# Functional Properties of Natural Products and Human Health

# Functional Properties of Natural Products and Human Health

Editors

**Daniela Rigano**
**Paola Bontempo**

MDPI

Basel • Beijing • Wuhan • Barcelona • Belgrade • Novi Sad • Cluj • Manchester

*Editors*
Daniela Rigano
University of Naples Federico II
Naples, Italy

Paola Bontempo
University of Campania
"Luigi Vanvitelli"
Naples, Italy

*Editorial Office*
MDPI
St. Alban-Anlage 66
4052 Basel, Switzerland

This is a reprint of articles from the Special Issue published online in the open access journal *Nutrients* (ISSN 2072-6643) (available at: https://www.mdpi.com/journal/nutrients/special_issues/Functional_Properties_Natural_Products_Human_Health).

For citation purposes, cite each article independently as indicated on the article page online and as indicated below:

| Lastname, A.A.; Lastname, B.B. Article Title. *Journal Name* **Year**, *Volume Number*, Page Range. |
|---|

**ISBN 978-3-0365-9096-7 (Hbk)**
**ISBN 978-3-0365-9097-4 (PDF)**
**doi.org/10.3390/books978-3-0365-9097-4**

# Contents

# About the Editors

**Daniela Rigano**

Daniela Rigano, Ph.D., is an Associate Professor in the Department of Pharmacy of the University of Naples Federico II. Her research interests include the isolation and structural characterization of secondary metabolites extracted from aromatic and food plants and the study of the properties of essential oils derived from plants of the Mediterranean region. The results of this activity have been documented in more than 130 publications in international scientific journals and in about 90 communications presented at national and international scientific conferences. Dr. Rigano has participated in different co-financed projects and is currently an Editor and Editorial Board Member of several international journals. She is a member of the Italian Society of Phytochemistry, the Italo-Latin American Society of Ethnomedicine, and the European Society of Phytochemistry. Prof. Rigano was ranked by the scientific journal *Plos Biology* as numbering among the top 2% of researchers in the world in 2021.

**Paola Bontempo**

Paola Bontempo, MD, PhD, is an Aggregate Professor of Clinical Pathology at the Department of Precision Medicine of the University of Campania Luigi Vanvitelli, Naples. Her research interests include pathology and oncology, but she particularly focuses on cancer, apoptosis, signal transduction, and epigenetics. She has been actively involved, for most of her scientific career, in studies concerning the mechanism(s) of proliferation differentiation and apoptosis. In particular, in the last few years, she focused on the analysis of cell cycle regulation/proliferation via steroid hormones and other extracellular stimuli and the elucidation of the underlying mechanisms. Recent relevant studies involved the epigenetic treatment of neoplastic disease and the identification and characterization of the antiproliferative and epigenetic effects of plant extracts and natural compounds. She has participated in different co-financed projects, and the results of her activity have been documented in many publications in international scientific journals. She is member of the Epigenetics Society and is an Editorial Board Member and Referee for several international journals.

nutrients

MDPI

*Editorial*

# Functional Properties of Natural Products and Human Health

**Paola Bontempo [1], Luigi De Masi [2] and Daniela Rigano [3,*]**

1 Department of Precision Medicine, University of Campania Luigi Vanvitelli, Via L. De Crecchio 7, 80138 Naples, Italy; paola.bontempo@unicampania.it
2 National Research Council (CNR), Institute of Biosciences and BioResources (IBBR), Via Università 133, 80055 Portici, Italy; luigi.demasi@ibbr.cnr.it
3 Department of Pharmacy, School of Medicine and Surgery, University of Naples Federico II, Via Domenico Montesano 49, 80131 Naples, Italy
* Correspondence: drigano@unina.it

**Citation:** Bontempo, P.; De Masi, L.; Rigano, D. Functional Properties of Natural Products and Human Health. *Nutrients* **2023**, *15*, 2961. https://doi.org/10.3390/nu15132961

Received: 20 June 2023
Accepted: 28 June 2023
Published: 29 June 2023



Natural products (NPs), broadly defined as chemicals produced by living organisms including microbes, marine organisms, animals, fungi and plants, are widely used as therapeutic agents for treating diseases and maintaining health and "wellness". NPs are regarded as an important repository for the development of potential novel drugs, as they may serve as either drug candidates or lead compounds for drug design. NPs have evolved over millions of years and acquired a unique chemical diversity, occupying a much greater chemical space than those created through synthetic chemistry methods, which consequently results in the diversity of their bioactivities and drug-like properties [1]. Plant-derived therapeutics, which may reach the pharmaceutic market as pure compounds or as complex mixtures containing thousands of different molecules, have several advantages, including wide availability, diverse pharmacological actions and a generally good profile of safety and tolerability. As novel therapeutic targets have emerged in recent times, the development of state-of-the-art technologies can be applied to drug discovery from natural sources. Therefore, there have been numerous reports from clinical studies testifying to the efficacy of medicinal plants and phytochemicals in ameliorating several human diseases such as cancers and inflammatory, cardiovascular, neurodegenerative, metabolic and infectious diseases. A plethora of basic in vitro and in vivo studies have also unraveled molecular mechanisms underlying NPs' health benefits [1].

In this Special Issue, eighteen original research articles are collected, of which one brief report and one review, on new perspectives on the chemical and functional properties of NPs, including the molecular mechanisms of the therapeutic effects of NPs in different models of human diseases.

Research on the plant resources allowed the isolation of a myriad of NPs with anti-inflammatory effects in vitro. Among these, Ma et al. [2] recently studied the bioactivity of the phenolic compound cryptochlorogenic acid (CCGA), a structural isomer of chlorogenic acid obtained from *Ageratina adenophora*, on the inflammatory response induced by lipopolysaccahride (LPS) in macrophages (RAW264.7), also clarifying the molecular mechanisms by which CCGA has shown in vitro anti-phlogosis potential. Moreover, *A. adenophora* is a toxic and invasive species in China (native to Central America), but nowadays it is considered a new bioresource for the discovery and production of specialized metabolites, such as CCGA.

Liu et al. [3] highlighted the beneficial effects of dietary procyanidins against pyroptosis, a type of inflammation-associated programmed cell death with a crucial role in liver diseases. These authors demonstrated that Procyanidin B2 (PCB2), a natural flavonoid known for its antioxidant and anti-inflammatory properties, attenuates pyroptosis induced in hepatocytes by nicotine, a natural alkaloid of tobacco that plays a fundamental role in the development of many diseases. The authors elucidated the role of nicotine in the pyroptotic death of hepatocytes and the mechanism by which PCB2 exerts its antipyroptotic and hepatoprotective effect through the activation of the peroxisome proliferator-activated receptor $\gamma$ (PPAR$\gamma$).

Phenolic compounds have been strongly linked with beneficial effects in many in vitro studies of human and animal models, as also confirmed by the other research described in this Special Issue.

Park et al. [4] demonstrated the ability of phenolic compounds to prevent the negative effects on the skin of blue light, also known as high-energy visible (HEV) light. Particularly, they studied the protective effects of yellow chaste weed (YCW) (*Helichrysum arenarium*) extract and its main components, apigenin and galangin, on HEV light-irradiated HaCaT keratinocytes from human skin. They found that YCW extract attenuated (i) the proliferation arrest of HaCaT cells and (ii) levels of reactive oxygen species (ROS). Furthermore, YCW extract, apigenin and galangin were able to modulate TRPV1/clusterin/FoxO3a and MAPK signaling. Therefore, they could be used as potential agents for counteracting the effects of HEV light in the protection of human skin.

Another phenolic compound, pinostrobin, a flavanone found in honey and propolis, was shown to be a potential melanogenic agent to be used in the treatment and prevention of vitiligo and other depigmentation disorders, for which, to date, the only therapeutic options are mostly based on immunosuppressive agents and UV light treatment [5]. This molecule has a stimulating effect on the activity of tyrosinase and on the production of melanin. Molecular docking analysis revealed that it possesses an appropriate molecular structure for effective interactions with major melanogenesis proteins.

Two papers in this Special Issue deal with the neuroprotective effect exerted by NPs. Pattarachotanant et al. [6] studied the effect of *Aquilaria crassna* (AC), a plant used in many traditional medicines, on the neurotoxicity and aging induced by high glucose. Through in vitro and in vivo studies, these authors demonstrated that AC leaf hexane extract (ACH) induces interesting neuroprotective activities in SH-SY5Y human neuroblastoma cells, including induction of neurite outgrowth and cell cycle regulation. In vivo studies on the nematode *Caenorhabditis elegans* have highlighted the beneficial effects of ACH on antioxidant activity, longevity and life span. The authors identified sitosterol and stigmasterol as the bioactive phytochemicals in ACH. Tan et al. [7] showed that the extract of *Pandanus amaryllifolius*, widely distributed in tropical and subtropical environments, has anti-amyloidogenic activity and promotes neuroprotective effects in amyloid-β-induced SH-SY5Y cells, through the restoration of cell viability, reduction in ROS and mitochondrial dysfunction. Furthermore, the unprecedented isolation of nicotinamide as a potential bioactive constituent, provides new insights into the promising potential of *P. amaryllifolius* extracts against Alzheimer's disease and suggests further exploration of other potential bioactive constituents.

Nowadays, when multidrug-resistant bacterial strains emerge and designing new antibiotics is a long process, rapidly developing a new possible line of infection treatment is essential. Clear evidence of the possible antimicrobial properties of NPs was shown by Nowicki et al., who evaluated the activity of dietary isothiocyanates, particularly sulforaphane and phenethyl isothiocyanate, against a *Shigella dysenteriae* infection in the *Galleria mellonella* larvae model [8]. These authors elucidated the mechanism behind the observed antibacterial effect, which involves induction of global stress response, stringent response, and decrease in Shiga toxin gene expression. The results here described constitute an important step in the perspective of using plant-derived NPs to treat bacterial infections.

Other research showed the potential of NPs to stimulate immune functions and to treat immune-related pathologies. Particularly, Ishiguro et al. [9] demonstrated that the water extract from the cell walls of *Chlorella sorokiniana*, a unicellular green alga used as a dietary supplement worldwide, could be useful to stimulate anti-microbial or anti-tumor immunity. The extract stimulated the growth of bone marrow cells and splenocytes, and increased cytokine mRNA associated with T-cell activation.

These, and other papers, clearly show that plant species represent an apparently inexhaustible resource of bioactive compounds. In spite of this, they can be susceptible of more or less wide variability in phytochemicals and, consequently, in the related bioactivities. The variability of the NP composition in the starting plant material due to environmental factors is a major challenge for the development of botanical drugs. An important line of research in this regard concerns the study of plant metabolome changes in response to variations in environmental conditions. The seasonality of plant harvesting is among the most important factors influencing the bioactive content, and so it deserves to be studied more in depth. In this regard, the in vitro study by Pucci et al. [10] showed that the extracts of common fig (*Ficus carica*) leaves collected in different seasons differently modulate lipid metabolism and adipogenesis in the cellular model of 3T3-L1 adipocytes. Importantly, dysfunctions related to adipogenesis and metabolism are very common in the population, so it would be advisable to utilize the fig autumnal extract, characterized by a higher coumarin content, to correctly downregulate the transcriptional pathway of adipogenesis, as shown in 3T3-L1 adipocytes.

The papers cited above describe noteworthy results, but they face limitations that, as most of the studies on NPs biological activities, are typically performed only through in vitro assessment. If the in vitro studies represent an essential prerequisite to advance the research on the biological effects of NPs, a complete understanding of the in vivo mechanism(s) is necessary and requires advanced studies.

Only a few natural products have been appropriately studied in models in vivo; among them, there is α-bisabolol, a monocyclic sesquiterpene derived from essential oils of many plants, whose pharmacological properties, including anticancer, antinociceptive, neuroprotective, cardioprotective and antimicrobial, were demonstrated based on both in vitro and in vivo studies [11]. Other interesting in vivo studies, described in this Special Issue, regard the potential role of NPs in the treatment of obesity and related diseases, a widespread public health issue all over the world. Increasing amounts of knowledge demonstrate that NPs can modulate metabolic syndrome and its risk factors with limited side effects. Investigations on the in vivo model of obese mice showed that sap of *Dendropanax trifidus*, a medicinal herb native to East Asia, can suppress body weight increase and stimulate energy metabolism in C2C12 muscle cells [12]. Estrogen receptor alpha is present in skeletal muscle cells, so sap containing estrogen as a major component could contribute to controlling body weight by modulation of cellular metabolism via this receptor. This study showed that cellular components could mediate glycolytic and metabolic changes within the cell, thus contributing to weight loss in mice treated with *D. trifidus* sap. Of particular importance in this respect is the gut, an endocrine organ producing the hormone peptide incretins, known for the positive modulatory effects on insulin production, glycemia, lipidemia, gastrointestinal motility, appetite and inflammation. Among incretins, glucose-dependent insulinotropic polypeptide (GIP) could represent a new potential target for treating diabetes, obesity and cardiovascular diseases. The seeds of tea (*Camellia sinensis*) contain saponins, bioactive compounds that have been found to regulate gastrointestinal motility and body weight. Their effects on GIP production, the underlying mechanism and signaling network were investigated in a study in vitro and in vivo [13]. This research showed how tea seed saponins are able to significantly increase GIP levels, both in the small intestine of normal mice and in the mouse intestinal endocrine STC-1 cells, through important mediators of intestinal metabolism.

Stress-induced alterations in the internal homeostasis maintained by the autonomic nervous system can induce hypertension and arteriosclerosis leading to a deterioration of cardiovascular health. Therefore, using the biological effects of NPs, research studies how to reduce the blood levels of cholesterol and prevent the progression of cardiovascular diseases. The findings of an innovative line of frontier research [14], reported in this Special Issue, showed the effects of essential oil (EO) from common basil (*Ocimum basilicum*) and its major compound linalool through olfactory stimulation of the nervous system on stress-induced dyslipidemia in the in vivo model of chronically stressed rats. The interesting results of this study revealed that the simple inhalation of basil EO (i) decreases body weight gain and blood levels of cholesterol, triglycerides, LDL, while (ii) increases HDL and ameliorates atherogenic index and cardiac risk factors. All these beneficial effects should be further explored as soon as possible in clinical trials. Another critical point in the studies about NPs' therapeutic properties is infact that very few have moved to clinical trials. In the study by Oue et al. [15], the effect of acute dietary $NO_3^-$ supplementation with beetroot juice on the venous vascular response and circulatory responses to exercise was analyzed on sixteen healthy young adults in a randomized crossover design. The results showed that the treatment leads to an increase in plasma of $NO_3^-$ concentration but does not change the venoconstriction in non-exercising limbs or the mean arterial pressure response to exercise.

In recent years, the popularity of fermented foods has increased in the Western world, largely due to the emerging evidence that secondary metabolites derived from fermentation-based products provide health benefits. To date, research indicated that fermented foods and beverage contains numerous bioactive substances, originating from the material used, but also resulting from the enzymatic transformations of organic compounds carried out by microorganisms or other techniques [16]. Two examples of research in this Special Issue showed that, as a result of the biochemical transformations of organic compounds by microorganisms, it is possible to obtain foods with not only extended shelf life and microbiological stability, but also a higher nutritional value. Particularly, Jakubczyk et al. [17] have studied the effects of fermentation time and type of Chinese tea (black, green, white and red) on the content of micronutrients in Kombucha, a low-alcohol beverage made by fermenting a sugared tea infusion with symbiotic cultures of bacteria and yeasts (SCOBY), commonly called "tea fungus". The authors demonstrated that the type of tea, as well as the days of fermentation, have a significant effect on the concentrations of selected minerals, and so Kombucha can represent a good supplement of micronutrients in the human diet. Song et al. [18] showed the in vivo bioactivities of fermented Chinese green tea and spores of the fungus *Eurotium cristatum*. Particularly, the methanol extract of the fermented tea and spores of *E. cristatum* both showed potent lipid-lowering activity in the blood of a high-fat-diet-induced hyperlipidemia model in golden hamsters and significantly reduced the accumulation of fat granules in the liver. Interestingly, the extract and *E. cristatum* spores share similar secondary metabolites, such as the new indole-containing diketopiperazine alkaloid, variecolorin P and four of its analogs.

Another important aspect to consider for a better exploitation of NPs therapeutic effects is that several factors could hamper the bioavailability and absorption of NPs in host cell systems and target sites, such as food matrix, environmental factors, molecule size and the association with gastrointestinal (GI) material. Nowadays, the remarkable progress in the use of encapsulation technologies is providing immense opportunities to provide stable and health-promoting compounds for different fields. In the study by Chen et al. [19], a chitosan-thioglycolic acid (CT) was developed as a physical barrier in the gastrointestinal tracts to inhibit nutrient uptake, exhibiting a superior mucoadhesive property compared to chitosan both in vitro and in vivo. Results show that CT could be exploited as a potential mucosal gel for obesity treatment, as hesperidin encapsulated in CT (CTH) not only effectively reduced mice body weight by 40.91% compared with the high-fat diet group, but also reduced the accumulation of body fat. Moreover, Enache et al. [20] designed two co-microencapsulated delivery systems, containing anthocyanins from cornelian cherry (*Cornus mas*) and the probiotic species *Lacticaseibacillus casei* 431®, into biopolymeric whey protein isolated-based materials, including casein and inulin. The selected microencapsulation matrices have unique properties, both from technological and nutritional points of view, and could be exploited to preserve the chemical stability during the shelf life of nutraceuticals or food products, and for a controlled release of bioactive compounds in a GI environment.

In conclusion, NPs represent a promising chemical pool for the discovery of novel drugs. Nevertheless, it is important to underline that issues such as NPs accessibility, sustainable supply and bioavailability require further study for NP-based drugs to continue making major contributions to human health and longevity. Furthermore, phytochemicals are generally classified as safe compounds as they are present in plants and produced naturally; however, the assessment of toxicity risk requires full consideration to eliminate possible toxic reactions and ensure not exceeding the recommended dose.

**Conflicts of Interest:** The authors declare no conflict of interest.

## References

1. Atanasov, A.G.; Zotchev, S.B.; Dirsch, V.M.; Supuran, C.T. Natural products in drug discovery: Advances and opportunities. *Nat. Rev. Drug Discov.* **2021**, *20*, 200–216. [CrossRef] [PubMed]
2. Ma, X.; Okyere, S.K.; Hu, L.; Wen, J.; Ren, Z.; Deng, J.; Hu, Y. Anti-Inflammatory Activity and Mechanism of Cryptochlorogenic Acid from *Ageratina adenophora*. *Nutrients* **2022**, *14*, 439. [CrossRef] [PubMed]
3. Liu, J.; Yao, Q.; Xie, X.; Cui, Q.; Jiang, T.; Zhao, Z.; Du, X.; Lai, B.; Xiao, L.; Wang, N. Procyanidin B2 Attenuates Nicotine-Induced Hepatocyte Pyroptosis through a PPARgamma-Dependent Mechanism. *Nutrients* **2022**, *14*, 1756. [CrossRef] [PubMed]
4. Park, J.Y.; Park, S.-H.; Oh, S.W.; Kwon, K.; Yu, E.; Choi, S.; Yang, S.; Han, S.B.; Jung, K.; Song, M.; et al. Yellow ChasteWeed and Its Components, Apigenin and Galangin, Affect Proliferation and Oxidative Stress in Blue Light-Irradiated HaCaT Cells. *Nutrients* **2022**, *14*, 1217. [CrossRef] [PubMed]
5. Yoon, J.; Youn, K.; Jun, M. Discovery of Pinostrobin as a Melanogenic Agent in cAMP/PKA and p38 MAPK Signaling Pathway. *Nutrients* **2022**, *14*, 3713. [CrossRef] [PubMed]
6. Pattarachotanant, N.; Sornkaew, N.; Warayanon, W.; Rangsinth, P.; Sillapachaiyaporn, C.; Vongthip, W.; Chuchawankul, S.; Prasansuklab, A.; Tencomnao, T. Aquilaria crassna Leaf Extract Ameliorates Glucose-Induced Neurotoxicity In Vitro and Improves Lifespan in Caenorhabditis elegans. *Nutrients* **2022**, *14*, 3668. [CrossRef] [PubMed]
7. Tan, M.; Ishikawa, H.; An, S. Pandanus amaryllifolius Exhibits In Vitro Anti-Amyloidogenic Activity and Promotes Neuroprotective Effects in Amyloid-beta-Induced SH-SY5Y Cells. *Nutrients* **2022**, *14*, 3962. [CrossRef] [PubMed]
8. Nowicki, D.; Krause, K.; Karczewska, M.; Szalewska-Pałasz, A. Evaluation of the Anti-Shigellosis Activity of Dietary Isothiocyanates in *Galleria mellonella* Larvae. *Nutrients* **2021**, *13*, 3967. [CrossRef] [PubMed]
9. Ishiguro, S.; Roth, M.; Welti, R.; Loyd, M.; Thakkar, R.; Phillips, M.; Robben, N.; Upreti, D.; Nakashima, A.; Suzuki, K.; et al. A Water Extract from *Chlorella sorokiniana* Cell Walls Stimulates Growth of Bone Marrow Cells and Splenocytes. *Nutrients* **2022**, *14*, 2901. [CrossRef] [PubMed]
10. Pucci, M.; Mandrone, M.; Chiocchio, I.; Sweeney, E.M.; Tirelli, E.; Uberti, D.; Memo, M.; Poli, F.; Mastinu, A.; Abate, G. Different Seasonal Collections of *Ficus carica* L. leaves Diversely Modulate Lipid Metabolism and Adipogenesis in 3T3-L1 Adipocytes. *Nutrients* **2022**, *14*, 2833. [CrossRef] [PubMed]
11. Eddin, L.; Jha, N.; Goyal, S.; Agrawal, Y.; Subramanya, S.; Bastaki, S.; Ojha, S. Health Benefits, Pharmacological Effects, Molecular Mechanisms, and Therapeutic Potential of α-Bisabolol. *Nutrients* **2022**, *14*, 1370. [CrossRef] [PubMed]
12. Lee, A.; Koh, E.; Kim, D.; Lee, N.; Cho, S.M.; Lee, Y.J.; Cho, I.-H.; Yang, H.-J. *Dendropanax trifidus* Sap-Mediated Suppression of Obese Mouse Body Weight and the Metabolic Changes Related with Estrogen Receptor Alpha and AMPK-ACC Pathways in Muscle Cells. *Nutrients* **2022**, *14*, 1098. [CrossRef] [PubMed]
13. Zhu, H.; Wang, K.; Chen, S.; Kang, J.; Guo, N.; Chen, H.; Liu, J.; Wu, Y.; He, P.; Tu, Y.; et al. Saponins from *Camellia sinensis* Seeds Stimulate GIP Secretion in Mice and STC-1 Cells via SGLT1 and TGR5. *Nutrients* **2022**, *14*, 3413. [CrossRef] [PubMed]
14. Kim, D.-S.; Hong, S.-J.; Yoon, S.; Jo, S.-M.; Jeong, H.; Youn, M.-Y.; Kim, Y.-J.; Kim, J.-K.; Shin, E.-C. Olfactory Stimulation with Volatile Aroma Compounds of Basil (*Ocimum basilicum* L.) Essential Oil and Linalool Ameliorates White Fat Accumulation and Dyslipidemia in Chronically Stressed Rats. *Nutrients* **2022**, *14*, 1822. [CrossRef] [PubMed]
15. Oue, A.; Iimura, Y.; Shinagawa, A.; Miyakoshi, Y.; Ota, M. Effect of Acute Dietary Nitrate Supplementation on the Venous Vascular Response to Static Exercise in Healthy Young Adults. *Nutrients* **2022**, *14*, 4464. [CrossRef] [PubMed]
16. Bontempo, P.; Stiuso, P.; Lama, S.; Napolitano, A.; Piacente, S.; Altucci, L.; Molinari, A.M.; De Masi, L.; Rigano, D. Metabolite Profile and In Vitro Beneficial Effects of Black Garlic (*Allium sativum* L.) Polar Extract. *Nutrients* **2021**, *13*, 2771. [CrossRef] [PubMed]
17. Jakubczyk, K.; Kupnicka, P.; Melkis, K.; Mielczarek, O.; Walczyńska, J.; Chlubek, D.; Janda-Milczarek, K. Effects of Fermentation Time and Type of Tea on the Content of Micronutrients in Kombucha Fermented Tea. *Nutrients* **2022**, *14*, 4828. [CrossRef] [PubMed]
18. Song, F.; Zhang, K.; Yang, J.; Wilson, A.; Chen, C.; Xu, X. The Hypolipidemic Characteristics of a Methanol Extract of Fermented Green Tea and Spore of *Eurotium cristatum* SXHBTBU1934 in Golden Hamsters. *Nutrients* **2023**, *15*, 1329. [CrossRef] [PubMed]
19. Chen, T.; Ho, Y.; Tang, R.; Ke, Y.; Lin, J.; Yang, I.; Lin, F. Thiolated Chitosan as an Intestinal Absorption Carrier with Hesperidin Encapsulation for Obesity Treatment. *Nutrients* **2021**, *13*, 4405. [CrossRef] [PubMed]
20. Enache, I.; Vasile, M.; Crăciunescu, O.; Prelipcean, A.; Oancea, A.; Enachi, E.; Barbu, V.; Stănciuc, N.; Vizireanu, C. Co-Microencapsulation of Anthocyanins from Cornelian Cherry (*Cornus mas* L.) Fruits and Lactic Acid Bacteria into Antioxidant and Anti-Proliferative Derivative Powders. *Nutrients* **2022**, *14*, 3458. [CrossRef] [PubMed]

*nutrients*

*Article*

# Anti-Inflammatory Activity and Mechanism of Cryptochlorogenic Acid from *Ageratina adenophora*

**Xiaoping Ma [1,2,†], Samuel Kumi Okyere [1,2,†], Liwen Hu [1,2], Juan Wen [1,2], Zhihua Ren [1,2], Junliang Deng [1,2] and Yanchun Hu [1,2,*]**

1 Key Laboratory of Animal Diseases and Environmental Hazards of Sichuan Province, College of Veterinary Medicine, Sichuan Agricultural University, Chengdu 611130, China; mxp886@sicau.edu.cn (X.M.); samuel20okyere@gmail.com (S.K.O.); huliwen197@163.com (L.H.); juanwen881010@163.com (J.W.); zhihua_ren@126.com (Z.R.); dengjl213@126.com (J.D.)

2 Key Laboratory of Animal Disease and Human Health of Sichuan Province, Sichuan Agricultural University, Chengdu 611130, China

* Correspondence: yanchunhu@sicau.edu.cn; Tel.: +86-2886291162

† These authors contributed equally to this work.

**Abstract:** *Ageratina adenophora* is an invasive plant known for its toxicity to livestock. Current research on this plant has shifted from toxicity prevention to the beneficial utilization of plant resources. This study was performed to investigate the effects and mechanisms of cryptochlorogenic acid (CCGA) isolated from *Ageratina adenophora* on the inflammatory responses induced by lipopolysaccharide (LPS) in RAW264.7 cells. RAW264.7 cells were pretreated with CCGA (200, 100, and 50 μg/mL) and subsequently stimulated with LPS (1 μg/mL) for 16 h. The cytotoxicity of CCGA was tested using the Cell Counting Kit (CCK8). The mechanism of action of CCGA in attenuating inflammation was also identified using enzyme-linked immunosorbent assay (ELISA), quantitative reverse transcription-polymerase chain reaction, and Western blot. The results showed that CCGA had a maximal safe concentration of 200 mg/mL. Moreover, CCGA reduced the level of nitric oxide (NO) and iNOS in LPS-induced RAW264.7 cells ($p < 0.01$). In addition, CCGA reduced the levels of pro-inflammatory cytokines (TNF-α, IL-1β, IL-6, and IL-8) and cyclooxygenase-2 (COX-2) in LPS-induced RAW264.7 cells at both the mRNA and protein levels ($p < 0.01$). CCGA prevented the activation of nuclear factor-kappa B (NF-kB) in LPS-induced RAW264.7 cells via the inhibition of IKK and IκB phosphorylation and the degradation of IκB proteins ($p < 0.01$). This finding indicated that CCGA isolated from *A. adenophora* may be a potential candidate for the treatment of inflammation-related diseases.

**Keywords:** *Ageratina adenophora*; cryptochlorogenic acid; anti-inflammatory; nuclear factor-kappa B pathway

**Citation:** Ma, X.; Okyere, S.K.; Hu, L.; Wen, J.; Ren, Z.; Deng, J.; Hu, Y. Anti-Inflammatory Activity and Mechanism of Cryptochlorogenic Acid from *Ageratina adenophora*. *Nutrients* **2022**, *14*, 439. https://doi.org/10.3390/nu14030439

Academic Editors: Daniela Rigano and Paola Bontempo

Received: 19 December 2021
Accepted: 17 January 2022
Published: 19 January 2022



## 1. Introduction

The chief tool used by most organisms against infection is inducing inflammatory responses [1]. However, elevated levels of inflammatory cytokines in the body circulation or in local inflammatory sites result in severe diseases [2–4]. Chronic inflammation is linked with various diseases, such as diabetes, depression, and cancer [5–13]. Among the inflammatory inducing agents, lipopolysaccharide (LPS) is reported to enhance the production of pro-inflammatory cytokines in macrophages, fibroblasts, and monocytes [14]. Furthermore, it has been reported that LPS-induced inflammatory reactions can occur via the NF-κB signaling pathway [15]. TNF-α cytokine actively takes part in systemic inflammation as well as immune cell regulation [16]. IL-1β is known for activating the plasmalemmal IL-1R1 receptor in various types of cells [17,18]. IL-6 is involved in numerous cellular activities, including cell proliferation, differentiation, and apoptosis [19]. IL-8 is a pro-inflammatory cytokine that is linked to tumor growth and progression [20]. Cyclooxygenase (COX) is an oxidoreductase enzyme noted for the conversion of arachidonic acid to prostaglandins, thereby initiating inflammatory reactions [21]. COX-2 is associated with PGE2 synthesis in numerous examples of inflammation [22].

Nitrite oxide (NO) is an inflammatory mediator of the nitric oxide synthase (NOS) family [23]. Three types of NOSs have been identified, namely, inducible NOS (iNOS), neuronal NOS (nNOS), and endothelial NOS (eNOS) [24]. Both nNOS and eNOS are constitutive NOSs, whereas iNOS is activated by cytokines to yield NO [25]. An increase in NO concentration in cells causes cytotoxicity and induces various inflammatory disorders [26,27].

Inflammation is linked with the NF-κB pathway. NF-κB is a nuclear transcriptional factor that is linked to both pro- and anti-inflammatory activities [28]. In dormant cells, the NF-κB dimers are joined by inhibitory molecules, such as IκB, to keep them stable in the cytoplasm. LPS activates IKK to initiate the phosphorylation of IκB in cells, which causes the degradation of IκBα protein and the activation of NF-κB translocation into the nucleus to bind target DNAs [29].

Inflammation is one of the key drivers of pathogenesis in many diseases; therefore, over the past years, numerous studies have identified and isolated many anti-inflammatory compounds from natural products, especially from plant sources, because they produce fewer side effects compared with commercial drugs. Among these compounds from natural products, polyphenolic compounds in plants are the most widely known to possess anti-inflammation activities [30], whereas phenolic acids are the most abundant among plant polyphenolic compounds. They have been isolated from different plant parts, such as the leaves and seeds of vegetables and the skins of fruits [31]. Among the phenolic acid group, chlorogenic acid is the most well-known and abundant compound known to possess beneficial biological properties, such as anticancer, anti-inflammatory, antioxidant, antimicrobial, and antiviral effects [32–35]. Cryptochlorogenic acid (CCGA) (Figure 1), a distinct isomer of chlorogenic acid with similar structural features, has also demonstrated potential anti-inflammatory activities in previous studies [36,37].

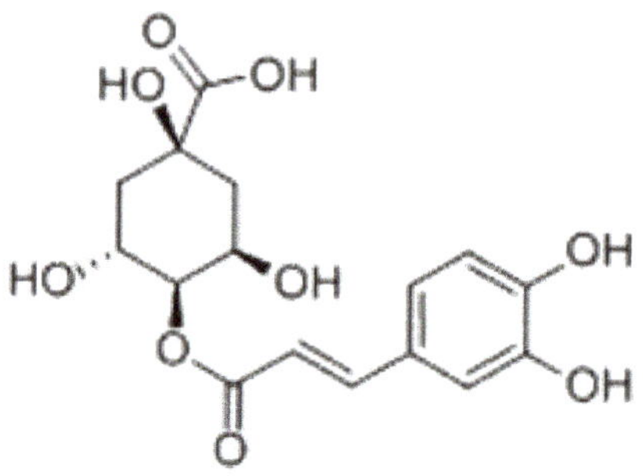


**Figure 1.** Chemical structure of cryptochlorogenic acid (CCGA).

*Ageratina adenophora* (Spreng.) R. M. King et H. Rob. is one of the most harmful invasive plants in China [38–41]. It has strong environmental adaptability and breeding ability, which can cause serious ecological disasters and affect the growth of the animal husbandry industry [42]. However, at present, the focus of research on *A. adenophora* has shifted from toxicity prevention to the biological utilization of the plant resource [43–45]; thus, researchers have identified several secondary metabolites with pharmacological properties, such as antibacterial, antitumor, anti-obesity, anticancer, and anti-inflammation effects, in this plant resource [46]. Cryptochlorogenic acid (CCA) is one of the major polyphenolic bioactive compounds isolated from *A. adenophora* [46]; however, its effects and mechanism in inflammation are still unknown. Therefore, this work was performed to explore the anti-inflammatory activity of CCGA isolated from *A. adenophora* in order to discover the pharmacological effects of CCGA as well as its molecular pathway.

## 2. Materials and Methods

### 2.1. Plant Sample, Chemicals, and Reagents

The collection, location, and identification of *Ageratina adenophora* for this study were reported in a study by Okyere et al. (2022). Cells: The RAW264.7 murine cell line was purchased from the Wuhan China Typical Culture Preservation Center (Wuhan, China), which offered Short Tandem Repeat (STR) analysis to confirm the authenticity of all cell lines. Lipopolysaccharides (LPS) were derived from *Escherichia coli* O111:4. High-glucose DMEM and CCK-8 were obtained from HyClone. The imported test kit for fetal bovine serum (GIBCO), the nitric oxide (NO) box, and ELISA kits for TNF-α, IL-1 β, IL-6, IL-8, IL-10, and COX-2 were purchased from Nanjing Jiancheng Company Ltd., Nanjing, China. The total RNA extraction kit and fluorescent quantitative PCR kit were obtained from the Beyotime Institute of Biotechnology (Beijing, China). Secondary antibodies were bought from the Beyotime Institute of Biotechnology (Beijing, China). All experiments were undertaken following the approved guidelines and experimental protocols by the Animal Care and Use Committee of Sichuan Agricultural University, China (no. 2019603005).

### 2.2. Extraction of CCGA

#### Preparation of Aqueous Extract from *Ageratina adenophora*

Fifty grams (50) of milled leaves was mixed in 100 mL of water and soaked for 12 h. The upper layer containing the organic solvent was collected and filtered by suction. After rotating evaporation at 105 kpa and 45 °C, the upper layer containing the organic solvent was dissolved in 1000 mL of extraction solution to begin the extraction process. The liquid ratio used for the extraction was methanol: ethyl acetate: water = 2:5:8 (*v/v/v*). After that, the lower part of the aqueous phase was collected and evaporated at 75 °C to obtain a yellowish-brown aqueous extract.

### 2.3. Separation of Cryptochlorogenic Acid by Macroporous Resin

#### 2.3.1. Macroporous Resin Pretreatment

HPD100 macroporous resin was soaked with 95% ethanol for 24 h. After full swelling, it was washed with ethanol and an appropriate amount of water until the effluent had no white turbidity. Afterward, it was soaked in distilled water until there was no smell of alcohol.

#### 2.3.2. Sample Loading and Elution System

The aqueous extract was dissolved in 95% ethanol, and the ratio of ethanol volume to extract was 10:1 (*v/w*). Using 5%, 15%, 65%, and 95% as eluent, the elution amount was 6 times of column volume. The eluent was then collected from below the column. The collected eluent and cryptochlorogenic acid standard were developed using the mixture of methanol and ethyl acetate in a ratio of 95:5 (*v/v*) as the layer-developing agent, and 2% ferric chloride solution was used as the color developing agent. The samples were collected according to the color-developing points for silica gel column chromatography.

### 2.4. Purification of Cryptochlorogenic Acid by Silica Gel Column Chromatography

Step 2.3 was repeated 5–6 times, and the eluate containing chlorogenic acid was combined for the sample of silica gel column chromatography. The silica gel powder was mixed with ethyl acetate and poured into a clean silica gel column. Twice the column volume of ethyl acetate was used as the eluent to stabilize the column, and the bubble-free column was used as the separation column. Samples were dissolved in ethyl acetate in a ratio of 4:1, and 20 mL of the sample was taken and slowly added to the silica gel column of the above cabinet. After loading, the valve was opened to allow the sample to enter the silica gel column. Using an ethyl acetate to methanol ratio of 98:2 ($v/v$) and a methanol to ethyl acetate ratio of 95:5 ($v/v$) as eluent, the elution amounts were 3 times the column volume and 6 times the column volume, respectively. The eluent containing cryptochlorogenic acid was collected and combined by thin layer chromatography for secondary column passing.

### *2.5. Identification of Monomer Purity by Ultra High Performance Liquid Chromatography*

#### 2.5.1. Chromatographic Conditions

In this analysis, an Agilent 1290 ultra-high liquid chromatograph (chromatographic column: XDB-C18(2.1 × 100 mm, 1.8 μm) was used to analyze chlorogenic acid using water (A) and methanol (B) as mobile phases, methanol–0.1% formic acid solution (30:70), a flow rate of 0.3 mL/min, a detection wavelength of 325 nm, and a column temperature of 28 °C. The sample injection volume was 1 μL, and the peak time was between 1.28 min and 12 h.

#### 2.5.2. Making the Standard Curve

The standard chlorogenic acid of 0.7 mg was accurately weighed and poured into a 3.5 mL volumetric flask; after that, it was dissolved in methanol, ultrasonic, and cooled. After cooling, 0.2 mg/mL concentration solution was prepared using methanol. The prepared solution was then diluted to different concentrations (0.01, 0.05, 0.10, 0.15, and 0.20 mg/mL), and 1 μL of the above standard solution and sample in sequence were loaded. The peak area was determined by ultra-high performance liquid chromatography (UHPLC) according to the chromatographic conditions in 2.5.1. The standard curve was drawn with the peak area as the ordinate and the standard concentration as the abscissa.

#### 2.5.3. Determination of Cryptochlorogenic Acid in *Ageratina adenophora*

An amount of 0.7 mg of the extract monomer was weighed and fixed into 3.5 mL of chromatographic methanol. UHPLC determination was conducted according to the chromatographic conditions in 2.5.1, and the content of cryptochlorogenic acid in the extracted monomer was calculated according to the standard curve.

### *2.6. Cell Viability Using CCK8 Method*

The cytotoxicity of CCGA was investigated using the CCK-8 (cell counting kit8) assay. RAW264.7 cells in the logarithmic growth phase were used for this experiment. After counting with a cell counting plate, 100 μL of cell suspension with $5 \times 10^4$ cells was seeded on a 96 well plate and cultured at 37 °C and 5% $CO_2$ for 24 h. After 24 h of culture at 37 °C and 5% $CO_2$, CCGA was released at 400, 200, 100, and 50 μg/mL concentrations for 24 h in DMEM. DMEM was added to the 96-well plate, blank group, and LPS group. After 4 h of continuous culture, 100 μL of cells without stimulation DMEM was added to the blank control group. The LPS group and experimental group were induced with LPS solution at a concentration of 1 μg/mL. for 16 h. After the replacement of fresh DMEM, 10 μL of CCK-8 reagent was added to each well and incubated at 37 °C for 4 h. Each group was set with 3 multiple wells and blank wells (with DMEM and CCK-8 solution, without cells). The absorbance was measured at a 450 nm wavelength. The data are represented as means ± standard deviation. Cell viability was calculated as:

$$\text{Cell viability (\%)} = (\text{treatment well OD-blank well OD}) / (\text{blank group OD-blank well OD}) \times 100\%$$

From the experimental results of CCK-8, the concentration groups of 200, 100, and 50 μg/mL CCGA were selected for further experiments. Therefore, we designed an experimental study of a blank control group, LPS group, and experimental group with concentrations of 200, 100, and 50 μg/mL. Each group was provided with 3 multiple wells. The experiment was performed in triplicate.

### *2.7. Measurement of Nitric Oxide in RAW264.7 Cells*

Nitric oxide levels were measured following the procedure by Ying et al. [25]. RAW264.7 cells were pretreated with CCGA at 50–200 μg/mL for 24 h; after that, cells were stimulated with LPS (1 μg/mL) for 16 h. Then, Griess reagent was used to estimate the NO production in the medium. For the blank groups, we used a sterile culture medium. The quantity of nitrite in the samples was measured by using the sodium nitrite standard curve as a reference point.

### 2.8. Enzyme-Linked Immunosorbent Assay (ELISA) for Detecting TNF-α, IL-1β, IL-6, IL-8, INOS, COX-2, and IL-10 Production

ELISA assay for detecting TNF-α, IL-1β, IL-6, IL-8, IL-10, iNOS, and COX-2 production were pretreated with CCGA for 24 h and then induced with LPS (100 ng/mL) for 16 h. The culture supernatant was estimated using commercially available ELISA kits. All procedures were accomplished by following the manufactures' instructions. The samples were analyzed in triplicate.

### 2.9. RNA Extraction and Real Time RT-PCR Analysis for Detecting TNF-α, IL-1β, IL-6, IL-8, IL-10, COX-2, and INOS Production

RAW264.7 cells were treated with different concentrations of CCGA for 24 h and stimulated with LPS. After 16 h, the supernatant was collected as the sample to be tested. Cells were blown with 1 mL Trizol and collected at −80 °C for RNA extraction. Afterward, RNA was extracted following the kit's instructions. The qRT-PCR was performed with β-actin as the control. The primers used are presented in Table 1. The relative expression of the genes was deduced by the $2^{-\Delta\Delta}$ CT method, where, ΔCT = CT target gene—CT internal reference gene, ΔΔCt = ΔCT treated samples −ΔCT control sample, and multiple change = $2^{-\Delta\Delta}$ C.

**Table 1.** Primers used for qRT-PCR.

| Genes | Primer Sequence (5′→3′) | References |
|---|---|---|
| TNF-α | F- AGCACAGAAAGCATGATCCG<br>R- ATGAGAGGGAGGCCATT | [10] |
| IL-1β | F- GTTCCCCAACTGGTACATCA<br>R- CCATACTTTAGGAAGACACGG | [11] |
| IL-6 | F- CTTCTTGGGACTGATGCTGGT G<br>R- CGCTGGCTTTGTCTTTCTTGTTA | [12] |
| IL-8 | F- CAAGGCTGGTCCATGCTCC<br>R- TGCTATCACTTCCTTTCTGTTG | [11] |
| IL-10 | F- GCTCTTACTGACTGGCATGAG<br>R- CGCAGCTCTAGGAGCATGTG | [13] |
| INOS | F- TCCTACACCACACCAAAC<br>R- CTCCAATCTCTGCCTATCC | [10] |
| COX | F- TGCTGTACAAGCAGTGGCAA<br>R- GCAGCCATTTCCTTCTCTCC | [10] |
| β-actin | F- CGGTTGGCCTTAGGGTTCAGGGGGG<br>R- GTGGGCCGCTCTAGGCACCA | [12] |

F, forward; R, reverse.

### 2.10. Transient Transfection and Luciferase Activity Assay

The transient transfection and luciferase activity assay was carried out following the methods and procedures of Ying et al. [25].

### 2.11. Western Blot Analysis

Using a scraper, cells were collected after thorough washing in ice-cold phosphate-buffered saline (PBS); after that, a protein extraction kit (Beyotime Institute of Biotechnology, Jiangsu, China) was used to extract the cytoplasmic proteins. The protein was resolved using SDS-PAGE and transferred to PVDF membranes. The membranes were incubated with blocking buffer (5% (*w*/*v*) skimmed milk powder in 1TBS containing 0.1% Tween-20 for 1 h, then incubated overnight at 4 °C with a primary antibody (Santa Cruz Biotechnology, Santa Cruz, CA, USA). Membranes were incubated for 1 h at room temperature with HRP-linked secondary antibodies after the membrane was washed. Then, the membranes were washed again and detected using an enhanced chemiluminescence (ECL) kit and exposed to X-ray films (Kodak, Shanghai, China).

### 2.12. Statistical Analysis

Statistical analyses were performed with SPSS software, version 20.0 (SPSS, Inc., Chicago, IL, USA). All data are presented as means ± SD. One-way analysis of variance (ANOVA) followed by Tukey's post hoc multiple comparison test was used and $p < 0.01$ was considered statistically significant.

## 3. Results

### 3.1. Detection of Cryptochlorogenic Acid by Thin Layer Chromatography

The results of the thin layer chromatography (TCL) are shown in Figure 2. The chromogenic points of the extracted monomers C and D in the thin layer chromatography are the same as those of A and B. Therefore, it was preliminarily inferred that the extracted monomers were cryptochlorogenic acid, and eluent with the same retention factors (RFs) was collected.

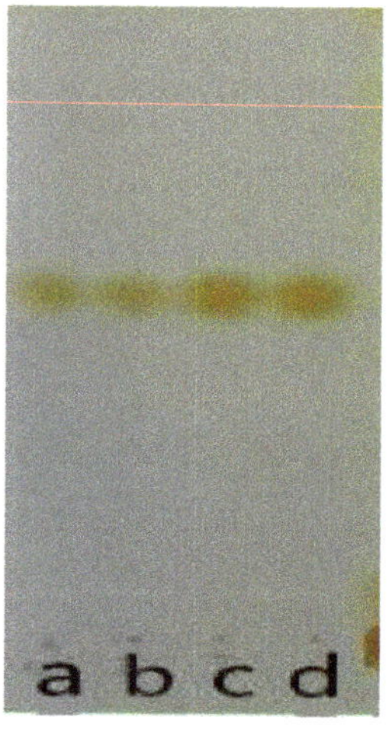


**Figure 2.** The results of TCL (a and b are standard substances; c and d are extractive monomers).

### 3.2. Determination of the Purity of Cryptochlorogenic Acid by Ultra High Performance Liquid Chromatography

#### 3.2.1. Standard Curve of Cryptochlorogenic Acid

As shown in Figure 3, according to UHPLC, the peak area of cryptochlorogenic acid standard was represented at the ordinate and the differential concentrations were represented at the abscissa. The graph showed the standard curve of cryptochlorogenic acid as y = 33039x + 94.402, where $R^2$ = 0.9995, indicating that there was a good linear relationship between cryptochlorogenic acid standard and its concentration.

#### 3.2.2. Determination of Cryptochlorogenic Acid in *Ageratina adenophora*

The UHPLC determination results of the cryptochlorogenic acid standard and cryptochlorogenic acid extracted from *A. adenophora* at 325 nm are shown in Figure 4. The retention time of the cryptochlorogenic acid standard is 1.276 min, and the retention time of monomer extracted from *A. adenophora* was 1.275 min. It can be inferred that the monomer extracted from *A. adenophora* was cryptochlorogenic acid. In addition, the peak area of cryptochlorogenic acid extracted from *A. adenophora* in Figure 2 was 6324.19, i.e., y = 6324.19; the concentration of cryptochlorogenic acid after dissolution was 0.189 mg/mL; and the specific chlorogenic acid mass was 0.66 mg. It was calculated that the purity of cryptochlorogenic acid extracted from *A. adenophora* was 94.3%. Therefore, we concluded that the amount of CCGA in 0.7 mg extract was 0.6 mg and the purity was 94.3%.

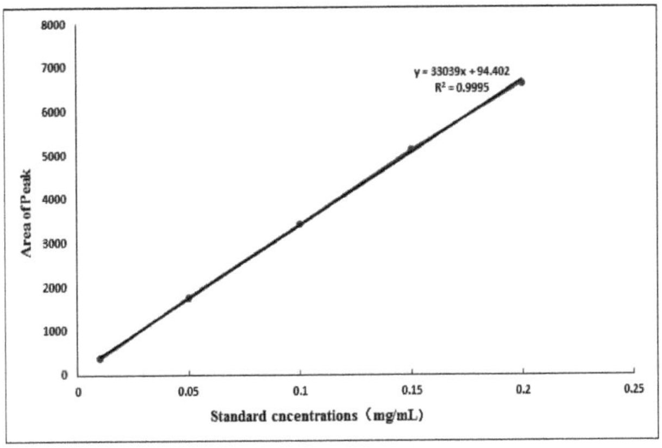


**Figure 3.** Standard linear regression curve.

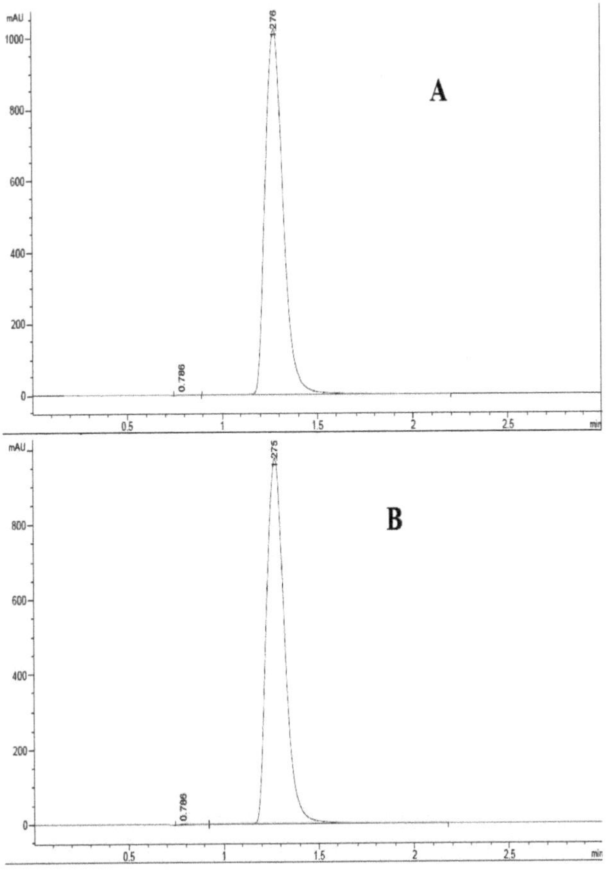


**Figure 4.** Determination of CCGA standard and extraction of CCGA from *A. adenophora* by UHPLC (**A**): Standard (0.2 mg/mL). (**B**): CCGA in *A. adenophora*.

### 3.3. Effects of CCGA on Cell Morphology and Cytotoxicity

The experimental results (Figure 5A) showed that after CCGA treatment and LPS induction, the cells in the blank control, 200 μg/mL, 100 μg/mL, and 50 μg/mL groups were smooth in appearance without pseudopodia, and some cells were compact in distribution and aggregated; however, the cells in the LPS and 400 μg/mL group were elongated with pseudopodia, and most of them were spindle-shaped.

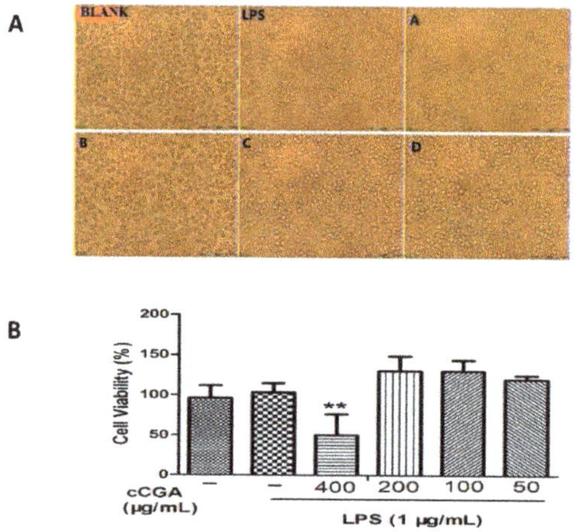


**Figure 5.** Cell morphology and viability after treatment with different concentrations of CCGA (BLANK: No LPS and CCGA treatments, LPS: Treatment with only LPS; A: 400 μg/mL CCGA + 1 μg/mL LPS; B: 200 μg/mL CCGA + 1 μg/mL LPS; C: 100 μg/mL CCGA + 1 μg/mL LPS; D: 50 μg/mL CCGA + 1 μg/mL LPS). (**A**) Photograph of viable cells after treatment with concentrations of CCGA (BLANK: No LPS and CCGA; LPS: Treatment with only LPS; A: 400 μg/mL CCGA + 1 μg/mL LPS; B: 200 μg/mL CCGA + 1 μg/mL LPS; C: 100 μg/mL CCGA + 1 μg/mL LPS; D: 50 μg/mL CCGA + 1 μg/mL LPS). (**B**) Viable cell count after treatment with concentrations of CCGA (A: 400 μg/mL; B: 200 μg/mL; C: 100 μg/mL; D: 50 μg/mL) Notes, ** $p < 0.01$ vs. LPS, $n = 9$ (mean ± standard deviation).

Furthermore, we observed that after 16 h, cells treated with CCGA at a concentration of 400 μg/mL were significantly lower than that of the LPS group (Figure 5B, $p < 0.01$), with a survival rate lower than 50%. Therefore, we concluded that CCGA at a concentration of 400 μg/mL was toxic to RAW264.7 cells. However, the 200, 100, and 50 μg/mL CCGA concentrations and the LPS groups had cell viability above 100%. Hence, we concluded that 1 μg/mL LPS and CCGA at concentrations of 200, 100, and 50 μg/mL did not show any obvious toxicity to RAW264.7 cells, and the maximum safe concentration of CCGA was 200 μg/mL. Therefore, CCGA concentrations of 200, 100, and 50 μg/mL were selected for subsequent tests.

### 3.4. Effect of CCGA on LPS-Induced Nitric Oxide (NO) and iNOS Production

In this study, we observed that CCGA could reduce the production of NO and the expression of iNOS in LPS-induced RAW264.7 cells compared with the LPS group (Figure 6A–C, $p < 0.01$).

### 3.5. The Regulatory Effect of CCGA from A. adenophora on Inflammation Related Cytokines in LPS-Induced RAW264.7 Cells

In this study, the mRNA and protein levels of pro-inflammatory cytokines (TNF- α, IL-1 β, IL-6, and IL-8) were elevated in the LPS group compared with the blank control group (Figure 7A–D and Figure 8A–D, $p < 0.01$). Furthermore, compared with the blank control group, anti-inflammatory cytokine IL-10 was downregulated in the LPS group

(Figures 7E and 8E, $p < 0.01$). In addition, the treatments groups reduced the expression levels of pro-inflammation cytokines (TNF- α, IL-1 β, IL-6, and IL-8) and increased the expression levels of IL-10 compared with the LPS group (Figure 7A–E and Figure 8A–E, $p < 0.01$).

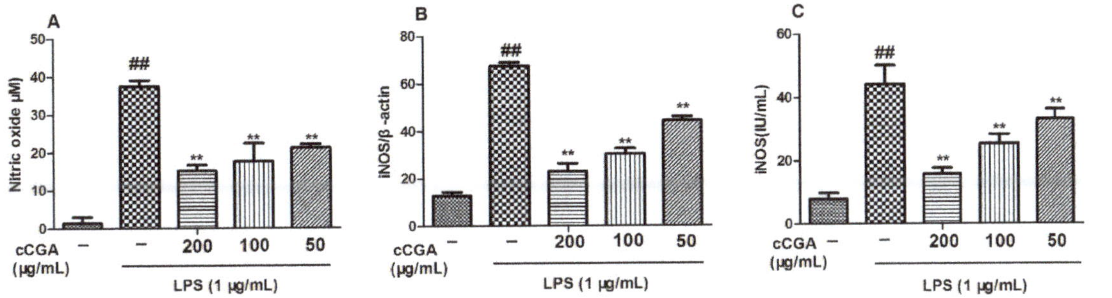


**Figure 6.** Effects of CCGA from *A. adenophora* on nitric oxide (NO) and iNOS production in LPS-induced RAW264.7 cells (**A**) Production level of NO after pretreatment with CCGA at 50–200 μg/mL for 24 h and then stimulation by LPS (1 μg/mL) for 16 h. (**B**) mRNA expression of iNOS after pretreatment with CCGA at 50–200 μg/mL for 24 h and then stimulation by LPS (1 μg/mL) for 16 h. (**C**) Protein levels of iNOS after pretreatment with CCGA at 50–200 μg/mL for 24 h and then stimulation by LPS (1 μg/mL) for 16 h. Notes: ## $p < 0.01$ vs. blank control group, ** $p < 0.01$ vs. LPS, $n = 3$ (mean ± standard deviation).

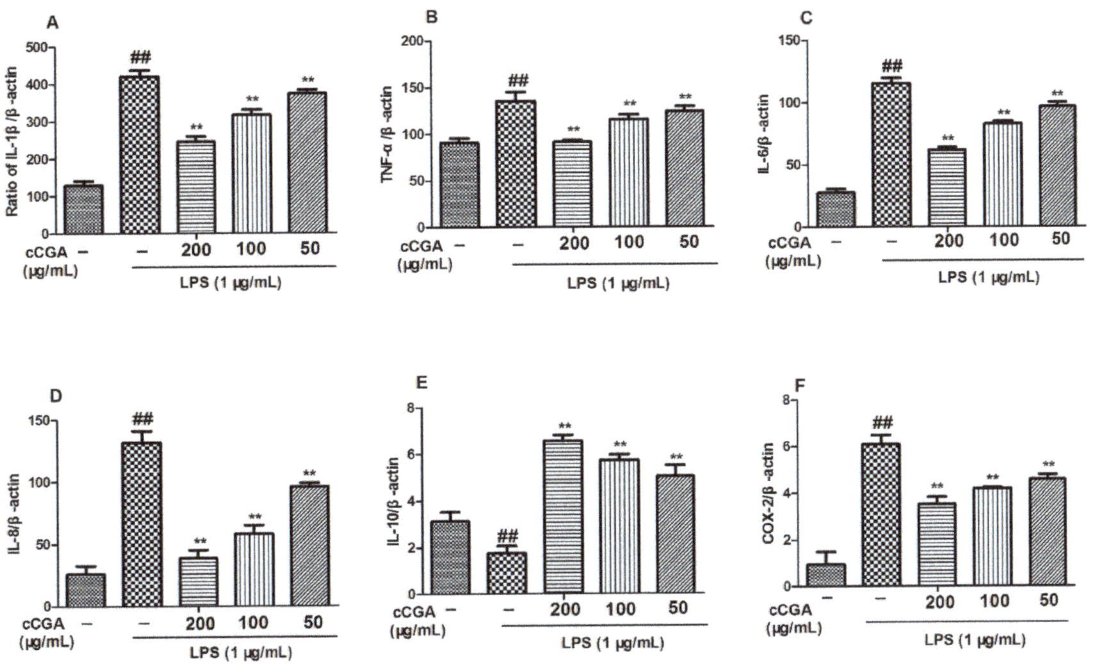


**Figure 7.** Effects of CCGA from *A. adenophora* on mRNA level of inflammatory-related cytokines in LPS-induced RAW264.7 cells. (**A–E**) mRNA level of inflammatory-related cytokines; (**F**) mRNA level of COX-2. Notes: ## $p < 0.01$ vs. blank control group, ** $p < 0.01$ vs. LPS, $n = 3$ (mean ± standard deviation).

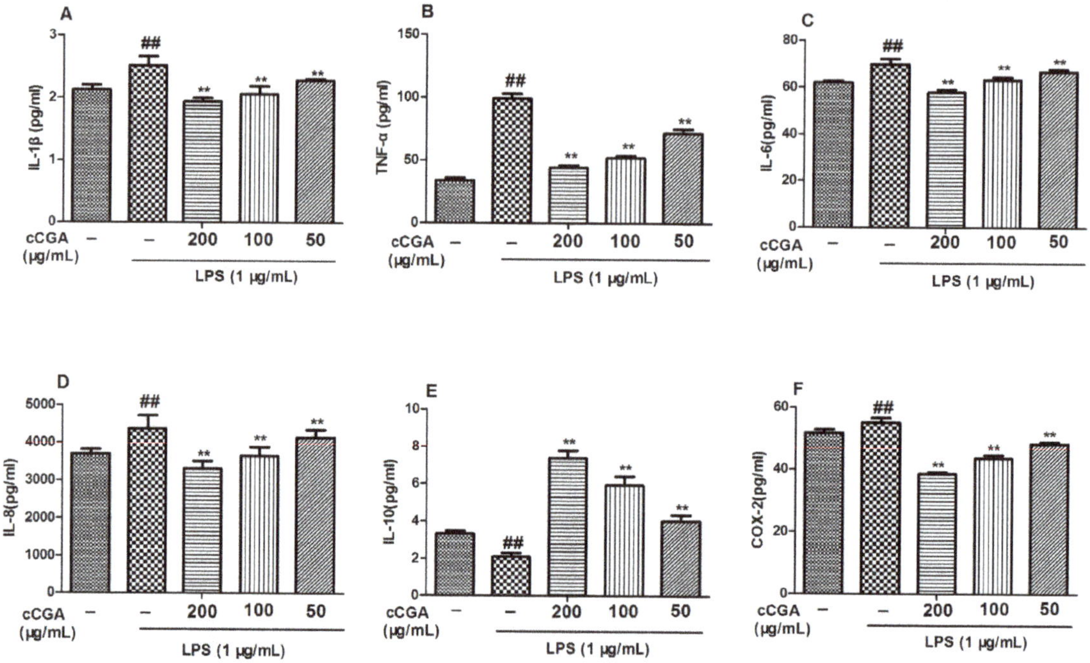


**Figure 8.** Effects of CCGA from *A. adenophora* on protein levels of inflammatory-related cytokines and COX-2 in LPS-induced RAW264.7 cells. (**A**–**E**) Protein levels of inflammatory-related cytokines. (**F**) Protein expression of COX-2. Notes: ## $p < 0.01$ vs. blank control group, ** $p < 0.01$ vs. LPS, $n = 9$ (mean ± standard deviation).

### 3.6. Effect of CCGA from A. adenophora on COX-2 Expression

According to the results of this study, the mRNA and protein levels of COX-2 in the LPS group were higher than those in the control blank group (Figures 7F and 8F, $p < 0.01$); however, CCGA could reduce the expression levels of COX-2 compared with the LPS group (Figures 5F and 7F, $p < 0.01$).

### 3.7. CCGA Inhibited LPS-Induced NF-kB Activation in RAW 264.7 Cells

To explore the anti-inflammatory mechanism of the CCGA, the NF-kB transcriptional activity was examined by the luciferase reporter gene assay. The results are represented in Figure 9A: CCGA reduced the LPS-induced NF-kB transcriptional activity in RAW 264.7 cells ($p < 0.01$). The main pathway known for the activation of NF-kB involves the phosphorylation of IKK and IκB, which activate the release and translocation of NF-kB into the nucleus. We further explored the cytoplasmic levels of p-IKK, IκB, and p-IκB using Western blot analysis to discover whether CCGA could regulate the NF-kB signaling pathway. We observed that CCGA prevented the phosphorylation of IKK and IκB and the degradation of the IκB protein after LPS stimulation (Figure 9B, $p < 0.01$).

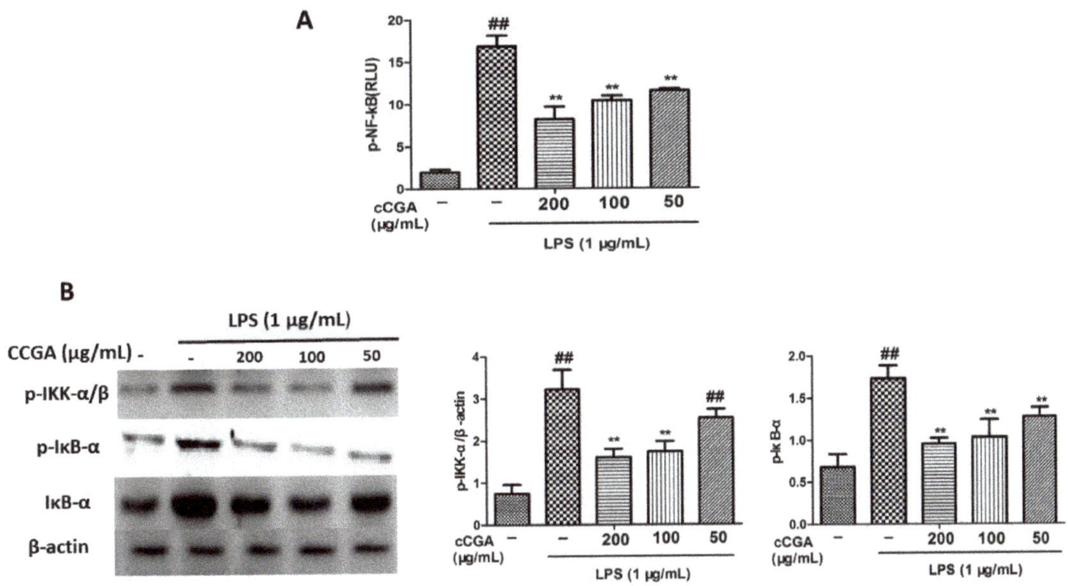


**Figure 9.** Effect of CCGA isolated from *A. adenophora* on phosphorylated proteins in NF-kB signaling pathway in LPS-induced RAW264.7 cells. (**A**) NF-kB transcriptional activity. Values are means ± standard deviation. (**B**) Protein levels of IkB-α, p-IkB-α, and p-IKK in RAW264.7 cells. Notes: ## $p < 0.01$ vs. blank control group, ** $p < 0.01$ vs. LPS, $n = 9$ (mean ± standard deviation).

## 4. Discussion

Plants contain secondary metabolites that are widely used in treating disease conditions [47]. These compounds are known to stimulate biochemical reactions in cells [48]. Pharmacological activities such as anti-aging, antioxidant, anti-cancer, and anti-inflammatory effects have been reported for these plant bioactive compounds [49]. Therefore, this study was conducted to ascertain the anti-inflammatory activity and mechanisms of cryptochlorogenic acid isolated from *A. adenophora,* an invasive and toxic plant. Our results demonstrate that CCGA at concentrations up to 200 µg/mL has better anti-inflammatory effects with no toxicity to cells; thus, the reduction in pro-inflammatory cytokine production was a result of CCGA activity and not cell death.

LPS activates inflammatory cytokines to terminate iNOS expression, thereby oxidizing L-arginine and increasing the production of NO [50]. NO, a free oxygen radical, is associated with the pathophysiology of several inflammatory diseases, as they act as cytotoxic agents in cells [51]. In activated macrophages, the production of NO may play a crucial role in the pathology of various acute and chronic inflammatory disorders [50]. Polyphenols decrease the production of NO in LPS induced inflammation. For example, a study by Moore et al. [52] reported that cranberry polyphenol could reduce the levels of NO in LPS-activated RAW 264.7 macrophages. Another study by Bigagli et al. [53] revealed that resveratrol and hydroxytyrosol synergistically inhibited the production of NO and PGE2 in LPS-stimulated RAW 264.7. In our study, we observed that LPS treatment elevated the NO and iNOS levels RAW264.7 cells; however, CCGA treatment could reduce NO production and downregulate the levels of iNOS in LPS-induced RAW264.7 cells. These results were similar to those of a study by Zhao et al. [37], which reported that CCGA at concentrations of 20, 40, and 80 µM reduced the production of NO and iNOS in LPS-induced RAW264.7 macrophages.

Numerous studies have reported a remarkable increase in pro-inflammatory cytokines in LPS-challenged macrophages and cells [54,55]. IL-1β, IL-6, and TNF-α are the most common cytokines associated with LPS-induced macrophages and have been reported to play vital roles in various biological functions, such as the control of the immune response, homeostasis, and inflammation [56]. In our study, the anti-inflammatory effect of CCGA on the mRNA and protein levels of inflammatory-related cytokines (TNF-α, IL-1β, IL-6, IL-8, COX-2, and IL-10) in LPS-induced RAW264.7 cells was elucidated. We observed that LPS treatment increased the production of inflammation-related cytokines (TNF- α, IL-1β, IL-6, IL-8, and COX-2) but reduced the levels of anti-inflammatory cytokine IL-10 in RAW264.7 cells. However, CCGA could prevent the production of these inflammatory cytokines (TNF- α, IL-1β, IL-6, IL-8, and COX-2) both at the mRNA and protein levels, whereas it enhanced the production of anti-inflammatory cytokines (IL-10) in RAW264.7 cells. These results are consistent with the study by Zhao et al. [37], which reported that the production of pro-inflammatory cytokines (TNF-α, IL-6, and COX-2) was reduced after treatment with 150 μM CCGA in LPS-induced inflammation in RAW264.7 macrophages. Similarly, Kim et al. [57] also reported that chlorogenic acid could suppress lipopolysaccharide-induced nitric oxide and interleukin-1β expression by hindering JAK2/STAT3 activation in RAW264.7 cells. Therefore, we conclude that CCGA prevented LPS-induced inflammation in RAW264.7 cells by inhibiting the production of inflammation-inducing cytokines.

NF-κB is the key tool for the regulation of pro-inflammatory enzymes and cytokines [58]. The initiation of the NF-κB pathway is regulated by IκB-α, whereas IκB kinase (IKK) regulates the degradation of IκB-α. [59]. Therefore, when IKK phosphorylates, it results in the degradation of IκB-α, which then activates NF-κB p65 to stimulate the production of pro-inflammatory factors in the cytoplasm [60,61]; hence, the suppression of the NF-κB signaling pathway is a promising strategy for the treatment of inflammatory disorders. Numerous studies have revealed several anti-inflammatory polyphenols that suppress NF-κB signaling [62]. Yang et al. [63] revealed that resveratrol inhibited the initiation of the NF-κB signaling pathway. In addition, another study by Serreli and Deiana [64] also reported that extra virgin olive oil polyphenols could inhibit NF-κB activation, thus reducing the levels of pro-inflammatory factors. Therefore, we investigated the effects of CCGA on the total phosphorylation of NF-κB family proteins (IKK and IκB) in RAW264.7 cells induced by LPS using Western blot. Our results showed that pretreatment with CCGA significantly blocked NF-kB promoter luciferase activity and downregulated the phosphorylation levels of IKK and IκB in LPS-induced RAW264.7 cells in a dose-dependent manner. Similar results were recorded in the study of Li et al. [65]. Therefore, we concluded that CCGA isolated from *A. adenophora* prevented inflammation by inhibiting the phosphorylation of NF-κB proteins. However, the exact molecular mechanism of the NF-κB signaling pathway inhibition by CCGA is not clear and thus requires further studies.

## 5. Conclusions

In conclusion, CCGA treatment reduced the levels of NO, iNOS, COX-2, and pro-inflammation cytokines in LPS-induced RAW264.7 cells. The anti-inflammatory properties of CCGA are achieved by blocking or suppressing the NF-κB signaling pathway in RAW 264.7 cells. Therefore, CCGA isolated from *Ageratina adenophora* can be used as a potential therapeutic agent for the treatment of inflammation-related diseases. However, there is a need for further studies on the exact activity of CCGA in inhibiting the NF-κB inflammation signaling pathway.

**Author Contributions:** X.M.: Conceptualization, Methodology, Writing—Original Draft, Writing—Review and Editing, Funding Acquisition. S.K.O., Conceptualization, Methodology, Writing—Original Draft, Writing—Review and Editing. L.H.: Conceptualization, Methodology, Writing—Original Draft, Software, Validation. J.W.: Methodology, Software, Validation. Z.R.: Writing—Software, Visualizing, Review, and Editing. J.D.: Writing—Validating, Resources, Review and Editing and supervision, Y.H.: Writing—Review and Editing, Supervision, Funding Acquisition. All authors have read and agreed to the published version of the manuscript.

**Funding:** This research was supported by Sichuan Province Science and Technology Support Program (Grant No. 2020YFS0337), School–Enterprise Cooperation Project (grant no. 002-2122339012).

**Institutional Review Board Statement:** All animal studies were approved by the Institutional Animal Care and Use Committee of Sichuan Agricultural University (certification no. 2019603005). All experiments were conducted in accordance with the Sichuan Agricultural University (SAU) Laboratory Animal Welfare and Ethics guidelines.

**Informed Consent Statement:** Not applicable.

**Data Availability Statement:** The data presented in this study are available on request from the corresponding author.

**Acknowledgments:** We would like to thank Nong Xiang and all the teaching staff at the College of Veterinary Medicine, Sichuan Agricultural University, Chengdu, for their advice and guidance in writing this paper.

**Conflicts of Interest:** We wish to confirm that there are no known conflict of interest associated with this publication, and there has been no significant financial support for this work that could have influenced its outcome.

## References

1. Mogensen, T.H. Pathogen recognition and inflammatory signaling in innate immune defenses. *Clin. Microbiol. Rev.* **2009**, *22*, 240–273. [CrossRef] [PubMed]
2. Serhan, C.N.; Savill, J. Resolution of inflammation: The beginning programs the end. *Nat. Immunol.* **2005**, *6*, 1191–1197. [CrossRef]
3. Hotamisligil, G.S. Inflammation, metaflammation and immunometabolic disorders. *Nature* **2017**, *542*, 177. [CrossRef] [PubMed]
4. Singh, N.; Baby, D.; Rajguru, J.P.; Patil, P.B.; Thakkannavar, S.S.; Pujari, V.B. Inflammation and cancer. *Ann. Afr. Med.* **2019**, *18*, 121–126. [CrossRef]
5. Catrysse, L.G.; van Loo, G. Inflammation and the metabolic syndrome: The tissuespecific functions of NF-κB. *Trends Cell Biol.* **2017**, *27*, 417–429. [CrossRef]
6. Eming, S.A.; Krieg, T.; Davidson, J.M. Inflammation in wound repair: Molecular and cellular mechanisms. *J. Investig. Dermatol.* **2007**, *127*, 514–525. [CrossRef] [PubMed]
7. Gilbert, D.L. Inflammation in Tic Disorders and Obsessive-Compulsive Disorder: Are PANS and PANDAS a Path Forward? *J. Child Neurol.* **2019**, *34*, 598–611. [CrossRef]
8. Kay, J.; Thadhani, E.; Samson, L.; Engelward, B. Inflammation-induced DNA damage, mutations and cancer. *DNA Repair* **2019**, *83*, 102673. [CrossRef]
9. Reinehr, T. Inflammatory markers in children and adolescents with type 2 diabetes mellitus. *Clin. Chim. Acta* **2019**, *496*, 100–107. [CrossRef]
10. McGeer, E.G.; McGeer, P.L. Inflammatory processes in Alzheimer's disease. *Prog. Neuro-Psychopharmacol. Biol. Psychiatry* **2003**, *27*, 741–749. [CrossRef]
11. Fingleton, B. Matrix metalloproteinases as regulators of inflammatory processes. *Biochimi. Biophys. (BBA)-Mol. Cell Res.* **2017**, *1864*, 2036–2042. [CrossRef]
12. Reddy, P.; Lent-Schochet, D.; Ramakrishnan, N.; McLaughlin, M.; Jialal, I. Metabolic syndrome is an inflammatory disorder: A conspiracy between adipose tissue and phagocytes. *Clin. Chim. Acta* **2019**, *496*, 35–44. [CrossRef]
13. Catarro, M.; Serrano, J.L.; Ramos, S.S.; Silvestre, S.; Almeida, P. Nimesulide analogues: From anti-inflammatory to antitumor agents. *Bioorg. Chem.* **2019**, *88*, 102966. [CrossRef] [PubMed]
14. Schumann, R.R.; Pfeil, D.; Lamping, N.; Kirschning, C.; Scherzinger, G.; Schlag, P.; Karawajew, L.; Herrmann, F. Lipopolysaccharide induces the rapid tyrosine phosphorylation of the mitogen-activated protein kinases erk-1 and p38 in cultured human vascular endothelial cells requiring the presence of soluble CD14. *Blood* **1996**, *87*, 2805–2814. [CrossRef] [PubMed]
15. Li, T.; Wu, Y.N.; Wang, H.; Ma, J.Y.; Zhai, S.S.; Duan, J. Dapk1 improves inflammation, oxidative stress and autophagy in LPS-induced acute lung injury via p38MAPK/NF-κB signaling pathway. *Mol. Immunol.* **2020**, *120*, 13–22. [CrossRef]
16. Mitoma, H.; Horiuchi, T.; Hatta, N.; Tsukamoto, H.; Harashima, S.; Kikuchi, Y.; Otsuka, J.; Okamura, S.; Fujita, S.; Harada, M. Infliximab induces potent anti-inflammatory responses by outside-toinside signals through transmembrane TNF-α. *Gastroenterology* **2005**, *128*, 376–392. [CrossRef] [PubMed]
17. Liu, X.; Nemeth, D.P.; McKim, D.B.; Zhu, L.; DiSabato, D.J.; Berdysz, O.; Gorantla, G.; Oliver, B.; Witcher, K.G.; Wang, Y.; et al. Cell-Type-Specific Interleukin 1 Receptor 1 Signaling in the Brain Regulates Distinct Neuroimmune Activities. *Immunity* **2019**, *50*, 317–333. [CrossRef] [PubMed]
18. Zhu, W.; London, N.R.; Gibson, C.C.; Davis, C.T.; Tong, Z.; Sorensen, L.K.; Shi, D.S.; Guo, J.; Smith, M.C.; Grossmann, A.H.; et al. Interleukin receptor activates a MYD88-ARNO-ARF6 cascade to disrupt vascular stability. *Nature* **2012**, *492*, 252–255. [CrossRef]
19. Petersen, A.; Pedersen, B. The role of IL-6 in mediating the anti-inflammatory. *J. Physiol. Pharmacol.* **2006**, *57*, 43–51.

20. Singh, J.K.; Simões, B.M.; Howell, S.J.; Farnie, G.; Clarke, R.B. Recent advances reveal IL-8 signaling as a potential key to targeting breast cancer stem cells. *Breast Cancer Res.* **2013**, *15*, 210. [CrossRef]
21. Zhu, X.; Yao, Y.; Yang, J.; Zhengxie, J.; Li, X.; Hu, S.; Zhang, A.; Dong, J.; Zhang, C.; Gan, G. COX-2-PGE2 signaling pathway con-tributes to hippocampal neuronal injury and cognitive impairment in PTZ-kindled epilepsy mice. *Int. Immunopharmacol.* **2020**, *87*, 106801. [CrossRef] [PubMed]
22. Lee, A.K.; Sung, S.H.; Kim, Y.C.; Kim, S.G. Inhibition of lipopolysaccharide inducible nitric oxide synthase, TNF-alpha and COX-2 expression by sauchinone effects on I-kappaBalpha phosphorylation, C/EBP and AP-1 activation. *Br. J. Pharmacol.* **2003**, *139*, 11–20. [CrossRef]
23. Spiller, F.; Oliveira Formiga, R.; Fernandes da Silva Coimbra, J.; Alves-Filho, J.C.; Cunha, T.M.; Cunha, F.Q. Targeting nitric oxide as a key modulator of sepsis, arthritis and pain. *Nitric Oxide* **2019**, *89*, 32–40. [CrossRef]
24. Ally, A.; Powell, I.; Ally, M.M.; Chaitoff, K.; Nauli, S.M. Role of neuronal nitric oxide synthase on cardiovascular functions in physiological and pathophysiological states. *Nitric Oxide* **2020**, *102*, 52–73. [CrossRef]
25. Ying, X.; Yu, K.; Chen, X.; Chen, H.; Hong, J.; Cheng, S.; Peng, L. Piperine inhibits LPS induced expression of inflammatory mediators in RAW 264.7 cells. *Cell Immunol.* **2013**, *285*, 49–54. [CrossRef]
26. Detmers, P.A.; Hernandez, M.; Mudgett, J.; Hassing, H.; Burton, C.; Mundt, S.; Chun, S.; Fletcher, D.; Card, D.J.; Lisnock, J.; et al. Deficiency in inducible nitric oxide synthase results in reduced atherosclerosis in apolipoprotein E-deficient mic. *J. Immunol.* **2000**, *16*, 3430–3435. [CrossRef] [PubMed]
27. Korhonen, R.; Lahti, A.; Kankaanranta, H.; Moilanen, E. Nitric oxide production and signaling in inflammation. *Curr. Drug Targets Inflamm. Allergy* **2005**, *4*, 471–479. [CrossRef]
28. Murtas, D.; Piras, F.; Minerba, L.; Ugalde, J.; Piga, M.; Maxia, C.; Perra, M.T.; Sirigu, P. Nuclear factor-κB expression is predictive of overall survival in patients with cutaneous melanoma. *Oncol. Lett.* **2010**, *1*, 633–639. [CrossRef] [PubMed]
29. Liu, H.; Talalay, P. Relevance of anti-inflammatory and antioxidant activities of exemestane and synergism with sulforaphane for disease prevention. *Proc. Natl. Acad. Sci. USA* **2013**, *110*, 19065–19070. [CrossRef] [PubMed]
30. Babu, P.V.; Liu, D.; Gilbert, E.R. Recent advances in understanding the anti-diabetic actions of dietary flavonoids. *J. Nutr. Biochem.* **2013**, *24*, 1777–1789. [CrossRef]
31. Kumar, N.; Goel, N. Phenolic acids: Natural versatile molecules with promising therapeutic applications. *Biotechnol. Rep.* **2019**, *24*, e00370. [CrossRef] [PubMed]
32. Sato, Y.; Itagaki, S.; Kurokawa, T.; Ogura, J.; Kobayashi, M.; Hirano, T.; Iseki, K. In vitro and in vivo antioxidant properties of chlorogenic acid and caffeic acid. *Int. J. Pharm.* **2011**, *403*, 136–138. [CrossRef]
33. Lou, Z.; Wang, H.; Zhu, S.; Ma, C.; Wang, Z. Antibacterial activity and mechanism of action of chlorogenic acid. *J. Food Sci.* **2011**, *76*, M398–M403. [CrossRef]
34. Dos Santos, M.D.; Almeida, M.C.; Lopes, N.P.; De Souza, G.E.P. Evaluation of the antiinflammatory, analgesic and antipyretic activities of the natural polyphenol chlorogenic acid. *Biol. Pharm. Bull.* **2006**, *29*, 2236–2240. [CrossRef]
35. Barahuie, F.; Hussein, M.Z.; Arulselvan, P.; Fakurazi, S.; Zainal, Z. Controlled in vitro release of the anticancer drug chlorogenic acid using magnesium/aluminiumlayered double hydroxide as a nanomatrix. *Sci. Adv. Mater.* **2016**, *8*, 501–513. [CrossRef]
36. Hwang, S.J.; Kim, Y.W.; Park, Y.; Lee, H.J.; Kim, K.W. Anti-inflammatory effects of chlorogenic acid in lipopolysaccharide-stimulated raw 264.7 cells. *Inflamm. Res.* **2014**, *63*, 81–90. [CrossRef]
37. Zhao, X.L.; Yu, L.; Zhang, S.D.; Ping, K.; Ni, H.Y.; Qin, X.Y.; Zhao, C.J.; Wang, W.; Efferth, T.; Fu, Y.J. Cryptochlorogenic acid attenu-ates LPS-induced inflam;matory response and oxidative stress via upregulation of the Nrf2/HO-1 signaling pathway in RAW 264.7 macrophages. *Int. Immunopharmacol.* **2020**, *83*, 106436. [CrossRef]
38. Ouyang, C.; Liu, X.; Yan, D.; Li, Y.; Wang, Q.; Cao, A.; Guo, M. Immunotoxicity assessment of cadinene sesquiterpenes from Eupatorium adenophorum in mice. *J. Integr. Agric.* **2016**, *15*, 2319–2325. [CrossRef]
39. Okyere, S.K.; Wen, J.; Cui, Y.; Xie, L.; Gao, P.; Wang, J.; Wang, S.; Hu, Y. Toxic mechanisms and pharmacological properties of euptox A, a toxic monomer from A. adenophora. *Fitoterapia* **2021**, *155*, 105032. [CrossRef]
40. Ren, Z.; Gao, P.; Okyere, S.K.; Cui, Y.; Wen, J.; Jing, B.; Deng, J.; Hu, Y. Ageratina adenophora inhibits spleen immune function in rats via the loss of the FRC Network and Th1-Th2 Cell Ratio elevation. *Toxins* **2021**, *13*, 309. [CrossRef] [PubMed]
41. Cui, Y.; Okyere, S.K.; Gao, P.; Wen, J.; Cao, S.; Wang, Y.; Deng, J.; Hu, Y. *Ageratina adenophora* Disrupts the Intestinal Structure and Immune Barrier Integrity in Rats. *Toxins* **2021**, *13*, 651. [CrossRef]
42. Okyere, S.K.; Mo, Q.; Pei, G.; Ren, Z.; Deng, J.; Hu, Y. Euptox A induces G0/G1 arrest and apoptosis of hepatocyte via ROS, mitochondrial dysfunction and caspases-dependent path- ways in vivo. *J. Toxicol. Sci.* **2020**, *45*, 661–671. [CrossRef]
43. Ren, Z.; Okyere, S.K.; Wen, J.; Xie, L.; Cui, Y.; Wang, S.; Wang, J.; Cao, S.; Shen, L.; Ma, X.; et al. An Overview: The Toxicity of *Ageratina adenophora* on Animals and Its Possible Interventions. *Int. J. Mol. Sci.* **2021**, *22*, 11581. [CrossRef] [PubMed]
44. Okyere, S.K.; Wen, J.; Cui, Y.; Xie, L.; Gao, P.; Zhang, M.; Wang, J.; Wang, S.; Ran, Y.; Ren, Z.; et al. Bacillus toyonensis SAU-19 and SAU-20 isolated from *Ageratina adenophora* alleviates the intestinal structure and integrity damage associated with gut dysbiosis in mice fed high fat mice. *Front. Microbiol.* 2021; *in press*. [CrossRef]
45. Okyere, S.K.; Xie, L.; Wen, J.; Ran, Y.; Ren, Z.; Deng, J.; Hu, Y. Bacillus toyonensis SAU-19 Ameliorates Hepatic Insulin Resistance in High-Fat Diet/Streptozocin-Induced Diabetic Mice. *Nutrients* **2021**, *13*, 4512. [CrossRef] [PubMed]
46. Poudel, R.; Neupane, N.P.; Mukeri, I.H.; Alok, S.; Verma, A. An updated review on invasive nature, phytochemical evaluation, & pharmacological activity of *Ageratina adenophora*. *Int. J. Pharm. Sci. Res.* **2020**, *119*, 2510–2520.

47. Parthasarathy, M.; Evan Prince, S. The potential effect of phytochemicals and herbal plant remedies for treating drug-induced hepatotoxicity: A review. *Mol. Biol. Rep.* **2021**, *48*, 4767–4788. [CrossRef]
48. Forni, C.; Facchiano, F.; Bartoli, M.; Pieretti, S.; Facchiano, A.; D'Arcangelo, D.; Norelli, S.; Valle, G.; Nisini, R.; Beninati, S.; et al. Beneficial Role of Phytochemicals on Oxidative Stress and Age-Related Diseases. *Biomed. Res. Int.* **2019**, *2019*, 8748253. [CrossRef]
49. George, B.P.; Chandran, R.; Abrahamse, H. Role of Phytochemicals in Cancer Chemoprevention: Insights. *Antioxidants* **2021**, *10*, 1455. [CrossRef]
50. Aktan, F. iNOS-mediated nitric oxide production and its regulation. *Life Sci.* **2004**, *75*, 639–653. [CrossRef]
51. Cirino, G.; Distrutti, E.; Wallace, J.L. Nitric oxide and inflammation. *Inflamm. Allergy Drug Targets* **2006**, *5*, 115–119. [CrossRef] [PubMed]
52. Moore, K.; Howard, L.; Brownmiller, C.; Gu, I.; Lee, S.O.; Mauromoustakos, A. Inhibitory effects of cranberry polyphenol and volatile extracts on nitric oxide production in LPS activated RAW 264.7 macrophages. *Food Funct.* **2019**, *10*, 7091–7102. [CrossRef] [PubMed]
53. Bigagli, E.; Cinci, L.; Paccosi, S.; Parenti, A.; D'Ambrosio, M.; Luceri., C. Nutritionally relevant concentrations of resveratrol and hydroxytyrosol mitigate oxidative burst of human granulocytes and monocytes and the production of pro-inflammatory mediators in LPS-stimulated RAW 264.7 macrophages. *Int. Immunopharmacol.* **2017**, *43*, 147–155. [CrossRef]
54. Hung, Y.-L.; Miyazaki, H.; Fang, S.-H.; Li, C.-Y.; Suzuki, K. The Structural Characteristics of Green Tea Polyphenols on Lipopolysaccharide-Stimulated RAW Cells. *J. Nutr. Biol.* **2018**, *2*, 151–157. [CrossRef]
55. Li, C.-Y.; Suzuki, K.; Hung, Y.-L.; Yang, M.-S.; Yu, C.-P.; Lin, S.-P.; Hou, Y.-C.; Fang, S.-H. Aloe metabolites prevent LPS-induced sepsis and inflammatory response by inhibiting mitogen-activated protein kinase activation. *Am. J. Chin. Med.* **2017**, *45*, 847–861. [CrossRef] [PubMed]
56. Yeom, M.; Kim, J.H.; Min, J.H.; Hwang, M.K.; Jung, H.S.; Sohn, Y. Xanthii fructus inhibits inflammatory responses in LPS-stimulated RAW 264.7 macrophages through suppressing NF-κB and JNK/p38 MAPK. *J. Ethnopharmacol.* **2015**, *176*, 394–401. [CrossRef]
57. Kim, S.H.; Park, S.Y.; Park, Y.L.; Myung, D.S.; Rew, J.S.; Joo, Y.E. Chlorogenic acid suppresses lipopolysaccharide-induced nitric oxide and interleukin-1β expression by inhibiting JAK2/STAT3 activation in RAW264.7 cells. *Mol. Med. Rep.* **2017**, *16*, 9224–9232. [CrossRef]
58. Kunnumakkara, A.B.; Shabnam, B.; Girisa, S.; Harsha, C.; Banik, K.; Devi, T.B.; Choudhury, R.; Sahu, H.; Parama, D.; Sailo, B.L.; et al. Inflammation, NF-κB, and Chronic Diseases: How are They Linked? *Crit. Rev. Immunol.* **2020**, *40*, 1–39. [CrossRef] [PubMed]
59. Endale, M.; Park, S.C.; Kim, S.; Kim, S.H.; Yang, Y.; Cho, J.Y.; Rhee, M.H. Quercetin disrupts tyrosine-phosphorylated phosphatidylinositol 3-kinase and myeloid differentiation factor-88 association, and inhibits MAPK/AP-1 and IKK/NF-κB-induced inflammatory mediators production in RAW 264.7 cells. *Immunobiology* **2013**, *218*, 1452–1467. [CrossRef]
60. Cheng, S.C.; Huang, W.C.; Pang, J.H.; Wu, Y.H.; Cheng, C.Y. Quercetin Inhibits the Production of IL-1β-Induced Inflammatory Cytokines and Chemokines in ARPE-19 Cells via the MAPK and NF-κB Signaling Pathways. *Int. J. Mol. Sci.* **2019**, *20*, 2957. [CrossRef] [PubMed]
61. Pontoriero, M.; Fiume, G.; Vecchio, E.; de Laurentiis, A.; Albano, F.; Iaccino, E.; Mimmi, S.; Pisano, A.; Agosti, V.; Giovannone, E.; et al. Activation of NF-κB in B cell receptor signaling through Bruton's tyrosine kinase-dependent phosphorylation of IκB-α. *J. Mol. Med.* **2019**, *97*, 675–690. [CrossRef] [PubMed]
62. Yahfoufi, N.; Alsadi, N.; Jambi, M.; Matar, C. The Immunomodulatory and Anti-Inflammatory Role of Polyphenols. *Nutrients* **2018**, *10*, 1618. [CrossRef] [PubMed]
63. Yang, X.; Xu, S.; Qian, Y.; Xiao, Q. Resveratrol regulates microglia M1/M2 polarization via PGC-1α in conditions of neuroinflammatory injury. *Brain Behav. Immun.* **2017**, *64*, 162–172. [CrossRef] [PubMed]
64. Serreli, G.; Deiana, M. Extra Virgin Olive Oil Polyphenols: Modulation of Cellular Pathways Related to Oxidant Species and Inflammation in Aging. *Cells* **2020**, *9*, 478. [CrossRef]
65. Li, Q.; Sun, J.; Mohammadtursun, N.; Wu, J.; Dong, J.; Li, L. Curcumin inhibits cigarette smoke-induced inflammation via modulating the PPARγ-NF-κB signaling pathway. *Food Funct.* **2019**, *10*, 7983–7994. [CrossRef]

nutrients

Article

# Procyanidin B2 Attenuates Nicotine-Induced Hepatocyte Pyroptosis through a PPARγ-Dependent Mechanism

Jia Liu [1,2], Qinyu Yao [1], Xinya Xie [1], Qi Cui [3], Tingting Jiang [3], Ziwei Zhao [3], Xiong Du [1], Baochang Lai [1], Lei Xiao [1,2] and Nanping Wang [4,*]

1 Cardiovascular Research Center, School of Basic Medical Sciences, Xi'an Jiaotong University, Xi'an 710061, China; jeven_liu@xjtu.edu.cn (J.L.); qinyu.yao@xjtu.edu.cn (Q.Y.); xiexinya916@stu.xjtu.edu.cn (X.X.); duxiong0414@stu.xjtu.edu.cn (X.D.); laibc@xjtu.edu.cn (B.L.); xiaolei0122@xjtu.edu.cn (L.X.)
2 Key Laboratory of Environment and Genes Related to Diseases, Xi'an Jiaotong University, Ministry of Education of China, Xi'an 710061, China
3 Advanced Institute for Medical Sciences, Dalian Medical University, Dalian 116044, China; qicui1019@163.com (Q.C.); jiangtdl@163.com (T.J.); curry15zw@163.com (Z.Z.)
4 Health Science Center, East China Normal University, Shanghai 200241, China
* Correspondence: npwang@hsc.ecnu.edu.cn; Tel.: +86-021-62235057

**Abstract:** Procyanidin B2 (PCB2), a natural flavonoid, has been demonstrated to exert anti-oxidation and anti-inflammatory effects on hepatic diseases. Increasing evidence shows the hepatoxicity of nicotine. However, whether PCB2 protects against nicotine-induced hepatoxicity and the underlying mechanisms remains uncharacterized. Here, we reported that nicotine promoted hepatocyte pyroptosis, as evidenced by the elevation of propidium iodide (PI)-positive cells, the activation of Caspase-1 and gasdermin D (GSDMD), the enhanced expression of NOD-like receptor containing pyrin domain 3 (NLRP3) and the increased release of lactate dehydrogenase (LDH), interleukin (IL)-1β and IL-18. The silencing of GSDMD by small interfering RNA (siRNA) efficiently inhibited the release of LDH and the secretion of IL-1β and IL-18. In addition, rosiglitazone (RGZ) prevented hepatocyte pyroptosis induced by nicotine. Furthermore, we showed that PCB2 attenuated nicotine-induced pyroptosis through the activation of peroxisome proliferator-activated receptor-γ (PPARγ) in hepatocytes. Moreover, administration of PCB2 ameliorated liver injury and hepatocyte pyroptosis in nicotine-treated mice. Hence, our findings demonstrated that PCB2 attenuated pyroptosis and liver damage in a PPARγ-dependent manner. Our results suggest a new mechanism by which PCB2 exerts its liver protective effects.

**Keywords:** Procyanidin B2; nicotine; pyroptosis; peroxisome proliferator-activated receptor-γ; hepatocyte

**Citation:** Liu, J.; Yao, Q.; Xie, X.; Cui, Q.; Jiang, T.; Zhao, Z.; Du, X.; Lai, B.; Xiao, L.; Wang, N. Procyanidin B2 Attenuates Nicotine-Induced Hepatocyte Pyroptosis through a PPARγ-Dependent Mechanism. *Nutrients* **2022**, *14*, 1756. https://doi.org/10.3390/nu14091756

Academic Editors: Daniela Rigano and Paola Bontempo

Received: 2 April 2022
Accepted: 21 April 2022
Published: 22 April 2022



## 1. Introduction

Tobacco smoking is reported to be one of the leading causes of preventable mortality [1]. The toxicity of tobacco combustion is attributed to a wide variety of products including smoke tars, nitrosamines, aldehydes and heavy metals [2]. Nicotine, a natural active alkaloid, is the addictive component of tobacco and e-cigarettes (an alternative to conventional cigarettes) [3]. It has been reported that nicotine plays a critical role in the development of many diseases, such as cardiovascular disorders and lung cancer [4]. Since clinical trials indicate that smoking is associated with the pathogenesis of hepatic diseases, including non-alcoholic fatty liver disease (NAFLD), hepatitis C and liver cancer [5–9], a quantity of studies focus on the hepatoxicity effect of nicotine. Bandyopadhyaya et al. reported that nicotine induced DNA damage and decreased the contents of total DNA in the liver tissues of Wistar rats, suggesting a nicotine-promoted genotoxicity [10]. In addition, EI-Sokkary et al. revealed the increased level of lipid peroxidation and the reduced activity of superoxide dismutase in the liver of chronic nicotine-administered Sprague Dawley rats [11]. Similarly, a recent study has reported that nicotine increases hepatic cytochrome p-450 level and oxidative stress, as well as the dilation and congestion of central veins in the liver of Sprague Dawley rats [12].

Pyroptosis is a type of the inflammation-associated programmed cell death, which is initially observed in *Salmonella*-infected macrophages [13]. Following studies revealed that the gasdermin family proteins function as pore-forming effectors, leading to cell membrane permeabilization and pyroptosis [14,15]. Recent studies have demonstrated the crucial role of pyroptosis in liver diseases. For instance, by using the hyperactive NLRP3 knock-in mice, Wree et al. revealed that the activation of hepatocyte NLRP3 inflammasome resulted in pyroptotic cell death and consequent liver inflammation and fibrosis [16]. Khanova et al. found that the activation of Caspase-11/4 induced the cleavage of GSDMD, thus triggering pyroptosis in alcoholic hepatitis liver in mice and patients [17]. It has been reported that GSDMD-mediated lipogenesis and NF-kB signaling play a key role in the development of non-alcoholic steatohepatitis; GSDMD-NT, the activation form of GSDMD, is positively correlated with liver fibrosis and NAFLD [18]. In addition, the activation of NLRP3 inflammasome leads to pyroptotic cell death of hepatocytes to release NLRP3 inflammasome particles, which are internalized by adjacent hepatic stellate cells, activating them to induce liver fibrosis [19]. Furthermore, the activation of Caspase-4 in Apaf-1 pyroptosome triggers the cleavage of GSDME, therefore inducing pyroptosis and subsequent cholestatic liver failure [20]. Recent reports have indicated that nicotine promotes pyroptosis in macrophages and endothelial cells [21–23]. However, whether nicotine could induce hepatocytes to undergo pyroptotic cell death and, if so, what the underlying molecular mechanisms are remain uncharacterized.

Procyanidins are a type of flavonoid polyphenols, which are widely distributed in plant food such as fruits, vegetables, beans and teas [24]. PCB2, composed by 2 flavan-3-ol (−)-epicatechin, is the most ubiquitous procyanidins and has been shown to possess pharmacological properties [25]. Accumulating evidence demonstrates the hepatoprotective effects of PCB2. For instance, Feng et al. have shown that PCB2 suppresses the development of liver fibrosis through the inhibition of the hedgehog signaling pathway [26]. PCB2 increases TFEB-mediated lysosomal activation and decreases oxidative stress, thereby ameliorating free fatty acids-induced hepatic steatosis [25]. It has been reported that PCB2 protects against carbon tetrachloride-caused hepatic damage via the enhancement of the antioxidative system and the downregulation of the inflammatory response [27]. In addition, our previous study has demonstrated that PCB2 suppresses the activation of NLRP3 inflammasome in endothelial cells, suggesting that PCB2 may have a potential in inhibiting pyroptosis [28]. However, whether PCB2 affects hepatocyte pyroptosis remains to be elucidated. To this end, we examined the role of nicotine in hepatocyte pyroptosis, and we also identified the potential effects of PCB2 on nicotine-triggered pyroptotic cell death and the underlying molecular mechanisms.

## 2. Materials and Methods

### 2.1. Reagents, Antibodies and Plasmids

Nicotine was from Cayman (Ann Arbor, MI, USA). PCB2, RGZ, GW9662, Hoechst 33342 and PI were from Sigma (St. Louis, MO, USA). Antibodies against Gasdermin D (39754), HMGB1 (6893), Cleaved-Caspase-1 (4199) and NLRP3 (15101) were from Cell Signaling Technology (Danvers, MA, USA). Antibody against Vinculin (V-4505) was obtained from Sigma. Flag-Gsdmd-FL (80950), Flag-Gsdmd-NT (80951) and Flag-Gsdmd-NT 4A (80952) were from Addgene (Watertown, MA, USA).

### 2.2. Animal Procedure

Male C57BL/6J mice aged 8 weeks were obtained from Vital River Laboratories (Beijing, China). Mice received one week of acclimatization and allocated into four groups at random: the vehicle group, the nicotine group, the PCB2 group and the nicotine plus PCB2 group ($n$ = 6 per group). PCB2 (50 mg/kg/day) or saline was daily administrated through oral gavage for 6 weeks. After 2 weeks of PCB2 administration, mice received once-daily intraperitoneal (IP) injections of nicotine (2 mg/kg/day) or saline for 4 weeks. The processes of animal care and experiments were authorized by the Animal Research Committee of Xi'an Jiaotong University (approval no. 2021-1723), and all animal care complied with the National Institutes of Health *Guide for the Care and Use of Laboratory Animals*. For sacrifice, mice were euthanized using the $CO_2$ method in accordance with the 2013 AVMA guidelines.

### 2.3. Cell Culture, Transfection and RNA Interference

Human hepatocyte LO2 cells were obtained from ATCC (Manassas, VA, USA) and cultured in DMEM and incubated with 100 μM nicotine for 24 h to induce pyroptosis. Plasmids and siRNA were transfected into LO2 cells by using Lipofectamine 3000 (Invitrogen, Carlsbad, CA, USA) following the manufacturer's instructions. The siRNA sequences targeting human GSDMD were as follows: siGSDMD-A, 5′-GUGUGUCAACCUGUCUAUCAATT-3′ (forward) and 5′-UUGAUAGACAGGUUGACACACTT-3′ (reverse); siGSDMD-B, 5′-CAGCACCUCAAUGAAUGUGUATT-3′ (forward) and 5′-UACACAUUCAUUGAG GUGCUGTT-3′ (reverse).

### 2.4. Immunoblotting

Cells and mouse liver tissues were homogenized and centrifuged at 12,000× *g* for 10 min to extract proteins. A BCA protein assay kit (ThermoFisher, Waltham, MA, USA) was used to calculate the protein concentrations. The proteins were separated by SDS-PAGE gels, transferred to PVDF membranes and immunoblotted with indicated antibodies.

### 2.5. Caspase-1 Activity Assay

The activity of Caspase-1 was determined by using Caspase-1 activity assay kit (Beyotime, Shanghai, China). Briefly, LO2 cells or 20 mg mouse liver tissues were homogenized at 4 °C. After centrifuge, Ac-YVAD-pNA (acetyl-Tyr-Val-Ala-Asp p-nitroanilide), a substrate of Caspase-1, were incubated with the supernatants at 37 °C for 2 h. The production of yellow pNA was measured at 405 nm with a spectrophotometer.

### 2.6. LDH Release Assay

The activity of LDH was determined by using an LDH cytotoxicity assay kit (ThermoFisher Scientific) following the manufacturer's instructions. The absorbance at 490 nm was measured using a spectrophotometer. Cytotoxicity was calculated against the absorbance of LDH releasing reagent treated cells.

### 2.7. Cell Death Assay

Necrotic cell death was examined with Hoechst 33342/PI staining. LO2 cells were plated in 6-well plate ($1 \times 10^5$ cells/well). The cells were incubated with a mixed solution of Hoechst 33342 (0.5 μg/mL) and PI (1 μg/mL) in a 37 °C incubator for 20 min. The fluorescence imaging was taken with a confocal microscope (E-C2; Nikon, Japan). ImageJ software was used for the quantification of PI-positive cells.

### 2.8. Enzyme-Linked Immunosorbent Assay (ELISA)

The concentrations of IL-1β and IL-18 in mouse serums and cultured medium of hepatocytes were determined by ELISA kits (Elabscience, Wuhan, China) following the manufacturer's protocols.

### 2.9. Cell Viability Assay

Cell viability was measured with CCK-8 assay and MTT assay. For CCK-8 assay, LO2 cells were cultured in 96-well plate, cell viability was determined using cell counting kit-8 (Sigma) in accordance with the manufacturer's instructions. Absorbance was measured at 450 nm using a spectrophotometer. For MTT assay, LO2 cells were seeded in 96-well plate, cell viability was determined using MTT cell proliferation and cytotoxicity assay kit (Beyotime) following the manufacturer's instructions.

*2.10. Statistical Analysis*

The statistical differences were analyzed with Student's *t*-test (evaluations of two groups) or one-way ANOVA (comparisons among multiple groups) followed by Bonferroni post-test and expressed as mean ± SD. $p < 0.05$ was considered statistically significant.

## 3. Results

*3.1. Nicotine Triggers Hepatocyte Pyroptosis through the Cleavage of GSDMD*

To examine the cytotoxicity of nicotine on hepatocytes, LO2 cells were exposed to nicotine at various concentrations. As shown in Figure 1A and Figure S1, nicotine triggered cell death in a dose-dependent manner. Next, we performed Hoechst 33342/PI double staining to discriminate between apoptotic cell death and necrotic cell death after hepatocytes exposed to nicotine. Our results showed that nicotine triggered necrotic cell death as indicated by the enhancement of PI positive staining rate (Figure 1B,C). In addition, LDH release was also enhanced in a concentration-dependent manner (Figure 1D). Both of pyroptosis and necroptosis are lytic types of necrotic cell death, but they require different membrane pore formation proteins [29]. To ascertain whether nicotine induces pyroptotic cell death, we treated LO2 cells with different doses of nicotine. The protein abundance of GSDMD-NT was increased upon the nicotine treatment (Figure 1E,F). Pyroptosis is mediated by inflammasome and Caspase activation, immunoblotting exhibited that nicotine up-regulated the protein abundances of NLRP3 and cleaved-Caspase-1, as well as high mobility group box protein 1 (HMGB1) (Figure 1E,F). Consistently, the activity of Caspase-1 was increased upon the nicotine exposure (Figure 1G). To further illustrate nicotine-promoted hepatocyte pyroptosis, we tested the levels of IL-1β and IL-18 in the conditioned medium by ELISA. The concentrations of these two inflammatory cytokines were also augmented following nicotine stimulation (Figure 1H,I). These results indicated that nicotine promoted hepatocyte pyroptosis in a concentration-dependent manner.

The gasdermins are a protein family causing membrane pore formation and lytic pro-inflammatory cell death [14]. To test whether the nicotine-induced pyroptosis is a GSDMD-dependent type of cell death, we silenced the expression of endogenous GSDMD in LO2 cells (Figure 2A,B). The depletion of GSDMD significantly attenuated cell viability impaired by nicotine (Figure 2C). Moreover, silencing of GSDMD inhibited the release of LDH and the generations of IL-1β and IL-18 (Figure 2D–F). It has been revealed that the cleavage of GSDMD plays a crucial role in pyroptotic cell death [30]. We showed that ectopic expression of mouse Gsdmd-NT (N-terminal cleavage product of Gsdmd-FL), but not Gsdmd-NT 4A (Gsdmd-NT mutant, loss of oligomerization and pore formation function [30]), significantly increased cell death in the context of nicotine treatment (Figure 2G–J). These data therefore revealed that nicotine-induced pyroptosis was dependent on the cleavage of GSDMD.

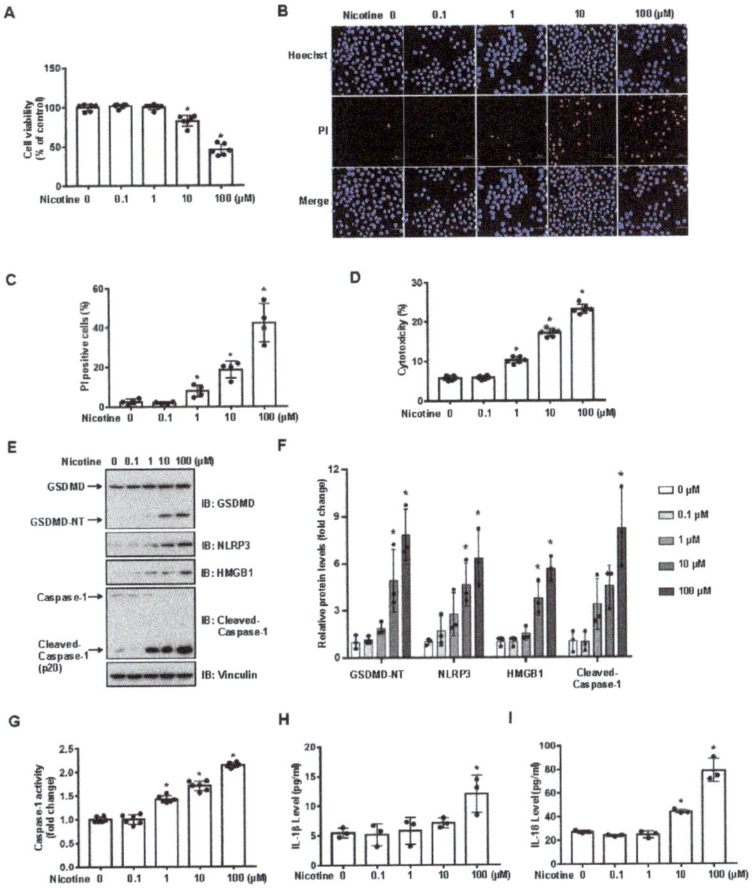


**Figure 1.** Nicotine induces pyroptosis in hepatocytes. LO2 cells were treated with nicotine at the indicated concentrations for 24 h. (**A**) Cell viability was detected by CCK-8 assay following exposure to nicotine ($n$ = 6). (**B**,**C**) Hoechst 33342/PI staining, fluorescence imaging (**B**) was acquired by confocal microscopy. Scale bar: 50 μm. The percentage of PI-positive cells (**C**) was analyzed by using ImageJ software ($n$ = 4). (**D**) Cytotoxicity was determined by LDH release assay following exposure to nicotine ($n$ = 6). (**E**) The proteins levels of GSDMD, NLRP3, HMGB1 and Cleaved-Caspase-1 was measured by using Western blotting. (**F**) Quantification of GSDMD-NT, NLRP3, HMGB1 and Cleaved-Caspase-1 levels as in (**E**) ($n$ = 3). (**G**) The activity of Caspase-1 was determined by using Caspase-1 activity assay kit ($n$ = 6). (**H**,**I**) The concentrations of IL-1β (**H**) and IL-18 (**I**) in the conditioned medium were determined by ELISA ($n$ = 6). Data shown represent the mean ± SD (*, $p < 0.05$).

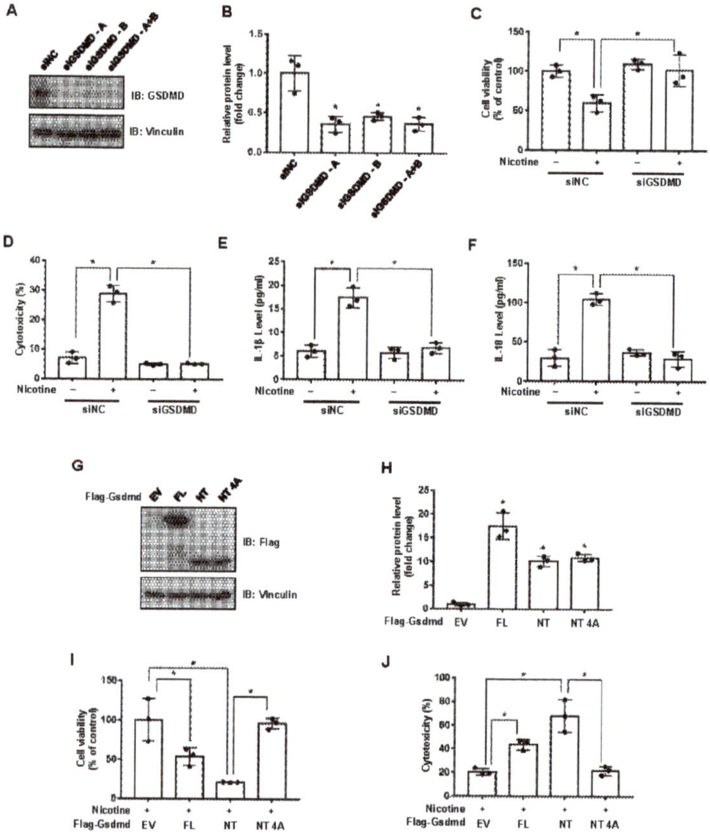


**Figure 2.** Nicotine induces pyroptosis through the cleavage of GSDMD. (**A**) LO2 cells were transfected with the indicated siRNA for 36 h, the protein levels of GSDMD were determined by using Western blotting. (**B**) Quantification of GSDMD protein level as in (**A**). (**C–F**) LO2 cells were transfected with the indicated siRNA for 24 h followed by the treatment with 100 μM nicotine for 24 h. Cell viability was determined by CCK-8 assay following exposure to nicotine (**C**). Cytotoxicity was determined by LDH release assay following exposure to nicotine (**D**). The concentrations of IL-1β (**E**) and IL-18 (**F**) in the conditioned medium were determined by ELISA. (**G**) Western blotting of cell lysates derived from LO2 cells transfected with the indicated plasmids for 36 h. (**H**) Quantification of protein levels as in (**G**). (**I**,**J**) LO2 cells transfected with the indicated plasmids for 24 h followed by the incubation with 100 μM nicotine for 24 h. Cell viability was detected by CCK-8 assay following exposure to nicotine (**I**). Cytotoxicity was determined by LDH release assay following exposure to nicotine (**J**). $n = 3$, data shown represent the mean ± SD (*, $p < 0.05$).

### 3.2. Activation of PPARγ Prevents Nicotine-Triggered Pyroptosis

PPARs regulate the expression of genes involved in glycolipid metabolism, differentiation and inflammation [31]. To investigate whether PPARs might attenuate nicotine-induced pyroptosis, LO2 cells were incubated with the agonist of PPARγ or PPARδ before nicotine treatment. As shown in Figure S2A,B and Figure 3C,D, RGZ (a selective agonist of PPARγ) but not GW501516 (a synthetic ligand of PPARδ), decreased the protein abundance of NLRP3 and the cleavage of GSDMD caused by nicotine. To confirm the hypothesis that RGZ suppresses pyroptosis through the activation of PPARγ. LO2 cells were exposed to GW9662, an antagonist of PPARγ, followed by the treatment of RGZ and nicotine. As shown in Figure 3A, GW9662 blocked the beneficial effect of RGZ on nicotine-impaired cell viability. In addition, in the presence of GW9662, RGZ failed to inhibit the elevated release of LDH (Figure 3B) and the activation of Caspase-1 induced by nicotine (Figure 3E). Moreover, nicotine-triggered secretion of inflammatory cytokines was reduced by RGZ whereas GW9662 successfully eliminated this effect (Figure 3F,G). These results indicated that PPARγ activation attenuated nicotine-induced hepatocyte pyroptosis.

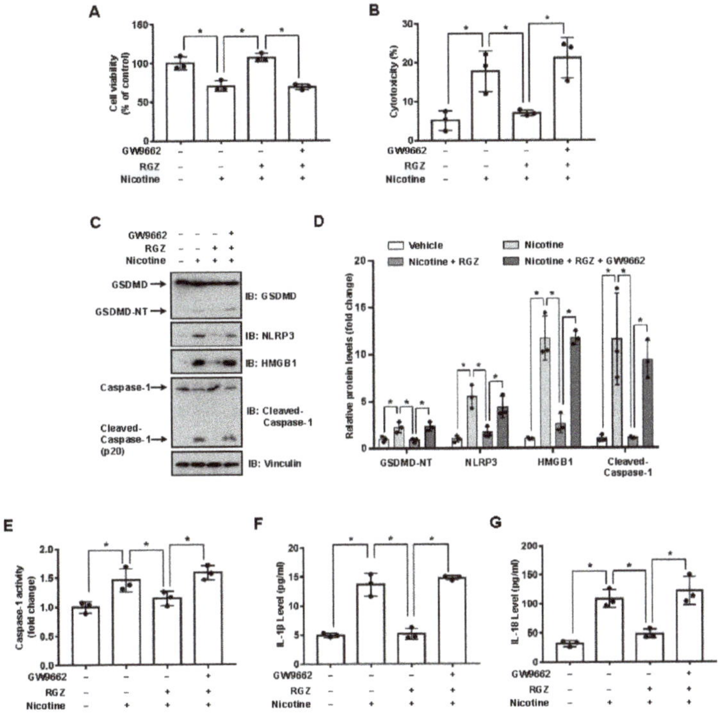

**Figure 3.** PPARγ activation attenuates nicotine-induced pyroptosis in hepatocytes. LO2 cells were treated with or without 10 μM GW9662 for 1 h followed by the treatment with 10 μM RGZ for 24 h. Cells were exposed to 100 μM nicotine for 24 h before harvesting. (**A**) Cell viability was detected by CCK-8 assay following exposure to nicotine. (**B**) Cytotoxicity was determined by LDH release assay following exposure to nicotine. (**C**) The proteins levels of GSDMD, NLRP3, HMGB1 and Cleaved-Caspase-1 by using Western blotting. (**D**) Quantification of GSDMD-NT, NLRP3, HMGB1 and Cleaved-Caspase-1 levels as in (**C**). (**E**) The activity of Caspase-1 was determined by using Caspase-1 activity assay kit. (**F**,**G**) The concentrations of IL-1β (**F**) and IL-18 (**G**) in the conditioned medium were determined by ELISA. $n = 3$, data shown represent the mean ± SD (*, $p < 0.05$).

### 3.3. PCB2 Attenuates Nicotine-Induced Pyroptosis in Hepatocytes

It has been revealed that the extracts or pure compounds derived from natural products exhibit anti-inflammatory properties [32]. Next, we decided to test whether natural products are advantageous in preventing nicotine-caused pyroptosis. By examining several natural products with anti-inflammatory activities, we observed that PCB2 mitigated nicotine-promoted the cleavage of GSDMD in LO2 hepatocytes (Figure S3). To further elucidate the possible role of PCB2 on nicotine-triggered hepatocyte pyroptosis, LO2 cells were incubated with PCB2 before the exposure to nicotine. As shown in Figure 4A, PCB2 ameliorated the impaired cell viability caused by nicotine treatment. In addition, the PI-positive cells and LDH release were increased after nicotine exposure, PCB2 attenuated the cell death provoked by nicotine (Figure 4B–D). Furthermore, immunoblotting showed that PCB2 decreased the cleavage of GSDMD and Caspase-1 triggered by nicotine (Figure 4E,F).

Moreover, PCB2 also diminished the activity of Caspase-1 (Figure 4G) and the concentration of pyroptosis response cytokines, including IL-1β and IL-18 (Figure 4H,I). Thus, we demonstrated that PCB2 effectively prevented LO2 cells from pyroptotic cell death triggered by nicotine.

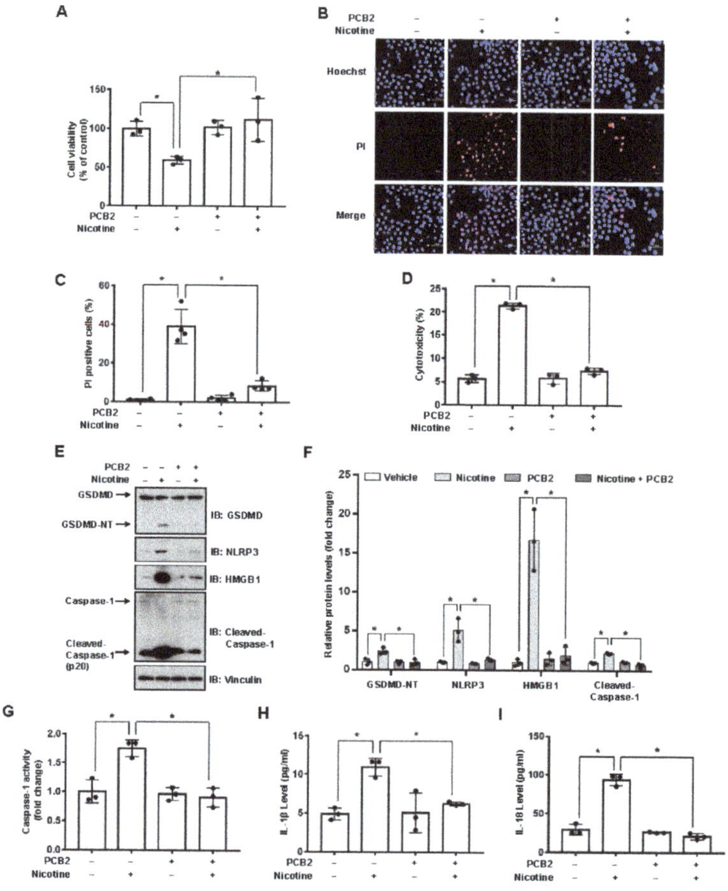

**Figure 4.** PCB2 prevents nicotine-induced pyroptosis in hepatocytes. LO2 cells were treated with 10 μM PCB2 for 24 h followed by the incubation with 100 μM nicotine for 24 h before harvesting. (**A**) Cell viability was detected by CCK-8 assay following exposure to nicotine ($n = 3$). (**B**,**C**) Hoechst 33342/PI staining, fluorescence imaging (**B**) was acquired by confocal microscopy. Scale bar: 50 μm. The percentage of PI-positive cells (**C**) was analyzed by using ImageJ software ($n = 4$). (**D**) Cytotoxicity was determined by LDH release assay following exposure to nicotine ($n = 3$). (**E**) The proteins levels of GSDMD, NLRP3, HMGB1 and Cleaved-Caspase-1 by using immunoblotting. (**F**) Quantification of GSDMD-NT, NLRP3, HMGB1 and Cleaved-Caspase-1 levels as in (**E**) ($n = 3$). (**G**) The activity of Caspase-1 was determined by using Caspase-1 activity assay kit ($n = 3$). (**H**,**I**) The concentrations of IL-1β (**H**) and IL-18 (**I**) in the conditioned medium were determined by ELISA ($n = 3$). Data shown represent the mean ± SD (*, $p < 0.05$).

### *3.4. PPARγ Antagonist Abrogates the Effects of PCB2 on Pyroptosis*

We recently showed that PCB2 induced macrophage M2 polarization through a PPARγ-dependent mechanism [33]. Thus, we hypothesized that PCB2 might mitigate pyroptosis via PPARγ activation. To this end, LO2 cells were pretreated with GW9662 and then with PCB2 before the exposure to nicotine in the presence of PCB2. We found that PCB2 failed to rescue nicotine-impaired cell viability in the presence of GW9662 (Figure 5A). Consistently, in the context of GW9662, PCB2 failed to decrease the number of PI-positive cells, the cleavage of GSDMD and Caspase-1, the activity of LDH and the release of inflammatory cytokines (Figure 5B–I). These results demonstrated that PCB2 attenuated nicotine-induced hepatocyte pyroptosis via a PPARγ-dependent mechanism.

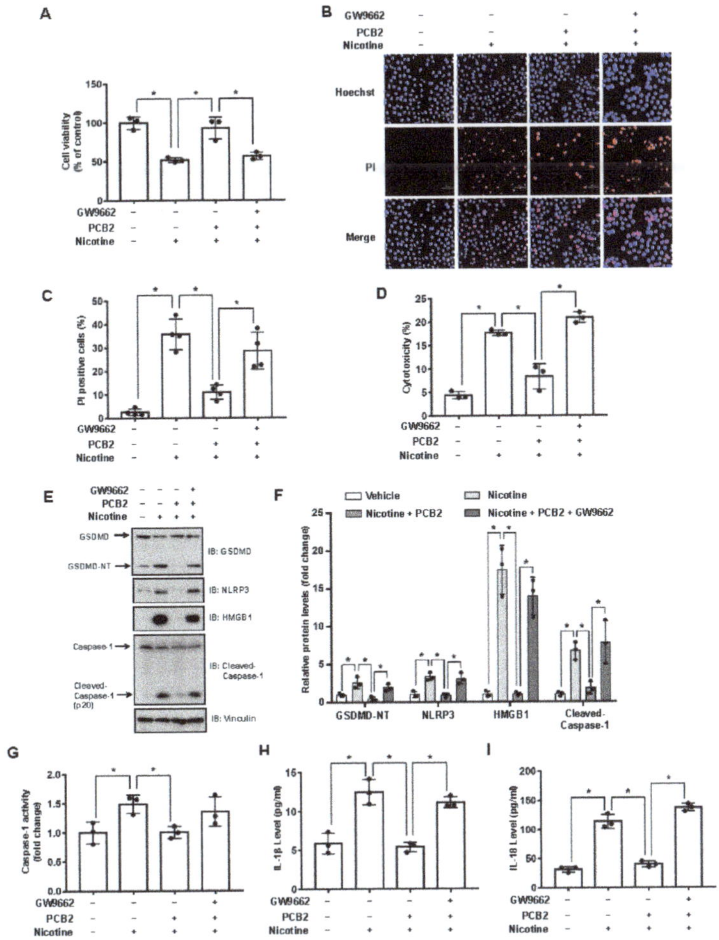


**Figure 5.** PCB2 prevents nicotine-induced pyroptosis through the activation of PPARγ. LO2 cells were treated with or without 10 μM GW9662 for 1 h followed by the treatment with 10 μM PCB2 for 24 h. Cells were exposed to 100 μM nicotine for 24 h before harvesting. (**A**) Cell viability was detected by CCK-8 assay following exposure to nicotine ($n = 3$). (**B**,**C**) Hoechst 33342/PI staining, fluorescence imaging (**B**) was acquired by confocal microscopy. Scale bar: 50 μm. The percentage of PI-positive cells (**C**) was analyzed by using ImageJ software ($n = 4$). (**D**) Cytotoxicity was determined by LDH release assay following exposure to nicotine ($n = 3$). (**E**) The proteins levels of GSDMD, NLRP3, HMGB1 and Cleaved-Caspase-1 by using immunoblotting. (**F**) Quantification of GSDMD-NT, NLRP3, HMGB1 and Cleaved-Caspase-1 levels as in (**E**) ($n = 3$). (**G**) The activity of Caspase-1 was determined by using Caspase-1 activity assay kit ($n = 3$). (**H**,**I**) The concentrations of IL-1β (**H**) and IL-18 (**I**) in the conditioned medium were determined by ELISA ($n = 3$). Data shown represent the mean ± SD (*, $p < 0.05$).

### 3.5. PCB2 Attenuates Liver Injury and Hepatocyte Pyroptosis in Mice

To validate these in vitro findings, we further examined the beneficial effects of PCB2 in mice (Figure 6A). Compared with the vehicle group, mice in the nicotine group had gained less weight; PCB2 ameliorated the weight loss (Figure S4). Liver injury markers, serum AST and ALT, were also analyzed. As shown in Figure 6B,C, compared with the vehicle group, the nicotine group had higher serum levels of AST and ALT, which were significantly increased by PCB2 administration. These results suggested that PCB2 attenuated weight loss and liver injury in nicotine-treated mice. Furthermore, immunoblotting showed that the protein abundance of GSDMD-NT, NLRP3 and HMGB1 were increased in the livers of nicotine group, whereas PCB2 administration to nicotine-treated mice decreased the expression of GSDMD-NT, NLRP3 and HMGB1 (Figure 6D,E). Consistently, nicotine increased the activity of Caspase-1 and the serum levels of inflammatory cytokines, PCB2 treatment mitigated these effects compared with nicotine group (Figure 6F–H). These data together supported the notion that PCB2 inhibited pyroptosis in the liver of nicotine-treated mice.

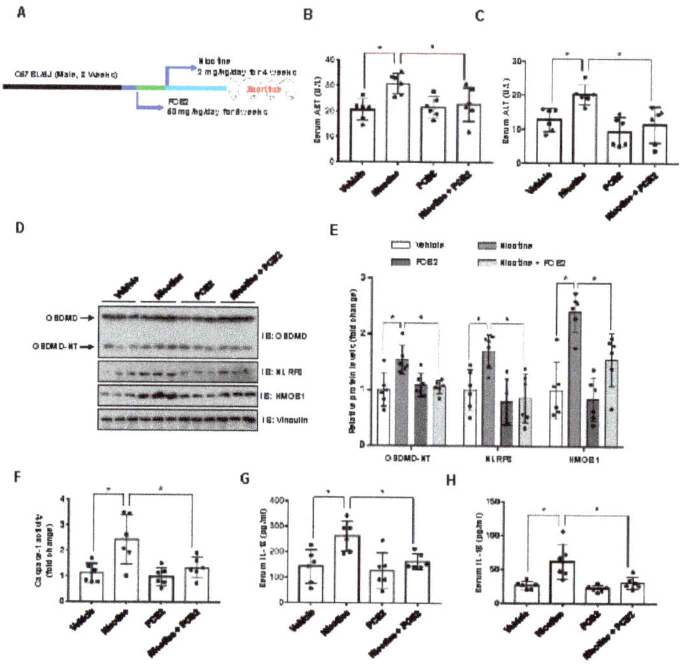

**Figure 6.** PCB2 ameliorates nicotine-induced pyroptosis and liver damage in mice. C57BL/6J mice were administrated with PCB2 (50 mg/kg/day) or saline for 6 weeks. After 2 weeks of PCB2 administration, mice were received daily IP injection of nicotine (2 mg/kg/day) or saline for 4 weeks. (**A**) A schematic illustration of the animal treatment procedure. (**B**) Serum levels of AST. (**C**) Serum levels of ALT. (**D**) Immunoblot analysis of the protein levels of GSDMD, NLRP3 and HMGB1 in liver tissues of the mice. (**E**) Quantification of GSDMD-NT, NLRP3 and HMGB1 levels as in (**D**). (**F**) The activity of Caspase-1 in liver tissues of the mice. (**G**,**H**) The concentrations of IL-1β (**G**) and IL-18 (**H**) in the serum of the mice. $n = 6$ per group; Data shown represent the mean ± SD (*, $p < 0.05$).

## 4. Discussion

The current study provided experimental evidence that PCB2 inhibited hepatocyte pyroptosis, thus ameliorating nicotine-induced liver injury. Our findings are as follows: (1) nicotine promoted the pyroptotic cell death of hepatocytes through the cleavage of GSDMD; (2) PCB2 attenuated hepatocyte pyroptosis and liver damage caused by nicotine; (3) the suppressive effect of PCB2 on pyroptosis was mediated via the activation of PPARγ.

Inflammasome complexes are assembled in response to various stimuli. Caspases are activated by inflammasomes or directly activated by the stimuli. The active Caspases process gasdermins to constitute cell membrane pores, triggering pyroptotic cell death and the release of cytosolic content such as LDH, HMGB1, IL-1β, IL-18 and even the entire inflammasome complexes. Several recent studies have indicated that nicotine-induced pyroptotic cell death contributes to the progression of atherosclerosis. Nicotine promotes atherosclerosis through the induction of NLRP3-dependent macrophage pyroptosis; HDAC6 depletion abolishes nicotine-caused pyroptosis by inhibiting NLRP3 transcription [22]. Another study in macrophages indicates that the increased level of ROS by nicotine induces the activation of TXNIP-NLRP3 inflammasome signaling, thereby promoting pyroptosis and exacerbating atherosclerosis [21]. In addition, the pro-pyroptosis effect of nicotine is attributed to its ability to up-regulate ROS generation and Caspase-1 activity in endothelial cells [23]. In this study, we revealed that nicotine induced pyroptosis in hepatocytes, featured by the increased level of PI positive rate, LDH release, IL-1β and IL-18, as well as the activation of GSDMD and Caspase-1 (Figure 1). These findings therefore demonstrated that nicotine triggers pyroptosis in hepatocytes.

The gasdermin family proteins are identified pyroptosis executors that cause cell membrane pore-forming and permeabilization. Structurally, gasdermins contain two distinct domains, which can be cleaved by Caspases. GSDMD and GSDME-induced pyroptosis have been well documented. The activation of GSDMD is regulated by Caspase-1 or Caspase-4/5/11 [34,35], apoptotic Caspase-8 also has been reported to induced GSDMD cleavage [36]. On the other hand, the cleavage of GSDME by Caspase-3 has been demonstrated to induce pyroptotic cell death [37]. It is noteworthy that the N-terminal of other gasdermin family members, such as GSDMA, GSDMB and GSDMC, also have a pore-formation function and can induce pyroptosis [38], although their upstream signaling remains largely uncharacterized. Our data showed that depletion of endogenous GSDMD inhibited nicotine-promoted pyroptosis (Figure 2A–F), indicating that nicotine induced hepatocyte pyroptosis via a GSDMD-dependent mechanism. In addition, ectopic expression of Gsdmd-NT increased the activity of LDH and decreased the viability of hepatocytes (Figure 2G–J), indicating that the cleavage of GSDMD plays a critical role in nicotine-caused pyroptosis. However, whether other gasdermins participate in nicotine-induced hepatocyte pyroptosis warrants further investigation.

Both acute and chronic liver injury may induce wound-healing response, leading to the formation of hepatic fibrosis even hepatic carcinoma [39]. It has been reported that chronic inhalation of electronic cigarettes vapor containing nicotine increases the level of circulating inflammatory cytokines and induces liver fibrosis in mice [40]. Compared with saline aerosol control, ApoE(−/−) mice exposed to electronic nicotine delivery systems exhibited hepatic lipid accumulation and oxidative stress, as well as increased hepatocyte apoptosis [3]. In addition, animal studies suggest that chronic nicotine administration leads to liver injury as indicated by the increased level of oxidative stress, DNA damage, transaminases and triglyceride accumulation in the liver [10–12,41,42]. Consistently, our data demonstrated that 4 weeks of nicotine exposure led to mouse liver damage (Figure 6B,C). Given the crucial role of pyroptosis in liver injury [16–19] and our in vitro and in vivo results that nicotine promoted hepatocyte pyroptosis, it is likely that nicotine-induced hepatocyte pyroptosis may contribute to the progression of liver injury.

A previous report has indicated that PPARγ plays a crucial role in protecting liver from diverse diseases such as hepatitis, liver fibrosis, liver cancer and NAFLD [43]. Although PPARγ agonist troglitazone showed severe hepatotoxicity, studies pointed out that the liver toxicity was associated with its exclusive side chain [44]. RGZ does not possess the hepatotoxic property of troglitazone and has no effect on inducing apoptosis of hepatocytes [45,46]. Conversely, RGZ is reported to prevent Schistosoma japonicum-induced liver fibrosis and reduce the serum level of inflammatory cytokines in mice [47]. Furthermore, RGZ attenuates the activation of NLRP3 inflammasome and subsequent Caspase-1 and IL-1β maturation in macrophage [48], suggesting RGZ may has a potential for the attenuation of pyroptosis. In this study, we demonstrated that RGZ reduced nicotine-caused hepatocyte pyroptotic cell death (Figure 3). More importantly, GW9662 prevented the effect of RGZ on nicotine-induced pyroptosis (Figure 3), indicating that the activation of PPARγ plays a significant role in ameliorating nicotine-induced hepatocyte pyroptosis.

Epidemiological studies indicate that fruit and vegetable consumptions are closely correlated with the reduced risk of hepatic diseases [49,50]. PCB2, a type of flavonoid polyphenols which is enriched in fruits and vegetables, has been shown to protect against liver diseases such as liver fibrosis and NAFLD [26,51]. The beneficial effects of PCB2 could be attributed to its anti-oxidation and ROS scavenging properties. However, other mechanisms might be involved in this process. We previously reported that PCB2 suppressed the detrimental effect of LPS through the inhibition of AP-1 in endothelial cells [28]. Consistently, our current study showed that PCB2 decreased nicotine-induced NLRP3 protein level and its downstream Caspase-1 maturation (Figure 4E,F). In addition, it has been reported that PCB2 promotes nuclear location of Nrf2 to induce the expression of PPARγ [52]. Another study points out that PPARγ suppresses NLRP3 inflammasome activation through binding with NLRP3 and disrupting the assemble of NLRP3 inflammasome complex [48]. We found that GW9662 inhibited the effect of PCB2 on nicotine-triggered pyroptosis (Figure 5), suggesting the inhibitory effect of PCB2 on NLRP3 inflammasome activation and subsequent pyroptosis is dependent on PPARγ activation. However, whether PCB2 promotes the interaction between PPARγ and NLRP3 to suppress pyroptosis warrants further investigation.

## 5. Conclusions

In conclusion, our results reveal a novel role of nicotine in hepatocyte pyroptotic cell death. Our findings also provide new insights into the mechanism by which PCB2 exerts its anti-pyroptotic effects and hepatic benefits through the activation of PPARγ, highlighting the beneficial effects of dietary procyanidins against hepatic diseases.

**Supplementary Materials:** The following supporting information can be downloaded at: https://www.mdpi.com/xxx/s1, Figure S1: Nicotine decreased cell viability of hepatocytes; Figure S2: Rosiglitazone (RGZ) prevents nicotine-induced pyroptosis in hepatocytes; Figure S3: Procyanidin B2 prevents nicotine-induced pyroptosis in hepatocytes; Figure S4: PCB2 ameliorated nicotine-induced weight loss.

**Author Contributions:** Conceptualization, J.L. and N.W.; methodology, J.L.; software, B.L.; validation, Q.Y.; formal analysis, L.X. and N.W.; investigation, J.L., Q.C., T.J., X.X., Z.Z. and X.D.; data curation, N.W.; writing—original draft preparation, J.L.; writing—review and editing, L.X. and N.W.; supervision, N.W.; funding acquisition, J.L., L.X. and N.W. All authors have read and agreed to the published version of the manuscript.

**Funding:** This research was funded by the National Natural Science Foundation of China (81770497, 82070829 and 81600389), the Ministry of Science and Technology (National Key R&D Program 2018YFA0800600) and the Special Financial Grant from the China Postdoctoral Science Foundation (2017T100757).

**Institutional Review Board Statement:** The animal study protocol was approved by the Animal Research Committee of Xi'an Jiaotong University (approval no. 2021-1723 and approved on 17 December 2021).

**Informed Consent Statement:** Not applicable.

**Data Availability Statement:** All relevant data are within the manuscript and its Supporting Information files.

**Conflicts of Interest:** The authors declare no conflict of interest.

## References

1. Frieden, T.R. A Safer, Healthier U.S.: The Centers for Disease Control and Prevention, 2009–2016. *Am. J. Prev. Med.* **2017**, *52*, 263–275. [CrossRef]
2. Long, G.A. Comparison of select analytes in exhaled aerosol from e-cigarettes with exhaled smoke from a conventional cigarette and exhaled breaths. *Int. J. Environ. Res. Public Health* **2014**, *11*, 11177–11191. [CrossRef]

3. Hasan, K.M.; Friedman, T.C.; Shao, X.; Parveen, M.; Sims, C.; Lee, D.L.; Espinoza-Derout, J.; Sinha-Hikim, I.; Sinha-Hikim, A.P. E-cigarettes and Western Diet: Important Metabolic Risk Factors for Hepatic Diseases. *Hepatology* **2019**, *69*, 2442–2454. [CrossRef]
4. Thorgeirsson, T.E.; Geller, F.; Sulem, P.; Rafnar, T.; Wiste, A.; Magnusson, K.P.; Manolescu, A.; Thorleifsson, G.; Stefansson, H.; Ingason, A.; et al. A variant associated with nicotine dependence, lung cancer and peripheral arterial disease. *Nature* **2008**, *452*, 638–642. [CrossRef]
5. Zimmermann, T.; Hueppe, D.; Mauss, S.; Buggisch, P.; Pfeiffer-Vornkahl, H.; Grimm, D.; Galle, P.R.; Alshuth, U. Effects of Smoking on Pegylated Interferon alpha 2a and First Generation Protease Inhibitor-based Antiviral Therapy in Naive Patients Infected with Hepatitis C Virus Genotype 1. *J. Gastrointest. Liver Dis. JGLD* **2016**, *25*, 15–24. [CrossRef]
6. Zein, C.O.; Unalp, A.; Colvin, R.; Liu, Y.C.; McCullough, A.J.; Nonalcoholic Steatohepatitis Clinical Research, N. Smoking and severity of hepatic fibrosis in nonalcoholic fatty liver disease. *J. Hepatol.* **2011**, *54*, 753–759. [CrossRef]
7. Zhao, J.K.; Wu, M.; Kim, C.H.; Jin, Z.Y.; Zhou, J.Y.; Han, R.Q.; Yang, J.; Zhang, X.F.; Wang, X.S.; Liu, A.M.; et al. Jiangsu Four Cancers Study: A large case-control study of lung, liver, stomach, and esophageal cancers in Jiangsu Province, China. *Eur. J. Cancer Prev. Off. J. Eur. Cancer Prev. Organ.* **2017**, *26*, 357–364. [CrossRef]
8. Karagozian, R.; Baker, E.; Houranieh, A.; Leavitt, D.; Baffy, G. Risk profile of hepatocellular carcinoma reveals dichotomy among US veterans. *J. Gastrointest. Cancer* **2013**, *44*, 318–324. [CrossRef]
9. El-Zayadi, A.; Selim, O.; Hamdy, H.; El-Tawil, A.; Badran, H.M.; Attia, M.; Saeed, A. Impact of cigarette smoking on response to interferon therapy in chronic hepatitis C Egyptian patients. *World J. Gastroenterol.* **2004**, *10*, 2963–2966. [CrossRef]
10. Bandyopadhyaya, G.; Sinha, S.; Chattopadhyay, B.D.; Chakraborty, A. Protective role of curcumin against nicotine-induced genotoxicity on rat liver under restricted dietary protein. *Eur. J. Pharmacol.* **2008**, *588*, 151–157. [CrossRef]
11. El-Sokkary, G.H.; Cuzzocrea, S.; Reiter, R.J. Effect of chronic nicotine administration on the rat lung and liver: Beneficial role of melatonin. *Toxicology* **2007**, *239*, 60–67. [CrossRef] [PubMed]
12. Khaled, S.; Makled, M.N.; Nader, M.A. Tiron protects against nicotine-induced lung and liver injury through antioxidant and anti-inflammatory actions in rats in vivo. *Life Sci.* **2020**, *260*, 118426. [CrossRef] [PubMed]
13. Brennan, M.A.; Cookson, B.T. Salmonella induces macrophage death by caspase-1-dependent necrosis. *Mol. Microbiol.* **2000**, *38*, 31–40. [CrossRef] [PubMed]
14. Broz, P.; Pelegrin, P.; Shao, F. The gasdermins, a protein family executing cell death and inflammation. *Nat. Rev. Immunol.* **2020**, *20*, 143–157. [CrossRef]
15. Shi, J.; Gao, W.; Shao, F. Pyroptosis: Gasdermin-Mediated Programmed Necrotic Cell Death. *Trends Biochem. Sci.* **2017**, *42*, 245–254. [CrossRef]
16. Wree, A.; Eguchi, A.; McGeough, M.D.; Pena, C.A.; Johnson, C.D.; Canbay, A.; Hoffman, H.M.; Feldstein, A.E. NLRP3 inflammasome activation results in hepatocyte pyroptosis, liver inflammation, and fibrosis in mice. *Hepatology* **2014**, *59*, 898–910. [CrossRef]
17. Khanova, E.; Wu, R.; Wang, W.; Yan, R.; Chen, Y.; French, S.W.; Llorente, C.; Pan, S.Q.; Yang, Q.; Li, Y.; et al. Pyroptosis by caspase11/4-gasdermin-D pathway in alcoholic hepatitis in mice and patients. *Hepatology* **2018**, *67*, 1737–1753. [CrossRef]
18. Xu, B.; Jiang, M.Z.; Chu, Y.; Wang, W.J.; Chen, D.; Li, X.W.; Zhang, Z.; Zhang, D.; Fan, D.M.; Nie, Y.Z.; et al. Gasdermin D plays a key role as a pyroptosis executor of non-alcoholic steatohepatitis in humans and mice. *J. Hepatol.* **2018**, *68*, 773–782. [CrossRef]
19. Gaul, S.; Leszczynska, A.; Alegre, F.; Kaufmann, B.; Johnson, C.D.; Adams, L.A.; Wree, A.; Damm, G.; Seehofer, D.; Calvente, C.J.; et al. Hepatocyte pyroptosis and release of inflammasome particles induce stellate cell activation and liver fibrosis. *J. Hepatol.* **2021**, *74*, 156–167. [CrossRef]
20. Xu, W.; Che, Y.; Zhang, Q.; Huang, H.; Ding, C.; Wang, Y.; Wang, G.; Cao, L.; Hao, H. Apaf-1 Pyroptosome Senses Mitochondrial Permeability Transition. *Cell Metab.* **2021**, *33*, 424–436.e410. [CrossRef]
21. Mao, C.Y.; Li, D.J.; Zhou, E.; Zhang, J.F.; Wang, C.Q.; Xue, C. Nicotine exacerbates atherosclerosis through a macrophage-mediated endothelial injury pathway. *Aging-US* **2021**, *13*, 7627–7643. [CrossRef]
22. Xu, S.; Chen, H.W.; Ni, H.E.; Dai, Q.Y. Targeting HDAC6 attenuates nicotine-induced macrophage pyroptosis via NF-kappa B/NLRP3 pathway. *Atherosclerosis* **2021**, *317*, 1–9. [CrossRef] [PubMed]
23. Wu, X.; Zhang, H.; Qi, W.; Zhang, Y.; Li, J.; Li, Z.; Lin, Y.; Bai, X.; Liu, X.; Chen, X.; et al. Nicotine promotes atherosclerosis via ROS-NLRP3-mediated endothelial cell pyroptosis. *Cell Death Dis.* **2018**, *9*, 171. [CrossRef] [PubMed]
24. Zhu, X.; Tian, X.; Yang, M.; Yu, Y.; Zhou, Y.; Gao, Y.; Zhang, L.; Li, Z.; Xiao, Y.; Moses, R.E.; et al. Procyanidin B2 Promotes Intestinal Injury Repair and Attenuates Colitis-Associated Tumorigenesis via Suppression of Oxidative Stress in Mice. *Antioxid. Redox Signal.* **2021**, *35*, 75–92. [CrossRef] [PubMed]
25. Su, H.M.; Li, Y.T.; Hu, D.W.; Xie, L.H.; Ke, H.H.; Zheng, X.D.; Chen, W. Procyanidin B2 ameliorates free fatty acids-induced hepatic steatosis through regulating TFEB-mediated lysosomal pathway and redox state. *Free Radic. Bio. Med.* **2018**, *126*, 269–286. [CrossRef]
26. Feng, J.; Wang, C.; Liu, T.; Li, J.; Wu, L.; Yu, Q.; Li, S.; Zhou, Y.; Zhang, J.; Chen, J.; et al. Procyanidin B2 inhibits the activation of hepatic stellate cells and angiogenesis via the Hedgehog pathway during liver fibrosis. *J. Cell. Mol. Med.* **2019**, *23*, 6479–6493. [CrossRef]
27. Yang, B.Y.; Zhang, X.Y.; Guan, S.W.; Hua, Z.C. Protective Effect of Procyanidin B2 against CCl4-Induced Acute Liver Injury in Mice. *Molecules* **2015**, *20*, 12250–12265. [CrossRef]

28. Yang, H.; Xiao, L.; Yuan, Y.; Luo, X.; Jiang, M.; Ni, J.; Wang, N. Procyanidin B2 inhibits NLRP3 inflammasome activation in human vascular endothelial cells. *Biochem. Pharmacol.* **2014**, *92*, 599–606. [CrossRef]
29. Frank, D.; Vince, J.E. Pyroptosis versus necroptosis: Similarities, differences, and crosstalk. *Cell Death Differ.* **2019**, *26*, 99–114. [CrossRef]
30. Liu, X.; Zhang, Z.; Ruan, J.; Pan, Y.; Magupalli, V.G.; Wu, H.; Lieberman, J. Inflammasome-activated gasdermin D causes pyroptosis by forming membrane pores. *Nature* **2016**, *535*, 153–158. [CrossRef]
31. Gross, B.; Pawlak, M.; Lefebvre, P.; Staels, B. PPARs in obesity-induced T2DM, dyslipidaemia and NAFLD. *Nat. Rev. Endocrinol.* **2017**, *13*, 36–49. [CrossRef]
32. Azab, A.; Nassar, A.; Azab, A.N. Anti-Inflammatory Activity of Natural Products. *Molecules* **2016**, *21*, 1321. [CrossRef]
33. Tian, Y.; Yang, C.; Yao, Q.; Qian, L.; Liu, J.; Xie, X.; Ma, W.; Nie, X.; Lai, B.; Xiao, L.; et al. Procyanidin B2 Activates PPARgamma to Induce M2 Polarization in Mouse Macrophages. *Front. Immunol.* **2019**, *10*, 1895. [CrossRef]
34. Shi, J.; Zhao, Y.; Wang, K.; Shi, X.; Wang, Y.; Huang, H.; Zhuang, Y.; Cai, T.; Wang, F.; Shao, F. Cleavage of GSDMD by inflammatory caspases determines pyroptotic cell death. *Nature* **2015**, *526*, 660–665. [CrossRef]
35. Kayagaki, N.; Stowe, I.B.; Lee, B.L.; O'Rourke, K.; Anderson, K.; Warming, S.; Cuellar, T.; Haley, B.; Roose-Girma, M.; Phung, Q.T.; et al. Caspase-11 cleaves gasdermin D for non-canonical inflammasome signalling. *Nature* **2015**, *526*, 666–671. [CrossRef]
36. Chen, K.W.; Demarco, B.; Heilig, R.; Shkarina, K.; Boettcher, A.; Farady, C.J.; Pelczar, P.; Broz, P. Extrinsic and intrinsic apoptosis activate pannexin-1 to drive NLRP3 inflammasome assembly. *EMBO J.* **2019**, *38*, 1638. [CrossRef]
37. Wang, Y.; Gao, W.; Shi, X.; Ding, J.; Liu, W.; He, H.; Wang, K.; Shao, F. Chemotherapy drugs induce pyroptosis through caspase-3 cleavage of a gasdermin. *Nature* **2017**, *547*, 99–103. [CrossRef]
38. Ding, J.; Wang, K.; Liu, W.; She, Y.; Sun, Q.; Shi, J.; Sun, H.; Wang, D.C.; Shao, F. Pore-forming activity and structural autoinhibition of the gasdermin family. *Nature* **2016**, *535*, 111–116. [CrossRef]
39. Hernandez-Gea, V.; Friedman, S.L. Pathogenesis of liver fibrosis. *Annu. Rev. Pathol.* **2011**, *6*, 425–456. [CrossRef]
40. Crotty Alexander, L.E.; Drummond, C.A.; Hepokoski, M.; Mathew, D.; Moshensky, A.; Willeford, A.; Das, S.; Singh, P.; Yong, Z.; Lee, J.H.; et al. Chronic inhalation of e-cigarette vapor containing nicotine disrupts airway barrier function and induces systemic inflammation and multiorgan fibrosis in mice. *Am. J. Physiol. Regul. Integr. Comp. Physiol.* **2018**, *314*, R834–R847. [CrossRef]
41. Kuru, P.; Bilgin, S.; Mentese, S.T.; Tazegul, G.; Ozgur, S.; Cilingir, O.T.; Akakin, D.; Yarat, A.; Kasimay, O. Ameliorative effect of chronic moderate exercise in smoke exposed or nicotine applied rats from acute stress. *Nicotine Tob. Res. Off. J. Soc. Res. Nicotine Tob.* **2015**, *17*, 559–565. [CrossRef]
42. Dangana, E.O.; Michael, O.S.; Omolekulo, T.E.; Areola, E.D.; Olatunji, L.A. Enhanced hepatic glycogen synthesis and suppressed adenosine deaminase activity by lithium attenuates hepatic triglyceride accumulation in nicotine-exposed rats. *Biomed. Pharmacother. Biomed. Pharmacother.* **2019**, *109*, 1417–1427. [CrossRef]
43. Wu, L.W.; Guo, C.Y.; Wu, J.Y. Therapeutic potential of PPAR gamma natural agonists in liver diseases. *J. Cell. Mol. Med.* **2020**, *24*, 2736–2748. [CrossRef]
44. Scheen, A.J. Thiazolidinediones and liver toxicity. *Diabetes Metab.* **2001**, *27*, 305–313. [PubMed]
45. Guo, L.; Zhang, L.; Sun, Y.M.; Muskhelishvili, L.; Blann, E.; Dial, S.; Shi, L.M.; Schroth, G.; Dragan, Y.P. Differences in hepatotoxicity and gene expression profiles by anti-diabetic PPAR gamma agonists on rat primary hepatocytes and human HepG2 cells. *Mol. Divers.* **2006**, *10*, 349–360. [CrossRef]
46. Bae, M.A.; Rhee, H.; Song, B.J. Troglitazone but not rosiglitazone induces G1 cell cycle arrest and apoptosis in human and rat hepatoma cell lines. *Toxicol. Lett.* **2003**, *139*, 67–75. [CrossRef]
47. Chen, H.; He, Y.W.; Liu, W.Q.; Zhang, J.H. Rosiglitazone prevents murine hepatic fibrosis induced by Schistosoma japonicum. *World J. Gastroenterol.* **2008**, *14*, 2905–2911. [CrossRef]
48. Yang, C.C.; Wu, C.H.; Lin, T.C.; Cheng, Y.N.; Chang, C.S.; Lee, K.T.; Tsai, P.J.; Tsai, Y.S. Inhibitory effect of PPARgamma on NLRP3 inflammasome activation. *Theranostics* **2021**, *11*, 2424–2441. [CrossRef]
49. Wu, Y.; Li, Y.; Giovannucci, E. Potential Impact of Time Trend of Lifestyle Risk Factors on Burden of Major Gastrointestinal Cancers in China. *Gastroenterology* **2021**, *161*, 1830–1841.e1838. [CrossRef]
50. Kim, S.A.; Shin, S. Fruit and vegetable consumption and non-alcoholic fatty liver disease among Korean adults: A prospective cohort study. *J. Epidemiol. Commun. H* **2020**, *74*, 1035–1042. [CrossRef]
51. Xing, Y.W.; Lei, G.T.; Wu, Q.H.; Jiang, Y.; Huang, M.X. Procyanidin B2 protects against diet-induced obesity and non-alcoholic fatty liver disease via the modulation of the gut microbiota in rabbits. *World J. Gastroenterol.* **2019**, *25*, 955–966. [CrossRef] [PubMed]
52. Liu, L.; Wang, R.; Xu, R.; Chu, Y.; Gu, W. Procyanidin B2 ameliorates endothelial dysfunction and impaired angiogenesis via the Nrf2/PPARgamma/sFlt-1 axis in preeclampsia. *Pharmacol. Res.* **2022**, 106127. [CrossRef] [PubMed]

*nutrients* MDPI

*Article*

# Yellow Chaste Weed and Its Components, Apigenin and Galangin, Affect Proliferation and Oxidative Stress in Blue Light-Irradiated HaCaT Cells

**Jung Yoen Park [1,†], See-Hyoung Park [2,†], Sae Woong Oh [1,†], Kitae Kwon [1], Eunbi Yu [1], Seoyoung Choi [1], Seoyoun Yang [1], Su Bin Han [1], Kwangsun Jung [3], Minkyung Song [4,*], Jae Youl Cho [5,*] and Jongsung Lee [1,*]**

1 Molecular Dermatology Laboratory, Department of Integrative Biotechnology, College of Biotechnology and Bioengineering, Sungkyunkwan University, Suwon City 16419, Korea; maria0502@skku.edu (J.Y.P.); hanzeeoo@skku.edu (S.W.O.); wesdwel@skku.edu (K.K.); yuebi95@skku.edu (E.Y.); csy2696@skku.edu (S.C.); chorim1004@skku.edu (S.Y.); subin816@skku.edu (S.B.H.)
2 Department of Bio and Chemical Engineering, Hongik University, Sejong City 30016, Korea; shpark74@hongik.ac.kr
3 Biocosmetics Laboratory, TOUN28 Inc., Seongnam 13449, Korea; jks8835@toun28.com
4 Integrative Research of T Cells Laboratory, Department of Integrative Biotechnology, College of Biotechnology and Bioengineering, Sungkyunkwan University, Suwon City 16419, Korea
5 Molecular Immunology Laboratory, Department of Integrative Biotechnology, College of Biotechnology and Bioengineering, Sungkyunkwan University, Suwon City 16419, Korea
* Correspondence: piscesmk@skku.edu (M.S.); jaecho@skku.edu (J.Y.C.); bioneer@skku.edu (J.L.); Tel.: +82-31-290-7861 (J.L.)
† These authors contributed equally to this work.

**Abstract:** While harmful effects of blue light on skin cells have been recently reported, there are few studies regarding natural products that alleviate its negative effects. Therefore, we investigated ameliorating effects of yellow chaste weed (YCW) (*Helichrysum arenarium*) extract and its components, apigenin and galangin, on blue light-irradiated HaCaT cells. In this study, we found that YCW extract improved the reduced proliferation of HaCaT cells induced by blue light-irradiation and reduced blue light-induced production of reactive oxygen species (ROS) levels. We also found that apigenin and galangin, the main components of YCW extract, showed the same activities as YCW extract. In experiments examining molecular mechanisms of YCW extract and its components such as apigenin and galangin, they all reduced expression of transient receptor potential vanilloid member 1 (TRPV1), its phosphorylation, and calcium ion ($Ca^{2+}$) influx induced by blue light irradiation. In addition, apigenin and galangin regulated phosphorylation of mitogen-activated protein kinases (MAPKs). They also reduced phosphorylation of mammalian sterile 20-like kinase-1/2 (MST-1/2), inducing phosphorylation of Akt (protein kinase B), one downstream molecule of MST-1/2. Moreover, apigenin and galangin promoted translocation of Forkhead box O3 (FoxO3a) from the nucleus to the cytosol by phosphorylating FoxO3a. Besides, apigenin and galangin interrupted blue light influences on expression of nuclear and secretory clusterin. Namely, they attenuated both upregulation of nuclear clusterin and downregulation of secretory clusterin induced by blue light irradiation. We also found that they downregulated apoptotic protein Bcl-2 associated X protein (Bax) and conversely upregulated anti-apoptotic protein B-cell lymphoma 2 (Bcl-2). Collectively, these findings indicate that YCW extract and its components, apigenin and galangin, antagonize the blue light-induced damage to the keratinocytes by regulating TRPV1/clusterin/FoxO3a and MAPK signaling.

**Keywords:** blue light; keratinocytes; TRPV1; calcium influx; ROS; clusterin; FoxO3a; cell proliferation; apoptosis

**Citation:** Park, J.Y.; Park, S.-H.; Oh, S.W.; Kwon, K.; Yu, E.; Choi, S.; Yang, S.; Han, S.B.; Jung, K.; Song, M.; et al. Yellow Chaste Weed and Its Components, Apigenin and Galangin, Affect Proliferation and Oxidative Stress in Blue Light-Irradiated HaCaT Cells. *Nutrients* **2022**, *14*, 1217. https://doi.org/10.3390/nu14061217

Academic Editors: Daniela Rigano and Anna Gramza-Michałowska

Received: 18 January 2022
Accepted: 10 March 2022
Published: 13 March 2022



## 1. Introduction

The skin is located at the outermost part of the human body, protecting us from a variety of external factors, including sun light, pollutants, and stresses. These various external factors induce many physiological changes in skin, such as aging, pigmentation/vitiligo, skin inflammation, and hair loss. Therefore, identifying the signal transduction caused by the external factors and developing treatments targeting them are important challenges in skin research [1–4].

Sunlight, one of the biggest factors affecting the skin, is largely composed of ultraviolet (UV), visible and infrared light, depending on energy and wavelength [5,6]. Among the various wavelengths, blue light (380–500 nm), also called high-energy visible (HEV) light, has especially received attention because of the increased use of electronic devices such as smartphones, laptops, computers, and TVs in everyday life [7,8]. Excessive exposure to blue light has been reported to induce several damages to skin cells, including oxidative stress, apoptosis, and reduced proliferation potential [9,10]. In particular, blue light induces several damages to the epidermis, including DNA damage and oxidative stress [11–13].

The transient receptor potential vanilloid member 1 (TRPV1) channel is a submember of the TRP cation channel family, expressed in kidney cells, bronchial epithelial cells, primary sensory neurons, and keratinocytes. TRPV1 is a non-selective cation receptor and responds to capsaicin. TRPV1 is activated by heat (>43 °C), low pH, ultraviolet, and blue light. Activated TRPV1 can also induce $Ca^{2+}$ influx in keratinocyte and mediate multiple signaling pathways. Blue light has been recently reported to upregulate and activate TRPV1, leading to ROS production and reduced cell proliferation [14]. In addition, in our previous report, we demonstrated that the effects of blue light on keratinocyte cell proliferation are mediated by upregulating TRPV1, a negative regulator of EGFR-FoxO3a signaling [10]. Blue light-induced production of ROS and TNF-$\alpha$ is also mediated through increased calcium influx via TRPV1 activation.

Calcium ions ($Ca^{2+}$) act as a secondary messenger for controlling various aspects of cell functions such as differentiation in keratinocytes and barrier homeostasis [15–17]. Calcium-sensing receptor and calcium-permeable channels expressed in skin could regulate various functions in skin barrier homeostasis [18–22]. Clusterin expression is also controlled by intracellular $Ca^{2+}$ influx [23,24]. Clusterin, a glycoprotein which is highly expressed and ubiquitously synthesized in the body, can be characterized into two isoforms: anti-apoptotic secretory CLU (sCLU) or apoptotic nuclear CLU (nCLU). These two forms of clusterin are controlled differently by $Ca^{2+}$ [25,26].

The Forkhead box O (FoxO) transcription factors include FoxO1, FoxO3a, FoxO4, and FoxO6 [27]. Notably, FoxO3a is well known to mediate cellular events, such as apoptosis, proliferation, aging, and longevity in various types of cells [27–31]. FoxO3a is the downstream of Akt, also known as protein kinase B (PKB), thereby the phosphorylation of the FoxO3a is modulated by Akt [32–34]. Phosphorylated FoxO3a by Akt induces binding with 14-3-3 protein, which moves FoxO3a from the nucleus to the cytoplasm [35]. Furthermore, in the presence of oxidative stress, FoxO3a is controlled by c-Jun N-terminal kinase (JNK)/mammalian sterile 20-like kinase 1 (MST 1) signaling pathways [36]. As JNK is activated by oxidative stress, the phosphorylation of 14-3-3 weakens the coherence with the FoxO3a, and subsequently isolated FoxO3a moves from cytoplasm to the nucleus, leading to the transactivation of the target genes [37].

In the aspect of preventing the adverse effects of blue light on the skin, there is a need to develop blue light-antagonizing agents, using natural or natural-derived ingredients or organic ingredients. In this study, we examined the possibility of yellow chaste weed (*Helichrysum arenarium*) and its components as agents to improve blue light-induced damage. The yellow chaste weed has been used as a traditional medicine since ancient times [38]. Apigenin (4′,5,7-trihydroxyflavone) and galangin (3,5,7-trihydroxy flavones) are the two main flavonoids of yellow chaste weed [39,40]. They have been reported to possess anti-cancer, antioxidant, and anti-inflammatory properties [41–50]. Although the biological properties of apigenin and galangin have already been extensively studied, it remains unclear how they affect blue light-irradiated HaCaT cells.

Therefore, in this study, we investigated antagonizing activities of yellow chaste weed and its components against blue light and its mechanisms of action in human keratinocytes, HaCaT cells.

## 2. Materials and Methods

### 2.1. Cell Culture and Materials

HaCaT cells, a human keratinocyte cell line, were obtained from ATCC (Manassas, VA, USA). The cells were cultured in Dulbecco's modified Eagle's medium (DMEM; SH30243.01, Hyclone, Logan, UT, USA) supplemented with 10% fetal bovine serum (FBS) and 1% of antibiotics (penicillin and streptomycin), at 37 °C in a humidified incubator (5% $CO^2$). The medium was changed every three days until the cells grew 70–80% confluence.

All types of antibodies were used in Western blot analysis. Antibodies for TRPV1 (1:1000 dilution, PA1-29421) and p-TRPV1 (Ser502) (1:1000 dilution, PA5-64860) were purchased from Invitrogen (Invitrogen, Waltham, MA, USA). Antibodies for JNK (1:2000 dilution, sc-572), p-JNK (1:1000 dilution, sc-6254), ERK 1/2 (1:2000, sc-292838), p-ERK 1/2 (Tyr 204) (1:2000 dilution, sc-101761), p38 MAPK (1:1000 dilution, sc-535), Lamin B1 (C-5) (1:500 dilution, sc-365962), Clusterin (1:1000 dilution, sc-166907), Clusterin-α (1:1000 dilution, sc-5289), Bcl-2 (1:1000 dilution, sc-7382), and Bax (1:1000 dilution, sc-7480) were purchased from Santa Cruz Biotechnology (Santa Cruz, Dallas, TX, USA). Antibodies for FoxO3a (1:2000 dilution, 2497S), p-FoxO3a (Ser253) (1:1000 dilution, 9466S), p-p38 MAPK (1:2000 dilution, 9216S), Akt (1:1000 dilution, 9272S), and p-Akt (Ser473) (1:1000 dilution, 12694S) were purchased from Cell Signaling Technology (CST, Danvers, MA, USA). Antibodies for Clusterin (1:1000 dilution, ab69644), α-tubulin (1:10,000 dilution, ab7291), and goat anti-mouse IgG (1:4000 dilution, Alexa fluor 488, ab150117) antibodies were ac-quired from Abcam (Abcam, Cambridge, UK). Antibodies for β-actin (1:4000 dilution, A5316), anti-rabbit immunoglobulin G (IgG) (1:4000 dilution, A0545), and anti-mouse IgG (1:4000 dilution, A9044) were purchased from Sigma–Aldrich (Sigma–Aldrich, St. Louis, MO, USA). Antibodies for p-MST 1 (Thr183)/MST 2 (Thr180) (1:1000 dilution, bs-3294R) were purchased from Bioss (Bioss Inc., Wo-burn, MA, USA).

### 2.2. YCW Extract, Apigenin and Galangin Pretreatment, and Blue Light Irradiation

YCW extract was obtained from Luvama Biolab Co., Ltd. (Seongnam, Korea). The powder of the extract was dissolved in dimethyl sulfoxide (DMSO) (472301, Sigma–Aldrich, St. Louis, MO, USA) for the experiments. Apigenin (SMB00702, Sigma–Aldrich, St. Louis, MO, USA), and galangin (sc-235240, Santa Cruz Biotechnology, Dallas, Texas, USA) were dissolved in DMSO. On the first day, HaCaT cells were treated with various concentrations of YCW extract (0.01, 0.05, 0.1%) or apigenin (10, 20, 30 μM) or galangin (1, 10, 20 μM) or vehicle (DMSO) for 24 h in phenol red-free culture medium (PRFCM) before blue light irradiation. On the second day, the culture media were removed, and the cells were incubated with fresh PRFCM. Then, the cells were irradiated with blue light with a photoreactor (CCP-4V, Luzchem, Ottawa, ON, Canada) that had an emission peak between 470–480 nm. The power density of LED blue light equipped in photoreactor was 76 W/$m^2$, and it was exposed for 30 min. The dose (76 W × 30 min/$m^2$ = 13.68 J/$cm^2$) of blue light irradiation used is equivalent to those received by the skin during summer after proximately 1500 s of sun exposure. After the first blue light irradiation (13.68 J/$cm^2$), cells were incubated in the presence of the indicated concentration of YCW extract, apigenin, and galangin for 24 h at 37 °C. On the third day, the same process as on the second day was repeated. In order to consider temperature as a control variable, the control group was also left in the dark at room temperature, while the experimental groups were irradiated.

### 2.3. Cell Counting Kit-8 Assay for Cytotoxic Analysis

The cytotoxicity effect of apigenin and galangin on HaCaT cells were measured by cell counting kit-8 assay (CCK-8; CK04-11, Dojindo, Japan). Cells were cultured in 6-well plates. After changing culture media, the cells were incubated for 24 h with apigenin and galangin. These processes were repeated twice more to make a total treatment time of 72 h.

After incubation, cells were washed with PBS and the media was changed to PRFCM. Next, the CCK-8 reagent (8 μL/well) was added to cells and incubated for 2 h at 37 °C. Then, the equal amount of supernatant (150 μL/well) was taken to measure the absorbance at 450 nm with microplate reader (Synergy HTX Multi-Mode Reader, BioTek, Winooski, VT, USA).

### 2.4. CellTiter-Glo® 2.0 Assay for Cell Proliferation Analysis

Cells were cultured on 96-well opaque wall. After 4 days of rough processing (24 h pretreatment and blue light irradiation × 2 days), plates were placed at room temperature for 30 min to equilibrate the plate and its contents to optimal temperature. The CellTiter-Glo® 2.0 reagent (G9242, Promega, Madison, WI, USA) equal to the volume of cell culture medium was presented in each well (100 μL). All the contents were mixed for 2 min on orbital shaker to induce cell lysis. Cells were incubated at room temperature for 10 min with light-blocked to stabilize the luminescent signal. Luminescence was recorded using a microplate reader (integration time of 0.5 s per well).

### 2.5. BrdU ELISA for Cell Proliferation Analysis

The cells were plated on 96 well black-wall/clear-bottom plates and proceeded 4 days repeated processing experiment. After the experiment, the cell proliferation was determined using the Cell Proliferation ELISA BrdU (colorimetric) kit (11647229001, Roche, Basel, Switzerland) in accordance with the manufacturer's instructions.

### 2.6. EdU Incorporation Imaging for Cell Proliferation Analysis

Click-iT® EdU Imaging Kits (C10337, Invitrogen, Waltham, MA, USA) were used to examine the cell proliferation. Cells were cultured in confocal plates. Apigenin 30 μM, galangin 20 μM, and vehicle (DMSO) were pretreated for 24 h on the first day. The next day, PRFCM were changed, cells were irradiated blue light (13.68 J/cm$^2$) and treated with indicated concentrations of substances. On the third day, same processes as on the second day were proceeded. Twelve hours after blue light (13.68 J/cm$^2$) exposure, cells replaced half of the media with an equal volume of EdU labeling solution (final concentration of 10 μM) and incubated for 12 h at 37 °C. From this time on, cells were washed three times with 1 mL of PBS with 0.1% Tween-20 (PBST) before every steps. After incubation, cells were fixed with 1 mL of 4% formaldehyde in PBS for 15 min at room temperature. Fixative was removed, 1 mL of 0.1% Triton X®-100 in PBS to permeabilize cells was added, and then incubated for 20 min at room temperature. After that, cells were blocked with 500 μL of 3% BSA in PBS for 1 h at room temperature. Next, cells were added with 200 μL of Click-iT® reaction cocktail and incubated for 30 min at room temperature in the dark. After EdU detection, DNA staining was carried out by adding 200 μL of Hoechst 33342 (Invitrogen, Waltham, MA, USA) in PBS, and incubated for 10 min at room temperature in the dark. Thereafter, cells were washed three times with PBS and wet mounted with PBS for slide preparation. Finally, the cells were observed using an LSM 700-laser scanning confocal microscope (Zeiss, Jena, Germany). To measure the intensity of immunofluorescence, images were taken on the same laser power and the mean value of signals was evaluated. Signals from the images were measured by using ZEN 2012 Blue (Zeiss, Jena, Germany) and ImageJ software (National Institutes of Health, Bethesda, MD, USA).

### 2.7. Flow Cytometry Analysis for Apoptosis

Cells were cultured in 60-mm plates and after 4 days of rough processing (24 h pretreatment and blue light irradiation × 2 days), suspended media were first collected, and cells were washed with 1 mL of PBS and collected on 15-mL tubes. Tubes were then centrifuged at 21,209× *g* for 3 min. Cells were harvested with 500 μL of trypsin and centrifuged at 1500 rpm for 3 min and removed supernatant. Cell pellets were resuspended with 100 μL of 1X annexin binding buffer with diethyl pyrocarbonate-treated double distilled water (DEPC-DW; C-9030, Bioneer, Daejeon, Korea). Cells were then stained with 100 μg/mL of propidium iodide (PI) and 5 μL of FITC-Annexin V using Dead Cell Apoptosis Kit with

Annexin V FITC and PI, for flow cytometry (V13242, Invitrogen, Waltham, MA, USA) and incubated for 15 min at room temperature in the dark. After the incubation, mix gently with 400 μL of 1X annexin binding buffer and keep the samples on ice. The stained cells were counted at least 10,000 cells and analyzed by flow cytometry (CytoFLEX, Beckman Coulter, Brea, CA, USA).

### *2.8. Fluo-4 NW Calcium Assay for Intracellular Calcium Influx Analysis*

Cells were cultured in 96-well black wall/clear bottom plates. Cells were pretreated with yellow chaste weed extract 0.1% or apigenin (10, 20, 30 μM) or galangin (1, 10, 20 μM) or vehicle (DMSO) for 1 h in PRFCM. Next, 100 μL of 1X Fluo-4 NW dye loading solution (F36205, Invitrogen, Waltham, MA, USA)/well were added to the cells and plates were incubated in the dark at 37 °C for 30 min, then at room temperature for an additional 30 min. After removing dye loading solution, 100 μL of assay buffer (1X HBSS, 20 mM HEPES) were added to cells and blue light (13.68 J/cm$^2$) were irradiated in plates except for control group. Then, cells were treated with indicated concentrations of yellow chaste weed extract or apigenin or galangin or vehicle (DMSO) dissolved in assay buffer. Fluorescence was measured using a microplate reader (excitation, 494 nm; emission, 516 nm).

### *2.9. DCF-DA Fluorescence Assay for ROS Production Analysis*

The cellular ROS were measured by 2′,7′-dichlorofluorescin diacetate (DCF-DA) assay kit (ab113851, Abcam, Cambridge, UK). The cells were plated on 96-well black-wall/clear-bottom plates. Cells were pretreated with apigenin (10, 20, 30 μM), galangin (1, 10, 20 μM), and vehicle (DMSO) for 24 h with culture media. They were rinsed twice in PBS and treated with 20 μM DCF-DA (freshly diluted in PBS) for 30 min at 37 °C with the light blocked. After incubation, they were replaced with PBS and blue light (30 min) were irradiated. Fluorescence signals were detected using a microplate reader (excitation, 485 nm: emission, 535 nm).

### *2.10. Western Blotting*

Cells were cultured in 60-mm plates and treated with indicated concentrations and incubation time of yellow chaste weed, apigenin, and galangin. After that, cells were har-vested and total proteins were extracted from the cells using RIPA buffer (9806S, Cell Signaling Technology, MA, USA), containing protease inhibitor cocktail (5872S, Cell Signaling Technology, Danvers, MA, USA). Nuclear and cytoplasmic fractions were extracted using NE-PER Nuclear and Cytoplasmic Extraction Reagents (78833, Thermo Fisher Scientific, Waltham, MA, USA), following manufacturer's instructions. Pierce™ BCA Protein Assay Kit (23227, Thermo Fisher Scientific, Waltham, MA, USA) was used to quantify ac-curate protein sample concentrations. Protein samples to be loaded were mixed with 4X sample buffer (161-1747, Bio-Rad, Hercules, CA, USA) and heated at 95 °C for 5 min. The protein samples were separated by sodium dodecyl sulfate-polyacrylamide gel electrophoresis (SDS-PAGE) and transferred onto polyvinylidene fluoride (PVDF) membrane (162-0177, Bio-Rad, Hercules, CA, USA). After transfer, the proteins were dyed with Ponceau S solution (P7170, Sigma–Aldrich, St. Louis, MO, USA), and sufficiently washed with Tris-buffered saline and 0.1% Tween-20 (TBST). Membranes were blocked for 2 h using 2% BSA and were incubated overnight with the primary antibodies at 4 °C. Then, the membranes were incubated for more than 1 h with the secondary antibodies at room temperature. Protein bands were visualized by Pierce™ ECL Western blotting substrate (23227, Thermo Fisher Scientific, Waltham, MA, USA).

### *2.11. Statistical Analysis*

The data are expressed as the mean ± standard error of the mean (SEM). Analyses of differences between two groups were performed using Student's *t*-test. The comparison between multiple groups was performed using one-way analysis of variance (ANOVA), followed by the Tukey's multiple-comparison test, for which the GraphPad Prism (5.0)

(GraphPad, La Jolla, CA, USA) software was used. Statistical significance was considered when the $p$-value was less than 0.05.

## 3. Results

### 3.1. Effect of YCW Extract on the Blue Light Irradiated HaCaT Cells

To examine the effects of YCW extract on the proliferation of HaCaT cells, cells were treated with various concentrations (0.01, 0.05, 0.1%) of YCW extract. When blue light-exposed HaCaT cells were treated with YCW extract, the proliferation was significantly increased (Figure 1A,B). The effect of blue light on the activation of TRPV1 in HaCaT cells was demonstrated by our previous study [10]. Next, we investigated the influence of YCW extract on expression and phosphorylation of TRPV1. After blue light irradiation, the expression of TRPV1 and phospho-TRPV1 increased apparently (Figure 1C,D). However, YCW extract-pretreated group has shown downregulation of TRPV1 and reduction of its phosphorylation (Figure 1C,D). TRPV1 is a non-selective cation channel and the activation of TRPV1 elevates calcium influx [51]. To evaluate the inhibition effect of YCW extract on $Ca^{2+}$ influx, cells were pretreated with YCW extract 0.1% for 1 h and then irradiated with blue light. As shown in Figure 1E, while blue light increased $Ca^{2+}$ influx through the TRPV1 channel, YCW extract reduced $Ca^{2+}$ influx induced by blue light irradiation. This result indicates that YCW extract enhances cell proliferation decreased in blue light exposed-HaCaT cells and suggests its effect is mediated through downregulation of TRPV1 and inhibition of its phosphorylation.

### 3.2. Effects of Apigenin and Galangin on the Cytotoxicity of HaCaT Cells and the Proliferation of Blue Light-Irradiated HaCaT Cells

HaCaT cells were treated with different concentrations (10–50 μM) of apigenin (Figure 2A) and (1–30 μM) of galangin (Figure 2B) for 3 days and the percentage of cell viability was determined (Figure 2C,D). As shown in Figure 2C, apigenin was cytotoxic at 40 μM and the cell viability of galangin was decreased at 30 μM. Therefore, the maximum final treatment concentrations for apigenin and galangin were 30 and 20 μM, respectively. To evaluate the effect of apigenin and galangin on proliferation of blue light-exposed keratinocytes, the cells were pretreated before blue light irradiation with 10, 20, and 30 μM of apigenin or 1, 10, and 20 μM of galangin for 24 h. As shown in Figure 2D, both apigenin and galangin enhanced the cell proliferation reduced by blue light. In the BrdU ELISA assay to confirm the proliferation effect of apigenin and galangin, both apigenin and galangin significantly attenuated the inhibitory effect of blue light on proliferation of HaCaT cells concentration-dependently (Figure 2E). Similarly, EdU incorporation assay was conducted to prove cellular proliferation. As shown in Figure 2F, compared to the control group, blue light exposure remarkably reduced the population of EdU-positive cells. However, this effect was suppressed by pretreatment of apigenin (30 μM) and galangin (20 μM). In addition, this effect of apigenin and galangin was demonstrated using flow cytometry analysis after staining with FITC-annexin V and PI. Annexin V-stained group represents apoptotic cells, while PI staining is used to detect dead cells in a population. As shown in Figure 2G, the proportion of living cells in blue light-irradiated group reduced to 81.19% when compared to a non-irradiated group (96.20%). However, treatment with apigenin or galangin enhanced the ratio of living cells. Together, these indicate that blue light-irradiated reduction of cell proliferation is attenuated by either apigenin or galangin.

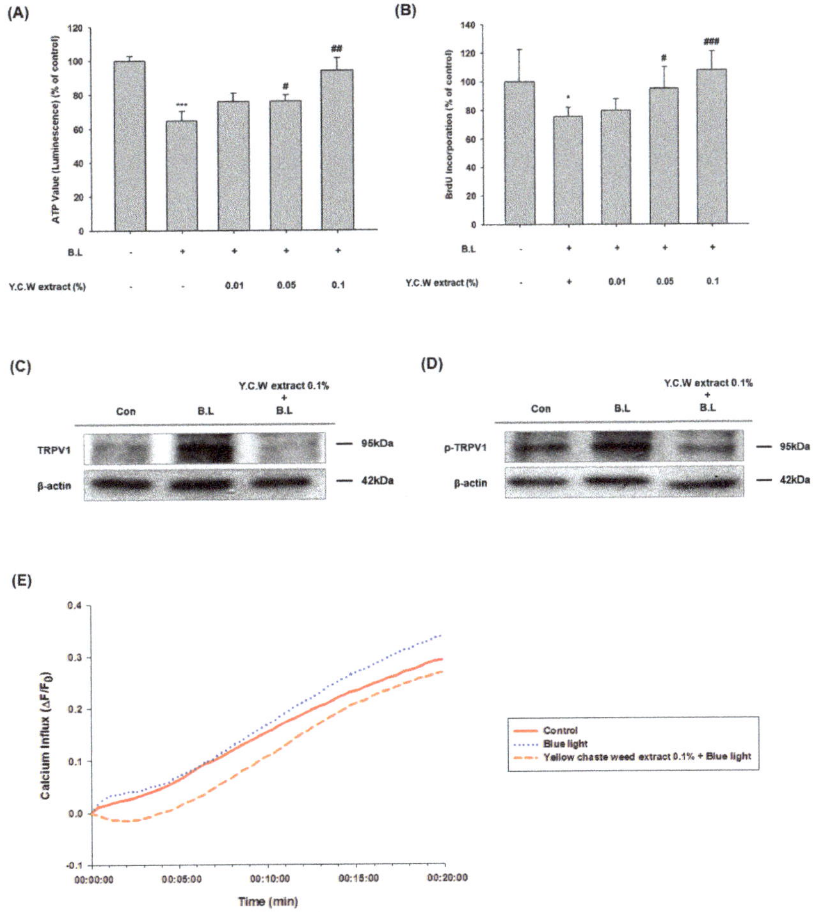


**Figure 1.** YCW extract increased cell proliferation via calcium dependent-TRPV1 signaling pathways in blue light irradiated HaCaT cells. (**A**) CellTiter-Glo® 2.0 assay of YCW extract (0.01–0.1%) pretreated with blue light-irradiated HaCaT cells. YCW extract was pretreated 24 h before irradiation of blue light (13.68 J/cm$^2$), and the same process was repeated on the next day. After 24 h incubation, an equal volume of CellTiter-Glo reagent was added to the cells. (**B**) Cell proliferation effect of YCW extract (0.01–0.1%) on blue light-irradiated HaCaT cells was measured using BrdU ELISA assay. Data are presented as the mean ± SEM of four independent experiments. Statistical significance of differences among the groups was assessed by one-way analysis of variance (ANOVA), followed by Tukey's multiple-comparison test, using the GraphPad Prism 5 software. * $p < 0.05$ vs. control group. * $p < 0.05$ vs. control, *** $p < 0.005$ vs. control, # $p < 0.05$ vs. blue light-irradiated group (B.L), ## $p < 0.01$ vs. B.L. ### $p < 0.005$ vs. B.L. (**C**) TRPV1 expression levels were determined by Western blotting at 24 h incubation after, pretreatment with YCW extract (0.1%) for and 2-day repetitive blue light irradiation (30 min, 76 W/m$^2$). (**D**) Phosphorylated-TRPV1 expression levels were determined by Western blotting at 90 min incubation after, 1 h YCW extract (0.1%) pretreatment and blue light irradiation (30 min, 76 W/m$^2$). The total proteins were extracted from the cells immediately after experimental conditions, and β-actin was used as a loading control. (**E**) $Ca^{2+}$ influx changes of blue light and with Y.C.W extract (0.1%) pretreatment by Fluo-4 NW assay. Cells were 1-h pretreated with YCW extract (0.1 %) and irradiated with blue light (10 min, 76 W/m$^2$). After the irradiation, fluorescence intensities were measured immediately over a certain period.

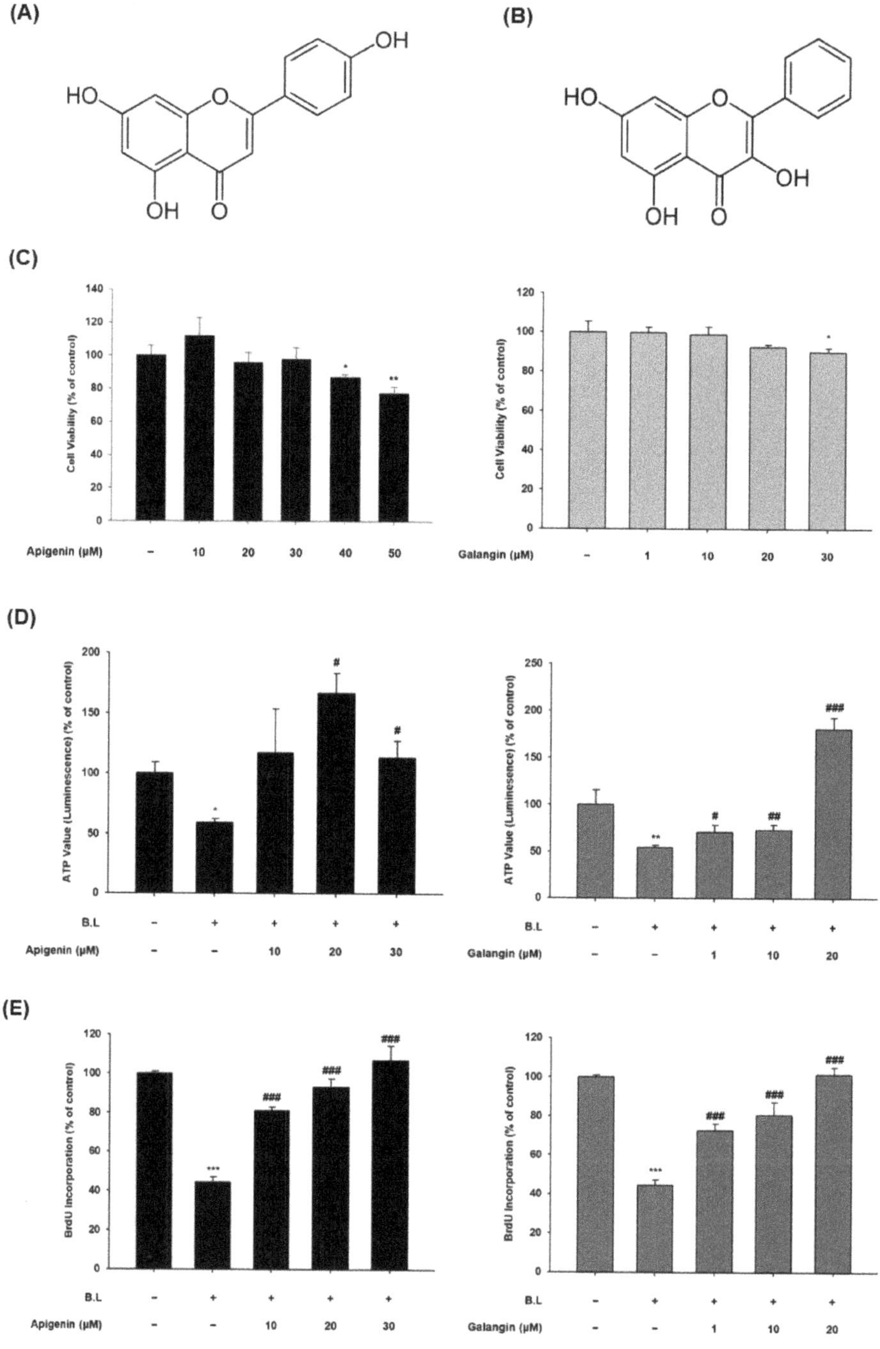


**Figure 2.** *Cont.*

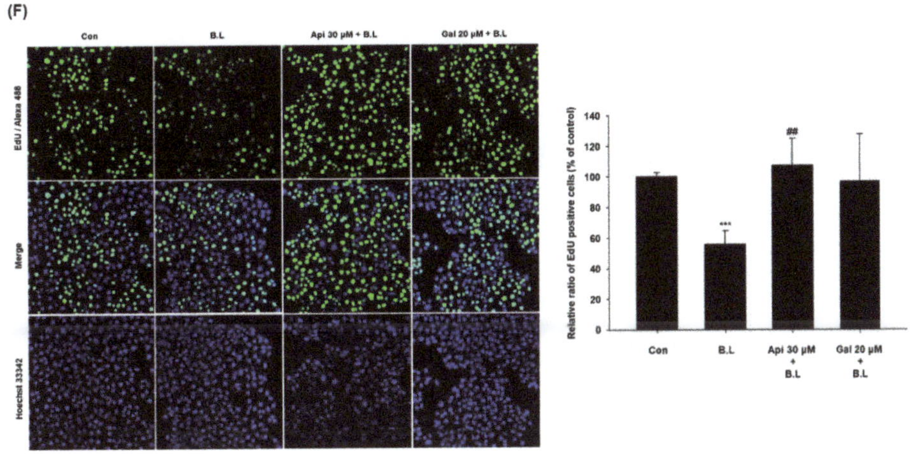


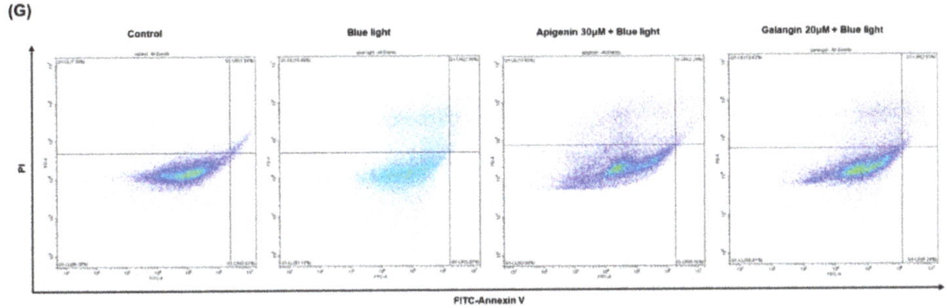


**Figure 2.** Apigenin and galangin, active components of YCW, increased proliferation in blue light-irradiated HaCaT cells. (**A**) Chemical structure of apigenin, (**B**) and galangin. (**C**) Cell cytotoxicity of apigenin (10–50 μM) and galangin (1–30 μM) were measured with CCK-8 assay. The effects of apigenin and galangin on proliferation of blue light (30 min, 76 W/m$^2$)-irradiated HaCaT cells were determined by CellTiter-Glo® 2.0 assay (**D**), BrdU ELISA assay (**E**), and EdU incorporation assay (**F**). Effect of apigenin (30 μM) or galangin (20 μM) on blue light-induced apoptosis in HaCaT cells was measured using flow cytometry analysis (**G**) * $p < 0.05$ vs. control, ** $p < 0.01$ vs. control, *** $p < 0.005$ vs. control, # $p < 0.05$ vs. B.L, ## $p < 0.01$ vs. B.L, ### $p < 0.005$ vs. B.L, V: viable cells, D: dead cells.

### 3.3. *Apigenin and Galangin Suppress TRPV1-Mediated Signaling Induced by Blue Light-Irradiation*

Previous studies in our laboratory have already shown that blue light upregulates TRPV1 and induce its phosphorylation [10]. Therefore, we examined effects of apigenin and galangin on blue light-induced upregulation of TRPV1. As shown in Figure 3A, various concentrations (10, 20, and 30 μM) of apigenin decreased the blue light-induced effect. Likewise, galangin also reduced the expression of TRPV1 induced by blue light irradiation (Figure 3A). In addition, TRPV1 phosphorylation contributes the activation of TRPV1 [52]. Therefore, we investigated effects of apigenin and galangin on TRPV1 phosphorylation. As shown in Figure 3B, blue light-induced phosphorylation of TRPV1 was reduced by both apigenin (30 μM) and galangin (20 μM). These data indicate that apigenin and galangin can antagonize against blue light effect by suppressing phosphorylation and expression of TRPV1.

Studies have shown that activation of the TRPV1 induces $Ca^{2+}$ entry into the cytosol [51]. Therefore, we examined the involvement of apigenin and galangin in the blue light-induced calcium influx. As shown in Figure 3C,D, while blue light irradiation increased the inflow of $Ca^{2+}$ compared to control group, apigenin and galangin reduced the calcium influx induced by blue light. These data indicate that apigenin and galangin antagonize TRPV1-mediated signaling induced by blue light irradiation.

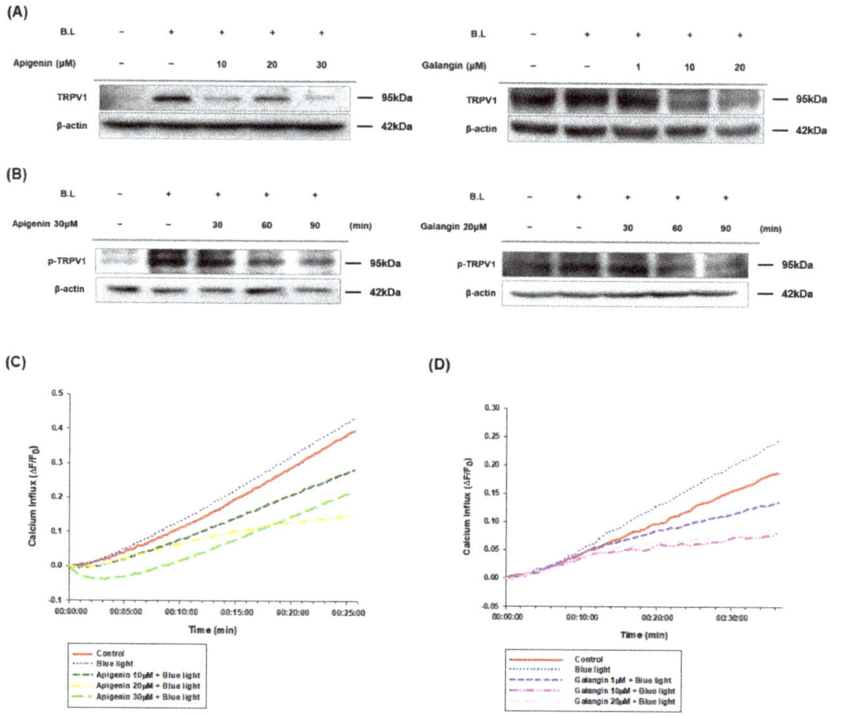


**Figure 3.** Apigenin and galangin reduced TRPV1 expression and its phosphorylation in blue light-irradiated HaCaT cells. (**A**,**B**) TRPV1 expression (**A**), and its phosphorylation levels (**B**) were determined by Western blotting. (**A**) Cells were incubated with apigenin (10, 20, 30 μM) and galangin (1, 10, 20 μM) for 24 h and were then irradiated with blue light (30 min, 76 W/m$^2$). After 24 h incubation, the cells were subjected to the same process twice and were finally subjected to Western blot analysis. (**B**) Phosphorylation levels of TRPV1 were determined by Western blotting at 30-, 60-, and 90-min incubation after pretreatment with apigenin (30 μM) or galangin (20 μM) for 1 h and subsequent blue light-irradiation (30 min, 76 W/m$^2$). The total proteins were extracted from the cells immediately after experimental conditions, and β-actin was used as a loading control. (**C**,**D**) Apigenin and galangin inhibited blue light-induced calcium influx in HaCaT cells. (**C**) $Ca^{2+}$ influx changes of blue light and with apigenin (10, 20, 30 μM) pretreatment, (**D**) with galangin (1, 10, 20 μM) pretreatment by Fluo-4 NW assay. Cells were pretreated with apigenin (10, 20, 30 μM) and galangin (1, 10, 20 μM) for 1 h and irradiated with blue light (10 min, 76 W/m$^2$). After the irradiation, fluorescence intensities were measured immediately over a certain period.

### *3.4. Apigenin and Galangin Reduce ROS Generation in Blue Light Irradiated HaCaT Cells*

Next, we examined whether pretreatment of apigenin or galangin ameliorates the accumulation of ROS in blue light irradiated HaCaT cells using a DCFDA dye. Treatment with apigenin or galangin significantly decreased endogenous ROS production induced by blue light irradiation in dose dependent manner, which was amplified by the blue light irradiation (Figure 4A,B). These data indicate that apigenin and galangin possess antioxidative activity.

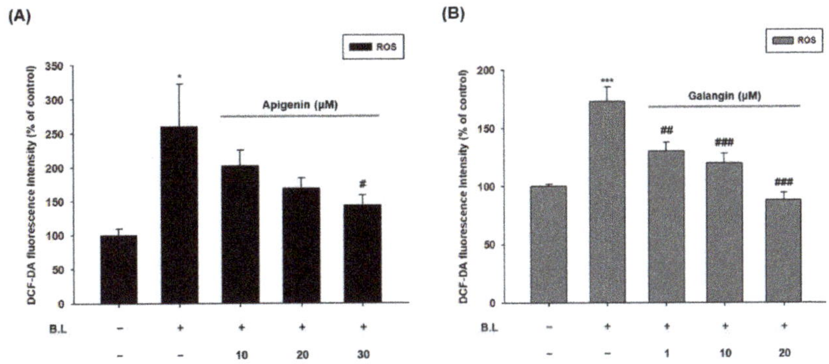


**Figure 4.** Apigenin and galangin decreased oxidative stress induced by blue light irradiation. (**A**,**B**) ROS production was determined by DCF-DA assay. Cells were incubated with apigenin (10, 20, 30 μM) (**A**) or galangin (1, 10, 20 μM) (**B**) for 24 h and were then irradiated with blue light (30 min, 76 W/m$^2$). After 24 h incubation, the cells were subjected to the same process twice and were finally subjected to DCF-DA assay. * $p < 0.05$ vs. control, *** $p < 0.005$, # $p < 0.05$ vs. B.L, ## $p < 0.01$%, ### $p < 0.005$%.

### *3.5. Apigenin and Galangin Antagonize against Blue Light by Regulating MAPKs and MST-1/2-Akt-FoxO3a Signaling*

Previous studies have reported that $Ca^{2+}$ regulates MAPK functions [53]. Since apigenin and galangin were verified to lower $Ca^{2+}$ flux by blue light irradiation, Western blot was performed to examine the effect of apigenin and galangin on MAPK activation. As shown in Figure 5A,B, apigenin and galangin attenuated the blue light effects on MAPKs. Specifically, apigenin and galangin reduced phosphorylation levels of JNK induced by blue light. On the contrary, while blue light reduced phosphorylation levels of ERK and p38 MAPK, this blue light effect was suppressed by apigenin and galangin treatment (Figure 5A,B). These results suggest that apigenin and galangin antagonize blue light effect by regulating MAPKs.

To investigate the role of FoxO3a, which is a downstream signal of MAPKs and regulator of cell proliferation, we examined effects of apigenin and galangin on phosphorylation levels and nuclear translocation of FoxO3a under blue light-irradiated condition. As shown in Figure 5C, while blue light irradiation reduced the levels of phosphorylated FoxO3a (Ser253), apigenin and galangin increased its phosphorylation levels. Additionally, FoxO3a was detected at higher level in the nuclear fractions of blue light irradiated group (Figure 5D). In contrast, apigenin and galangin treatment reduced the nuclear translocation of FoxO3a (Figure 5D). These results indicate that apigenin and galangin reduce nuclear translocation of FoxO3a by phosphorylating FoxO3a.

Mst-1/2 is known to phosphorylate FoxO3a and dissociate FoxO3a from 14-3-3 protein, which lead to nuclear translocation of FoxO3a [54]. To examine effect of apigenin and galangin on MST-1/2 activation, Western blot analysis for phosphorylated MST-1/2 was carried out. As shown in Figure 5E,F, while blue light irradiation caused significant increase in phosphorylated MST-1/2, apigenin and galangin reduced its phosphorylation levels. This data indicates that apigenin and galangin regulate MST-1/2- FoxO3a signaling. Furthermore, to investigate relationship between Akt and both FoxO3a and MST-1/2 signaling, we performed Western blot analysis for phosphorylation forms of them. As shown in Figure 5G,H, while blue light decreased the phosphorylation levels of Akt, apigenin and galangin induced the phosphorylation of Akt. These results indicate that apigenin and galangin antagonize against blue light by regulating MST-1/2-Akt-FoxO3a signaling.

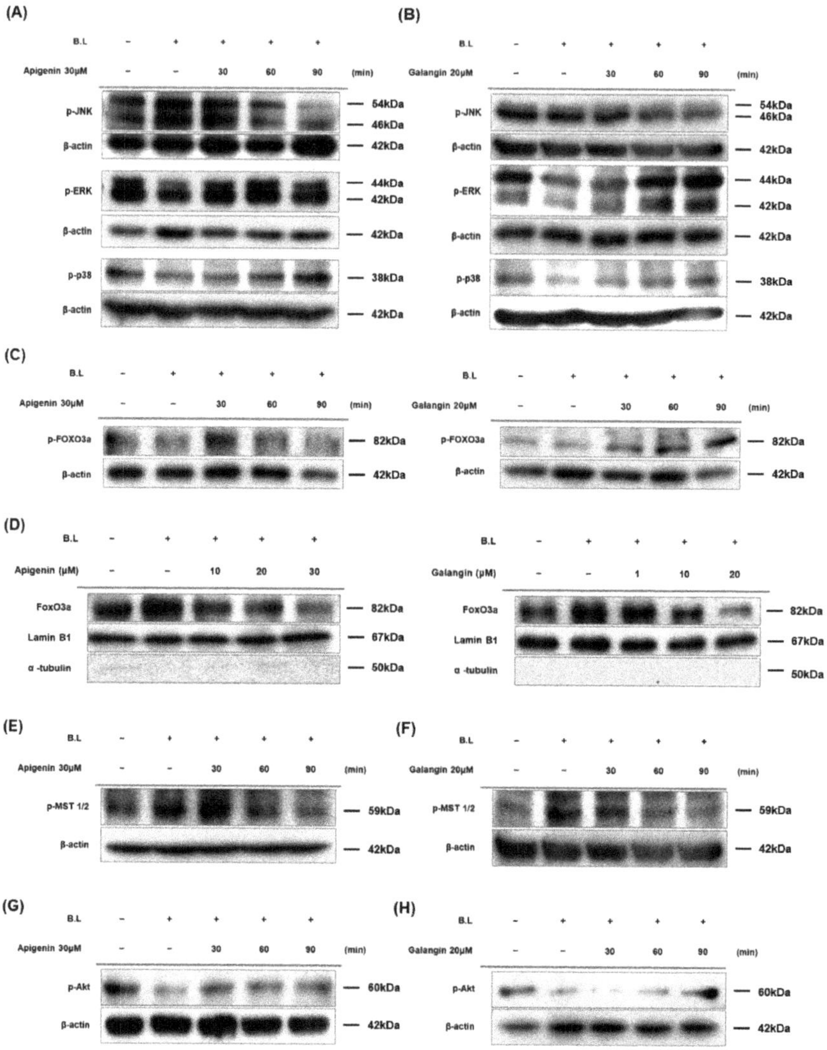

**Figure 5.** Apigenin and galangin antagonize against blue light by regulating MAPKs and MST-1/2-Akt-FoxO3a signaling. (**A**,**B**) Apigenin and galangin regulate phosphorylation of MAPKs in blue light-irradiated HaCaT cells. (**A**) Phosphorylation levels of MAPKs were determined by Western blotting at 30-, 60-, and 90-min incubation after, pretreatment with apigenin (30 μM) or galangin (20 μM) for 1 h and subsequent blue light irradiation (30 min, 76 W/m$^2$). The total proteins were extracted from the cells immediately after experimental conditions, and β-actin was used as a loading control. (**C**,**D**) Apigenin and galangin inhibited FoxO3a nuclear translocation by FoxO3a phosphorylation. (**C**) Phosphorylation levels of FoxO3a were determined by Western blotting at 30-, 60-, and 90-min incubation after, pretreatment with apigenin (30 μM) or galangin (20 μM) for 1 h and subsequent blue light irradiation (30 min, 76 W/m$^2$). The total proteins were extracted from the cells immediately after experimental conditions, and β-actin was used as a loading control. (**D**) The nuclear translocation of FoxO3a were determined by Western blotting at 24 h incubation after, 24 h apigenin (10, 20, 30 μM) and galangin (1, 10, 20 μM) pretreatment and 2-day repetitive blue light irradiation (30 min, 76 W/m$^2$). The nuclear fraction was extracted from the cells immediately after experimental conditions by using NE-PER™ Nuclear and Cytoplasmic Extraction reagents. Lamin B1

was used as a loading control of nucleus fraction aliquot. (**E**,**F**) Apigenin and galangin suppress phosphorylation of MST-1/2. (**E**) Phosphorylation levels of MST-1/2 were determined by Western blotting at 30-, 60-, and 90-min incubation after, pretreatment with apigenin (30 μM) (**E**) or galangin (20 μM) (**F**) for 1 h and subsequent blue light irradiation (30 min, 76 W/m$^2$). The total proteins were extracted from the cells immediately after experimental conditions, and β-actin was used as a loading control. (**G**,**H**) Apigenin and galangin activated Akt in blue light irradiated HaCaT cells. Phosphorylation levels of Akt were determined by Western blotting at 30-, 60-, and 90-min incubation after, pretreatment with apigenin (30 μM) (**G**), and galangin (20 μM) (**H**) for 1 h and subsequent blue light irradiation (30 min, 76 W/m$^2$). The total proteins were extracted from the cells immediately after experimental conditions, and β-actin was used as a loading control.

### 3.6. *Apigenin and Galangin Regulate Expression of Clusterin, Bax, and Bcl-2*

Clusterin, Bcl-2 associated X protein (Bax), and B-cell lymphoma-2 (Bcl-2) have been well known to be involved in cell survival and apoptosis. We examined whether apigenin and galangin affect blue light-induced expression of two clusterin isoforms (nuclear clusterin and secretory clusterin), Bax and Bcl-2. As shown in Figure 6A,B, while blue light irradiation increased the expression of nuclear clusterin (pro-apoptotic protein), protein levels of secretory clusterin (pro-survival protein) were decreased. However, apigenin and galangin decreased expression of nCLU and increased expression of sCLU. In addition, as shown in Figure 6C,D, blue light irradiation increased protein levels of Bax (apoptotic protein) and reduced protein levels of Bcl-2 (anti-apoptotic protein). These blue light effects on Bax and Bcl-2 was attenuated by apigenin and galangin. These data indicate that apigenin and galangin exert antagonizing activity by in-creasing expression of cell survival-related genes.

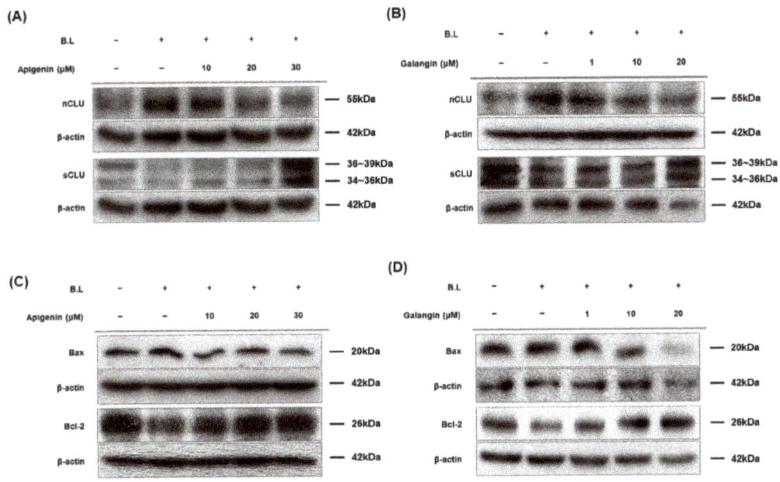


**Figure 6.** Apigenin and galangin regulated expression of nuclear and secretory clusterin as well as Bax/Bcl-2 ratio in blue light-irradiated HaCaT cells. (**A**,**B**) Protein levels of nuclear clusterin (nCLU) and secretory clusterin (sCLU) were determined by Western blot analysis. Cells were incubated with apigenin (10, 20, 30 μM) (**A**) or galangin (1, 10, 20 μM) (**B**) for 24 h and were then irradiated with blue light (30 min, 76 W/m$^2$). After 24 h incubation, the cells were subjected to the same process twice and were finally subjected to Western blot analysis. The total proteins were extracted from the cells immediately after experimental conditions, and β-actin was used as a loading control. (**C**,**D**) Protein levels of Bax, apoptotic protein, and Bcl-2, anti-apoptotic protein, were determined by Western blotting. Cells were incubated with apigenin (10, 20, 30 μM) (**C**) or galangin (1, 10, 20 μM) (**D**) for 24 h and were then irradiated with blue light (30 min, 76 W/m$^2$). After 24 h incubation, the cells were subjected to the same process twice and were finally subjected to Western blot analysis. The total proteins were extracted from the cells immediately after experimental conditions, and β-actin was used as a loading control.

## 4. Discussion

In everyday life, a large amount of blue light is irradiated to our skin through not only sunlight, but also electronic devices. The blue light is well known to reduce cell proliferation and induce oxidative damage, which in turn could cause photoaging [13,55–58]. Nevertheless, little is known about the mechanism for cell damage caused by blue light. In addition, only few substances have been reported to prevent the adverse effects of blue light. In this study, yellow chaste weed (YCW), and its main components such as apigenin and galangin were found to exert anti-blue light effect through regulation of TRPV1-mediated signaling.

First, our results showed that the YCW (*Helichrysum arenarium*), an herbaceous perennial plant, enhanced cell proliferation of human keratinocytes decreased by blue light irradiation. Next, according to our previous study, cellular damage caused by blue light irradiation in human keratinocyte is due to TRPV1 regulation [10]. TRPV1 cation channel can be activated by various stimuli and ligands such as heat (>43 °C), acid, capsaicin, and other vanilloid compounds (e.g., endocannabinoid, anandamide) [59]. Moreover, many studies have elucidated that TRPV1 expression is related to photoaging of the skin by UV [60–62]. Similar to UV results in these studies, our report has also proved that the expression of TRPV1 was increased by blue light irradiation [10]. Likewise, in this paper, we found that the TRPV1 expression elevated by the blue light was downregulated by high concentration of YCW extract (Figure 1C). In addition, phosphorylation of TRPV1 mediate channel's activation [63]. Our data showed that blue light enhanced responsiveness of TRPV1 and yellow chaste weed extract dephosphorylated TRPV1 (Figure 1D). Further, TRPV1 channel opening stimulates $Ca^{2+}$ entry into cytosol [14]. Blue light exposure increased $Ca^{2+}$ influx, which was suppressed by YCW extract (Figure 1E). These data suggest that YCW extract could antagonize negative effects of blue light irradiation.

Apigenin and galangin, main flavonoids of YCW, showed the same effects as YCW extract. Specifically, while blue light inhibited keratinocyte proliferation, this blue light effect was attenuated by apigenin and galangin (Figure 2). Apigenin and galangin reduced expression and phosphorylation of TRPV1 as well as intracellular $Ca^{2+}$ influx induced by blue light irradiation (Figure 3). These data suggest that apigenin and galangin has anti-blue light activity by regulating TRPV1 signaling.

ROS are produced during normal intracellular activation and are associated with various biological processes, including cell differentiation, gene expression, and response to cytokines [64,65]. Therefore, maintaining homeostasis of these ROS is critical for cell growth and survival. Oxidative stress is increased due to the imbalance between the production of ROS and the antioxidant reaction to remove it [64,66]. Thus, ROS has so far been recognized as a substance that simply oxidizes proteins, DNA, and lipids, causing cell necrosis, but also plays an important role in certain cytokines or growth factors signaling as an essential second messenger within cells [67,68]. In our study, we demonstrated that YCW extract, apigenin, and galangin can reduce intracellular ROS levels increased by blue light irradiation, suggesting that they possess anti-oxidative activity (Figure 4).

In addition, under various external stimuli, such as oxidative stress, mitogen-activated protein kinases (MAPKs) are involved in cell growth, apoptosis, and differentiation. MAPK subfamily consists of c-Jun amino-terminal kinase (JNK), extracellular signal-regulated kinase (ERK), and p38 kinase, and regulates the expression of various genes by phosphorylating transcriptional regulatory factors [69–71]. Activated JNK stimulates an apoptotic signaling by regulating expression of p53-dependent genes, such as Bcl-2 associated X protein (Bax) and p53-upregulated modulator of apoptosis (PUMA) [72,73]. JNK can also induce apoptosis by suppressing BH3-only family of Bcl-2 proteins [74]. In this study, apigenin and galangin reduced levels of phosphorylated-JNK (p-JNK), which can lead to anti-apoptotic signal (Figure 5A,B). In addition, both ERK and p38 kinase are known to promote cell survival, proliferation, and development [75–77]. In this study, apigenin and galangin also increased phosphorylation of ERK and p38 kinase which were reduced by blue light irradiation (Figure 5A,B).

MAPKs are known to regulate FoxO3a phosphorylation [78–80]. For example, under stress condition, activated JNK suppresses Akt activities which dephosphorylate FoxO3a and promotes FoxO3a nuclear translocation [78]. Specifically, phosphorylated Ser473 Akt can phosphorylate FoxO3a at Ser 253, leading to FOXO3a export from nucleus to cytosol and decreases its transactivation activity [32–34]. In addition, MST-1/2 is a positive regulator of FoxO3a activity [54]. MST-1/2 activation translocate FoxO3a from cytosol to nucleus and increase its transcription. Akt and MST-1/2 signaling negatively crosslink each other, and suppression of either Akt or MST-1/2 inhibits cell proliferation [81–83]. Activated p-Akt moves into nucleus, which phosphorylates FOXO3a and release from DNA to lower transcription activities [84,85]. The p-FoxO3a is exported out of the nucleus by binding to the 14-3-3 protein and remains inactive in the cytosol [86]. The FoxO3a transcription factors in the nucleus regulate target genes related to cell death, cell cycle arrest and oxidative stress [27]. In this study, we found that apigenin and galangin reduced phosphorylation of MST-1/2, and increased phosphorylation of Akt, leading to phosphorylation of FoxO3a and its export from nucleus (Figure 5C–H).

Nuclear and secretory clusterin respectively carry out apoptotic or anti-apoptotic functions, by regulating Ku70, Bax, Bcl-2, and caspase activities [23,26,87–89]. In this study, blue light irradiation increased nuclear clusterin and reduced secretory clusterin, leading to reduced cell growth. Apigenin and galangin antagonized the blue light effect on expression of two isoforms of clusterin, consequently promoting cell proliferation (Figure 6). Collectively, these results indicate that apigenin and galangin exert anti-blue light effect by enhancing cell proliferation and decreasing ROS production through regulation of TRPV1 signaling and its related genes shown in Figure 7. However, since the connection between the molecules involved in the blue light-induced signaling was not clear, further study is needed to demonstrate it.

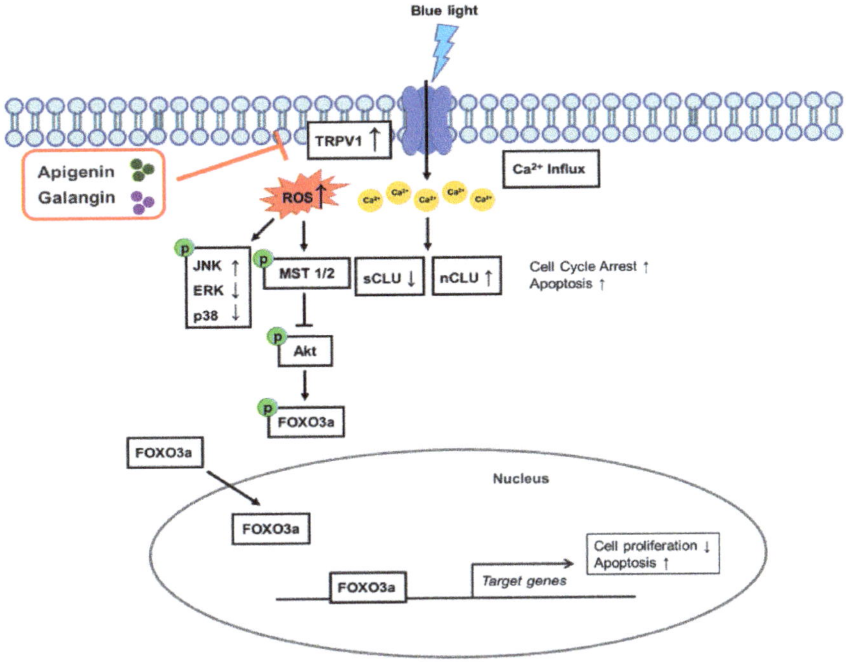


**Figure 7.** Schematic diagram showing action mechanism of apigenin and galangin in blue light-irradiated human keratinocytes.

## 5. Conclusions

In conclusion, these findings indicate that apigenin and galangin, active components of YCW, can ameliorate negative effects of blue light through TRPV1/clusterin/FoxO3a and MAPK signaling pathways, suggesting the possibility of apigenin and galangin as an anti-blue light agent for human skin.

**Author Contributions:** J.Y.P., S.-H.P., S.W.O., M.S., J.Y.C. and J.L., conceptualization, methodology, formal analysis, writing of the original draft; S.W.O., K.K., E.Y., S.C., S.Y., K.J. and S.B.H., conceptualization, methodology; M.S., J.Y.C. and J.L., funding acquisition, conceptualization, methodology, supervision, writing review, and editing. All authors have read and agreed to the published version of the manuscript.

**Funding:** This research was supported by the Medical Device Technology Development Program (Grant No.: 20008861) funded by the Ministry of Trade, Industry, and Energy (MOTIE), Republic of Korea, and a grant from the Basic Science Research Program through the National Research Foundation of Korea (NRF) funded by the Ministry of Science and Technology Information and Communication (Grant No. 2020R1F1A1067731).

**Institutional Review Board Statement:** Not applicable.

**Informed Consent Statement:** Not applicable.

**Data Availability Statement:** The data used to support the findings of this study are available from the corresponding author upon request.

**Conflicts of Interest:** The authors declare no conflict of interest. The funders had no role in the design of the study; in the collection, analyses, or interpretation of data; in the writing of the manuscript; or in the decision to publish the results.

## References

1. Halliday, G.M.; Norval, M.; Byrne, S.N.; Huang, X.X.; Wolf, P. The effects of sunlight on the skin. *Drug Discov. Today Dis. Mech.* **2008**, *5*, e201–e209. [CrossRef]
2. Nguyen, A.V.; Soulika, A.M. The dynamics of the skin's immune system. *Int. J. Mol. Sci.* **2019**, *20*, 1811. [CrossRef] [PubMed]
3. Proksch, E.; Fölster-Holst, R.; Jensen, J.-M. Skin barrier function, epidermal proliferation and differentiation in eczema. *J. Dermatol. Sci.* **2006**, *43*, 159–169. [CrossRef] [PubMed]
4. Khavkin, J.; Ellis, D.A. Aging skin: Histology, physiology, and pathology. *Facial Plast. Surg. Clin.* **2011**, *19*, 229–234. [CrossRef] [PubMed]
5. Liebel, F.; Kaur, S.; Ruvolo, E.; Kollias, N.; Southall, M.D. Irradiation of skin with visible light induces reactive oxygen species and matrix-degrading enzymes. *J. Investig. Dermatol.* **2012**, *132*, 1901–1907. [CrossRef] [PubMed]
6. Byrne, C.; Subramanian, G.; Pillai, S.C. Recent advances in photocatalysis for environmental applications. *J. Environ. Chem. Eng.* **2018**, *6*, 3531–3555. [CrossRef]
7. Schalka, S.; de Paula Corrêa, M.; Sawada, L.Y.; Canale, C.C.; de Andrade, T.N. A novel method for evaluating sun visible light protection factor and pigmentation protection factor of sunscreens. *Clin. Cosmet. Investig. Dermatol.* **2019**, *12*, 605. [CrossRef] [PubMed]
8. Campiche, R.; Curpen, S.J.; Lutchmanen-Kolanthan, V.; Gougeon, S.; Cherel, M.; Laurent, G.; Gempeler, M.; Schuetz, R. Pigmentation effects of blue light irradiation on skin and how to protect against them. *Int. J. Cosmet. Sci.* **2020**, *42*, 399–406. [CrossRef] [PubMed]
9. Clement, M.; Daniel, G.; Trelles, M. Optimising the design of a broad-band light source for the treatment of skin. *J. Cosmet. Laser Ther.* **2005**, *7*, 177–189. [CrossRef] [PubMed]
10. Yoo, J.A.; Yu, E.; Park, S.-H.; Oh, S.W.; Kwon, K.; Park, S.J.; Kim, H.; Yang, S.; Park, J.Y.; Cho, J.Y. Blue Light Irradiation Induces Human Keratinocyte Cell Damage via Transient Receptor Potential Vanilloid 1 (TRPV1) Regulation. *Oxid. Med. Cell. Longev.* **2020**, *2020*, 8871745. [CrossRef] [PubMed]
11. Nakashima, Y.; Ohta, S.; Wolf, A.M. Blue light-induced oxidative stress in live skin. *Free Radic. Biol. Med.* **2017**, *108*, 300–310. [CrossRef] [PubMed]
12. Chamayou-Robert, C.; DiGiorgio, C.; Brack, O.; Doucet, O. Blue light induces DNA damage in normal human skin keratinocytes. *Photodermatol. Photoimmunol. Photomed.* **2022**, *38*, 69–75. [CrossRef] [PubMed]
13. Godley, B.F.; Shamsi, F.A.; Liang, F.Q.; Jarrett, S.G.; Davies, S.; Boulton, M. Blue light induces mitochondrial DNA damage and free radical production in epithelial cells. *J. Biol. Chem.* **2005**, *280*, 21061–21066. [CrossRef] [PubMed]
14. Tominaga, M.; Tominaga, T. Structure and function of TRPV1. *Pflügers Arch.* **2005**, *451*, 143–150. [CrossRef] [PubMed]

15. Bikle, D.D.; Xie, Z.; Tu, C.-L. Calcium regulation of keratinocyte differentiation. *Expert Rev. Endocrinol. Metab.* **2012**, *7*, 461–472. [CrossRef] [PubMed]
16. Lee, S.E.; Lee, S.H. Skin barrier and calcium. *Ann. Dermatol.* **2018**, *30*, 265–275. [CrossRef] [PubMed]
17. Murata, T.; Honda, T.; Egawa, G.; Yamamoto, Y.; Ichijo, R.; Toyoshima, F.; Dainichi, T.; Kabashima, K. Transient elevation of cytoplasmic calcium ion concentration at a single cell level precedes morphological changes of epidermal keratinocytes during cornification. *Sci. Rep.* **2018**, *8*, 6610. [CrossRef] [PubMed]
18. Behne, M.J.; Sanchez, S.; Barry, N.P.; Kirschner, N.; Meyer, W.; Mauro, T.M.; Moll, I.; Gratton, E. Major translocation of calcium upon epidermal barrier insult: Imaging and quantification via FLIM/Fourier vector analysis. *Arch. Dermatol. Res.* **2011**, *303*, 103–115. [CrossRef] [PubMed]
19. Celli, A.; Sanchez, S.; Behne, M.; Hazlett, T.; Gratton, E.; Mauro, T. The epidermal $Ca^{2+}$ gradient: Measurement using the phasor representation of fluorescent lifetime imaging. *Biophys. J.* **2010**, *98*, 911–921. [CrossRef] [PubMed]
20. Celli, A.; Tu, C.L.; Lee, E.; Bikle, D.D.; Mauro, T.M. Decreased Calcium-Sensing Receptor Expression Controls Calcium Signaling and Cell-To-Cell Adhesion Defects in Aged Skin. *J. Investig. Dermatol.* **2021**, *141*, 2577–2586. [CrossRef]
21. Celli, A.; Crumrine, D.; Meyer, J.M.; Mauro, T.M. Endoplasmic reticulum calcium regulates epidermal barrier response and desmosomal structure. *J. Investig. Dermatol.* **2016**, *136*, 1840–1847. [CrossRef] [PubMed]
22. Magno, A.L.; Ward, B.K.; Ratajczak, T. The calcium-sensing receptor: A molecular perspective. *Endocr. Rev.* **2011**, *32*, 3–30. [CrossRef] [PubMed]
23. Caccamo, A.E.; Scaltriti, M.; Caporali, A.; D'arca, D.; Scorcioni, F.; Astancolle, S.; Mangiola, M.; Bettuzzi, S. Cell detachment and apoptosis induction of immortalized human prostate epithelial cells are associated with early accumulation of a 45 kDa nuclear isoform of clusterin. *Biochem. J.* **2004**, *382*, 157–168. [CrossRef] [PubMed]
24. Pajak, B.; Orzechowski, A. Ethylenediaminetetraacetic acid affects subcellular expression of clusterin protein in human colon adenocarcinoma COLO 205 cell line. *Anti-Cancer Drugs* **2007**, *18*, 55–63. [CrossRef] [PubMed]
25. Shannan, B.; Seifert, M.; Boothman, D.; Tilgen, W.; Reichrath, J. Clusterin and DNA repair: A new function in cancer for a key player in apoptosis and cell cycle control. *J. Mol. Histol.* **2006**, *37*, 183–188. [CrossRef] [PubMed]
26. Leskov, K.S.; Klokov, D.Y.; Li, J.; Kinsella, T.J.; Boothman, D.A. Synthesis and functional analyses of nuclear clusterin, a cell death protein. *J. Biol. Chem.* **2003**, *278*, 11590–11600. [CrossRef] [PubMed]
27. Calnan, D.R.; Brunet, A. The FoxO code. *Oncogene* **2008**, *27*, 2276–2288. [CrossRef] [PubMed]
28. McClelland Descalzo, D.L.; Satoorian, T.S.; Walker, L.M.; Sparks, N.R.; Pulyanina, P.Y.; Zur Nieden, N.I. Glucose-Induced Oxidative Stress Reduces Proliferation in Embryonic Stem Cells via FOXO3A/beta-Catenin-Dependent Transcription of p21(cip1). *Stem Cell Rep.* **2016**, *7*, 55–68. [CrossRef] [PubMed]
29. McGowan, S.E.; McCoy, D.M. Platelet-derived growth factor-A regulates lung fibroblast S-phase entry through p27(kip1) and FoxO3a. *Respir. Res.* **2013**, *14*, 68. [CrossRef]
30. Joseph, J.; Ametepe, E.S.; Haribabu, N.; Agbayani, G.; Krishnan, L.; Blais, A.; Sad, S. Inhibition of ROS and upregulation of inflammatory cytokines by FoxO3a promotes survival against Salmonella typhimurium. *Nat. Commun.* **2016**, *7*, 12748. [CrossRef]
31. Fluteau, A.; Ince, P.G.; Minett, T.; Matthews, F.E.; Brayne, C.; Garwood, C.J.; Ratcliffe, L.E.; Morgan, S.; Heath, P.R.; Shaw, P.J.; et al. The nuclear retention of transcription factor FOXO3a correlates with a DNA damage response and increased glutamine synthetase expression by astrocytes suggesting a neuroprotective role in the ageing brain. *Neurosci. Lett.* **2015**, *609*, 11–17. [CrossRef] [PubMed]
32. Brunet, A.; Bonni, A.; Zigmond, M.J.; Lin, M.Z.; Juo, P.; Hu, L.S.; Anderson, M.J.; Arden, K.C.; Blenis, J.; Greenberg, M.E. Akt promotes cell survival by phosphorylating and inhibiting a Forkhead transcription factor. *Cell* **1999**, *96*, 857–868. [CrossRef]
33. Dobson, M.; Ramakrishnan, G.; Ma, S.; Kaplun, L.; Balan, V.; Fridman, R.; Tzivion, G. Bimodal regulation of FoxO3 by AKT and 14-3-3. *Biochim. Biophys. Acta* **2011**, *1813*, 1453–1464. [CrossRef] [PubMed]
34. Shukla, S.; Bhaskaran, N.; Maclennan, G.T.; Gupta, S. Deregulation of FoxO3a accelerates prostate cancer progression in TRAMP mice. *Prostate* **2013**, *73*, 1507–1517. [CrossRef] [PubMed]
35. Tzivion, G.; Dobson, M.; Ramakrishnan, G. FoxO transcription factors; Regulation by AKT and 14-3-3 proteins. *Biochim. Biophys. Acta* **2011**, *1813*, 1938–1945. [CrossRef] [PubMed]
36. Khan, M.; Rutten, B.P.F.; Kim, M.O. MST1 Regulates Neuronal Cell Death via JNK/Casp3 Signaling Pathway in HFD Mouse Brain and HT22 Cells. *Int. J. Mol. Sci.* **2019**, *20*, 2504. [CrossRef] [PubMed]
37. Weng, Q.; Liu, Z.; Li, B.; Liu, K.; Wu, W.; Liu, H. Oxidative Stress Induces Mouse Follicular Granulosa Cells Apoptosis via JNK/FoxO1 Pathway. *PLoS ONE* **2016**, *11*, e0167869. [CrossRef]
38. Eshbakova, K.; Aisa, H. Components of Helichrysum arenarium. *Chem. Nat. Compd.* **2009**, *45*, 929–930. [CrossRef]
39. Grinev, V.; Shirokov, A.; Navolokin, N.; Polukonova, N.; Kurchatova, M.; Durnova, N.; Bucharskaya, A.; Maslyakova, G. Polyphenolic compounds of a new biologically active extract from immortelle sandy flowers (*Helichrysum arenarium* (L.) Moench.). *Russ. J. Bioorganic Chem.* **2016**, *42*, 770–776. [CrossRef]
40. Czinner, E.; Kery, A.; Hagymási, K.; Blázovics, A.; Lugasi, A.; Szoke, E.; Lemberkovics, E. Biologically active compounds of *Helichrysum arenarium* (L.) Moench. *Eur. J. Drug Metab. Pharmacokinet.* **1999**, *24*, 309–313. [CrossRef]
41. Ghițu, A.; Schwiebs, A.; Radeke, H.H.; Avram, S.; Zupko, I.; Bor, A.; Pavel, I.Z.; Dehelean, C.A.; Oprean, C.; Bojin, F.; et al. A Comprehensive Assessment of Apigenin as an Antiproliferative, Proapoptotic, Antiangiogenic and Immunomodulatory Phytocompound. *Nutrients* **2019**, *11*, 858. [CrossRef] [PubMed]

42. Che, D.N.; Cho, B.O.; Kim, J.-S.; Shin, J.Y.; Kang, H.J.; Jang, S.I. Effect of Luteolin and Apigenin on the Production of Il-31 and Il-33 in Lipopolysaccharides-Activated Microglia Cells and Their Mechanism of Action. *Nutrients* **2020**, *12*, 811. [CrossRef] [PubMed]
43. Zhang, H.; Birch, J.; Pei, J.; Mohamed Ahmed, I.A.; Yang, H.; Dias, G.; Abd El-Aty, A.M.; Bekhit, A.E.-D. Identification of Six Phytochemical Compounds from Asparagus officinalis L. Root Cultivars from New Zealand and China Using UAE-SPE-UPLC-MS/MS: Effects of Extracts on $H_2O_2$-Induced Oxidative Stress. *Nutrients* **2019**, *11*, 107. [CrossRef] [PubMed]
44. Wen, M.; Wu, J.; Luo, H.; Zhang, H. Galangin induces autophagy through upregulation of p53 in HepG2 cells. *Pharmacology* **2012**, *89*, 247–255. [CrossRef] [PubMed]
45. Kim, D.A.; Jeon, Y.K.; Nam, M.J. Galangin induces apoptosis in gastric cancer cells via regulation of ubiquitin carboxy-terminal hydrolase isozyme L1 and glutathione S-transferase P. *Food Chem. Toxicol.* **2012**, *50*, 684–688. [CrossRef] [PubMed]
46. Zhang, H.-T.; Luo, H.; Wu, J.; Lan, L.-B.; Fan, D.-H.; Zhu, K.-D.; Chen, X.-Y.; Wen, M.; Liu, H.-M. Galangin induces apoptosis of hepatocellular carcinoma cells via the mitochondrial pathway. *World J. Gastroenterol. WJG* **2010**, *16*, 3377–3384. [CrossRef] [PubMed]
47. Zhang, W.; Tang, B.; Huang, Q.; Hua, Z. Galangin inhibits tumor growth and metastasis of B16F10 melanoma. *J. Cell. Biochem.* **2013**, *114*, 152–161. [CrossRef] [PubMed]
48. Tolomeo, M.; Grimaudo, S.; Di Cristina, A.; Pipitone, R.M.; Dusonchet, L.; Meli, M.; Crosta, L.; Gebbia, N.; Invidiata, F.P.; Titone, L. Galangin increases the cytotoxic activity of imatinib mesylate in imatinib-sensitive and imatinib-resistant Bcr-Abl expressing leukemia cells. *Cancer Lett.* **2008**, *265*, 289–297. [CrossRef] [PubMed]
49. Ha, T.K.; Kim, M.E.; Yoon, J.H.; Bae, S.J.; Yeom, J.; Lee, J.S. Galangin induces human colon cancer cell death via the mitochondrial dysfunction and caspase-dependent pathway. *Exp. Biol. Med.* **2013**, *238*, 1047–1054. [CrossRef] [PubMed]
50. Wang, X.; Gong, G.; Yang, W.; Li, Y.; Jiang, M.; Li, L. Antifibrotic activity of galangin, a novel function evaluated in animal liver fibrosis model. *Environ. Toxicol. Pharmacol.* **2013**, *36*, 288–295. [CrossRef] [PubMed]
51. Zhai, K.; Liskova, A.; Kubatka, P.; Büsselberg, D. Calcium entry through TRPV1: A potential target for the regulation of proliferation and apoptosis in cancerous and healthy cells. *Int. J. Mol. Sci.* **2020**, *21*, 4177. [CrossRef] [PubMed]
52. Pingle, S.; Matta, J.; Ahern, G. Capsaicin receptor: TRPV1 a promiscuous TRP channel. In *Transient Receptor Potential (TRP) Channels; Handbook of Experimental Pharmacology*; Springer: Berlin/Heidelberg, Germany, 2007; Volume 179, pp. 155–171.
53. Agell, N.; Bachs, O.; Rocamora, N.; Villalonga, P. Modulation of the Ras/Raf/MEK/ERK pathway by $Ca^{2+}$, and calmodulin. *Cell. Signal.* **2002**, *14*, 649–654. [CrossRef]
54. Lehtinen, M.K.; Yuan, Z.; Boag, P.R.; Yang, Y.; Villen, J.; Becker, E.B.; DiBacco, S.; de la Iglesia, N.; Gygi, S.; Blackwell, T.K.; et al. A conserved MST-FOXO signaling pathway mediates oxidative-stress responses and extends life span. *Cell* **2006**, *125*, 987–1001. [CrossRef] [PubMed]
55. Jakhar, D.; Kaul, S.; Kaur, I. Increased usage of smartphones during COVID-19: Is that blue light causing skin damage? *J. Cosmet. Dermatol.* **2020**, *19*, 2466–2467. [CrossRef] [PubMed]
56. Oplander, C.; Hidding, S.; Werners, F.B.; Born, M.; Pallua, N.; Suschek, C.V. Effects of blue light irradiation on human dermal fibroblasts. *J. Photochem. Photobiol. B* **2011**, *103*, 118–125. [CrossRef] [PubMed]
57. Oplander, C.; Deck, A.; Volkmar, C.M.; Kirsch, M.; Liebmann, J.; Born, M.; van Abeelen, F.; van Faassen, E.E.; Kroncke, K.D.; Windolf, J.; et al. Mechanism and biological relevance of blue-light (420–453 nm)-induced nonenzymatic nitric oxide generation from photolabile nitric oxide derivates in human skin in vitro and in vivo. *Free Radic. Biol. Med.* **2013**, *65*, 1363–1377. [CrossRef] [PubMed]
58. Liebmann, J.; Born, M.; Kolb-Bachofen, V. Blue-light irradiation regulates proliferation and differentiation in human skin cells. *J. Investig. Dermatol.* **2010**, *130*, 259–269. [CrossRef] [PubMed]
59. Caterina, M.J.; Julius, D. The vanilloid receptor: A molecular gateway to the pain pathway. *Annu. Rev. Neurosci.* **2001**, *24*, 487–517. [CrossRef] [PubMed]
60. Li, W.H.; Lee, Y.M.; Kim, J.Y.; Kang, S.; Kim, S.; Kim, K.H.; Park, C.H.; Chung, J.H. Transient receptor potential vanilloid-1 mediates heat-shock-induced matrix metalloproteinase-1 expression in human epidermal keratinocytes. *J. Investig. Dermatol.* **2007**, *127*, 2328–2335. [CrossRef] [PubMed]
61. Lee, Y.M.; Kim, Y.K.; Kim, K.H.; Park, S.J.; Kim, S.J.; Chung, J.H. A novel role for the TRPV1 channel in UV-induced matrix metalloproteinase (MMP)-1 expression in HaCaT cells. *J. Cell. Physiol.* **2009**, *219*, 766–775. [CrossRef]
62. Kusumaningrum, N.; Lee, D.H.; Yoon, H.S.; Park, C.H.; Chung, J.H. Ultraviolet light-induced gasdermin C expression is mediated via TRPV1/calcium/calcineurin/NFATc1 signaling. *Int. J. Mol. Med.* **2018**, *42*, 2859–2866. [CrossRef] [PubMed]
63. Levine, J.D.; Alessandri-Haber, N. TRP channels: Targets for the relief of pain. *Biochim. Biophys. Acta* **2007**, *1772*, 989–1003. [CrossRef] [PubMed]
64. Rhee, S.G. Cell signaling. $H_2O_2$, a necessary evil for cell signaling. *Science* **2006**, *312*, 1882–1883. [CrossRef] [PubMed]
65. Rhee, S.G.; Bae, Y.S.; Lee, S.R.; Kwon, J. Hydrogen peroxide: A key messenger that modulates protein phosphorylation through cysteine oxidation. *Sci. STKE* **2000**, *2000*, pe1. [CrossRef] [PubMed]
66. Kim, S.H.; Kim, H. Inhibitory Effect of Astaxanthin on Oxidative Stress-Induced Mitochondrial Dysfunction-A Mini-Review. *Nutrients* **2018**, *10*, 1137. [CrossRef] [PubMed]
67. Rhee, S.G.; Kang, S.W.; Jeong, W.; Chang, T.S.; Yang, K.S.; Woo, H.A. Intracellular messenger function of hydrogen peroxide and its regulation by peroxiredoxins. *Curr. Opin. Cell Biol.* **2005**, *17*, 183–189. [CrossRef] [PubMed]

68. Rhee, S.G.; Chang, T.S.; Bae, Y.S.; Lee, S.R.; Kang, S.W. Cellular regulation by hydrogen peroxide. *J. Am. Soc. Nephrol.* **2003**, *14*, S211–S215. [CrossRef] [PubMed]
69. Fialkow, L.; Chan, C.K.; Rotin, D.; Grinstein, S.; Downey, G.P. Activation of the mitogen-activated protein kinase signaling pathway in neutrophils. Role of oxidants. *J. Biol. Chem.* **1994**, *269*, 31234–31242. [CrossRef]
70. Guyton, K.Z.; Liu, Y.; Gorospe, M.; Xu, Q.; Holbrook, N.J. Activation of mitogen-activated protein kinase by $H_2O_2$. Role in cell survival following oxidant injury. *J. Biol. Chem.* **1996**, *271*, 4138–4142. [CrossRef] [PubMed]
71. Kang, S.W. Role of reactive oxygen species in cell death pathways. *Hanyang Med. Rev.* **2013**, *33*, 77–82. [CrossRef]
72. Davis, R.J. Signal transduction by the JNK group of MAP kinases. *Cell* **2000**, *103*, 239–252. [CrossRef]
73. Chang, L.; Karin, M. Mammalian MAP kinase signalling cascades. *Nature* **2001**, *410*, 37–40. [CrossRef] [PubMed]
74. Dhanasekaran, D.N.; Reddy, E.P. JNK signaling in apoptosis. *Oncogene* **2008**, *27*, 6245–6251. [CrossRef] [PubMed]
75. Wang, J.; He, C.; Zhou, T.; Huang, Z.; Zhou, L.; Liu, X. NGF increases VEGF expression and promotes cell proliferation via ERK1/2 and AKT signaling in Muller cells. *Mol. Vis.* **2016**, *22*, 254–263. [PubMed]
76. Liu, S.; Gao, F.; Wen, L.; Ouyang, M.; Wang, Y.; Wang, Q.; Luo, L.; Jian, Z. Osteocalcin Induces Proliferation via Positive Activation of the PI3K/Akt, P38 MAPK Pathways and Promotes Differentiation Through Activation of the GPRC6A-ERK1/2 Pathway in C2C12 Myoblast Cells. *Cell Physiol. Biochem.* **2017**, *43*, 1100–1112. [CrossRef] [PubMed]
77. Sharma, G.D.; He, J.; Bazan, H.E. p38 and ERK1/2 coordinate cellular migration and proliferation in epithelial wound healing: Evidence of cross-talk activation between MAP kinase cascades. *J. Biol. Chem.* **2003**, *278*, 21989–21997. [CrossRef] [PubMed]
78. Wang, X.; Chen, W.R.; Xing, D. A pathway from JNK through decreased ERK and Akt activities for FOXO3a nuclear translocation in response to UV irradiation. *J. Cell. Physiol.* **2012**, *227*, 1168–1178. [CrossRef] [PubMed]
79. Tikhanovich, I.; Kuravi, S.; Campbell, R.V.; Kharbanda, K.K.; Artigues, A.; Villar, M.T.; Weinman, S.A. Regulation of FOXO3 by phosphorylation and methylation in hepatitis C virus infection and alcohol exposure. *Hepatology* **2014**, *59*, 58–70. [CrossRef]
80. Soulez, M.; Tanguay, P.L.; Dô, F.; Dort, J.; Crist, C.; Kotlyarov, A.; Gaestel, M.; Ferron, M.; Dumont, N.A.; Meloche, S. ERK3-MK5 signaling regulates myogenic differentiation and muscle regeneration by promoting FoxO3 degradation. *J. Cell. Physiol.* **2022**. [CrossRef]
81. Bi, W.; Xiao, L.; Jia, Y.; Wu, J.; Xie, Q.; Ren, J.; Ji, G.; Yuan, Z. c-Jun N-terminal kinase enhances MST1-mediated pro-apoptotic signaling through phosphorylation at serine 82. *J. Biol. Chem.* **2010**, *285*, 6259–6264. [CrossRef]
82. Yuan, Z.; Kim, D.; Shu, S.; Wu, J.; Guo, J.; Xiao, L.; Kaneko, S.; Coppola, D.; Cheng, J.Q. Phosphoinositide 3-kinase/Akt inhibits MST1-mediated pro-apoptotic signaling through phosphorylation of threonine 120. *J. Biol. Chem.* **2010**, *285*, 3815–3824. [CrossRef] [PubMed]
83. Chao, Y.; Wang, Y.; Liu, X.; Ma, P.; Shi, Y.; Gao, J.; Shi, Q.; Hu, J.; Yu, R.; Zhou, X. Mst1 regulates glioma cell proliferation via the AKT/mTOR signaling pathway. *J. Neurooncol.* **2015**, *121*, 279–288. [CrossRef] [PubMed]
84. Greer, E.L.; Brunet, A. FOXO transcription factors at the interface between longevity and tumor suppression. *Oncogene* **2005**, *24*, 7410–7425. [CrossRef] [PubMed]
85. Brownawell, A.M.; Kops, G.J.; Macara, I.G.; Burgering, B.M. Inhibition of nuclear import by protein kinase B (Akt) regulates the subcellular distribution and activity of the forkhead transcription factor AFX. *Mol. Cell. Biol.* **2001**, *21*, 3534–3546. [CrossRef] [PubMed]
86. Matsuzaki, H.; Daitoku, H.; Hatta, M.; Tanaka, K.; Fukamizu, A. Insulin-induced phosphorylation of FKHR (Foxo1) targets to proteasomal degradation. *Proc. Natl. Acad. Sci. USA* **2003**, *100*, 11285–11290. [CrossRef] [PubMed]
87. Pucci, S.; Bonanno, E.; Pichiorri, F.; Angeloni, C.; Spagnoli, L.G. Modulation of different clusterin isoforms in human colon tumorigenesis. *Oncogene* **2004**, *23*, 2298–2304. [CrossRef] [PubMed]
88. Caccamo, A.E.; Scaltriti, M.; Caporali, A.; D'Arca, D.; Corti, A.; Corvetta, D.; Sala, A.; Bettuzzi, S. $Ca^{2+}$ depletion induces nuclear clusterin, a novel effector of apoptosis in immortalized human prostate cells. *Cell Death Differ.* **2005**, *12*, 101–104. [CrossRef] [PubMed]
89. Artemaki, P.I.; Sklirou, A.D.; Kontos, C.K.; Liosi, A.A.; Gianniou, D.D.; Papadopoulos, I.N.; Trougakos, I.P.; Scorilas, A. High clusterin (CLU) mRNA expression levels in tumors of colorectal cancer patients predict a poor prognostic outcome. *Clin. Biochem.* **2020**, *75*, 62–69. [CrossRef] [PubMed]

*nutrients* MDPI

*Article*

# Discovery of Pinostrobin as a Melanogenic Agent in cAMP/PKA and p38 MAPK Signaling Pathway

**Jeong-Hyun Yoon [1,†], Kumju Youn [2,†] and Mira Jun [1,2,3,*]**

1 Department of Health Sciences, The Graduate School of Dong-A University, Busan 49315, Korea
2 Department of Food Science and Nutrition, Dong-A University, Busan 49315, Korea
3 Center for Food & Bio Innovation, Dong-A University, Busan 49315, Korea
* Correspondence: mjun@dau.ac.kr; Tel.: +82-51-200-7323; Fax: +82-51-200-7535
† These authors contributed equally to this work.

**Abstract:** Melanogenesis is the process of melanin synthesis to protect the skin against ultraviolet radiation and other external stresses. The loss of skin pigmentation is closely related to depigmented skin disorders. The melanogenic effects of pinostrobin, an active flavanone found in honey, were evaluated. B16F10 cells were used for melanin content, tyrosinase activity, and the expression of melanogenesis-related markers. Moreover, computational simulations were performed to predict docking and pharmacokinetics. Pinostrobin increased melanin levels and tyrosinase activity by stimulating the expression of melanogenic regulatory factors including tyrosinase, tyrosinase-related protein (TRP) 1 and microphthalmia transcription factor (MITF). Specifically, the phosphorylation of cAMP response element binding (CREB) involved in the MITF activation was augmented by pinostrobin. Moreover, the compound upregulated the β-catenin by cAMP/PKA-mediated GSK-3β inactivation. Co-treatment with a PKA inhibitor, inhibited melanin production, tyrosinase activity, and expression of MITF, *p*-CREB, *p*-GSK-3β and *p*-β-catenin, demonstrating that pinostrobin-stimulated melanogenesis was closely related to cAMP/PKA signaling pathway. Furthermore, the combination of pinostrobin and a specific p38 inhibitor, showed that MITF upregulation by pinostrobin was partly associated with the p38 signaling pathway. Docking simulation exhibited that the oxygen group at C-4 and the hydroxyl group at C-5 of pinostrobin may play an essential role in melanogenesis. In silico analysis revealed that pinostrobin had the optimal pharmacokinetic profiles including gastrointestinal absorption, skin permeability, and inhibition of cytochrome (CYP) enzymes. From the present results, it might be suggested that pinostrobin could be useful as a potent and safe melanogenic agent in the depigmentation disorder, vitiligo.

**Keywords:** melanogenesis; pinostrobin; MITF; melanogenesis-related enzyme; cAMP/PKA; p38

**Citation:** Yoon, J.-H.; Youn, K.; Jun, M. Discovery of Pinostrobin as a Melanogenic Agent in cAMP/PKA and p38 MAPK Signaling Pathway. *Nutrients* **2022**, *14*, 3713. https://doi.org/10.3390/nu14183713

Academic Editors: Daniela Rigano and Paola Bontempo

Received: 20 July 2022
Accepted: 6 September 2022
Published: 9 September 2022



## 1. Introduction

Human skin plays a major role in immunologic responses providing a physical barrier that may affect the physiological status of the body. Moreover, the skin acts as an immune system and, through its pigments, provides a protective mechanism against ultraviolet radiation (UVR) [1]. Melanin is responsible for the hair, eyes and skin, pigmentation, which is produced by melanosomes in melanocytes. Melanocytes transfer melanin via their dendrites to adjacent keratinocytes, where they form the melanin caps that reduce UV-stimulated epidermal DNA damage [2]. The loss of skin pigmentation can cause compromised cutaneous immunity, resulting in conditions, such as vitiligo, an acquired skin pigmentation disorder. The main cause of the depigmentation disease is the destruction of melanocytes and the impediment of melanogenesis pathways [3].

Melanogenesis, a biosynthetic pathway for melanin production, is mediated by three specific enzymes, tyrosinase, tyrosinase related protein 1 (TRP 1) and TRP 2 [4]. Tyrosinase, the rate-limiting enzyme in melanin synthesis, catalysis the hydroxylation of tyrosine into dihydroxyphenylalanine (DOPA) and the further oxidation of DOPA into DOPA quinone. Aside from tyrosinase, TRP 1 and TRP 2 reside in melanosomes and also regulate melanin production [5]. In mammals, two types of melanin are produced, pheomelanin (yellow/red) and eumelanin (brown/black). Tyrosinase is a common enzyme required for both types of melanin synthesis, whereas TRP 1 and TRP 2 are more specific for eumelanin formation. These tyrosinase family genes are controlled by microphthalmia transcription factor (MITF), a master regulator of melanocyte proliferation and survival [6].

When exposed to UVR, α-melanocyte stimulating hormone (α-MSH) is released and binds to melanin cortin-1 receptor (MC1R), resulting in the activation of cyclic adenosine monophosphate (cAMP) [7]. cAMP subsequently stimulates cAMP-dependent protein kinase A (PKA) and the catalytic subunit of PKA translocates to the nucleus where it phosphorylates cAMP-response element binding (CREB). The phosphorylated active CREB further binds MITF, which in turn stimulates the transcription of the key melanogenic enzymes [8].

β-Catenin signaling is one of the pathways involved in MITF expression, which contributes to melanogenesis. In normal conditions, the level of cytoplasmic β-catenin is kept low via multiprotein complexes, such as Axin, APC, GSK-3 and CK1 mediated degradation [9]. Previous studies demonstrated that activating the cAMP/PKA pathway stimulates the phosphorylation of GSK-3β at Ser 9, facilitating β-catenin accumulation [10,11]. The increased β-catenin is transported into the nucleus in the form of a complex with the lymphoid-enhancing factor-1/T cell factor (LEF-TCF) transcription factor, and then promotes MITF expression [12].

Mitogen-activated protein kinases (MAPKs) including extracellular responsive kinase (ERK), c-Jun *N*-terminal kinase (JNK), and p38 MAPK are major signaling molecules associated with the regulation of melanogenesis [13]. The phosphorylated p38 activates MITF to ultimately stimulate melanin synthesis, while the activation of ERK 1/2 and JNK leads to a decrease in melanogenesis via MITF degradation [14].

Natural products with diverse structures are well recognized as important sources in the development of safe and effective therapeutic agents for diseases including cancers, inflammation, dementia, vitiligo, etc. [15]. In our ongoing research of exploring natural melanogenic activators, 50 well-known natural compounds from foods were screened for stimulating effects on tyrosinase activity and cellular melanin production. Among them, pinostrobin, a major flavonoid found in honey, propolis, and finger roots, had the highest activity in producing melanin compared with the other flavonoids, such as quercetin, catechin, resveratrol, hesperidin, naringenin, etc. In the present study, the melanogenic activity of pinostrobin and its possible underlying cAMP/PKA and MAPK signaling pathways were elucidated for the first time.

## 2. Materials and Methods

### *2.1. Chemicals and Reagents*

Pinostrobin (5-hydroxy-7-methoxyflavanone) (≥99%), MTT reagent, H89, PD98059, SB203580, and L-DOPA were purchased from Sigma-Aldrich (St. Louis, MO, USA). *p*-CREB (Ser133), *p*-GSK-3β (Ser9), *p*-β-catenin (Ser675), *p*-ERK, *p*-JNK, *p*-p38, and p38 antibodies were purchased from Cell Signaling Technology Inc. (Beverly, MA, USA), whereas antibodies for tyrosinase, TRP 1, TRP 2, MITF, GSK-3β, JNK, ERK, β-actin, mouse IgG-HRP, and rabbit antibody IgG-HRP were obtained from Santa Cruz Biotechnology Inc. (Santa Cruz, CA, USA).

### *2.2. Cell Lines and Pinostrobin Treatment*

B16F10 cell was purchased from the Korean Cell Line Bank (Seoul, Korea) and cultured in DMEM supplemented with 10% FBS and 1% penicillin/streptomycin (all from Hyclone, Logan, UT, USA) at 37 °C in a humid atmosphere of 5% $CO_2$. A stock solution of pinostrobin was prepared by dissolving DMSO. Control cells were treated with equal volumes of DMSO.

### 2.3. Cell Viability Assay

The cells ($3 \times 10^3$ cells/well) were seeded into 96-well and cultured for 24 h. The medium was then replaced with fresh medium containing pinostrobin at various concentrations (25 μM, 50 μM, 100 μM, or 200 μM) and incubated for 3 days. Next, the cells were treated with MTT solution (5 mg/mL) at 37 °C for 3 h. Then, the MTT solution was removed and DMSO was added to dissolve. Absorbance was measured at 570 nm using a microplate spectrophotometer (ELX808, BioTek, Winooski, VT, USA).

### 2.4. Measurement of Melanin Content

Melanocytes were exposed to pinostrobin for 3 days of culture and washed. The cells were excised with trypsin-EDTA and the culture medium was removed by centrifugation. The pellet was then incubated in 1N NaOH solution at 80 °C for 2 h to dissolve melanin. The melanin contents were measured by an ELISA reader (405 nm) and presented as a percentage of those of untreated controls [16].

### 2.5. Evaluation of Tyrosinase Activity

The cells ($2.0 \times 10^5$) were seeded in 6-well plates and stimulated with pinostrobin for 3 days. The extracted protein content was assessed using a BCA Protein Assay kit (Thermo Scientific Pierce, Rockford, IL, USA). L-DOPA (10 mM) was added to each well of a 96-well plate. Then, potassium phosphate buffer (0.1 M, pH 6.8) and cell protein lysate (1 μg/μL) were added. The reaction mix was incubated for 1 h at 37 °C and then measured at 490 nm [17].

### 2.6. Western Blot Analysis

After pinostrobin treatment, the cells were harvested and lysed. The cell lysates (25 μg) were mixed with 5X Laemmli buffer (Elpis Biotech, Daejeon, Korea) and boiled for 5 min. The samples were divided using sodium dodecyl sulfate-polyacrylamide gels and transferred to a polyvinylidene difluoride membrane (Amersham Biosciences, Piscataway, NJ, USA). The membranes were blocked in 5% skim milk and incubated for 24 h at 4 °C with diluted primary antibodies: β-actin (1:10,000), MITF (1:500), tyrosinase (1:500), TRP 1 (1:2000), TRP 2 (1:1000), *p*-CREB (1:1000), *p*-GSK-3β (1:2000), GSK-3β (1:1000), *p*-β-catenin (1:1000) *p*-ERK (1:3000), ERK (1:3000), *p*-p38 (1:2000), p38 (1:2000), *p*-JNK (1:1000), and JNK (1:1000). Then, the membranes were then further incubated with a secondary antibody corresponding to the primary antibody. The protein band was detected using an ECL kit (Advansta Inc, Menlo Park, CA, USA) and photographed with the Atto EZ-capture system (Tokyo, Japan). The intensities of the protein bands were performed using the ImageJ software (NIH, Bethesda, MD, USA). Each protein detection had three replicates.

### 2.7. In Silico Pharmacokinetic Prediction and Docking Simulation

SwissADME (http://www.swissadme.ch, accessed on 19 July 2022 Swiss Institute of Bioinformatics, Lausanne, Switzerland) is a web-based tool to generate physicochemical descriptors, such as molecular weight, molecular refractivity, count of specific atom types, polar surface area, lipophilicity and water solubility. It also calculates pharmacokinetic properties including human gastrointestinal absorption, caco-2 permeability, skin permeability coefficient and interaction of molecules with cytochromes p450 (CYP). The SMILES (Simplified Molecular Input Line Entry System) format of pinostrobin generated from ChemSketch software 2012 version (ACD/Labs., Toronto, ON, Canada) was entered in the SwissADME [18].

The 3D structure of pinostrobin (73201) was obtained from PubChem. The crystal structure of PKA (ID No. 1CX4), and p38 (ID No. 3ZS5) was retrieved from the Protein Data Bank (PDB). Docking simulation was performed using AutoDock Vina 1.1.2 (The Scripps Research Institute, La Jolla, CA, USA) [19]. The best docking mode was the highest ligand-binding affinity. RMSD values of the lower bound and the upper bound were zero. The dimensions of the box were set to 40 Å × 40 Å × 40 Å, which is large enough to cover the entire binding site, and the default exhaustiveness value was 4. Poses were visually assessed and analyzed through PyMOL 2.5.0 version (Schrödinger, Inc., New York, NY, USA). The pharmacophore of hydrogen bonds and Van der Waals interactions were conducted using the Ligplot+ program version 2.2 (EMBL-EBI, Hinxton, Cambridgeshire, UK).

### 2.8. Statistical Analysis

Statistical analysis was conducted with ANOVA followed by Tukey's multiple comparisons test. * $p < 0.05$, ** $p < 0.01$, and *** $p < 0.001$ were considered statistically significant differences.

## 3. Results

### 3.1. Effects of Pinostrobin on Melanin Production and Tyrosinase Activity in B16F10 Cells

The structure of pinostrobin purchased from Sigma-Aldrich was revealed in Figure 1a. As shown in Figure 1b, it did not exhibit any cytotoxicity in the range from 25 to 200 μM in B16F10 cells. For further experiments, doses below 100 μM were selected. The compound significantly increased the melanin content at all tested concentrations (Figure 1c). In particular, the melanin production was noticeably elevated more than 2.3 times after 100 μM of pinostrobin pretreatment.

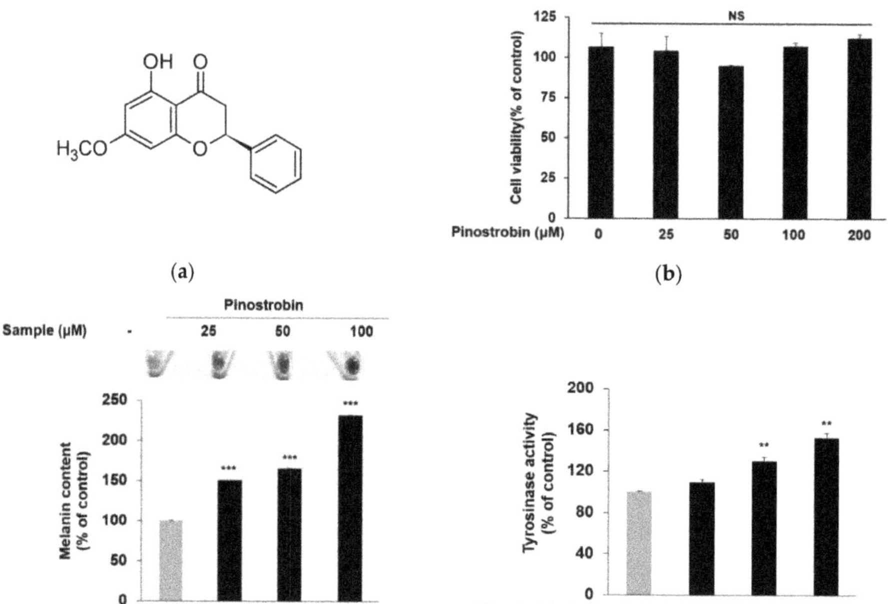


**Figure 1.** Effect of pinostrobin on cell viability, intracellular melanin contents, tyrosinase activity in B16F10 cells. (**a**) Chemical structure of pinostrobin. (**b**) After treatment with various concentrations of pinostrobin for 72 h, the MTT assay was performed to analyze the cell viability. (**c**,**d**) After treatment under the same conditions used for the determination of cell viability, cells were collected and lysed to measure intracellular melanin content and tyrosinase activity. Cell viability, intracellular melanin contents, and tyrosinase activity in control cells were regarded as 100%. *** $p < 0.001$ and ** $p < 0.01$ compared with control group. NS; not significant.

Since tyrosinase is the rate-limiting enzyme for regulating melanin production, increasing the activity of tyrosinase has become a target for the development of melanogenic agents. As shown in Figure 1d, pinostrobin exhibited a somewhat weaker effect at 25 μM, but considerably augmented intracellular tyrosinase activity at 50 μM and 100 μM (130.1 ± 3.9% and 152.7 ± 4.5%, respectively).

### *3.2. Pinostrobin Regulated the Protein Expression of Melanogenic Enzymes and MITF*

To determine whether stimulation of tyrosinase activity by pinostrobin was caused by the regulation of the tyrosinase gene family, Western blotting was performed. As shown in Figure 2a–c, the expression of tyrosinase and TRP 1 was significantly augmented by the pretreatment with pinostrobin at all concentrations, in particular, it increased to 1.6-fold and 2.1-fold, respectively, with the highest concentration of pinostrobin. On the contrary, pinostrobin showed no effects on TRP 2 expression (Figure 2d).

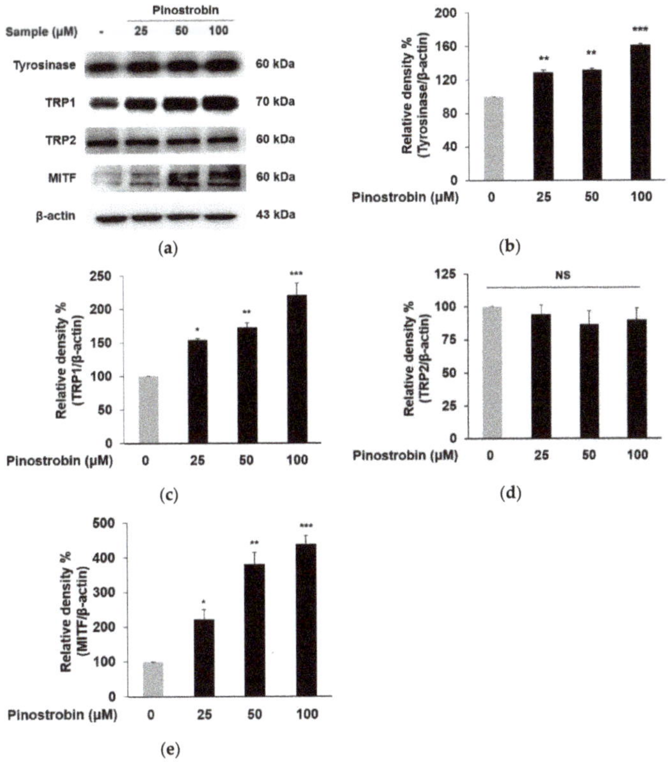


**Figure 2.** Effect of pinostrobin on the protein levels of tyrosinase, TRP 1, TRP 2, and MITF in B16F10 cells. Cells were treated with pinostrobin at various concentrations for 24 h. (**a**) Tyrosinase, TRP 1, TRP 2, and MITF, were analyzed by Western blotting. (**b–e**) Blots were quantified using ImageJ software. β-actin is used as a loading control for quantitative Western blotting. The untreated cells were regarded as 100%. Values are expressed represent as the mean±SD of three independent experiments. *** $p < 0.001$, ** $p < 0.01$ and * $p < 0.05$ compared with control group. NS; not significant.

Since MITF plays a crucial role in melanogenesis as a key transcriptional modulator of tyrosinase and TRPs, MITF expression was further investigated [6]. As depicted in Figure 2a,e, the tested compound significantly induced MITF expression to 2.2-fold, 3.8-fold, and 4.4-fold at a concentration of 25–100 μM, respectively, which clearly demonstrated that the melanogenic effect of pinostrobin was promoted by MITF-dependent activation of tyrosinase and TRP 1.

### 3.3. Pinostrobin Mediated Melanogenesis by Upregulating CREB, GSK-3β and β-Catenin

CREB, GSK-3β, and β-catenin are transcription factors that are closely related to MITF-dependent melanogenesis [20]. CREB activates MITF transcription, resulting in increasing melanin synthesis. In addition, phosphorylated GSK-3β induces β-catenin translocation, where it upregulates MITF and thus promotes melanogenic gene expression [9].

To define the mechanisms underlying the melanin synthetic property of pinostrobin, changes in the upstream activator of MITF induced by pinostrobin were examined. As depicted in Figure 3a,b, pinostrobin at 50 and 100 μM promoted the phosphorylation of CREB more than two-fold compared to the control group. The compound obviously elevated GSK-3β phosphorylation without significantly affecting total GSK-3β expression (Figure 3a,c). Particularly, the expression level of phosphorylated GSK-3β was dramatically elevated (>2 folds) after pretreatment with pinostrobin at 50 and 100 μM. Consistent with the results of GSK-3β, the level of phosphorylated β-catenin was significantly augmented in pinostrobin-treated cells (Figure 3a,d).

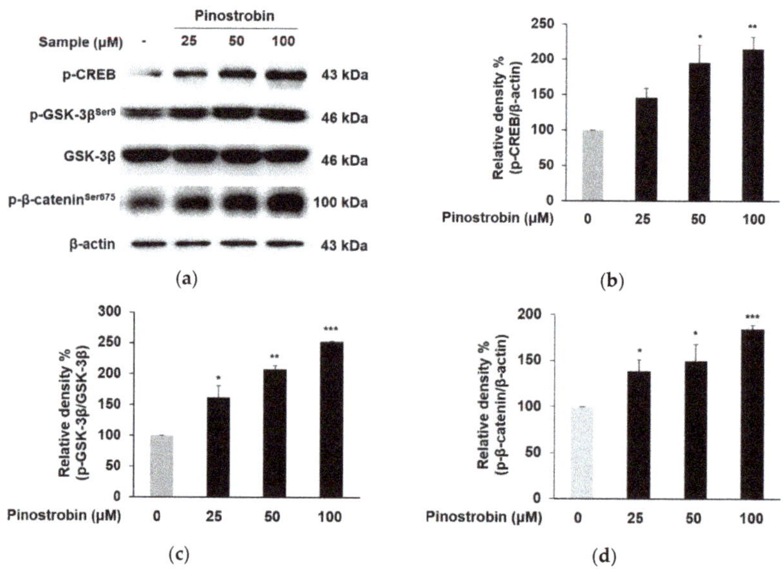


**Figure 3.** Effect of pinostrobin on the phosphorylation of CREB, GSK-3β, and β-catenin in B16F10 cells. Cells were treated with pinostrobin at various concentrations for 24 h. (**a**) *p*-CREB, *p*-GSK-3β, and β-catenin, were analyzed by Western blotting. (**b–d**) Blots were quantified using ImageJ software. Protein levels were normalized to the corresponding loading controls. The untreated cells were regarded as 100%. Values are expressed represent as the mean ± SD of three independent experiments. *** $p < 0.001$ ** $p < 0.01$, and * $p < 0.05$ compared with control group.

### 3.4. Pinostrobin Controlled the Melanogenesis via PKA Dependent Signaling Pathway

To verify the involvement of the cAMP/PKA cascade in pinostrobin-mediated melanogenesis, the specific PKA inhibitor (H89) was used. As shown in Figure 4a,b, H89 at 10 μM markedly reduced the level of melanin contents and tyrosinase activity compared with the H89 untreated control. Co-treatment of pinostrobin and H89 significantly reduced the pinostrobin-induced melanin synthesis and tyrosinase activity ($p < 0.001$). In addition, pinostrobin in combination with H89 obviously decreased the expression of tyrosinase, TRP 1, and MITF more than only pinostrobin did, as illustrated in Figure 4c–f. Moreover, upregulated phosphorylation of CREB, GSK-3β, and β-catenin induced by pinostrobin at 100 μM was markedly attenuated when co-treated with H89 (Figure 4g–j), implying that the cAMP/PKA signaling pathway is directly related to pinostrobin-mediated melanogenesis.

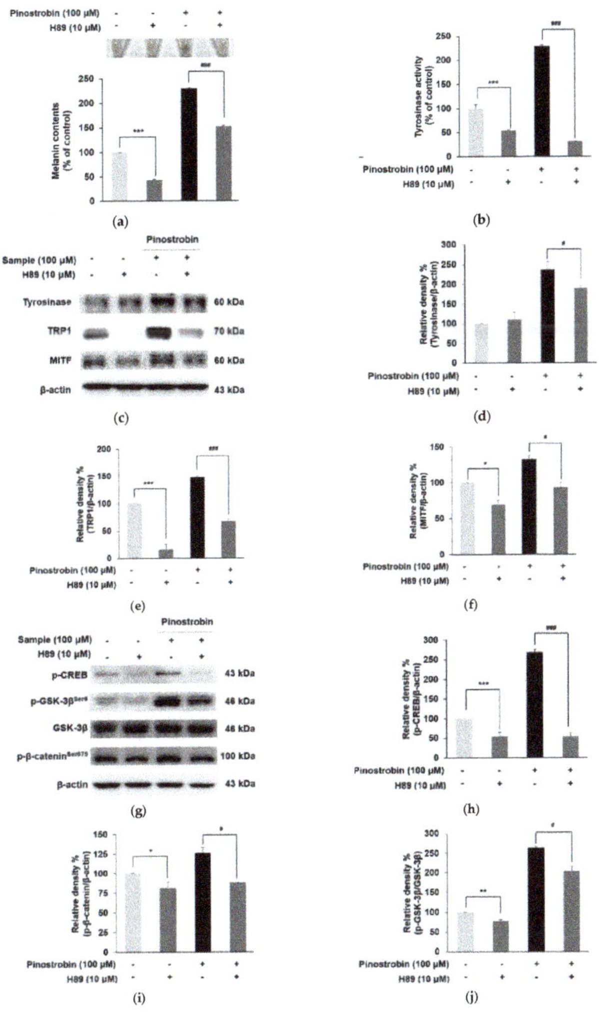

**Figure 4.** Effect of pinostrobin and H89 (PKA inhibitor) on PKA signaling pathway in B16F10 cells. Cells were pre−treated with H89 for 30 min prior to exposure to pinostrobin (100 μM) for 72 h to evaluate (**a**) the intracellular melanin contents, and (**b**) tyrosinase activity. (**c**–**f**) Cells were treated with H89 (10 μM) for 30 min, followed by treatment with pinostrobin (100 μM) for 24 h. Subsequently, tyrosinase, TRP 1, and MITF were analyzed by Western blotting. (**g**–**j**) Western blotting result of *p*-CREB, *p*-GSK-3β, and *p*-β-catenin expression with co-treatment of H89 and pinostrobin. Blots were quantified using ImageJ software. Protein levels were normalized to the corresponding loading controls. The untreated cells were regarded as 100%. Values are expressed represent as the mean ± SD of three independent experiments. *** $p < 0.001$, ** $p < 0.01$, and * $p < 0.05$ compared with control group; ### $p < 0.001$ and # $p < 0.05$ compared with pinostrobin group without H89.

### 3.5. Pinostrobin Regulated p38 and ERK Signaling Pathway

MAPKs were demonstrated as one of the upstream pathways of melanogenesis by regulating MITF activation [14]. The phosphorylation of p38, ERK, and JNK was evaluated to elucidate the mechanism underlying the MITF-mediated melanogenesis property of pinostrobin. As shown in Figure 5a–c, the levels of p38 phosphorylation were remarkably elevated, while the ERK phosphorylation was significantly reduced in pinostrobin-treated cells as compared to those of the control. However, the compound had no effect on phosphorylated JNK (Figure 5a,d). These results indicated that both p38 activation and ERK suppression were partly related to melanogenesis by pinostrobin.

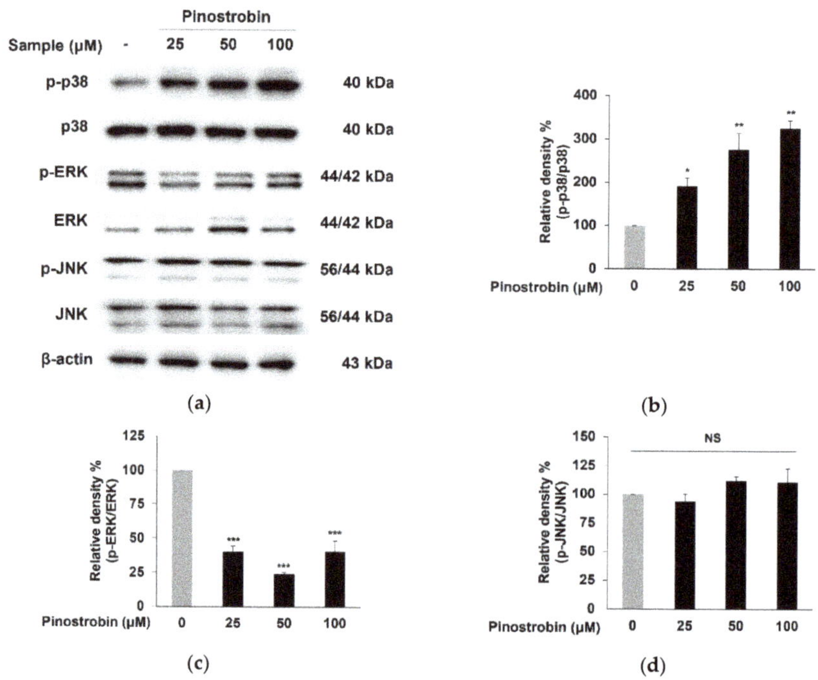


**Figure 5.** Effect of pinostrobin on the phosphorylation of p38, ERK, and JNK in B16F10 cells. (**a**–**d**) Cells were treated with pinostrobin at various concentrations for 24 h and then harvested. *p*-p38, *p*-ERK, and *p*-JNK, were analyzed by Western blotting. Blots were quantified using ImageJ software. Protein levels were normalized to the corresponding loading controls. The untreated cells were regarded as 100%. Values are expressed represent as the mean ± SD of three independent experiments. *** $p < 0.001$ ** $p < 0.01$, and * $p < 0.05$ compared with control group. NS; not significant.

### 3.6. Pinostrobin Stimulated Melanogenesis via p38 Signaling Pathway

To confirm the involvement of p38 and ERK phosphorylation in the melanogenic effect of pinostrobin, the expression of melanogenic-related markers was evaluated in the presence of pinostrobin and MAPK inhibitors. When the p38 signaling pathway was blocked by co-treating the cells with a specific p38 inhibitor SB203580 and pinostrobin, the melanin contents and tyrosinase activity considerably decreased compared to cells treated with pinostrobin alone (Figure 6a,b). Moreover, the combination of pinostrobin and SB203580 noticeably reduced the expression of MITF and its downstream targets tyrosinase as well as the p38 expression induced by pinostrobin, suggesting that the compound increased tyrosinase activity and expression via p38-mediated upregulation of MITF (Figure 6c–h). As illustrated in Figure 6i,j, ERK inhibitor (PD98059) significantly induced melanin production and tyrosinase activation. However, the co-treatment of pinostrobin and PD98059 showed no effect on melanin contents and tyrosinase activity, indicating the contributions of ERK signaling are relatively weaker than p38 in the MAPK pathway. These results demonstrated that the p38 pathway was closely associated with the MITF-mediated melanogenesis by pinostrobin.

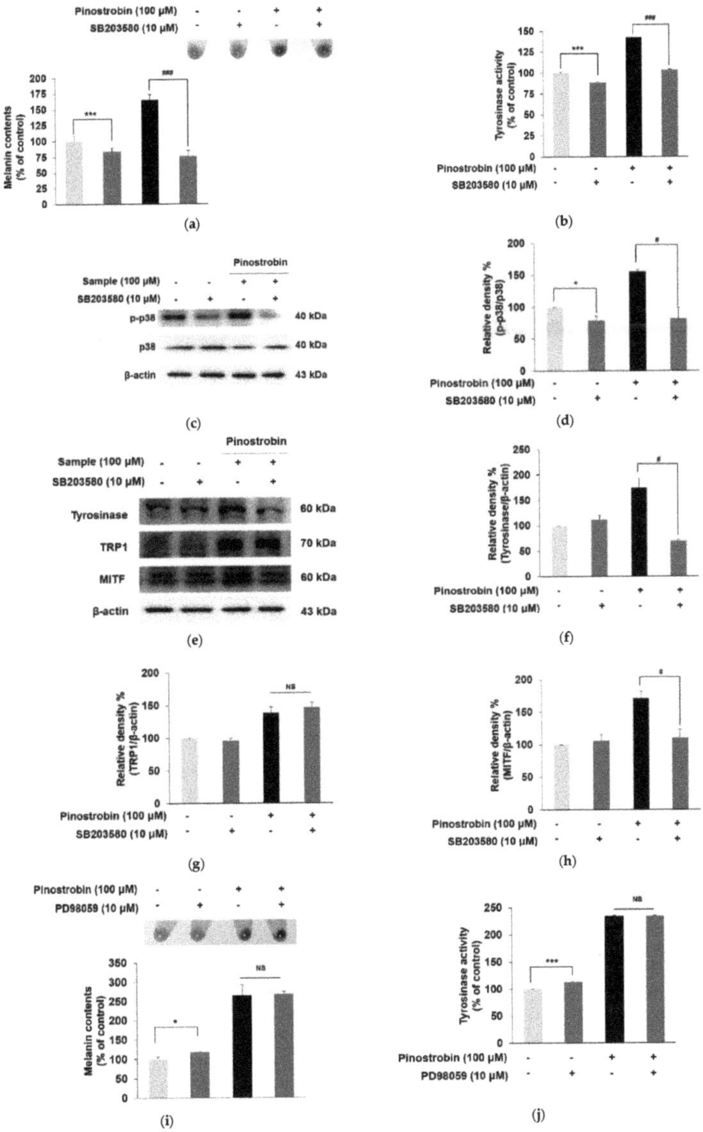


**Figure 6.** Effect of pinostrobin with SB203580 (p38 inhibitor) or PD98059 (ERK inhibitor) on MAPK signaling pathway in B16F10 cells. Cells were pre-treated with SB203580 (10 μM) and pinostrobin (100 μM) for 72 h to evaluate (**a**) the intracellular melanin contents, and (**b**) tyrosinase activity. B16F10 cells were treated with SB203580 (10 μM) for 30 min, followed by treatment with pinostrobin (100 μM) for 24 h. Subsequently, the expression of (**c**,**d**) *p*-p38, (**e**,**f**) tyrosinase, (**g**) TRP 1 and (**h**) MITF was analyzed by Western blotting. B16F10 cells were co-treated with PD98059 (10 μM) and pinostrobin (100 μM) for 72 h to evaluate (**i**) the intracellular melanin contents, and (**j**) tyrosinase activity. Blots were quantified using ImageJ software. Protein levels were normalized to the corresponding loading controls. The untreated cells were regarded as 100%. Values are expressed represent as the mean ± SD of three independent experiments. *** $p < 0.001$, and * $p < 0.05$ compared with control group; ### $p < 0.001$ and # $p < 0.05$ compared with pinostrobin group without SB203580 or PD98059. NS; not significant.

To validate the contribution of the PKA signaling pathway to pinostrobin-induced p38 phosphorylation, B16F10 cells were co-treated with PKA inhibitor and pinostrobin (Figure 7a,b). H89 alone decreased the phosphorylation level of p38 MAPK, whereas co-treatment with pinostrobin and H89 did not change the expression of *p*-p38, suggesting that pinostrobin-induced PKA and p38 signaling pathways were independent.

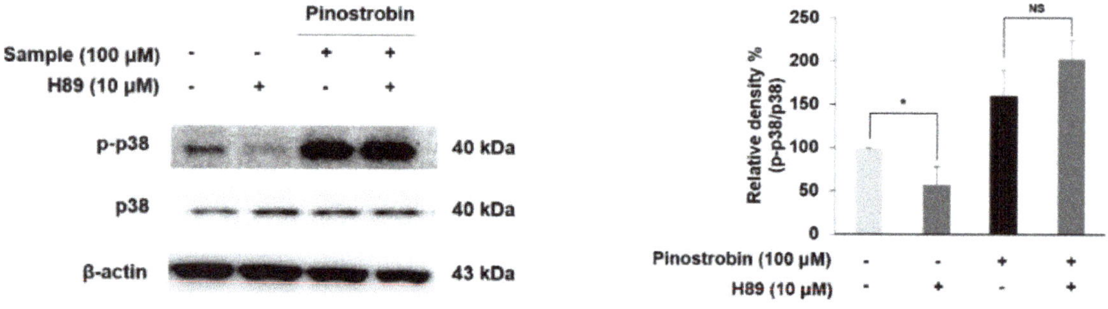


(a) (b)

**Figure 7.** Effect of pinostrobin with H89 on phosphorylated p38 expression in B16F10 cells. (**a**) Cells were pre−treated with H89 for 30 min prior to exposure pinostrobin (100 μM) for 24 h and then harvested. (**b**) Blots were quantified using ImageJ software. Protein levels were normalized to the corresponding loading controls. The untreated cells were regarded as 100%. Values are expressed represent as the mean ± SD of three independent experiments. * $p < 0.05$ compared with control group. NS; not significant.

### *3.7. Molecular Docking Simulation of Pinostrobin*

Table 1 and Figure 8 showed the molecular docking results of pinostrobin with biomarkers in the cAMP/PKA and p38 MAPK signaling pathway. Two hydrogen bonds were found between pinostrobin and PKA residues with the lowest binding energy of -7.3 kcal/mol. The Asp214 and Gln377 residues in PKA participated in hydrogen interactions with the hydroxyl group at C-5 and oxygen group at C-4 of pinostrobin (bonding distances of 2.84 Å and 3.15 Å, respectively). The binding site for the PKA-pinostrobin complex was formed by van Der Waals interactions with Asp149, Leu151, Tyr213, Arg216, Gly217, Asp309, and Asp375 (Table 1 and Figure 8a).

**Table 1.** Docking simulation of pinostrobin with upstream targets involved in melanogenesis.

| Target Protein | Binding Energy (kcal/mol) | No. of H- Bonds | H-Bonding Residues | H-Bond Length (Å) | van Der Waals Residues |
|---|---|---|---|---|---|
| PKA | -7.3 | 2 | Asp214,<br>Gln377 | 3.15<br>2.84 | Asp149, Leu151, Tyr213, Arg216, Gly217, Asp309, Asp375 |
| p38 | -7.6 | 2 | Ser56,<br>His64 | 2.82<br>3.03 | Ala34, Tyr35, Arg67, Thr68, Glu71, Asp168 |

In the p38-pinostrobin complex, the lowest binding energy between pinostrobin and p38 was −7.6 kcal/mol (Table 1 and Figure 8b). Six residues of p38, such as Ala34, Tyr35, Arg67, Thr68, Glu71, and Asp168 were demonstrated to participate in van Der Waals interactions with the compound. Interestingly, the hydroxyl group at C-5 of pinostrobin simultaneously bound at Ser56, and His64 residues of p38 with bonding distances of 2.82 Å and 3.03 Å, respectively.

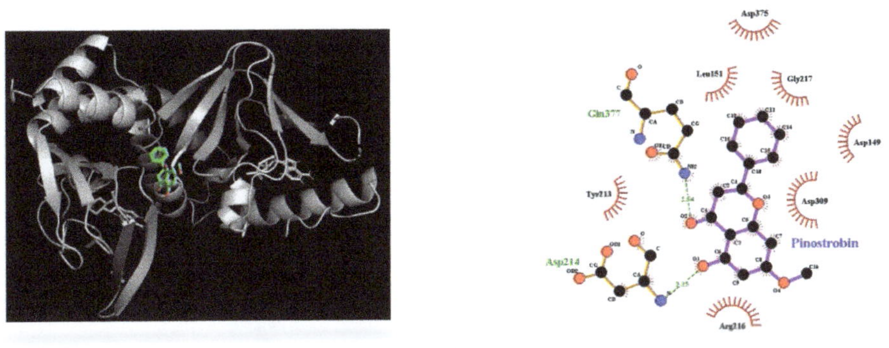


(a)

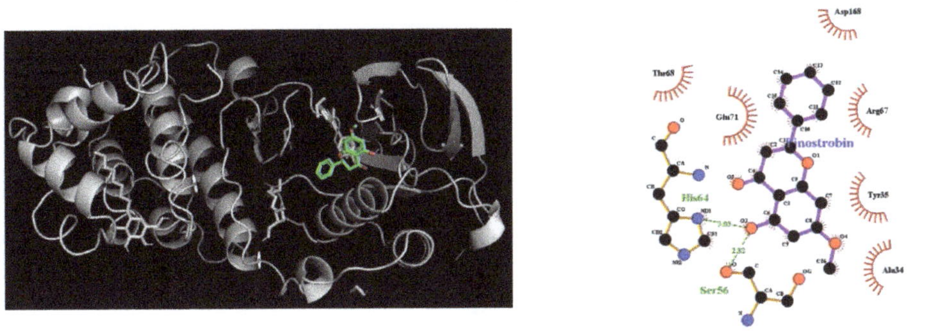

(b)

**Figure 8.** Molecular docking interactions of pinostrobin with (**a**) PKA and (**b**) p38; surface view, and interaction map, such as hydrogen (dotted line in green) and hydrophobic bonding (red dashed semicircle) between pinostrobin with upstream targets involved in melanogenesis.

### 3.8. *In Silico SwissADME Profile of Pinostrobin*

Pharmacokinetic parameters are needed to understand the efficacy and safety of the target compound in vivo. Designing novel candidates requires substantial attention to their pharmacokinetic properties including intestinal absorption, penetration of the skin, and inhibition of CYP isoforms.

As shown in Table 2, pinostrobin was predicted to be absorbed in the gastrointestinal (GI) tract and penetrated the stratum corneum, the rate-limiting barrier for skin penetration. In addition, the compound was predicted as an inhibitor of several CYP enzymes including CYP1A2 and CYP2C19.

**Table 2.** Pharmacokinetic properties of pinostrobin.

| Properties | Predicted Value | Standard Value |
|---|---|---|
| Permeability of Caco-2 (log Papp in $10^{-6}$ cm/s) | 1.3 | Caco-2 permeability >0.90 (high) |
| Human intestinal absorption (%) | 93.8 | <30% (poorly absorbed) |
| Skin Permeability (log Kp) | −2.8 | Log Kp >−2.5 (low) |
| CYP3A4 inhibitor | No | |
| CYP1A2 inhibitor | Yes | |
| CYP2C19 inhibitor | Yes | |
| CYP2C9 inhibitor | No | |
| CYP2D6 inhibitor | No | |

## 4. Discussion

Vitiligo is the most common disease of acquired depigmentation adversely affecting patient quality of life [21]. It has been reported that around 1% of the world population has vitiligo, regardless of skin type or gender [22]. The etiological factor of vitiligo is still obscure and several different hypotheses have been suggested, which involve a combination of susceptibility genes and environmental factors contributing to the autoimmune destruction of melanocytes [23]. To date, the therapeutic options for vitiligo are limited and are mostly based on the use of immunosuppressive agents including topical corticosteroids, and UV light treatment. However, those treatments not only lack sustained efficacy, but also have side effects, such as scales, skin atrophy, pustules, telangiectasia, and local endosymbiosis proliferation [24]. Recently, Janus kinase (JAK)–signal transducer and activator of transcription (STAT) have been shown to be involved in the pathogenesis of vitiligo [25]. Coumarin-neurotransmitter derivatives have been reported as promising inhibitors of JAK–STAT signaling. [26]. Moreover, clinical trials have shown that JAK inhibitors including tofacitinib, ruxolitinib, and baricitinibare recovered the pigment loss via alleviating the response of autoimmune and inflammation [25].

Safe and natural compounds that affect melanogenesis may be considered potential preventive and/or therapeutic agents for depigmentation skin disorders [27]. Pinostrobin is the bioactive flavanone found in honey, finger root ginger, etc. [28,29]. Pinostrobin is a chiral flavonoid with two enantiomers. However, in most cases, chiral compounds are produced in nature in optically pure forms, where only one enantiomer is biosynthesized in the producing organism [30,31]. For pinostrobin, only the (-)-isomer of pinostrobin is produced by natural sources, such as honey, propolis, and finger roots [32], where the synthesis of pinostrobin often results in a racemic mixture (50:50) of the two enantiomers. The compound has been reported to have various biological activities, such as antioxidant, anti-inflammatory, anti-leukemia, anti-tumor, and neuroprotective properties [33–35].

A novel activity of pinostrobin, the melanogenic property, was demonstrated in the present study. Melanogenesis is a complex process that is related to more than 120 genes [36]. Among the genes, the tyrosinase family enzymes including tyrosinase, TRP 1, and TRP 2 have been recognized as the key regulators of melanogenesis [37]. Melanin is biosynthesized in the melanosome by the action of the tyrosinase and TRPs, which are transcriptionally regulated by MITF. Pinostrobin elevated melanin synthesis and tyrosinase activity by promoting the expression of melanogenic regulatory factors including tyrosinase, TRP 1, and MITF.

The cAMP/PKA phosphorylates the CREB, which is known to be an activator of MITF expression [11]. The activation of the cAMP/PKA pathway also stimulated β-catenin activation through GSK-3β inactivation, leading to increased MITF expression [38]. As a result, the compound enhanced melanogenesis via activating cAMP/PKA pathways and its downstream targets including CREB, GSK-3β, and β-catenin, thus upregulation of MITF protein levels subsequent activation of tyrosinase and TRP 1 enzymes. However, pre-treatment with pinostrobin in the presence of a PKA inhibitor considerably decreased the melanogenic effect of pinostrobin, demonstrating that MITF upregulation is mediated by the cAMP/PKA signaling pathway.

Previous studies have demonstrated that melanogenesis is mediated by the regulation of MITF activation via phosphorylation of p38 MAPK [13,39]. In the present study, pinostrobin augmented the phosphorylation level of p38, whereas the combination of the compound and SB203580 reduced the increasing effect of pinostrobin on the melanin synthesis, tyrosinase activity, expression of MITF and its downstream target tyrosinase, proving that p38 is strongly associated with the melanogenesis effect of pinostrobin.

Compared with the extensive research on the beneficial effects of flavonoids including antioxidant, anti-cancer, and anti-inflammatory activities, relatively little study has been done on their melanogenic effects. Several flavonoids have been found to be involved in melanogenesis [40–44]. Cirsimaritin found in rosemary, exerted a profound effect on melanin synthesis by a similar mechanism as shown with pinostrobin [40]. Major citrus flavonoids, such as hesperetin and naringenin had stimulatory effects on melanin production [41–43]. However, hesperidin suppressed melanin synthesis in normal human melanocytes as well as in B16F10 cells despite its structural similarity [44]. Considering the structures of the above melanogenic compounds, it can be suggested that the C-5 hydroxyl group and C-7 methoxy group may partially play an important role in melanogenic activity.

Molecular docking simulation is a well-established and widely used method in the drug discovery process, which predicts ligand–target interactions at the molecular level, allowing the identification of novel compounds of therapeutic interest. This approach also explores binding affinity and structure-relationship by evaluating critical phenomena involved in the intermolecular recognition process. In fact, the results of the in silico docking simulation revealed that the oxygen group at C-4 and the hydroxyl group at C-5 of pinostrobin were responsible for molecular interaction, suggesting these functional groups were determinants of its biological function. Moreover, it revealed that pinostrobin entrenched specific interactions with the key amino acid residues of melanogenic related markers via hydrogen bonding and van der Waals interactions, demonstrating the compound has an appropriate molecular structure to promote melanin synthesis.

Pharmacokinetic profiling, such as absorption, distribution, metabolism, and excretion (ADME) processes, is important to understanding the in vivo behavior and mechanism of action of a candidate. To date, there have been several studies regarding the pharmacokinetic properties of pinostrobin in rats [45–47]. Pinostrobin was mainly distributed in the gastrointestinal tract in rats after a single oral administration [45]. The compound in the parent form exhibited less than 1.6% elimination into the urine, bile and feces, suggesting that pinostrobin was mostly metabolized in vivo and played a role in different organs [45,46]. In addition, pinostrobin at 1–100 mg/kg for 7 days showed no mutagenic effect in Wister rats, further confirming the safety consumption of the compound [47]. Although there has been no experimental study on the skin permeation of pinostrobin so far, our computational prediction of the skin permeability coefficient (Log Kp) by the octanol–water partition coefficient and molecular size exhibited that pinostrobin can penetrate the skin [48]. The CYPs are a superfamily of heme-containing isoenzymes located mainly in the intestine and liver which play a vital role in drug metabolism [49]. Evaluating the potential of a compound to inhibit a CYPs is important, as the co-administration of compounds may result in one or both inhibiting the other's metabolism. In the present study using in silico prediction, pinostrobin was found to be an inhibitor of CYP1A2 and CYP2C19. It should be avoided as an alternative supplement during the administration of drugs (substrate of CYP1A2 and CYP2C19) as it is likely to cause food–drug interactions. These results may help to understand CYP-mediated food–drug interactions and evaluate the safety profile of natural products used therapeutically.

## 5. Conclusions

The present study revealed, for the first time, that pinostrobin exerted a stimulatory effect on tyrosinase activity and melanin production without cytotoxicity by inducing expression of tyrosinase, TRP 1 and MITF. The compound increased phosphorylation of CREB and GSK-3β, and subsequently promoted activation of β-catenin, suggesting that pinostrobin induced melanogenesis via regulating the cAMP/PKA signaling pathway. The upregulation of MITF and its downstream targets by pinostrobin was also associated with the p38 MAPK signaling pathway. Furthermore, molecular docking analysis revealed the pinostrobin possessed an appropriate molecular structure for potent interactions with major melanogenesis proteins. In silico prediction also found that the compound had optimal pharmacokinetic properties, but further study is needed to validate skin permeability in biological evaluation. Therefore, it can be suggested that our compound could be used as a novel, potent and safe melanogenic agent for the prevention and/or treatment of vitiligo.

**Author Contributions:** J.-H.Y.; Data curation, Validation, Writing—Original Draft, K.Y.; Data curation, Formal analysis, Software, Writing—Original Draft, M.J.; Supervision, Data curation, Writing—Review and Editing. All authors have read and agreed to the published version of the manuscript.

**Funding:** This research was funded by Dong-A University.

**Institutional Review Board Statement:** Not applicable.

**Informed Consent Statement:** Not applicable.

**Data Availability Statement:** Not applicable.

**Conflicts of Interest:** The authors declare no conflict of interest.

## References


1. Madronich, S.; McKenzie, R.L.; Björn, L.O.; Caldwell, M.M. Changes in biologically active ultraviolet radiation reaching the Earth's surface. *J. Photochem. Photobiol. B* **1998**, *46*, 5–19. [CrossRef]
2. Ando, H.; Niki, Y.; Ito, M.; Akiyama, K.; Matsui, M.S.; Yarosh, D.B.; Ichihashi, M. Melanosomes are transferred from melanocytes to keratinocytes through the processes of packaging, release, uptake, and dispersion. *J. Invest. Dermatol.* **2012**, *132*, 1222–1229. [CrossRef] [PubMed]
3. Niu, C.; Aisa, H.A. Upregulation of melanogenesis and tyrosinase activity: Potential agents for vitiligo. *Molecules* **2017**, *22*, 1303. [CrossRef] [PubMed]
4. del Marmol, V.; Beermann, F. Tyrosinase and related proteins in mammalian pigmentation. *FEBS Lett.* **1996**, *381*, 165–168. [CrossRef]
5. Naish-Byfield, S.; Riley, P.A. Tyrosinase autoactivation and the problem of the lag period. *Pigment Cell Res.* **1998**, *11*, 127–133. [CrossRef] [PubMed]
6. Gaggioli, C.; Buscà, R.; Abbe, P.; Ortonne, J.P.; Ballotti, R. Microphthalmia-associated transcription factor (MITF) is required but is not sufficient to induce the expression of melanogenic genes. *Pigment Cell Res.* **2003**, *16*, 374–382. [CrossRef]
7. Millington, G.W. Proopiomelanocortin (POMC): The cutaneous roles of its melanocortin products and receptors. *Clin. Exp. Dermatol.* **2006**, *31*, 407–412. [CrossRef] [PubMed]
8. Bertolotto, C.; Abbe, P.; Hemesath, T.J.; Bille, K.; Fisher, D.E.; Ortonne, J.P.; Ballotti, R. Microphtalmia gene product as a signal transducer in cAMP-induced differentiation of melanocytes. *J. Cell Biol.* **1998**, *142*, 827–835. [CrossRef]
9. Bellei, B.; Pitisci, A.; Catricala, C.; Larue, L.; Picardo, M. Wnt/beta-catenin signaling is stimulated by alpha-melanocyte-stimulating hormone in melanoma and melanocyte cells: Implication in cell differentiation. *Pigment. Cell. Melanoma Res.* **2011**, *24*, 309–325. [CrossRef]
10. Hino, S.; Tanji, C.; Nakayama, K.I.; Kikuchi, A. Phosphorylation of beta-catenin by cyclic AMP-dependent protein kinase stabilizes beta-catenin through inhibition of its ubiquitination. *Mol. Cell. Biol.* **2005**, *25*, 9063–9072. [CrossRef]
11. Tsang, T.F.; Chan, B.; Tai, W.C.; Huang, G.; Wang, J.; Li, X.; Jiang, Z.H.; Hsiao, W.L.W. Gynostemma pentaphyllum saponins induce melanogenesis and activate cAMP/PKA and Wnt/beta-catenin signaling pathways. *Phytomedicine* **2019**, *60*, 153008. [CrossRef] [PubMed]
12. MacDonald, B.T.; Tamai, K.; He, X. Wnt/β-catenin signaling: Components, mechanisms, and diseases. *Dev. Cell* **2009**, *17*, 9–26. [CrossRef] [PubMed]
13. D'Mello, S.A.; Finlay, G.J.; Baguley, B.C.; Askarian-Amiri, M.E. Signaling Pathways in Melanogenesis. *Int. J. Mol. Sci.* **2016**, *17*, 1144. [CrossRef] [PubMed]
14. Karunarathne, W.A.H.M.; Molagoda, I.M.N.; Kim, M.S.; Choi, Y.H.; Oren, M.; Park, E.K.; Kim, G.Y. Flumequine-Mediated Upregulation of p38 MAPK and JNK Results in Melanogenesis in B16F10 Cells and Zebrafish Larvae. *Biomolecules* **2019**, *9*, 596. [CrossRef]
15. Mohd Zaid, N.A.; Sekar, M.; Bonam, S.R.; Gan, S.H.; Lum, P.T.; Begum, M.Y.; Mat Rani, N.N.I.; Vaijanathappa, J.; Wu, Y.S.; Subramaniyan, V.; et al. Promising Natural Products in New Drug Design, Development, and Therapy for Skin Disorders: An Overview of Scientific Evidence and Understanding Their Mechanism of Action. *Drug Des. Dev. Ther.* **2022**, *16*, 23–66. [CrossRef]
16. Wang, Y.C.; Haung, X.Y.; Chiu, C.C.; Lin, M.Y.; Lin, W.H.; Chang, W.T.; Tseng, C.C.; Wang, H.M.D. Inhibitions of melanogenesis via Phyllanthus emblica fruit extract powder in B16F10 cells. *Food Biosci.* **2019**, *28*, 177–182. [CrossRef]
17. Wu, Q.Y.; Wong, Z.C.; Wang, C.; Fung, A.H.; Wong, E.O.Y.; Chan, G.K.L.; Dong, T.T.X.; Chen, Y.; Tsim, K.W.K. Isoorientin derived from Gentiana veitchiorum Hemsl. flowers inhibits melanogenesis by down-regulating MITF-induced tyrosinase expression. *Phytomedicine* **2019**, *57*, 129–136. [CrossRef]
18. Jia, C.Y.; Li, J.Y.; Hao, G.F.; Yang, G.F. A drug-likeness toolbox facilitates ADMET study in drug discovery. *Drug Discov. Today* **2020**, *25*, 248–258. [CrossRef]
19. Trott, O.; Olson, A.J. AutoDock Vina: Improving the speed and accuracy of docking with a new scoring function, efficient optimization, and multithreading. *J. Comput. Chem.* **2010**, *31*, 455–461. [CrossRef]
20. Hartman, M.L.; Czyz, M. MITF in melanoma: Mechanisms behind its expression and activity. *Cell. Mol. Life Sci.* **2015**, *72*, 1249–1260. [CrossRef]
21. Picardo, M.; Dell'Anna, M.L.; Ezzedine, K.; Hamzavi, I.; Harris, J.E.; Parsad, D.; Taieb, A. Vitiligo. *Nat. Rev. Dis. Primers* **2015**, *1*, 15011. [CrossRef] [PubMed]
22. Mohamed, A.; Hassan, R. Concise review of recent studies in vitiligo. *Qatar Med. J.* **2013**, *2*, 10.

23. Spritz, R.A. The genetics of generalized vitiligo and associated autoimmune diseases. *J. Dermatol. Sci.* **2006**, *41*, 3–10. [CrossRef] [PubMed]
24. Migayron, L.; Boniface, K.; Seneschal, J. Vitiligo, From Physiopathology to Emerging Treatments: A Review. *Dermatol. Ther.* **2020**, *10*, 1185–1198. [CrossRef] [PubMed]
25. Niu, C.; Xie, H.; Aisa, H.A. Janus Kinase Inhibitors: A Review of Their Application in the Vitiligo. *Mini Rev. Med. Chem.* **2021**, *21*, 3203–3218. [CrossRef]
26. Niu, C.; Zang, D.; Aisa, H.A. Study of Novel Furocoumarin Derivatives on Anti-Vitiligo Activity, Molecular Docking and Mechanism of Action. *Int. J. Mol. Sci.* **2022**, *23*, 7959. [CrossRef]
27. Gauthier, Y.; Cario-Andre, M.; Lepreux, S.; Pain, C.; Taieb, A. Melanocyte detachment after skin friction in non lesional skin of patients with generalized vitiligo. *Br. J. Dermatol.* **2003**, *148*, 95–101. [CrossRef]
28. Fahey, J.F.; Stephenson, K.K. Pinostrobin from honey and Thai ginger (Boesenbergia pandurata): A potent flavonoid inducer of mammalian phase 2 chemoprotective and antioxidant enzymes. *J. Agric. Food Chem.* **2002**, *50*, 7472–7476. [CrossRef]
29. Youn, K.; Jun, M. Biological evaluation and docking analysis of potent BACE1 inhibitors from Boesenbergia rotunda. *Nutrients* **2019**, *11*, 662. [CrossRef]
30. Miller, K.A.; Tsukamoto, S.; Williams, R.M. Asymmetric total syntheses of (+)- and (-)-versicolamide B and biosynthetic implications. *Nat. Chem.* **2009**, *1*, 63–68. [CrossRef]
31. Finefield, J.M.; Sherman, D.H.; Kreitman, M.; Williams, R.M. Enantiomeric natural products: Occurrence and biogenesis. *Angew. Chem. Int. Ed. Engl.* **2012**, *51*, 4802–4836. [CrossRef]
32. Patel, N.K.; Jaiswal, G.; Bhutani, K.K. A review on biological sources, chemistry and pharmacological activities of pinostrobin. *Nat. Prod. Res.* **2016**, *30*, 2017–2027. [CrossRef] [PubMed]
33. Pobłocka-Olech, L.; Inkielewicz-Stepniak, I.; Krauze-Baranowska, M. Anti-inflammatory and antioxidative effects of the buds from different species of Populus in human gingival fibroblast cells: Role of bioflavanones. *Phytomedicine* **2019**, *56*, 1–9. [CrossRef] [PubMed]
34. Smolarz, H.D.; Mendyk, E.; Bogucka-Kocka, A.; Kockic, J. Pinostrobin-An anti-leukemic flavonoid from Polygonum lapathifolium L. ssp nodosum (Pers.) Dans. *Z. Naturforsch. C J. Biosci.* **2006**, *61*, 64–68. [CrossRef] [PubMed]
35. Xian, Y.F.; Ip, S.P.; Lin, Z.X.; Mao, Q.Q.; Su, Z.R.; Lai, X.P. Protective effects of pinostrobin on beta-amyloid-induced neurotoxicity in PC12 cells. *Cell. Mol. Neurobiol.* **2012**, *32*, 1223–1230. [CrossRef]
36. Yamaguchi, Y.; Brenner, M.; Hearing, V.J. The regulation of skin pigmentation. *J. Biol. Chem.* **2007**, *282*, 27557–27561. [CrossRef]
37. Costin, G.E.; Hearing, V.J. Human skin pigmentation: Melanocytes modulate skin color in reponse to stress. *FASEB J.* **2007**, *27*, 976–994. [CrossRef]
38. Suzuki, A.; Ozono, K.; Kubota, H.; Tachikawa, K.; Michigami, T. PTH/cAMP/PKA signaling facilitates canonical Wnt signaling via inactivation of glycogen synthase kinase-3beta in osteoblastic Saos-2 cells. *J. Cell. Biochem.* **2008**, *104*, 304–317. [CrossRef]
39. Kang, S.J.; Choi, B.R.; Lee, E.K.; Kim, S.H.; Yi, H.Y.; Park, H.R.; Song, C.H.; Lee, Y.J.; Ku, S.K. Inhibitory Effect of Dried Pomegranate Concentration Powder on Melanogenesis in B16F10 Melanoma Cells; Involvement of p38 and PKA Signaling Pathways. *Int. J. Mol. Sci.* **2015**, *16*, 24219–24242. [CrossRef]
40. Kim, H.J.; Kim, I.S.; Dong, Y.; Lee, I.S.; Kim, J.S.; Kim, J.S.; Woo, J.T.; Cha, B.Y. Melanogenesis-inducing effect of cirsimaritin through increases in microphthalmia-associated transcription factor and tyrosinase expression. *Int. J. Mol. Sci.* **2015**, *16*, 8772–8788. [CrossRef]
41. Ohguchi, K.; Akao, Y.; Nozawa, Y. Stimulation of melanogenesis by the citrus flavonoid naringenin in mouse B16 melanoma cells. *Biosci. Biotechnol. Biochem.* **2006**, *70*, 1499–1501. [CrossRef] [PubMed]
42. Huang, Y.; Yang, C.; Chiou, Y. Citrus flavanone naringenin enhances melanogenesis through the activation of Wnt/β-catenin signalling in mouse melanoma cells. *Phytomedicine* **2011**, *18*, 1244–1249. [CrossRef] [PubMed]
43. Huang, Y.C.; Liu, K.C.; Chiou, Y.L. Melanogenesis of murine melanoma cells induced by hesperetin, a Citrus hydrolysate-derived flavonoid. *Food Chem. Toxicol.* **2012**, *50*, 653–659. [CrossRef] [PubMed]
44. Lee, H.J.; Lee, W.J.; Chang, S.E.; Lee, G.Y. Hesperidin, a popular antioxidant inhibits melanogenesis via Erk1/2 mediated MITF degradation. *Int. J. Mol. Sci.* **2015**, *16*, 18384–18395. [CrossRef] [PubMed]
45. Sun, X.; Liu, X.; Chen, S. The Pharmacokinetics, Tissue Distribution, Metabolism, and Excretion of Pinostrobin in Rats: Ultra-High-Performance Liquid Chromatography Coupled with Linear Trap Quadrupole Orbitrap Mass Spectrometry Studies. *Front. Pharmacol.* **2020**, *11*, 574638. [CrossRef]
46. Sayre, C.L.; Alrushaid, S.; Martinez, S.E.; Anderson, H.D.; Davies, N.M. Pre-clinical pharmacokinetic and pharmacodynamic characterization of selected chiral flavonoids: Pinocembrin and pinostrobin. *J. Pharm. Pharmaceut. Sci.* **2015**, *18*, 368–396. [CrossRef]
47. Charoensin, S.; Punvittayagul, C.; Pompimon, W.; Mevatee, U.; Wongpoomchai, R. Toxicological and clastogenic evaluation of pinocembrin and pinostrobin isolated from Boesenbergia pandurata in Wistar rats. *Thai J. Toxicol.* **2010**, *25*, 29–40.
48. Potts, R.O.; Guy, R.H. Predicting skin permeability. *Pharm. Res.* **1992**, *9*, 663–669. [CrossRef]
49. Ogu, C.C.; Maxa, J.L. Drug interactions due to cytochrome P450. In *Baylor University Medical Center Proceedings*; Taylor & Francis: Abingdon, UK, 2000; Volume 13, pp. 421–423.

nutrients MDPI

Article

# *Aquilaria crassna* Leaf Extract Ameliorates Glucose-Induced Neurotoxicity In Vitro and Improves Lifespan in *Caenorhabditis elegans*

**Nattaporn Pattarachotanant [1,2], Nilubon Sornkaew [1,2], Watis Warayanon [1,2], Panthakarn Rangsinth [3], Chanin Sillapachaiyaporn [1,4], Wudtipong Vongthip [1,4], Siriporn Chuchawankul [3], Anchalee Prasansuklab [1,5] and Tewin Tencomnao [1,2,*]**

1 Natural Products for Neuroprotection and Anti-Ageing (Neur-Age Natura) Research Unit, Faculty of Allied Health Sciences, Chulalongkorn University, Bangkok 10330, Thailand
2 Department of Clinical Chemistry, Faculty of Allied Health Sciences, Chulalongkorn University, Bangkok 10330, Thailand
3 Department of Transfusion Medicine and Clinical Microbiology, Faculty of Allied Health Sciences, Chulalongkorn University, Bangkok 10330, Thailand
4 Program in Clinical Biochemistry and Molecular Medicine, Department of Clinical Chemistry, Faculty of Allied Health Sciences, Chulalongkorn University, Bangkok 10330, Thailand
5 College of Public Health Sciences, Chulalongkorn University, Bangkok 10330, Thailand
* Correspondence: tewin.t@chula.ac.th; Tel.: +66-2-218-1533

**Abstract:** Hyperglycemia is one of the important causes of neurodegenerative disorders and aging. *Aquilaria crassna* Pierre ex Lec (AC) has been widely used to relieve various health ailments. However, the neuroprotective and anti-aging effects against high glucose induction have not been investigated. This study aimed to investigate the effects of hexane extract of AC leaves (ACH) in vitro using human neuroblastoma SH-SY5Y cells and in vivo using nematode *Caenorhabditis elegans*. SH-SY5Y cells and *C. elegans* were pre-exposed with high glucose, followed by ACH treatment. To investigate neuroprotective activities, neurite outgrowth and cell cycle progression were determined in SH-SY5Y cells. In addition, *C. elegans* was used to determine ACH effects on antioxidant activity, longevity, and healthspan. In addition, ACH phytochemicals were analyzed and the possible active compounds were identified using a molecular docking study. ACH exerted neuroprotective effects by inducing neurite outgrowth via upregulating growth-associated protein 43 and teneurin-4 expression and normalizing cell cycle progression through the regulation of cyclin D1 and SIRT1 expression. Furthermore, ACH prolonged lifespan, improved body size, body length, and brood size, and reduced intracellular ROS accumulation in high glucose-induced *C. elegans* via the activation of gene expression in the DAF-16/FoxO pathway. Finally, phytochemicals of ACH were analyzed and revealed that β-sitosterol and stigmasterol were the possible active constituents in inhibiting insulin-like growth factor 1 receptor (IGFR). The results of this study establish ACH as an alternative medicine to defend against high glucose effects on neurotoxicity and aging.

**Keywords:** neurite outgrowth; GAP-43; Teneurin-4; Cyclin D1; SIRT1; *daf-16*; *aqp-1*

**Citation:** Pattarachotanant, N.; Sornkaew, N.; Warayanon, W.; Rangsinth, P.; Sillapachaiyaporn, C.; Vongthip, W.; Chuchawankul, S.; Prasansuklab, A.; Tencomnao, T. *Aquilaria crassna* Leaf Extract Ameliorates Glucose-Induced Neurotoxicity In Vitro and Improves Lifespan in *Caenorhabditis elegans*. *Nutrients* **2022**, *14*, 3668. https://doi.org/10.3390/nu14173668

Academic Editor: Paolo Brambilla

Received: 14 July 2022
Accepted: 26 August 2022
Published: 5 September 2022



## 1. Introduction

Neurodegenerative disease (ND) comprises a group of neuronal disorders such as Alzheimer's disease [1], Parkinson's disease (PD) [2], Huntington's disease, etc. ND results in the damage of neuronal structure and function and the inhibition of neuronal differentiation. Hyperglycemia is a key factor causing neuronal damage and neuronal differentiation inhibition [1,3]. Previous studies have indicated that high glucose concentration induced SH-SY5Y neuronal cell damage through the induction of cell cycle arrest via the downregulation of cyclin D1, phospho-cell division cycle 2 (pcdc2), and phospho-Retinoblastoma (pRb) proteins. Moreover, high glucose concentration reduced the healthspan and lifespan of *Caenorhabditis elegans* in a round worm model via the increase of reactive oxygen species (ROS) formation and advanced glycation end product (AGEs) modification of mitochondrial proteins in a daf-2 independent manner [4].

*Aquilaria crassna* Pierre ex Lec (AC) belongs to the Thymelaeaceae family and is well known as agarwood. AC is a fragrant wood that has many ties with religious history, rituals, and ceremonies [5]. It is widely used as an ingredient in food, including medicinal wine in traditional Chinese and Korean medicines [6], biscuits, herbal soups, and instant noodles [7]. For pharmacological activities, plant materials from AC have been reported to exert different bioactivities, including the inhibition of inflammatory cytokines (TNF-$\alpha$ and IL-1$\alpha$) [8] as well as anti-ischemic (cardioprotective) [9,10], antianaphylaxis [11], hepatoprotective [12], anti-cancer [13], and antioxidant [13,14] properties.

Previous studies have shown that the extract of AC leaf demonstrates neuroprotection and neuritogenesis in vitro [15] as well as antihyperglycemia and glucose-uptake enhancement activities in streptozotocin-induced diabetic rats [16]. Nevertheless, the pharmacological activities and underlying mechanisms of AC against hyperglycemia-associated neurodegenerative diseases and the effect on healthspan and lifespan need to be clarified.

The present study aimed to investigate the neuroprotective and anti-aging effects of hexane extract AC (ACH) and its underlying mechanisms on high glucose-induced neurotoxicity using human neuroblastoma cells (SH-SY5Y) and high glucose-induced life span and healthspan reduction in a *C. elegans* model. The neuroprotective properties of ACH were evaluated through the ability of this extract to exert neuritogenesis potential, including neurite outgrowth formation, and to induce cell cycle normalization. The anti-aging effect of ACH in *C. elegans* was provided by the ability to prolong life span and improve body size, body length, and brood size. In addition, possible active phytochemicals in AC were also analyzed by in silico molecular docking, gas chromatography-mass spectrometry/mass spectrometry (GC-MS/MS), reverse-phase high-pressure liquid chromatography (RP-HPLC), and nuclear magnetic resonance (NMR). Therefore, this experiment provided experimental evidence of the applications of ACH in the prevention or treatment of neurodegeneration and aging associated with high glucose conditions.

## 2. Materials and Methods

### 2.1. Chemicals and Reagents

Dimethyl sulfoxide (DMSO), hexane, and ethanol were purchased from Merck KGaA (Darmstadt, Germany). Phenylmethyl sulphonyl fluoride (PMSF) was purchased from United States Biological (Cleveland, OH, USA). Kodak processing chemicals were used for autoradiography films. The Amersham ECL select Western blotting detection reagent was purchased from Merck KGaA (Darmstadt, Germany). Dulbecco's modified Eagle medium (DMEM)/low glucose, fetal bovine serum (FBS), and penicillin-streptomycin solution (10,000 units/mL of penicillin and 10,000 µg/mL of streptomycin) were purchased from Thermo Scientific HyClone (Logan, UT, USA). A solution of 30% acrylamide/bis-acrylamide (37.5:1) was purchased from Bio-Rad Laboratories (Hercules, CA, USA). Ammonium persulfate (APS) was purchased from Millipore Sigma (Burlington, MA, USA). The monoclonal rabbit Cyclin D1 (92G2, cat#2978) and $\beta$ actin (13E5, cat#4970) were purchased from Cell Signaling Technology (Beverly, MA, USA). The monoclonal mouse SIRT1 (B-7, sc-74465) was purchased from Santa Cruz Biotechnology (Dallas, TX, USA). The polyclonal sheep IgG Teneurin-4 (AF6320) and sheep IgG HRP-conjugated antibody (HAF016) were from R&D systems, Inc. (Minneapolis, MN, USA) and the monoclonal rabbit GAP43 (ab75810) antibody from Abcam (Cambridge, UK).

Propidium iodide (PI) was purchased from Biolegend (San Diego, CA, USA). 3-(4,5-dimethylthiazol-2-yl)-2,5-diphenyltetrazoliumbromide (MTT) was purchased from Bio Basic Inc. (Markham, ON, Canada). Trizol reagent was purchased from Thermo Fisher Scientific (Waltham, MA, USA).

### 2.2. Plant Extraction

Leaves of *Aquilaria crassna* (agarwood) were dried under shade and ground into fine powder. Successive extraction was performed by the maceration method using hexane (40 g of dried plant: 400 mL of solvent) at room temperature for 72 h. The supernatant was collected, filtrated, and evaporated by the rotary evaporator at 45 °C. The phytochemical constituents were stabilized by keeping dried crude extract at −20 °C and protected from light exposure The extract was dissolved in DMSO as stock solution (100 mg/mL), passed through a 0.2 µm filter, and stored at −20 °C until use. Hexane extraction of *Aquilaria crassna* was designated as ACH.

### 2.3. Cell Line

SH-SY5Y cells were purchased from a cell line service (Heidelberg, Germany) and were cultured in DMEM/low glucose (HyClone, Logan, UT, USA) containing 10% FBS and antibiotics including 100 U/mL penicillin and 100 µg/mL streptomycin at 37 °C in a humidified atmosphere at 5% $CO_2$.

### 2.4. Antioxidant Determination

These assays were modified for a microplate format, as previously described [17,18].

#### 2.4.1. Folin–Ciocalteu Phenol Assay (FCP)

The extracts (50 µL) and 10% Folin–Ciocalteu phenol reagent (50 µL) were mixed and incubated in the dark at room temperature (RT) for 30 min. A sodium carbonate ($Na_2CO_3$) solution (50 µL) was added, mixed, and incubated in the dark at RT for 20 min. Reaction absorbance was measured using the Enspire® Multimode Plate Reader (Perkin-Elmer) at 760 nm. Gallic acid was used as the standard. The amount of phenolic compound was in a Gallic acid equivalent (GE) mg/g of dry weight.

#### 2.4.2. Determination of Total Flavonoid

The extracts (50 µL) were mixed with the solution containing 150 µL of ethanol, 10 µL of 1M Sodium acetate (NaOAc), and 10 µL of Aluminum Chloride ($AlCl_3$). The mixture was incubated in the dark at RT for 40 min and measured at 415 nm. Quercetin was used as the standard. The content of flavonoid was in Quercetin equivalent (QE) mg/g of dry weight.

#### 2.4.3. Radical Scavenging Activity Assays

First, working 2,2′-azino-bis (3-ethylbenzthiazoline-6-sulphonic acid) (ABTS•$^+$) ($OD_{734}$ = 0.7–0.8) solution was prepared by the addition of 2.45 mM $K_2S_2O_8$ to 7 mM ABTS (ratio 2:3) and incubated at 4 °C for 16–18 h. In addition, 0.2 mM of 2,2-diphenyl-1-picryl-hydrazyl-hydrate (DPPH•) ($OD_{517}$ = 0.8–0.9) was diluted in ethanol.

The extract (1 mg/mL) was reacted with DPPH• or ABTS•$^+$ and incubated at RT for 15 or 30 min, respectively. Absorbance was measured at 517 nm and 734 nm, respectively. Ascorbic acid or Vitamin C was used as the standard for both assays. The antioxidant capacity had Vitamin C equivalent antioxidant capacity (VCEAC) in mg/g of dry weight.

### 2.5. MTT Assay

To verify which ACH concentrations were nontoxic, SH-SY5Y cells were treated with ACH extract at the concentrations of 10, 25, 50 and 100 µg/mL for 24 h. Then, 20 µL of 5 mg/mL MTT was added to each well and incubated at 37 °C for 4 h. After incubation, 10% SDS in 0.01 NHCl (50 µL) was added and incubated overnight to dissolve the formazan crystal. Finally, the absorbance at 570 nm with a spectrometer was performed. The percentage of cell viability compared to the control group was a nontoxic concentration [19].

### 2.6. Intracellular Reactive Oxygen Species (ROS) Assay

For cell culture, cells were seeded at 10,000 cells/well in 96-well black plates with a clear bottom and cultured at 37 °C for 24 h. Having been incubated, cells were treated with high glucose (50 mM) for 48 h. Subsequently, cells were post-treated with ACH extract at the concentration of 10, 25, 50, and 100 μg/mL for 24 h. The ROS level was measured by following the published protocol [20]. After treatment, 5 μM (diluted in serum-free culture medium) of non-fluorescent 2′,7′-dichloro-dihydrofluorescein diacetate ($H_2$DCFDA) was loaded, incubated for 45 min, and then washed three times with PBS. The fluorescence was measured with an excitation wavelength of 485 nm and an emission wavelength of 535 nm.

### 2.7. Neurite Outgrowth Assay and Scoring of Neurite Length and Neurite-Bearing Cells

SH-SY5Y cells were cultured in 10% FBS culture medium for 24 h and induced the differentiation with 1% FBS culture medium and 10 μM retinoic acid (RA) for 24 h. During RA-induced differentiation, the effect of high glucose (50 mM) on neurite outgrowth inhibition was observed following addition for 48 h. Likewise, the protective effect on high glucose-induced neurite outgrowth inhibition of ACH was detected after cells were treated with ACH (10 and 50 μg/mL) for 24 h.

Cells were imaged by a differential interference contrast (DIC) microscope for 3 independent experiments. For neurite scoring, a cell that had at least 1 neurite extension longer than the diameter of its cell body was scored as a neurite-bearing cell. Cell clusters contained more than 5 cells were excluded [21]. Otherwise, the neurite length was obtained by the measurement of the longest neurite in each image using the software Motic Images Plus 3.0 [22].

### 2.8. Cell Cycle Analysis

SH-SY5Y cells were treated with 50 mM glucose for 48 h. Having been incubated, cells were post-treated with ACH for 24 h. Then, cells were fixed with precooled absolute ethanol at −20 °C for at least 2 h and stained with propidium iodide containing RNase A at 37 °C for 15 min in the dark [20]. Cell cycle progression was analyzed using a BD FACSCalibur flow cytometer (BD Biosciences, San Jose, CA, USA).

### 2.9. Western Blot Analysis

To identify the pathway through which ACH extract attenuates high glucose-induced neurite outgrowth inhibition and cell cycle progression delay, SH-SY5Y cells were treated as explained earlier. Total protein was extracted using NP-40 lysis buffer. Western blotting was performed using the standard experimental procedure [20] with primary antibodies, including cyclin D1 (1:2000), SIRT1 (1:5000), GAP-43 (1:5000), Teneurin-4 (1:2000), and β-actin (1:5000). Chemiluminescence detection reagent was used to develop HRP-linked secondary antibody bands. These bands were obtained using ImageJ software.

### 2.10. *C. elegans* Strain, Maintenance, Synchronization, and Treatment

Wild type *C. elegans* or Bristol N2 were cultivated on nematode growth medium (NGM) agar and maintained at 20 °C on living *Escherichia coli* (OP50) with an $OD_{600}$ of 1.5. However, in the high glucose-treated experiment using dead bacteria, *E. coli* OP50 were incubated at 65 °C for at least 30 min to stop bacteria growth [23].

*C. elegans* and *E. coli* OP50 were obtained from the Caenorhabditis Genetics Center (University of Minnesota, Twin Cities, MN, USA). To exclude the antimicrobial activity of ACH against *E. coli* OP50, an agar diffusion test was conducted.

Age synchronization of the worms was as previously described [18]. For all experiments, N2 synchronized L1 larvae were kept on NGM agar plates containing dead *E. coli* OP50 ($OD_{600}$ = 1.5) and divided into 4 groups. The first was treated with 0.1% DMSO (solvent control group). The second was treated with 50 mM glucose. Groups three and four were co-treated with 50 mM glucose and ACH at the concentrations of 10 and 50 μg/mL, respectively.

### 2.11. Brood Size, Body Length, and Body Size

These experiments were modified from the published protocol [18,24]. For brood size assay, L1 larvae stage worms in each group containing *E. coli* OP50 and treatment were cultured at 20 °C. After 48 h, L4 larvae was individually sorted and transferred to the new NGM containing *E. coli* OP50 and the treatment. L4 larvae were allowed to grow into adults and lay eggs. Under a dissecting microscope, eggs were observed and counted. While waiting for the adult worms to stop laying eggs, eggs were separated from adult worms every day.

For body length and size assay, at least 20 adult day 1 worms in each group were imaged by a 10× objective lens of a bright-field microscope. The body length and size of adult worms was analyzed by the software Motic Images Plus 3.0.

### 2.12. Intracellular ROS Accumulation

N2 synchronized L1 larvae were treated and cultured. At adult day 1, worms were stained with 50 µM $H_2$DCFDA and incubated away from light for 45 min to 1 h at 20 °C. Having been stained, worms were washed with M9 and mounted on a glass slide. Next, 10 mM sodium azide was added to paralyze the worms [18]. At least 30 worms were photographed using a confocal microscope (Olympus FluoView FV10i). Fluorescent intensity was measured and analyzed using the software ImageJ from National Institutes of Health (Bethesda, MD, USA).

### 2.13. Lifespan Assay

N2 synchronized L1 larvae were cultured on different NGM containing *E. coli* OP50 treated with 0.1% DMSO, 50 mM glucose, or ACH. At the L4 stage, worms were transferred to new NGM every day to separate them from their new population and to avoid starvation.

### 2.14. RT-qPCR

To identify the pathway through which ACH extract attenuates high glucose-induced lifespan and healthspan reduction, L1 larvae stage worms were treated and cultured for 72 h. Further, adult day 1 worms were collected and RNA was extracted by Trizol reagent following the manufacturer's instructions. Reverse transcription followed the recommended protocol of Maxime RT PreMix Kit. The qPCR was accomplished in CFX Real time PCR, and the measured fluorescent signals indicated the PCR results. qPCR conditions were 95 °C for 15 min, denaturation at 95 °C for 15 s for 45 cycles, and primer annealing/extension at 55 °C for 30 s. The gene-specific primers were *daf-16*, *sod-3*, *aqp-1*, and *act-1*. The sequences of these RNA primers were shown in Table 1.

**Table 1.** The sequence of RNA primers.

| Primer | Forward 5′-3′ | Reverse 5′-3′ |
|---|---|---|
| *daf-16* | TTTCCGTCCCCGAACTCAA | ATTCGCCAACCCATGATGG |
| *sod-3* | TTCGAAAGGGAATCTAAAAGAAG | GCCAAGTTGGTCCAGAAGATAG |
| *aqp-1* | TTTTGGCAAGGAACCTCATC | GCTGTTGCCATAACTGCAAA |
| *act-1* | AGACAATGGATCCGGAATGT | CATCCCAGTTGGTGACGATA |

### 2.15. Phytochemical Constituent Analysis of Extracts by Gas Chromatograph-Mass Spectrometer/Mass Spectrometer (GC-MS/MS) Analysis

The components of ACH were analyzed at the Scientific and Technological Research Equipment Center (STREC) (Chulalongkorn University, Bangkok, Thailand), according to the standard procedure [20]. The GC-MS Triple Quad system was an Agilent 7890B GC system coupled with an Agilent 7000C MS and a capillary column characterized as HP-5MS 5% Phenyl Methyl Siloxane, length 30 m, i.d. 0.25 mm, and phase thickness 0.25 µm. The GC was operated with helium as the carrier gas at the rate of 1 mL/min. The inlet had a temperature of 250 °C, pressure set to 8.2317 psi, and 1 µL injection. The GC oven was initially at 60 °C before rising to 300 °C with a linear gradient of 3 °C/min and kept at 300 °C for 5 min. The total run time was 85 min. Approximately 20 mg of the extract was dissolved in 1 mL of hexane.

The obtained spectra were used to calculate the retention indices and authentic mass spectra data, and these were compared with those stored in the NIST Mass Spectrometry Data Center to identify phytochemical constituents. With the purpose of obtaining the retention indices, a series of *n*-alkane calibration standards (C8-C40) was run in the same condition and calculated with the Kovats Index [25,26].

### 2.16. Plant Extract Isolation and Identification

ACH extract was isolated by column chromatography using the condition hexane:ethyl acetate (EtOAc) (90:10). The eluates were examined by thin layer chromatography (TLC), and combined fractions were obtained. Interesting fractions were subjected to column chromatography on silica gel using hexane:EtOAc (90:10) to isolate their compounds, and their structures were elucidated by using $^1$H-NMR.

$^1$H-NMR were recorded on a Bruker AVANCE 400 FT-NMR spectrometer operating at 400 ($^1$H) MHz. Column chromatography was carried out using Merck silica gel 60 (finer than 0.063 mm). For TLC, Merck precoated silica gel 60 $F_{254}$ plates were used. Spots on TLC were detected under UV light and by spraying with anisaldehyde-$H_2SO_4$ reagent followed by heating.

### 2.17. Reverse-Phase High-Pressure Liquid Chromatography (RP-HPLC)

The presence of active components in fractions of ACH was analyzed using reverse-phase high-performance liquid chromatography (RP–HPLC). The HPLC system consisted of a binary solvent delivery pump (Scientific, Spectra SYSTEM™ P1000 isocratic pump properl), column compartment, and Photo Diode Array detector. A Shodex HPLC Column Silica C18-4E, 120 Å, 5 μm, 4.6 × 250 mm was used as the stationary phase, and the peaks were obtained using Lab solutions software. The separation of compounds was performed by the isocratic mobile phase using methanol:acetonitrile in the ratio (90:10) with a flow rate of 1.0 mL min$^{-1}$, and the column was set at ambient temperature. The samples and standards were injected using SGE syringe into the injection loop and were detected at 208 nm. The presence of stigmasterol at 28.697 min and β-sitosterol at 32.447 min were identified by comparing its retention time with ACH.

### 2.18. Molecular Docking

#### 2.18.1. Ligand Preparation

All chemical structures of the phytochemicals were obtained from the PubChem database. We minimized the energy, cleaned the geometry, and generated the PDB format of all chemical structures using BIOVIA Discovery Studio Visualizer (version 20.1) (San Diego, CA, USA). Next, we converted the file format to the protein databank, partial charge (Q), and atom type (T) or PDBQT using AutoDockTools-1.5.6 software (The Scripps Research Institute, San Diego, CA, USA) [27].

#### 2.18.2. Protein Preparation

The X-ray crystallographic structures of insulin-like growth factor 1 receptor (IGFR) (PDB ID: 5FXS) [28] were retrieved from the RCSB Protein Data Bank. Protein structures were processed using the Prepare Protein Set Up in AutoDockTools (version 1.5.6) software. The complex structure composed of all water molecules and the original ligand was excluded. Next, the complex protein structure was added with all the missing hydrogens and Kollman charges and then converted to PDBQT file format as the inputs for the docking study [29].

2.18.3. Molecular Docking

The docking analyses were performed according to the previous report [29]. In brief, the AutoDock 4.2 software package supported by AutodockTools 1.5.6 was used. The Lamarckian Genetic Algorithm with default parameters was used to perform the protein–ligand interaction studies, and the results were further visualized using the Discovery Studio Visualizer (BIOVIA, San Diego, CA, USA).

### *2.19. Statistical Analysis*

All data were presented as the mean ± standard error mean (SEM). Means were from at least 3 independent experiments. All data were analyzed by a one-way analysis of variance (ANOVA) followed by a post hoc Tukey test ($p$-value < 0.05) using low glucose-treated cells as the control group.

## **3. Results**

### *3.1. Antioxidant Properties and Total Phenolic and Flavonoid Contents*

The ACH extraction yield was 2.98% ($w/w$). The phenolic and flavonoid analysis showed that ACH contained 6.75 ± 1.29 mg GAE/g dry weight for total phenolic content and 2.49 ± 1.51 mg QE/g dry weight for total flavonoid content. The antioxidant activity was tested using DPPH and ABTS assays. The data of antioxidant capacity were stated in mg of vitamin C equivalent antioxidant capacity (VCEAC)/g dry weight. ACH had antioxidant capacities of 4.96 ± 0.49 mg VCEAC/g dry weight in DPPH and 3.07 ± 0.98 mg VCEAC/g dry weight in ABTS. Moreover, ACH possessed the free radical scavenging capacities of 3.55 ± 1.64% and 5.07 ± 0.73% against DPPH and ABTS radicals, respectively.

### *3.2. Effects of ACH on Cell Viability and High Glucose-Induced Reactive Oxygen Species (ROS)*

We found that ACH extract at the concentrations of 10–50 μg/mL had no significant effect on cell viability (Figure 1A). Next, to examine the effects of ACH on high glucose-induced intracellular ROS, the experiment was modified from the previous protocol to investigate the treatment of SH-SY5Y cells with 50 mM glucose alone for 48 h. After high glucose treatment, cells were added with the extracts and cultured for 24 h. The H2DCFDA assay revealed that 10–50 μg/mL of concentrations showed a significant reducing effect on the intracellular ROS accumulation ($^{\#}$ $p < 0.05$ vs. 50 mM glucose) (Figure 1B). Moreover, the effects of ACH on high glucose-induced cytotoxicity were also examined. In the MTT assay, SH-SY5Y cells were pre-exposed with 50 mM glucose, followed by ACH treatment. As shown in Figure 1C, we found that glucose at the concentration of 50 mM could significantly decrease cell viability, while ACH (10–50 μg/mL) treatment alleviated the effect of glucose-induced cytotoxicity and significantly increased cell viability.

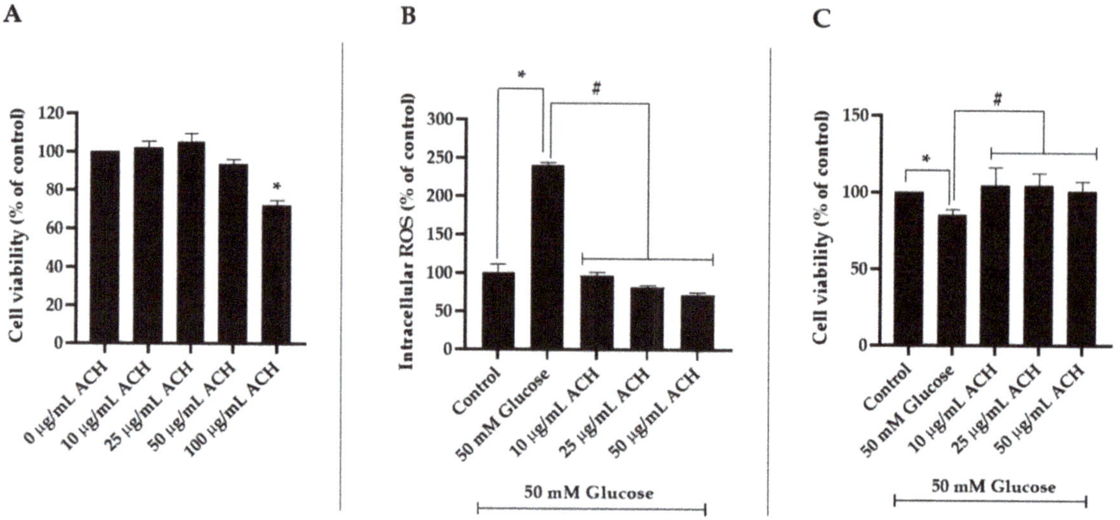


**Figure 1.** The effect of different concentrations of ACH on SH-SY5Y cell viability (**A**). MTT assay was used to clarify cell viability. The effect of ACH extracts on high glucose-induced intracellular ROS accumulation was performed using H2DCFDA assay, and the relative intracellular ROS accumulation level is shown in (**B**). MTT assay showed cell viability in cells pre-exposed with high glucose and followed by ACH treatment (**C**). Data are presented as the means ± SEM, * $p < 0.05$ vs. control and # $p < 0.05$ vs. 50 mM glucose.

### *3.3. Effects of ACH on Neurite Outgrowth*

Neuronal differentiation is one of the neuroprotective factors that plays an important role in neuronal development and the formation of synapses [30]. Neurite outgrowth is the process representing the neuronal differentiation. The activity of ACH on neurite outgrowth formation in high glucose-treated cells was investigated. To create the neurite outgrowth model, SH-SY5Y was cultured in serum starvation, which was low glucose DMEM supplemented with 1% FBS to avoid overgrowth of cells; 10 μM retinoic acid (RA) was added for 24 h [22]. In response to serum starvation and retinoic acid, SH-SY5Y will begin neurite outgrowth formation. Figure 2A shows the comparison of the neurite outgrowth formation between the untreated and RA-treated groups. We found that RA treatment could induce neurite outgrowth, as evidenced by the increased expression of growth-associated protein 43 (GAP-43), a marker of neurite outgrowth, and the increased expression of Teneurin-4, a transmembrane protein that plays a role in the neurogenesis, is highly expressed in the central nervous system, and mediates neurite outgrowth [31,32]. Confirmed by Western blot analysis (Figure 2D–F), the expression of GAP-43 and Teneurin-4 in RA-treated group was increased, and relative GAP-43 and Teneurin-4 expression was 2.94 ± 0.22-fold change and 11.28 ± 0.11-fold change, respectively ($^{+}$ $p < 0.05$ vs. 10 μM RA).

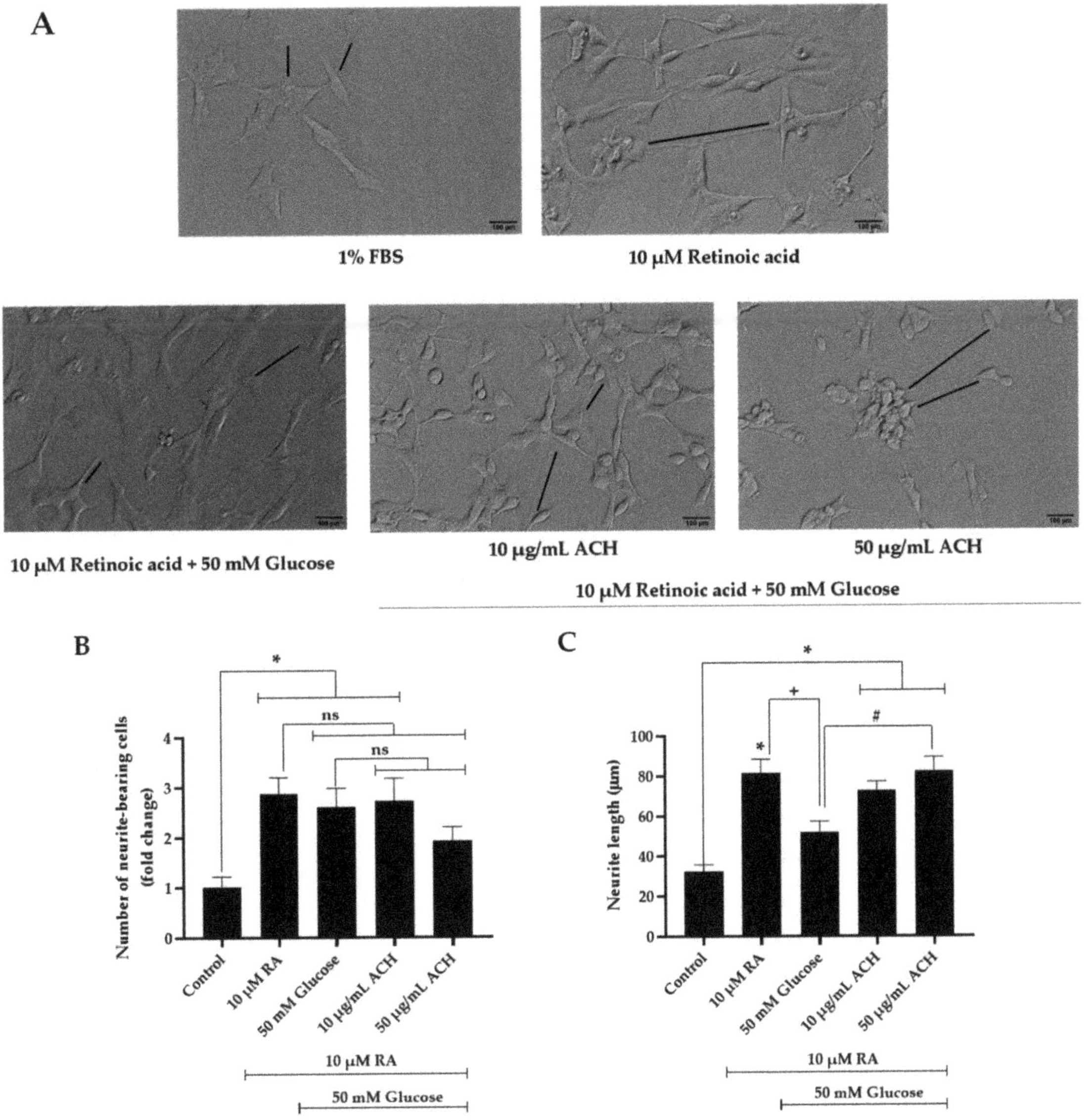


**Figure 2.** *Cont.*

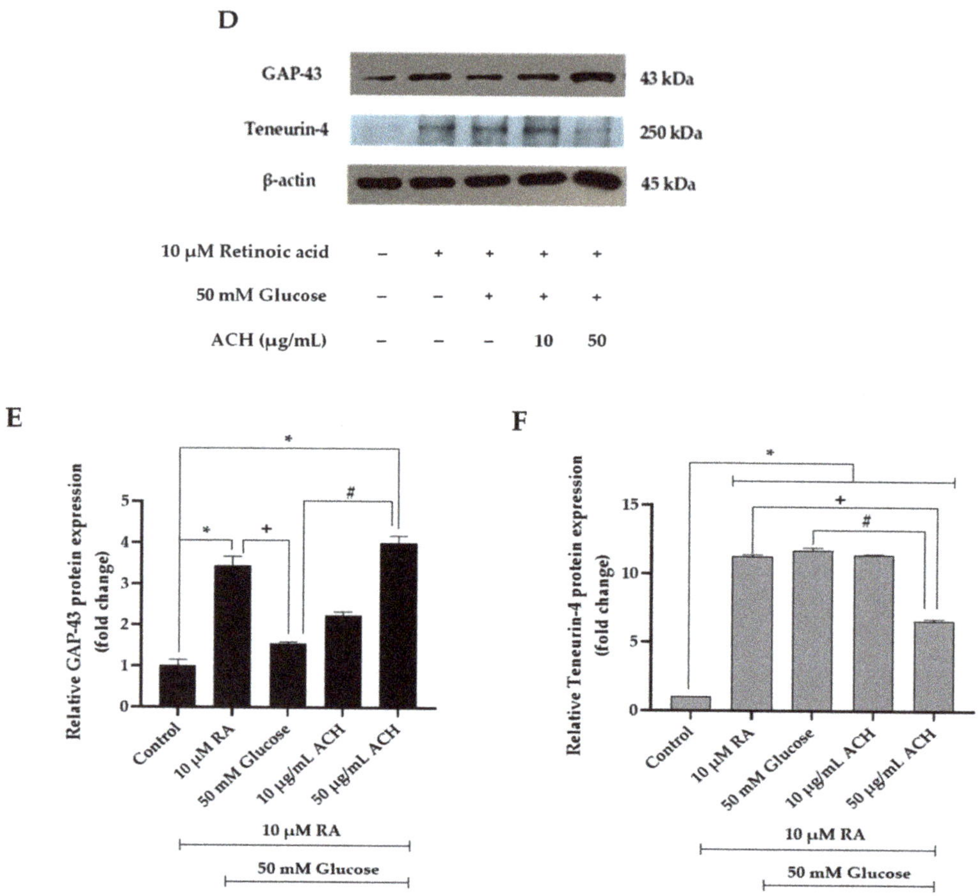


**Figure 2.** The neuroprotective effect of ACH on neuronal differentiation. The neurite outgrowth process was observed under a differential interference contrast (DIC) microscope at 10× magnification (black lines represented the measurement of neurite outgrowth) (**A**). The number of bearing cells and neurite length are shown in (**B**,**C**), respectively. GAP-43 and Teneurin-4 expression is shown in a representative Western blot (**D**). Normalized values of both GAP-43 and Teneurin-4 against β-actin (**E**,**F**). All data are presented as the mean ± SEM, * $p < 0.05$ vs. control; $^{+}$ $p < 0.05$ vs. 10 μM RA; $^{\#}$ $p < 0.05$ vs. 50 mM glucose; ns = not significant.

High glucose concentration is a critical factor that could trigger neuronal damage and inhibit neurite outgrowth [1,3]. To examine the effect of high glucose on neurite outgrowth, cells were cultured in high glucose conditions for 48 h. We found that 50 mM of glucose could significantly reduce the neurite outgrowth formation and neurite length when compared to the RA-treated group ($^{+}$ $p < 0.05$) (Figure 2A,C) and GAP-43 expression (1.54 ± 0.04-fold change or 55.15% reduction compared to RA-treated group, Figure 2D,E). Next, we examined the effect of the extracts on the elongation of neuronal processes. Treatment of the cells with 10 or 50 μg/mL ACH was performed. Figure 2A showed that ACH extract could induce the neurite outgrowth in high glucose-treated cells, and Figure 2D–F shows that ACH could also increase the GAP-43 expression in a concentration-dependent manner (2.23 ± 0.09-fold change and 4.01 ± 0.17-fold change, respectively). In addition, Teneurin-4 expression was not altered in both the high glucose-treated group and 10 μg/mL ACH-treated group, while 50 μg/mL ACH could cause a considerable reduction of Teneurin-4 expression (6.61 ± 0.09-fold change or 41.37% reduction compared to RA-treated group). Finally, the number of bearing cells and neurite length are shown in Figure 2B,C, respectively. The number of neurite-bearing cells and neurite length were significantly increased in retinoic acid–treated cells. The number was not altered in both the high glucose-treated group and 10 μg/mL ACH-treated group. However, Figure 2A shows the cell cluster containing more than five cells when treated with ACH (50 μg/mL). These cell clusters were excluded for scoring as neurite-bearing cells; therefore, the number of neurite-bearing cells in the ACH-treated group was decreased (not significant) when compared to the RA-treated group. On the contrary, both concentrations of ACH could significantly increase the length of neurites in a dose-dependent manner compared to the control group, but ACH at the concentration of 50 μg/mL could significantly increase the length of neurites compared to the RA-treated group.

### *3.4. Effects of ACH on Cell Cycle Delay*

High glucose level may cause neurotoxicity by interfering with cell cycle progression. A flow cytometer was used for the cell cycle diagram and cell numbers. Figure 3A,B demonstrates the activity of cell cycle distribution and the data generated from the flow cytometer. The percentage of cell numbers in the G1 phase in the high glucose-treated group was significantly higher than the control group (43.79% ± 1.83 for control and 60.03% ± 0.35; for high glucose-treated group), which causes cells to arrest in the resting phase (G0/G1). Post-treatment of ACH extract could attenuate the high glucose effect and normalize cell cycle progression. The percentage of cells in both the 10 and 50 μg/mL ACH treatments was significantly lower than the high glucose treatment (48.68% ± 2.56 and 49.86% ± 2.49, respectively).

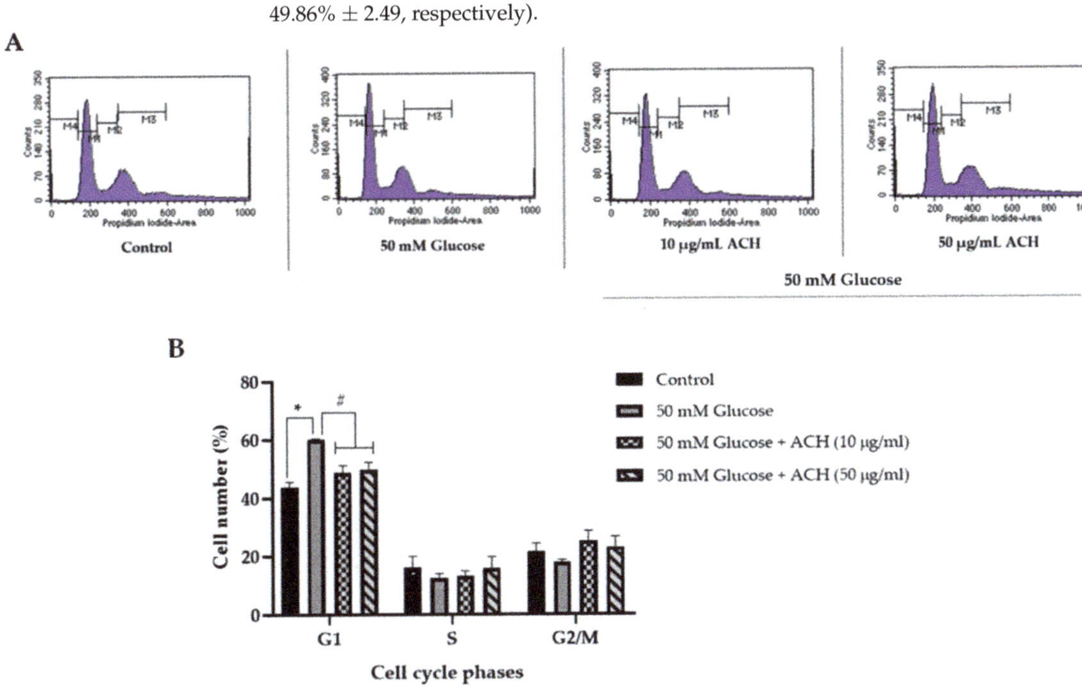


**Figure 3.** *Cont.*

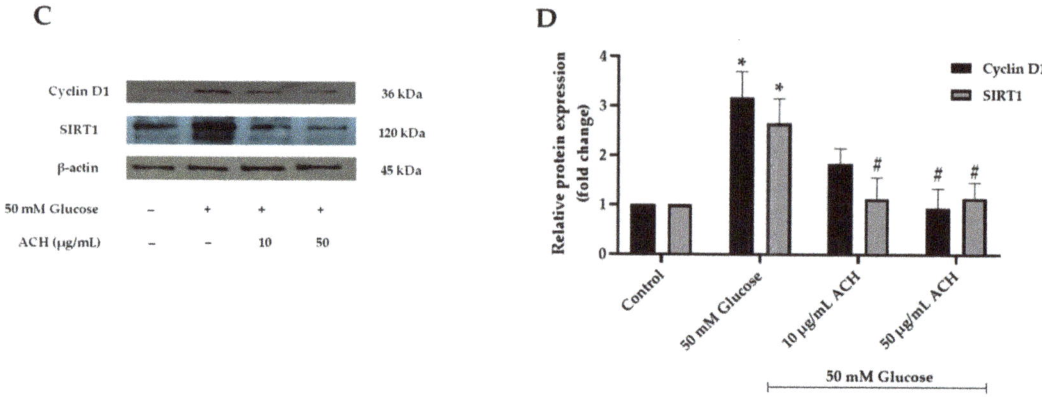


**Figure 3.** The neuroprotection of ACH on high glucose-induced cell cycle delay. Quantitative determination based on propidium iodide (PI) staining was carried out using a flow cytometer. Cell histogram (**A**) and the percentage of cell numbers (**B**) are shown. Cyclin D1 and sirtuin 1 (SIRT1) are as shown in a representative Western blot (**C**). Normalized values of both cyclin D1 and SIRT1 against β-actin (**D**). The mean ± SEM values of normalized cyclin D1 and SIRT1 expression were obtained from three independent experiments, * $p < 0.05$ vs. control; # $p < 0.05$ vs. 50 mM glucose.

We also investigated the expression of two cell cycle-associated proteins, cyclin D1 and SIRT1, using Western blot. Cyclin D1 is the G1 cyclin regulatory partner to control cell cycle progression. As in our previous study, upregulation of cyclin D1 caused cell cycle arrest at G1. Figure 3C,D shows the expression of cyclin D1 was significantly increased when cells were treated with 50 mM glucose alone (* $p < 0.05$ vs. control). The relative cyclin D1 expression was 3.18 ± 1.04 (* $p < 0.05$). When post-treated with ACH, cyclin D1 was significantly reduced in response to treatment with only 50 μg/mL ACH (0.95 ± 0.78; # $p < 0.05$ vs. high glucose alone).

As shown in Figure 3C,D, the relative SIRT1 expression was significantly increased in response to treatment with high glucose over time (2.66 ± 0.85; * $p < 0.05$ vs. control). Both 10 and 50 μg/mL ACH could normalize cell cycle progression by downregulating SIRT1 expression compared to the 50 mM glucose group (1.13 ± 0.75 and 1.15 ± 0.54; # $p < 0.05$ vs. high glucose alone, respectively).

### *3.5. ACH Extract Attenuated the High Glucose-Induced Reduction of Body Length and Size and Brood Size*

At adult day 1 stage, we found differences in both body size and body length following exposure to high glucose levels. Morphological changes were imaged by a 10× objective lens of a bright-field microscope (Figure 4A). The statistical difference between 0.1% DMSO (control group) and the 50 mM glucose-treated group is seen in Figure 4B–D. The body size and length were significantly lower than control group (401.46 ± 10.44 μm$^2$ and 1166.89 ± 13.24 μm; * $p < 0.05$). ACH extract co-treatment could work against the effect of high glucose by significantly improving both body size and length compared with high glucose alone. As seen in Figure 4B,C, the body size in 10 and 50 μg/mL co-treatment was 608.13 ± 7.95 μm$^2$ and 561.03 ± 10.11 μm$^2$; # $p < 0.05$, respectively. The body length was 1354.38 ± 10.51 μm, # $p < 0.05$, and 1333.14 ± 13.67 μm, # $p < 0.05$, in 10 and 50 μg/mL ACH, respectively.

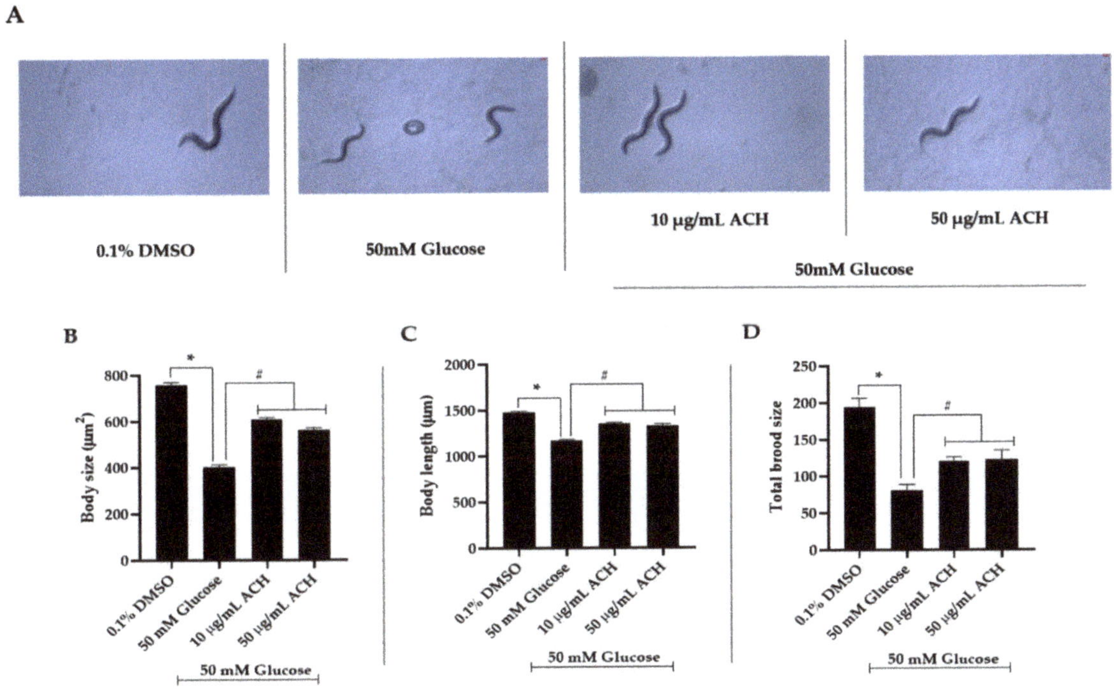


**Figure 4.** ACH extracts attenuated the high glucose-induced reduction of body length and size and brood size. Bright-field microscope images of *C. elegans* were taken using a 10x objective, and representative images are shown (**A**). The body size and length were measured from at least 20 adult day 1 worms. The number of eggs hatching were counted. The mean ± SEM values of body size (**B**), body length (**C**) and brood size (**D**) are shown, * $p < 0.05$ vs. control; # $p < 0.05$ vs. 50 mM glucose-fed worms.

Moreover, we found that high glucose-fed worms showed a significant lower number of eggs laid than the control group (Figure 4D) (80.51 ± 8.10; * $p < 0.05$ vs. control). The total brood size in 10 and 50 μg/mL ACH-fed worms was significantly increased as 119.81 ± 6.04 and 121.90 ± 12.96; # $p < 0.05$ vs. 50 mM glucose-fed worms, respectively.

### 3.6. ACH Extract against High Glucose-Induced Oxidative Stress in C. elegans

The effect of ACH extracts on intracellular ROS production was investigated. Treatment of *C. elegans* with high glucose alone induced intracellular ROS production (127.00 ± 5.71%). However, cultured with both concentrations of ACH, it could attenuate the accumulation of intracellular ROS in a dose-dependent manner (Figure 5A,B) compared with the high glucose-fed worms (102.97 ± 1.62% and 91.46 ± 1.59% in 10 and 50 μg/mL ACH-fed worms, respectively). These results confirmed that ACH extract protects against high glucose-induced healthspan reduction by suppressing ROS formation.

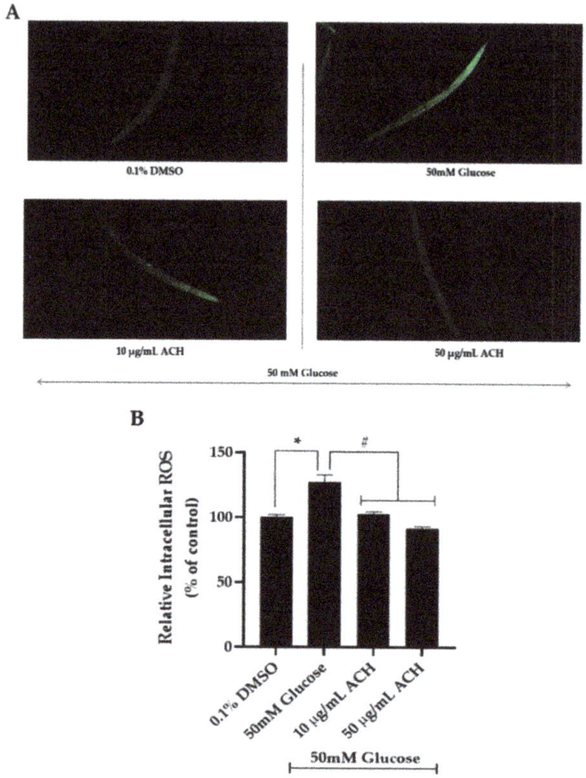


**Figure 5.** The protective effect of ACH extracts on high glucose-induced oxidative stress in *C. elegans*. Intracellular ROS accumulation was performed using H2DCFDA assay and imaged by a confocal microscope (**A**). Relative intracellular ROS accumulation level is shown in (**B**). Data are presented as the means ± SEM, * $p < 0.05$ vs. control; # $p < 0.05$ vs. high glucose-fed worms.

### 3.7. ACH Extracts Extend Lifespan in High Glucose-Fed Worms

To investigate the effect of ACH extracts on high glucose-induced lifespan reduction, N2 synchronized L1 larvae were cultured in high glucose alone or high glucose combined with ACH extracts. The results indicated that the mean lifespan of high glucose-fed worms was shorter than 0.1% DMSO-treated worms (approximate 31%). Moreover, ACH could increase the mean lifespan of high glucose-fed worms in a concentration-dependent manner (20.13% and 40.75% compared to mean lifespan of control group). All results are shown in Figure 6 and Table 2.

**Table 2.** Results and statistical analyses of lifespan of *C. elegans* treated with high glucose and ACH.

| Groups | Mean Lifespan | | *p*-Value (vs. Control) | *p*-Value (vs. 50 mM Glucose) | Number of Worms |
|---|---|---|---|---|---|
| | Day ± SEM | % Increase (vs. 50 mM Glucose) | | | |
| 0.1% DMSO | 18.55 ± 0.42 | 31.00 | - | 0.0001 | 102 |
| 50 mM Glucose | 14.16 ± 0.40 | - | 0.0001 | - | 118 |
| 50 mM Glucose + 10 μg/mL ACH | 17.01 ± 0.47 | 20.13 | 0.0379 | 0.0001 | 93 |
| 50 mM Glucose + 50 μg/mL ACH | 19.93 ± 0.43 | 40.75 | 0.0192 | 0.0001 | 91 |

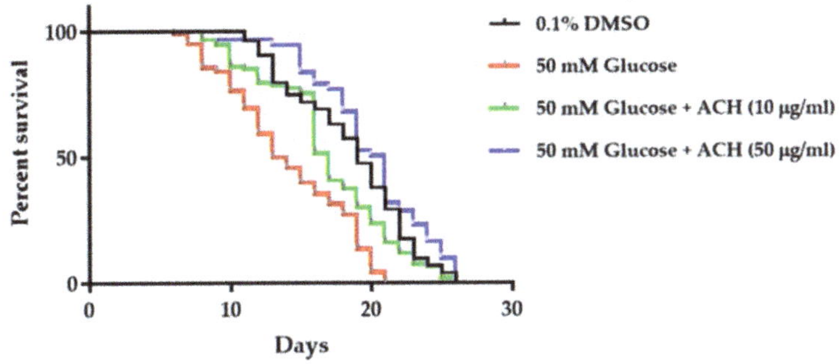


**Figure 6.** The effect of high glucose and ACH crude extracts on *C. elegans* lifespan.

### 3.8. ACH Extracts Mediated Extension of Lifespan and Healthspan in High Glucose-Fed Worms through DAF-16/FoxO and aqp-1

To investigate the effect of ACH extract on mRNA expression, *C. elegans* was cultured and treated as explained earlier. Compared to the vehicle control, we found that high glucose (50 mM) could decrease mRNA expression of *daf-16* (0.65 ± 0.1-fold; * $p < 0.05$), *sod-3*, and *aqp-1*. All mRNA expression could be increased after ACH treatment. The highest change was found in the *sod-3* mRNA expression in 50 μg/mL ACH-fed worms (# $p < 0.05$ vs. high glucose-fed worms). All ACH concentrations significantly increased both *daf-16* and *aqp-1* expression in the concentration-dependent manner. All results are shown in Figure 7.

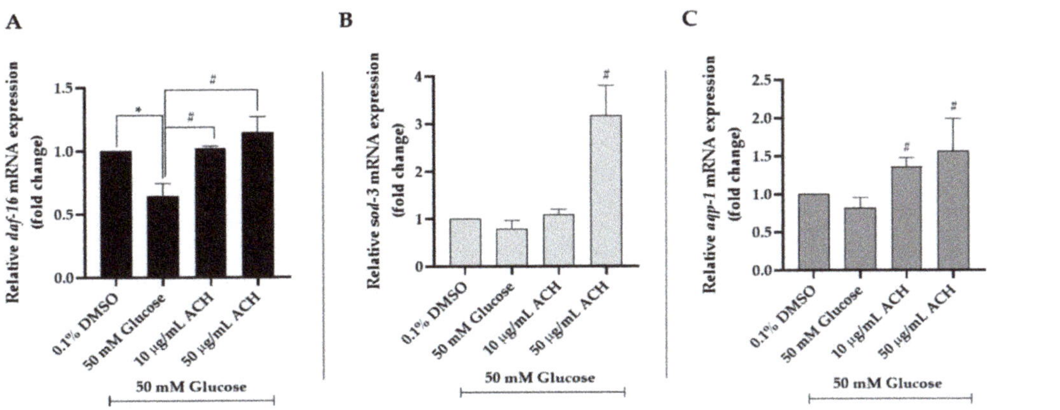


**Figure 7.** The mRNA expression: (**A**) *daf-16*, (**B**) *sod-3*, and (**C**) *aqp-1* genes of ACH treatment on high glucose-fed worms. All data were presented as the mean ± SEM), * $p < 0.05$ vs. control; # $p < 0.05$ vs. high glucose-fed worms.

### 3.9. Phytochemical Constituents of ACH

The GC-MS/MS chromatogram of ACH extract showed 10 major peaks, and the components corresponding to the peaks were determined as follows: 24-methylenecycloartan-3-one (14.17%), squalene (13.55%), D:A-friedooleanan-3-ol, (3.α.)–(10.21%), tritriacontane (9.38%), vitamin E (8.94%), β-amyrin (5.17%), 9,19-cyclolanostan-3-ol,24,24-epoxymethano-,acetate (3.74%), lupenone (lup-20(29)-en-3-one) (3.31%), hentriacontane (2.68%), olean-12-en-3-one (2.23%), and other compounds showed low of percentage of peak area (Table 3).

**Table 3.** Proposed phytochemical constituents in ACH extract compared with the National Institute of Standards and Technology (NIST) database.

| Compound | RT | Area (%) | MF | MW |
|---|---|---|---|---|
| tetradecane | 22.867 | 0.15 | $C_{14}H_{30}$ | 198 |
| hexadecane | 30.730 | 0.12 | $C_{16}H_{34}$ | 226 |
| phytol | 39.231 | 0.3 | $C_{20}H_{40}O$ | 296 |
| n-hexadecanoic acid | 43.3 | 0.9 | $C_{16}H_{32}O_2$ | 256 |
| oleic acid | 48.638 | 0.57 | $C_{18}H_{34}O_2$ | 282 |
| oleamide (9-Octadecenamide, (Z)-) | 54.799 | 0.97 | $C_{18}H_{35}NO$ | 281 |
| squalene | 66.506 | 13.55 | $C_{30}H_{50}$ | 410 |
| nonacosane | 68.036 | 0.49 | $C_{29}H_{60}$ | 408 |
| 9,19-cyclolanost-24-en-3-ol, acetate, (3.β.) | 68.794 | 0.38 | $C_{32}H_{52}O_2$ | 468 |
| 2,2,4-trimethyl-3-(3,8,12,16-tetramethyl-heptadeca-3,7,11,15-tetraenyl)-cyclohexanol | 69.037 | 0.4 | $C_{30}H_{52}O$ | 428 |
| Y-tocopherol | 71.233 | 0.69 | $C_{28}H_{48}O_2$ | 416 |
| hentriacontane | 72.371 | 2.68 | $C_{31}H_{64}$ | 436 |
| vitamin E | 73 | 8.94 | $C_{29}H_{50}O_2$ | 430 |
| stigmasterol | 75.305 | 1.17 | $C_{29}H_{48}O$ | 412 |
| D-friedoolean-14-en-3-one | 76.122 | 1.25 | $C_{30}H_{48}O$ | 424 |
| tritriacontane | 76.463 | 9.38 | $C_{33}H_{68}$ | 464 |
| β-amyrin | 76.547 | 5.17 | $C_{30}H_{50}O$ | 426 |
| olean-12-en-3-one | 77.411 | 2.23 | $C_{30}H_{48}O$ | 424 |
| lupenone (lup-20(29)-en-3-one) | 77.498 | 3.31 | $C_{30}H_{48}O$ | 424 |
| α-amyrin | 77.878 | 1.94 | $C_{30}H_{50}O$ | 426 |
| lup-20(29)-en-3-ol, acetate, (3.β.) | 78.077 | 0.93 | $C_{32}H_{52}O_2$ | 468 |
| D:A-friedooleanan-7-one, 3-hydroxy | 78.41 | 1.19 | $C_{30}H_{50}O_2$ | 442 |
| ursa-9(11),12-dien-3-ol | 78.65 | 1.16 | $C_{30}H_{48}O$ | 424 |
| 9,19-cyclolanostan-3-ol,24,24-epoxymethano, acetate | 78.796 | 3.74 | $C_{33}H_{54}O_3$ | 498 |
| betulin | 79.05 | 1.00 | $C_{31}H_{52}O$ | 440 |
| friedelan-3-one | 79.473 | 0.42 | $C_{30}H_{50}O$ | 426 |
| D:A-friedooleanan-3-ol, (3.α.) | 80.004 | 10.21 | $C_{30}H_{52}O$ | 428 |
| pentatriacontane | 80.197 | 1.42 | $C_{35}H_{72}$ | 492 |
| 24-methylenecycloartan-3-one | 80.474 | 14.17 | $C_{31}H_{50}O$ | 438 |

RT: retention time; MF: molecular formula; MW: molecular weight.

### *3.10. The Ability of ACH-Derived Phytochemical Constituents as Inhibitors of IGFR Using an In Silico Approach*

Next, we investigated the competence of ACH on IGFR inhibition and identified which phytochemicals were more efficient than the binding of each positive control. At least 27 phytochemicals obtained from GC-MS/MS were analyzed by molecular docking analysis.

In the current experiment, EGCG and resveratrol were used as the positive control in the molecular docking study. EGCG has been previously reported as an inhibitor of the activity of IGFR [33]. In addition, resveratrol exerted suppression on IGF-1 [34–36].

In Table 4, the binding energy of −9.88, −6.54, and −6.57 kcal/mol was exerted by the original ligand, EGCG, and resveratrol, respectively.

**Table 4.** Docking results of the compounds with IGFR.

| No. | Compound | MW | Binding Energy (kcal/mol) | Inhibition Constant | Amino Acid Interaction | | |
|---|---|---|---|---|---|---|---|
| | | | | | Hydrogen Bond | Hydrophobic Bond | Electrostatic Bond |
| | EGCG (positive control) | 458.4 | −6.54 | 16.18 μM | ASP1086<br>SER1089<br>GLU1015 | LEU1005 (2)<br>ARG1084 (2)<br>GLY1085 | |
| | Resveratrol (positive control) | 228.24 | −6.57 | 15.39 μM | MET1082<br>MET1082<br>GLU1080<br>GLN1007 | VAL1013<br>ALA1031<br>MET1156 | |
| 1 | 24-Methylenecycloartan-3-one | 438.7 | −5.13 | 172.22 μM | GLY1085 | LEU1005<br>VAL1013 (2)<br>ALA1031<br>LYS1033<br>MET1079<br>LEU1081 (2)<br>MET1082<br>MET1142<br>MET1156 (2) | |
| 2 | Squalene | 410.7 | −6.37 | 21.57 μM | | ARG1003<br>LEU1005 (3)<br>VAL1013<br>ALA1031 (2)<br>LYS1033<br>VAL1063<br>MET1079<br>LEU1081 (2)<br>MET1142 (2)<br>MET1156 (2) | |
| 3 | D:A-Friedooleanan-3-ol, (3.alpha.)- | 428.7 | −5.90 | 47.43 μM | GLU1080 | VAL1013<br>ALA1031<br>ALA1031<br>MET1142<br>VAL1063<br>MET1079<br>MET1142<br>VAL1013<br>LYS1033<br>MET1079<br>VAL1013<br>MET1156 | |
| 4 | Friedelan-3-one | 426.7 | −7.88 | 1.66 μM | | VAL1013 (3)<br>ALA1031 (3)<br>LYS1033 (2)<br>MET1079<br>MET1142<br>MET1156 | |
| 5 | Stigmasterol | 412.7 | −9.32 | 146.8 nM | ARG1003 | LEU1005 (3)<br>VAL1013 (2)<br>LEU1081<br>MET1142 (2) | |
| 6 | Tritriacontane | 464.9 | −3.51 | 2.69 mM | | LEU1005 (3)<br>ALA1031 (2)<br>LYS1033<br>MET1079<br>MET1142 (4)<br>MET1156 (2)<br>ILE1160 | |
| 7 | Vitamin E | 430.7 | −7.92 | 1.56 μM | GLY1085<br>ASP1086<br>SER1089 | VAL1013 (3)<br>ALA1031 (2)<br>LYS1033<br>MET1079<br>MET1082<br>MET1142 (2)<br>MET1156<br>ILE1160 | **ASP1086** |

**Table 4.** *Cont.*

| No. | Compound | MW | Binding Energy (kcal/mol) | Inhibition Constant | Amino Acid Interaction | | |
|---|---|---|---|---|---|---|---|
| | | | | | Hydrogen Bond | Hydrophobic Bond | Electrostatic Bond |
| 8 | D-Friedoolean-14-en-3-one | 424.7 | −8.88 | 355.27 nM | | LEU1005 (4)<br>VAL1013<br>ALA1031<br>LEU1081 (2)<br>MET1142 | |
| 9 | 9,19-Cyclolanostan-3-ol,24,24-epoxymethano-, acetate | 498.8 | −7.30 | 4.49 µM | SER1089<br>GLY1085<br>MET1082 | LEU1005<br>VAL1013 (2)<br>ALA1031 (2)<br>LYS1033<br>MET1142 (2)<br>MET1156 (2)<br>ILE1160<br>MET1079 (2)<br>LEU1081 | |
| 10 | D:A-Friedooleanan-7-one, 3-hydroxy- | 442.7 | −5.37 | 116.19 µM | GLU1080 | LEU1005<br>VAL1013<br>ALA1031<br>MET1079<br>MET1082<br>MET1142 (2) | |
| 11 | Beta-amyrin | 426.7 | −9.02 | 245.77 nM | THR1083 | LEU1005<br>VAL1013 (4)<br>ALA1031<br>LYS1033 (3)<br>MET1142 (2)<br>MET1156 (2) | |
| 12 | Olean-12-en-3-one | 424.7 | −9.77 | 68.61 nM | | LEU1005<br>VAL1013 (3)<br>ALA1031 (3)<br>LYS1033 (2)<br>MET1079 (2)<br>MET1142 (3)<br>MET1156 | |
| 13 | Lup-20(29)-en-3-ol, acetate, (3.beta.)- | 468.8 | −7.36 | 4.02 µM | | LEU1005<br>VAL1013<br>ALA1031 (3)<br>VAL1063<br>MET1079<br>LEU1081<br>MET1142 (3) | |
| 14 | Lupenone (Lup-20(29)-en-3-one) | 424.7 | −9.56 | 97.57 nM | SER1089 | LEU1005<br>VAL1013<br>ALA1031 (2)<br>LYS1033 (2)<br>MET1142 (3)<br>MET1156 (2)<br>VAL1063<br>MET1079<br>LEU1081<br>MET1082 | |
| 15 | Gamma-Tocopherol | 416.7 | −7.75 | 2.08 µM | | LEU1005 (3)<br>VAL1013 (2)<br>LYS1033<br>MET1079<br>LEU1081<br>MET1142<br>MET1156 (2) | |
| 16 | Hentriacontane | 436.8 | −3.39 | 3.25 mM | | LEU1005 (2)<br>VAL1013 (2)<br>ALA1031 (2)<br>LYS1033<br>LEU1081<br>MET1142 (2)<br>MET1156<br>ILE1160 (2)<br>TYR1161 | |
| 17 | Oleamide (9-Octadecenamide, (Z)-) | 281.5 | −4.51 | 490.34 µM | GLU1080 | LEU1005 (2)<br>VAL1013 (3)<br>ALA1031<br>LYS1033 (2)<br>MET1079<br>MET1156 | |

**Table 4.** *Cont.*

| No. | Compound | MW | Binding Energy (kcal/mol) | Inhibition Constant | Amino Acid Interaction | | |
|---|---|---|---|---|---|---|---|
| | | | | | Hydrogen Bond | Hydrophobic Bond | Electrostatic Bond |
| 18 | Alpha-amyrin | 426.7 | −9.21 | 178.76 nM | | LEU1005<br>VAL1013 (2)<br>ALA1031 (3)<br>LYS1033<br>MET1079<br>MET1142 (2)<br>MET1156 | |
| 19 | Pentatriacontane | 492.9 | −1.59 | 68.88 mM | | LEU1005 (5)<br>VAL1013<br>ALA1031<br>ARG1084<br>MET1142 | |
| 20 | Ursa-9(11),12-dien-3-ol | 424.7 | −8.53 | 554.63 nM | | VAL1013 (2)<br>ALA1031<br>LYS1033<br>ILE1160<br>VAL1063<br>LEU1081<br>MET1082 (2)<br>MET1142 (2)<br>MET1156 | |
| 21 | Betulin | 442.7 | −8.86 | 322.86 nM | | LEU1005 (2)<br>ALA1031 (3)<br>VAL1063<br>MET1079<br>MET1082<br>MET1142 (4)<br>MET1156 | |
| 22 | n-Hexadecanoic acid | 256.42 | −3.76 | 1.75 mM | MET1082<br>GLU1080 | LEU1005<br>VAL1013 (3)<br>ALA1031<br>LYS1033 (2)<br>MET1079<br>MET1156 (2) | |
| 23 | Oleic acid | 282.5 | −4.43 | 564.46 µM | SER1089<br>ASP1086 | VAL1013 (2)<br>ALA1031 (2)<br>LYS1033<br>MET1142 (2)<br>MET1156 (2) | |
| 24 | Nonacosane | 408.8 | −3.85 | 1.52 mM | | LEU1005<br>VAL1013 (3)<br>ALA1031 (2)<br>LYS1033 (2)<br>VAL1063<br>MET1079 (2)<br>MET1142<br>MET1156 (2) | |
| 25 | 2,2,4-Trimethyl-3-(3,8,12,16-tetramethyl-heptadeca-3,7,11,15-tetraenyl)-cyclohexanol | 428.7 | −6.75 | 11.23 µM | SER1089 | LEU1005 (2)<br>VAL1013 (2)<br>ALA1031 (3)<br>LYS1033<br>MET1079<br>MET1142 (2)<br>MET1156 (2) | |
| 26 | 9,19-Cyclolanost-24-en-3-ol, acetate, (3.beta.)- | 468.8 | −5.29 | 132.45 µM | | LEU1005 (3)<br>ALA1031 (2)<br>MET1079<br>ARG1084<br>TYR1090<br>MET1142 (2) | |
| 27 | Phytol | 296.5 | −5.17 | 161.26 µM | ASP1086 | LEU1005 (2)<br>VAL1013 (3)<br>ALA1031 (2)<br>LYS1033 (2)<br>LEU1081<br>MET1082<br>MET1142 (2)<br>MET1156 | |

Based on the docking results in Table 4, five phytochemicals showed outstanding inhibition against IGFR, with higher binding energy (higher than −9 kcal/mol) compared to the others; noticeably, their binding energy was also higher than both positive controls. The five phytochemical constituents included olean-12-en-3-one, lupenone, stigmasterol, α-amyrin, and β-amyrin. Interestingly, 2D diagrams of the five phytochemicals, original ligand, and both positive controls are shown in Figure S1 in the Supplementary Materials for IGFR.

For the method validation, 2-[4-[4-[[(6Z)-5-chloranyl-6-pyrazolo [1,5-a]pyridin-3-ylid ene-1H-pyrimidin-2-yl]amino]-3,5-dimethyl-pyrazol-1-yl]piperidin-1-yl]-N,N-dimethyl-eth anamide, a reported inhibitor of the IGFR (5FXS) crystal structure, was removed and re-docked into the original active cavity of the protein. The results showed that the original inhibitor was capably re-docked into the similar location and orientation of the native crystal structure with RMSD 2.48 Å (less than 2.5 Å is considered a near-native solution, and 2–3 Å is acceptable for docking [doi.org/10.1002/prot.24214, doi.org/10.3390/molecules23051038]). Moreover, the predicted binding energy was −9.88 kcal/mol, demonstrating the acceptable reproducibility of analysis. Protein–ligand interactions showed that re-docking conformation of the original interacted with key amino acids found in the co-crystallized structure. The re-docking showed that the original ligand formed a hydrogen bond with two amino acids found in the co-crystallized complex: GLU1080 and MET1082. In addition, it shared amino acid interaction with five out of six amino acids: LEU1005, VAL1013, ALA1031, LYS1033 and LEU1081, by hydrophobic bonding. Therefore, these results indicate that the protocol used in this study is reliable and could be applied for further predictions.

### *3.11. Isolation and Chemical Characterization of Active Compounds in ACH Extract*

In order for active compounds to be derived from ACH to investigate the protective effect against high glucose-induced toxicity, ACH crude extract (1.2 g) was isolated by column chromatography using the condition hexane:ethyl acetate (EtOAc) (90:10). As a result, the eluates were examined by TLC, and seven combined fractions (ACH1-ACH7) were obtained.

We further investigated the effect of fraction ACH on high glucose-induced neurotoxicity. Neurite outgrowth-promoting activities of fraction ACH were evaluated in SH-SY5Y cells. We found that only ACH3 could exert remarkable neuroprotection by attenuating high glucose-induced neurite outgrowth inhibition (Figure 8A). The protection of this fraction on cell viability in high glucose concentrations was determined. Only ACH3 at the concentrations of 25 and 50 µg/mL protected SH-SY5Y cell viability from high glucose-induced damage (Figure 8B). In addition, Western blot results (Figure 8C,D) showed that ACH3 (50 µg/mL) could attenuate high glucose-inhibited neurite outgrowth through the induction of GAP-43 and Teneurin-4 expression. Together, we propose that this shows that fraction ACH3 is a probable active fraction.

Next, fraction ACH3 was subjected to column chromatography on silica gel using hexane:EtOAc (90:10). Compound **1** (10.1 mg) and **2** (50.2 mg) were afforded from ACH3. Their structures were further elucidated using $^1$H-NMR.

The results indicated that a mixture of Compounds **1** and **2** was stigmastrol and β-sitosterol. The structure of these compounds is shown in Figure S2 in the Supplementary Material. The spectrum of these compounds obtained from $^1$H-NMR is shown in Figure S3 in the Supplementary Material. Chemical information on these isolated compounds is as follows.

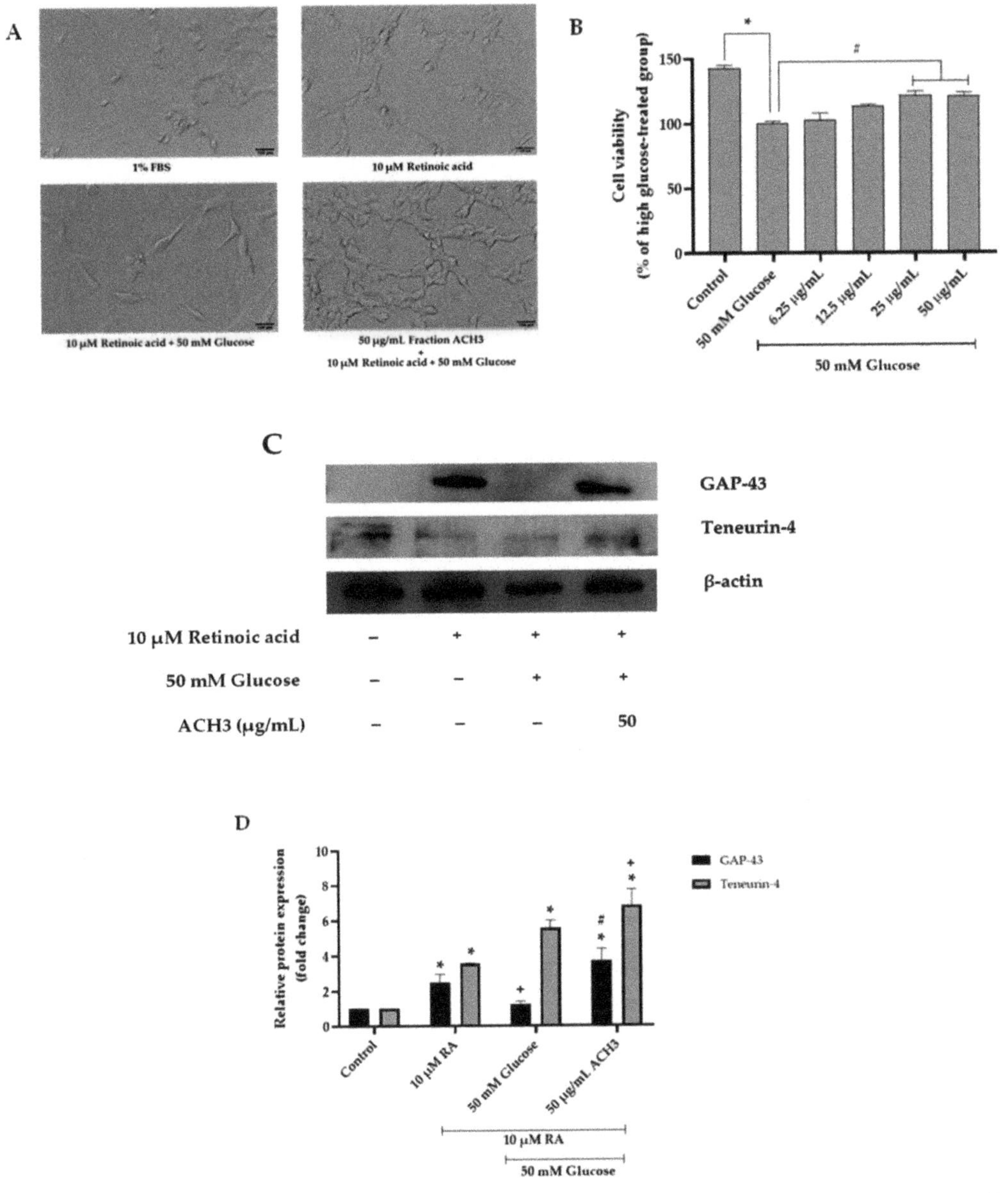


**Figure 8.** The effect of fraction ACH3 on neurite outgrowth (**A**). MTT assay was performed to clarify the effect of fraction ACH3 on cell viability (**B**). GAP-43 and Teneurin4 expression shown in a representative Western blot (**C**). Normalized values of both GAP-43 and Teneurin-4 against β-actin (**D**). All data are presented as the mean ± SEM, * $p < 0.05$ vs. control; $^{+}$ $p < 0.05$ vs. 10 μM RA; $^{\#}$ $p < 0.05$ vs. high glucose-treated cells.

### 3.12. *Compound* **1** *and* **2** *(A Mixture of Stigmastrol and β-Sitosterol)*

White powder; $^1$H-NMR ($CDCl_3$, 400 MHz): 0.65 (*s*, 3H), 0.80 (*d*, 3H, *J* = 6.4), 0.82 (*d*, 3H, *J* = 6.4), 0.84 (*t*, 3H, *J* = 7.2), 0.98 (*d*, 3H, *J* = 6.5), 1.02 (*s*, 3H), 3.50 (*tdd*, 1H, *J* = 4.5, 4.2, 3.8), 4.98 (*m*, 1H), 5.14 (*m*, 1H), 5.32 (*t*, 1H, *J* = 6.4).

Finally, the presence of β-sitosterol and stigmasterol was evaluated by using reverse-phase high-performance liquid chromatography (RP–HPLC). The HPLC fingerprints are shown in Figure S4 in the Supplementary Material. The amount of β-sitosterol and stigmasterol in ACH extract was 54.11 ± 0.5 ppm and 22.47 ± 0.42 ppm, respectively.

## 4. Discussion

*Aquilaria crassna* (AC) is a fragrant and medicinal plant used in many traditional medicines. We studied the protective effect of AC on high glucose-induced neurotoxicity and aging. To create a high glucose-induced neurotoxicity model, 50 mM glucose was selected, as a previous report [4] showed that *C. elegans* culturing under high glucose condition (50 mM) resulted in a glucose concentration in whole-body extracts of approximate 15 mM. At 10–15 mM glucose, it resembles the glucose concentrations in diabetic patients under poor glucose control. Furthermore, 50 mM glucose could induce nitric oxide production, protein kinase C activation, myo-inositol metabolism alteration, and defective tissue perfusion in high glucose-treated SH-SY5Y cells. The above-mentioned process involved the pathogenesis of diabetic complications, including neuropathy [37].

First, ACH supported the neuroprotective effect on high glucose-inhibited neurite outgrowth. The neuronal polarization is a crucial step that consists of two importantly different processes via a long axon and several short dendrites. These processes are associated with brain functions such as memory, learning, and emotion [38]. The profound outgrowth of axons was used as evidence of neuronal polarization [39]. In addition, GAP-43 and Teneurin-4 proteins were used as the important markers of neuronal polarization in this study. During the early stage of neurite outgrowth, filopodia are formed through the protrusion of small cell membranes. Teneurin-4 is a transmembrane protein and a positive regulator that functions in filopodia-like protrusion formation and neurite outgrowth [31]. Additionally, GAP-43 is another positive regulator during neurite outgrowth development [40]. The ACH extract on neurite outgrowth results (Figure 2) indicated that ACH could attenuate high glucose-inhibited neurite outgrowth. Interestingly (Figure 2B,C), we found that high glucose could only reduce GAP-43 expression; it did not alter Teneurin-4 expression. This could be described as high glucose reducing the neurite outgrowth development but not affecting the filopodia-like formation. The level of 50 μg/mL ACH could cause the considerable reduction of Teneurin-4 expression; this concentration of ACH could induce neurite outgrowth development but did not increase the number of filopodia-like formations. These data were in accordance with the number of neurite-bearing cells (or filopodia-like formations) (Figure 2D) and the neurite length (Figure 2E).

For cell cycle results (Figure 3A,B), high glucose causes cell cycle arrest at the G1 phase, and ACH normalized the cell cycle progress in high glucose-treated cells by significantly decreasing the percentage of cell number at the G1 phase. To confirm the effect of ACH on the cell cycle, the expression of proteins cyclin D1 and SIRT1 was detected by Western blot. SIRT1, also known as NAD-dependent deacetylase sirtuin-1, plays an important role in normalizing cell cycle progression, inhibiting cell senescence, and extending the lifespan of organisms [41–43]. Our previous study [20] confirmed the function of SIRT1 on high glucose-induced cell cycle delay and in response to glucose starvation. When the glucose level in cells was low, SIRT1 expression in the nucleus was activated and upregulated by the induction of phosphorylated GAPDH. In contrast, Figure 3C,D showed the different results. Western blot and normalized SIRT1 values revealed that treating cells with 50 mM glucose for 48 h could significantly increase SIRT1 expression. This process shows that the other important factor in addition to the glucose level was the time-dependent manner. SIRT1 expression was transiently downregulated by high glucose treatment. Then, cells were treated with high glucose over time. SIRT1 expression level gradually increased from 24 h to 48 h [44]. Our result (Figure 1A) indicated that high glucose could cause cellular oxidative stress. SIRT1 is triggered and redistributed in the mechanism responsible for maintaining cell homeostasis under oxidative stress [45,46]. Therefore, it could be described as why the level of SIRT1 expression was significantly increased in high glucose-treated cells.

*Caenorhabditis elegans* or *C. elegans* can be widely used as a model for the study of the molecular target and mechanisms affected by the pathological glucose concentrations. *C. elegans* has its short lifespan, and easily modifiable genome, and a simple glucose receptor system [47]. To examine the effect of high glucose concentration on the development of *C. elegans*, L1 larvae stage worms were cultured in NGM agar supplemented with *E. coli* OP50 and 50 mM glucose. This glucose concentration resulted in the intracellular glucose concentration of 14 to 15 mM and was sufficient to achieve significant effects on the development, offspring, and lifespan of *C. elegans*.

It has been documented that long-lasting high glucose conditions result in the substantial accumulation of the modification of mitochondrial proteins and a steady increase in ROS formation [48]. ROS plays an important role in the reduction of lifespan and healthspan in *C. elegans*. It well known that under stress conditions, a large amount of ROS will be released. ROS was associated with implications for pathogenesis of many diseases; in addition, it is one of the major causes of aging. Our previous research publications indicated that a certain concentration of antioxidants will contribute to the longevity of *C. elegans* [49].

To explain the mechanisms by which high glucose induced the reduction of lifespan and ACH extracts attenuated the toxicity of high glucose, we combined our results and found that high glucose caused lifespan reduction by interfering with worm development (such as body length and size), reducing in the number of progeny or eggs, and increasing the accumulation of intracellular ROS.

High glucose concentration disrupted the activation of genes associated with the protective mechanisms and caused shortened lifespan and reduced healthspan of *C. elegans*. Insulin and its close homolog insulin-like growth factor 1 (IGF-1) could bind to the tyrisine-kinase receptor, resulting in the inhibition of the forkhead boxO (FoxO) transcription factor. This transcription factor plays a vital role in many cellular processes, such as stress resistance, energy metabolism, apoptosis, cell cycle arrest, and organism longevity [50,51]. The FoxO family member *daf-16* and its transcription target gene, including *sod-3*, are important key factors contributing to the mediation of lifespan extension, metabolism, and stress response [27,52]. In addition, *daf-16* could act as a cumulative way for other genes to have an influence on lifespan [53]. As recently shown, the lifespan of *C. elegans* was significantly shortened by a high glucose-enriched diet through inhibiting the transcription factor *daf-16*. In addition, the role in response to glucose-induced lifespan shortening was associated with the reduction of the activity of the downstream aquaporin gene *aqp-1* [54]. This resulted in the *aqp-1* gene being noteworthy. The authors found a large impact on lifespan related to the loss of *aqp-1*. The *aqp-1* is a regulatory gene required for *daf-16*-regulated gene expression, including normalization of *sod-3* expression. Therefore, *aqp-1* might be a feedback regulator of insulin/IGF-1 signaling.

These reasons are consistent with our current results. ACH extract could attenuate the effect of glucose on shortening *C. elegans* lifespan through improving the mRNA expression of the glucose-responsive *daf-16* target gene, *sod-3*, and *aqp-1*. Collectively, ACH extract might be an interesting neuroprotectant and anti-aging agent. As shown in the above-mentioned experimental results, high glucose shortens the lifespan and reduces the healthspan of *C. elegans*, and ACH extract can attenuate the influence of high glucose. Conserved from *C. elegans* to mammals, many studies have indicated that glucose stimulated insulin secretion in *C. elegans* and is related to insulin-like growth factor 1 receptor (IGFR) and insulin/IGF-1 signaling [50,54,55]. Reducing the GFR and IGF-1 signaling pathway slows the aging process, doubles lifespan, and improves healthspan [56,57].

Next, various phytochemicals contained in ACH were analyzed and identified by GC-MS/MS. Moreover, the binding affinity of the phytochemicals of ACH against IGFR was analyzed by in silico analysis. The results revealed that five phytochemicals of olean-12-en-3-one, lupenone, stigmasterol, α-amyrin, and β-amyrin were more efficient than the binding of each positive control (EGCG or resveratrol).

Analyzed and confirmed by NMR and RP-HPLC, fraction ACH3 consisted of both active compounds: β-sitosterol and stigmasterol. Phytosterol is a class of natural products found in food, cosmetics, and medicinal plants such as AC. Along with over 200 types, stigmasterol and β-sitosterol are the most extensive in many plants [58]. There were many studies indicating that phytosterol might play an important role in the prevention of neurodegenerative disorders. Stigmasterol exerted neuroprotection against oxidative stress-induced neurodegeneration by upregulating FoxO, catalase, anti-apoptotic protein B-cell lymphoma 2 (Bcl-2), and SIRT1 expression in neurons [59]. Moreover, stigmasterol possesses neuroprotective activities against oxidative stress-indued ischemic injury and autophagy [60]. Stigmasterol could also decrease the activity of beta-secretase enzyme (BACE1) and the β-secretase cleavage of amyloid precursor protein (APP) [61]. On the other hand, β-sitosterol could inhibit neuroinflammation in neurodegenerative diseases through repressing pro-inflammatory markers, including interleukin-6, inducible nitric oxide, cyclooxygenase-2, and the phosphorylation of nuclear factor kappa B [62]. β-sitosterol prevented mitochondria dysfunction by increasing the mitochondrial ATP concentrations and mitochondrial potential [63]. Moreover, β-sitosterol could increase antioxidant enzymes through activating the estrogen receptor/PI3-kinase pathway, exerting anti-acetyl choline esterase (AChE) and anti-butyl choline esterase (BChE) inhibitory potentials in both in in vitro and in vivo models, and also playing a role as a free radical scavenger by regulating the glutathione level [64]. In our previous study, stigmasterol was a probable active component of *Momordica charantia* extract for the prevention for Polycyclic aromatic hydrocarbons (PAHs)-induced neurotoxicity [19]. β-sitosterol and stigmasterol have also been reported to possess the strongest neurite outgrowth-promoting activities of four sterol compounds in neuronal cell culture [65]. In accordance with this study, fraction ACH3 (50 μg/mL) exerted the neuroprotection of high glucose-inhibited neurite outgrowth. Therefore, β-sitosterol and stigmasterol are potentially interesting neuroprotectants derived from the ACH extract.

## 5. Conclusions

In summary, the AC leaf is a beneficial plant with rich bioactive compounds. ACH represents neuroprotection from high glucose-induced neuronal cell damage, including the induction of GAP-43/Teneurin-4-mediated neurite outgrowth and cell cycle normalization via the cyclin D1/SIRT1 signaling pathway. ACH also exerts oxidative resistance properties, healthspan improvement, and lifespan extension via the *daf-16*/FoxO and *aqp-1* pathway. At least 27 phytochemical compounds were identified by GC-MS/MS. Moreover, molecular docking analysis and NMR reveal that β-sitosterol and stigmasterol are bioactive phytochemicals in ACH. Collectively, ACH could be developed as an agent for the protection of high glucose-induced neurotoxicity and aging.

**Supplementary Materials:** The following supporting information can be downloaded at: https://www.mdpi.com/article/10.3390/nu14173668/s1, Table S1: The results of method validation between co-crystalized ligand at the original active site of IGFR (PDB ID: 5FXS); Figure S1: The results of the molecular docking study of IGFR were represented by 2D diagram of phytochemical-receptor interactions; Figure S2: Structure of compound **1** (stigmasterol) and compound **2** (β-sitosterol) from fraction ACH3; Figure S3: $^1$H-NMR spectrum of a mixture of β-sitosterol and stigmastrol (compound **1** and **2**) in $CDCl_3$; Figure S4: Reverse-phase HPLC analysis of standard β-sitosterol, stigmasterol and the extract of ACH showing the presence of both β-sitosterol and stigmasterol.

**Author Contributions:** Conceptualization, S.C., A.P. and T.T.; validation, N.P. and A.P.; formal analysis, N.S. and N.P.; investigation, W.V., W.W., N.S., P.R., C.S. and N.P.; data curation, N.P.; writing—original draft preparation, N.S. and N.P.; supervision, S.C., A.P. and T.T.; project administration, T.T.; funding acquisition, T.T. All authors have read and agreed to the published version of the manuscript.

**Funding:** This Research is funded by the Thailand Science research and Innovation Fund Chulalongkorn University (CU_FRB65_hea (77)_172_37_02).

**Institutional Review Board Statement:** Not applicable.

**Informed Consent Statement:** Not applicable.

**Data Availability Statement:** Not applicable.

**Acknowledgments:** N.P. wishes to thank the Second Century Fund (C2F), Chulalongkorn University, Bangkok, Thailand. The authors would like to acknowledge Noppadol Sa-Ard-Iam of the Immunology Research Center, Faculty of Dentistry, Chulalongkorn University, Thailand, for his help in flow cytometry analysis. We would like to express our gratitude to Nakorn Niamnont of the Department of Chemistry, King Mongkut's University of Technology Thonburi, Bangkok, Thailand for allowing the authors to access the NMR facility. Last but not least, we would like to express our sincere appreciation to Supakorn Yuenyong-wannachot of Siam Agarwood (2020) Co., Ltd., Thailand for providing the knowledge and information of agarwood leading to this research project.

**Conflicts of Interest:** The authors declare no conflict of interest.

## Abbreviations

| | |
|---|---|
| AC | *Aquilaria crassna* |
| ACH | AC hexane extract |
| GAP-43 | Growth-associated protein 43 |
| SIRT1 | Surtuin-1 |
| GAPDH | Glyceraldehyde-3-phosphate dehydrogenase |
| DAF-16/FoxO | Forkhead box protein O |
| *sod-3* | Superoxide dismutase-3 |
| *aqp-1* | Aquaporin-1 |
| ROS | Reactive oxygen species |
| $H_2$DCFDA | 2,7-Dichlorofluorescein diacetate |
| ABTS | 2,2-Azino-bis(3-ethylbenzothiazoline-6-sulfonic acid) |
| DPPH | Diammonium salt, 2,2-diphenyl-1-picrylhydrazyl |
| DMSO | Dimethyl sulfoxide |
| GC-MS/MS | Gas chromatography mass spectrometry/mass spectrometry |
| NMR | Nuclear magnetic resonance |
| RP-HPLC | Reverse-phase high-performance liquid chromatography |
| TLC | Thin layer chromatography |
| VCEAC | Vitamin C equivalent antioxidant capacity |
| IGFR | Insulin-like growth factor 1 receptor |
| EGCG | Epigallocatechin gallate |

## References

1. Schartner, E.; Sabbir, M.G.; Saleh, A.; Silva, R.V.; Chowdhury, S.R.; Smith, D.R.; Fernyhough, P. High glucose concentration suppresses a SIRT2 regulated pathway that enhances neurite outgrowth in cultured adult sensory neurons. *Exp. Neurol.* **2018**, *309*, 134–147. [CrossRef] [PubMed]
2. Saal, K.A.; Galter, D.; Roeber, S.; Bähr, M.; Tönges, L.; Lingor, P. Altered Expression of Growth Associated Protein-43 and Rho Kinase in Human Patients with Parkinson's Disease. *Brain Pathol.* **2017**, *27*, 13–25. [CrossRef] [PubMed]
3. Singh, J.N.; Negi, G.; Kharatmal, S.B.; Mule, N.K.; Sharma, D.; Sharma, S.S. Short-term extracellular glucose exposure alters neurite outgrowth and intracellular reactive oxygen species without altering viability in neuronal cells. *Biology* **2017**, *55*, 648–654.
4. Schlotterer, A.; Kukudov, G.; Bozorgmehr, F.; Hutter, H.; Du, X.; Oikonomou, D.; Ibrahim, Y.; Pfisterer, F.; Rabbani, N.; Thornalley, P. *C. elegans* as model for the study of high glucose-mediated life span reduction. *Diabetes* **2009**, *58*, 2450–2456. [CrossRef]
5. Barden, A.; Anak, N.A.; Mulliken, T.; Song, M. Heart of the matter: Agarwood use and trade and CITES implementation for *Aquilaria malaccensis*. In *Traffic International*; Traffic: Cambridge, UK, 2000; pp. 1–52.
6. Persoon, G.A.; Beek, H. Growing 'the wood of the gods': Agarwood production in southeast Asia. In *Smallholder Tree Growing for Rural Development and Environmental Services*; Springer: Berlin/Heidelberg, Germany, 2008; pp. 245–262.
7. Hashim, Y.Z.H.-Y.; Kerr, P.G.; Abbas, P.; Salleh, H.M. *Aquilaria* spp.(agarwood) as source of health beneficial compounds: A review of traditional use, phytochemistry and pharmacology. *J. Ethnopharmacol.* **2016**, *189*, 331–360. [CrossRef]
8. Peng, D.-Q.; Yu, Z.-X.; Wang, C.-H.; Gong, B.; Liu, Y.-Y.; Wei, J.-H. Chemical Constituents and Anti-Inflammatory Effect of Incense Smoke from Agarwood Determined by GC-MS. *Int. J. Anal. Chem.* **2020**, *2020*, 4575030. [CrossRef]
9. Jermsri, P.; Jiraviriyakul, A.; Unajak, S.; Kumphune, S. Effect of *Aquilaria crassna* crude extract on simulated ischemia induced cardiac cell death. *Int. J. Pharm. Bio Sci.* **2012**, *3*, 604–613.

10. Jermsri, P.; Kumphune, S. Ethylacetate extract of *Aquilaria crassna* preserve actin cytoskeleton on simulated ischemia induced cardiac cell death. *J. Med. Plants Res.* **2012**, *6*, 4057–4062.
11. Kim, Y.; Lee, E.; Lee, Y.; Kim, H.; Song, B.; Lee, E.; Kim, H. Effect of the aqueous extract of *Aquilaria agallocha* stems on the immediate hypersensitivity reactions. *J. Ethnopharmacol.* **1997**, *58*, 31–38. [CrossRef]
12. Vakati, K.; Rahman, H.; Eswaraiah, M.C.; Dutta, A. Evaluation of hepatoprotective activity of ethanolic extract of *Aquilaria agallocha* leaves (EEAA) against CCl4 induced hepatic damage in rat. *Sch. J. App. Med. Sci.* **2013**, *1*, 9–12.
13. Kamonwannasit, S.; Nantapong, N.; Kumkrai, P.; Luecha, P.; Kupittayanant, S.; Chudapongse, N. Antibacterial activity of *Aquilaria crassna* leaf extract against *Staphylococcus epidermidis* by disruption of cell wall. *Ann. Clin. Microbiol. Antimicrob.* **2013**, *12*, 1–7. [CrossRef]
14. Dahham, S.S.; Ahamed, M.B.K.; Saghir, S.M.; Alsuede, F.S.; Iqbal, M.A.; Majid, A.M.S.A. Bioactive essential oils from *Aquilaria crassna* for cancer prevention and treatment. *Glob. J. Adv. Pure Appl. Sci.* **2014**, *4*, 26–31.
15. Supasuteekul, C.; Tadtong, S.; Putalun, W.; Tanaka, H.; Likhitwitayawuid, K.; Tengamnuay, P.; Sritularak, B. Neuritogenic and neuroprotective constituents from *Aquilaria crassna* leaves. *J. Food Biochem.* **2017**, *41*, e12365. [CrossRef]
16. Pranakhon, R.; Pannangpetch, P.; Aromdee, C. Antihyperglycemic activity of agarwood leaf extracts in STZ-induced diabetic rats and glucose uptake enhancement activity in rat adipocytes. *Songklanakarin J. Sci. Technol.* **2011**, *33*, 405–410.
17. Prasansuklab, A.; Meemon, K.; Sobhon, P.; Tencomnao, T. Ethanolic extract of *Streblus asper* leaves protects against glutamate-induced toxicity in HT22 hippocampal neuronal cells and extends lifespan of *Caenorhabditis elegans*. *BMC Complement. Altern. Med.* **2017**, *17*, 551. [CrossRef]
18. Rangsinth, P.; Prasansuklab, A.; Duangjan, C.; Gu, X.; Meemon, K.; Wink, M.; Tencomnao, T. Leaf extract of *Caesalpinia mimosoides* enhances oxidative stress resistance and prolongs lifespan in *Caenorhabditis elegans*. *BMC Complement. Altern. Med.* **2019**, *19*, 164. [CrossRef]
19. Pattarachotanant, N.; Prasansuklab, A.; Tencomnao, T. *Momordica charantia* L. Extract Protects Hippocampal Neuronal Cells against PAHs-Induced Neurotoxicity: Possible Active Constituents Include Stigmasterol and Vitamin E. *Nutrients* **2021**, *13*, 2368. [CrossRef]
20. Pattarachotanant, N.; Tencomnao, T. Citrus hystrix extracts protect human neuronal cells against high glucose-induced senescence. *Pharmaceuticals* **2020**, *13*, 283. [CrossRef]
21. Wong, K.-H.; Sabaratnam, V.; Naidu, M.; Keynes, R. Activity of aqueous extracts of lion's mane mushroom *Hericium erinaceus* (Bull.: Friday) Pers.(Aphyllophoromycetideae) on the neural cell line NG108-15. *Int. J. Med. Mushrooms* **2007**, *9*, 57–65. [CrossRef]
22. Nasu, R.; Furukawa, A.; Suzuki, K.; Takeuchi, M.; Koriyama, Y. The effect of glyceraldehyde-derived advanced glycation end products on β-tubulin-inhibited neurite outgrowth in sh-sy5y human neuroblastoma cells. *Nutrients* **2020**, *12*, 2958. [CrossRef]
23. Ren, M.; Feng, H.; Fu, Y.; Land, M.; Rubin, C.S. Protein kinase D is an essential regulator of *C. elegans* innate immunity. *Immunity* **2009**, *30*, 521–532. [CrossRef]
24. Duangjan, C.; Rangsinth, P.; Gu, X.; Wink, M.; Tencomnao, T. Lifespan extending and oxidative stress resistance properties of a leaf extracts from *Anacardium occidentale* L. in *Caenorhabditis elegans*. *Oxidative Med. Cell. Longev.* **2019**, *2019*, 9012396. [CrossRef] [PubMed]
25. Kao, W.-Y.; Hsiang, C.-Y.; Ho, S.-C.; Ho, T.-Y.; Lee, K.-T. Chemical profiles of incense smoke ingredients from agarwood by headspace gas chromatography-tandem mass spectrometry. *Molecules* **2018**, *23*, 2969. [CrossRef]
26. Yusof, S.; Tajuddin, S.N.; Mansor, R.; Cheng, P.W.; Ramli, A.N.M. Gas chromatography analysis of artificially inoculated agarwood compounds related to high quality agarwood from Malaysia plantation. *Chem. Adv. Mater.* **2018**, *3*, 60–66.
27. Rangsinth, P.; Duangjan, C.; Sillapachaiyaporn, C.; Isidoro, C.; Prasansuklab, A.; Tencomnao, T. *Caesalpinia mimosoides* Leaf Extract Promotes Neurite Outgrowth and Inhibits BACE1 Activity in Mutant APP-Overexpressing Neuronal Neuro2a Cells. *Pharmaceuticals* **2021**, *14*, 901. [CrossRef]
28. Tao, M.; Li, R.; Zhang, Z.; Wu, T.; Xu, T.; Zogona, D.; Huang, Y.; Pan, S.; Xu, X. Vitexin and Isovitexin Act Through Inhibition of Insulin Receptor to Promote Longevity and Fitness in *Caenorhabditis elegans*. *Mol. Nutr. Food Res.* **2022**, *66*, e2100845. [CrossRef] [PubMed]
29. Rangsinth, P.; Sillapachaiyaporn, C.; Nilkhet, S.; Tencomnao, T.; Ung, A.T.; Chuchawankul, S. Mushroom-derived bioactive compounds potentially serve as the inhibitors of SARS-CoV-2 main protease: An in silico approach. *J. Tradit. Complement. Med.* **2021**, *11*, 158–172. [CrossRef]
30. Lam, C.T.; Gong, A.G.; Lam, K.Y.; Zhang, L.M.; Chen, J.-P.; Dong, T.T.; Lin, H.-Q.; Tsim, K.W. Jujube-containing herbal decoctions induce neuronal differentiation and the expression of anti-oxidant enzymes in cultured PC12 cells. *J. Ethnopharmacol.* **2016**, *188*, 275–283. [CrossRef]
31. Suzuki, N.; Numakawa, T.; Chou, J.; de Vega, S.; Mizuniwa, C.; Sekimoto, K.; Adachi, N.; Kunugi, H.; Arikawa-Hirasawa, E.; Yamada, Y. Teneurin-4 promotes cellular protrusion formation and neurite outgrowth through focal adhesion kinase signaling. *FASEB J.* **2014**, *28*, 1386–1397. [CrossRef]
32. Zhang, S.; Duangjan, C.; Tencomnao, T.; Liu, J.; Lin, J.; Wink, M. Neuroprotective effects of oolong tea extracts against glutamate-induced toxicity in cultured neuronal cells and β-amyloid-induced toxicity in *Caenorhabditis elegans*. *Food Funct.* **2020**, *11*, 8179–8192. [CrossRef]
33. Shimizu, M.; Deguchi, A.; Hara, Y.; Moriwaki, H.; Weinstein, I.B. EGCG inhibits activation of the insulin-like growth factor-1 receptor in human colon cancer cells. *Biochem. Biophys. Res. Commun.* **2005**, *334*, 947–953. [CrossRef]

34. Arablou, T.; Delbandi, A.A.; Khodaverdi, S.; Arefi, S.; Kolahdouz-Mohammadi, R.; Heidari, S.; Mohammadi, T.; Aryaeian, N. Resveratrol reduces the expression of insulin-like growth factor-1 and hepatocyte growth factor in stromal cells of women with endometriosis compared with nonendometriotic women. *Phytother. Res.* **2019**, *33*, 1044–1054. [CrossRef]
35. Hioki, T.; Kawabata, T.; Sakai, G.; Fujita, K.; Kuroyanagi, G.; Matsushima-Nishiwaki, R.; Kim, W.; Otsuka, T.; Iida, H.; Tokuda, H. Resveratrol suppresses insulin-like growth factor I-induced osteoblast migration: Attenuation of the p44/p42 MAP kinase pathway. *Biosci. Biotechnol. Biochem.* **2020**, *84*, 2428–2439. [CrossRef]
36. Vanamala, J.; Reddivari, L.; Radhakrishnan, S.; Tarver, C. Resveratrol suppresses IGF-1 induced human colon cancer cell proliferation and elevates apoptosis via suppression of IGF-1R/Wnt and activation of p53 signaling pathways. *BMC Cancer* **2010**, *10*, 238. [CrossRef]
37. Shindo, H.; Thomas, T.P.; Larkin, D.D.; Karihaloo, A.K.; Inada, H.; Onaya, T.; Stevens, M.J.; Greene, D.A. Modulation of basal nitric oxide-dependent cyclic-GMP production by ambient glucose, myo-inositol, and protein kinase C in SH-SY5Y human neuroblastoma cells. *J. Clin. Investig.* **1996**, *97*, 736–745. [CrossRef]
38. Takano, T.; Funahashi, Y.; Kaibuchi, K. Neuronal polarity: Positive and negative feedback signals. *Front. Cell Dev. Biol.* **2019**, *7*, 69. [CrossRef] [PubMed]
39. Yamamoto, H.; Demura, T.; Morita, M.; Banker, G.A.; Tanii, T.; Nakamura, S. Differential neurite outgrowth is required for axon specification by cultured hippocampal neurons. *J. Neurochem.* **2012**, *123*, 904–910. [CrossRef]
40. Benowitz, L.I.; Routtenberg, A. GAP-43: An intrinsic determinant of neuronal development and plasticity. *Trends Neurosci.* **1997**, *20*, 84–91. [CrossRef]
41. Lee, S.-H.; Lee, J.-H.; Lee, H.-Y.; Min, K.-J. Sirtuin signaling in cellular senescence and aging. *BMB Rep.* **2019**, *52*, 24. [CrossRef]
42. Lamichane, S.; Baek, S.H.; Kim, Y.-J.; Park, J.H.; Dahal Lamichane, B.; Jang, W.B.; Ji, S.; Lee, N.K.; Dehua, L.; Kim, D.Y. MHY2233 attenuates replicative cellular senescence in human endothelial progenitor cells via SIRT1 signaling. *Oxidative Med. Cell. Longev.* **2019**, *2019*, 6492029. [CrossRef] [PubMed]
43. Wang, Y.; Liang, Y.; Vanhoutte, P.M. SIRT1 and AMPK in regulating mammalian senescence: A critical review and a working model. *FEBS Lett.* **2011**, *585*, 986–994. [CrossRef] [PubMed]
44. Jia, Y.; Zheng, Z.; Wang, Y.; Zhou, Q.; Cai, W.; Jia, W.; Yang, L.; Dong, M.; Zhu, X.; Su, L. SIRT1 is a regulator in high glucose-induced inflammatory response in RAW264. 7 cells. *PLoS ONE* **2015**, *10*, e0120849.
45. Oberdoerffer, P.; Michan, S.; McVay, M.; Mostoslavsky, R.; Vann, J.; Park, S.-K.; Hartlerode, A.; Stegmuller, J.; Hafner, A.; Loerch, P. SIRT1 redistribution on chromatin promotes genomic stability but alters gene expression during aging. *Cell* **2008**, *135*, 907–918. [CrossRef] [PubMed]
46. Singh, C.K.; Chhabra, G.; Ndiaye, M.A.; Garcia-Peterson, L.M.; Mack, N.J.; Ahmad, N. The role of sirtuins in antioxidant and redox signaling. *Antioxid. Redox Signal.* **2018**, *28*, 643–661. [CrossRef]
47. Kimura, K.D.; Tissenbaum, H.A.; Liu, Y.; Ruvkun, G. daf-2, an insulin receptor-like gene that regulates longevity and diapause in *Caenorhabditis elegans*. *Science* **1997**, *277*, 942–946. [CrossRef]
48. Mendler, M.; Schlotterer, A.; Morcos, M.; Nawroth, P. Understanding diabetic polyneuropathy and longevity: What can we learn from the nematode *Caenorhabditis elegans*? *Exp. Clin. Endocrinol. Diabetes* **2012**, *120*, 182–183. [CrossRef]
49. Prasanth, M.I.; Brimson, J.M.; Chuchawankul, S.; Sukprasansap, M.; Tencomnao, T. Antiaging, stress resistance, and neuroprotective efficacies of *Cleistocalyx nervosum* var. *paniala* fruit extracts using *Caenorhabditis elegans* model. *Oxidative Med. Cell. Longev.* **2019**, *2019*, 7024785. [CrossRef]
50. Katic, M.; Kahn, C. The role of insulin and IGF-1 signaling in longevity. *Cell. Mol. Life Sci. CMLS* **2005**, *62*, 320–343. [CrossRef]
51. Salih, D.A.; Brunet, A. FoxO transcription factors in the maintenance of cellular homeostasis during aging. *Curr. Opin. Cell Biol.* **2008**, *20*, 126–136. [CrossRef]
52. Jensen, V.L.; Gallo, M.; Riddle, D.L. Targets of DAF-16 involved in *Caenorhabditis elegans* adult longevity and dauer formation. *Exp. Gerontol.* **2006**, *41*, 922–927. [CrossRef]
53. Murphy, C.T.; McCarroll, S.A.; Bargmann, C.I.; Fraser, A.; Kamath, R.S.; Ahringer, J.; Li, H.; Kenyon, C. Genes that act downstream of DAF-16 to influence the lifespan of *Caenorhabditis elegans*. *Nature* **2003**, *424*, 277–283. [CrossRef]
54. Lee, S.-J.; Murphy, C.T.; Kenyon, C. Glucose shortens the life span of *C. elegans* by downregulating DAF-16/FOXO activity and aquaporin gene expression. *Cell Metab.* **2009**, *10*, 379–391. [CrossRef]
55. Barbieri, M.; Bonafè, M.; Franceschi, C.; Paolisso, G. Insulin/IGF-I-signaling pathway: An evolutionarily conserved mechanism of longevity from yeast to humans. *Am. J. Physiol. Endocrinol. Metab.* **2003**, *285*, E1064–E1071. [CrossRef]
56. Kenyon, C.; Chang, J.; Gensch, E.; Rudner, A.; Tabtiang, R. A *C. elegans* mutant that lives twice as long as wild type. *Nature* **1993**, *366*, 461–464. [CrossRef]
57. Mao, K.; Quipildor, G.F.; Tabrizian, T.; Novaj, A.; Guan, F.; Walters, R.O.; Delahaye, F.; Hubbard, G.B.; Ikeno, Y.; Ejima, K. Late-life targeting of the IGF-1 receptor improves healthspan and lifespan in female mice. *Nat. Commun.* **2018**, *9*, 1–12. [CrossRef]
58. Moreau, R.A.; Nyström, L.; Whitaker, B.D.; Winkler-Moser, J.K.; Baer, D.J.; Gebauer, S.K.; Hicks, K.B. Phytosterols and their derivatives: Structural diversity, distribution, metabolism, analysis, and health-promoting uses. *Prog. Lipid Res.* **2018**, *70*, 35–61. [CrossRef]
59. Pratiwi, R.; Nantasenamat, C.; Ruankham, W.; Suwanjang, W.; Prachayasittikul, V.; Prachayasittikul, S.; Phopin, K. Mechanisms and neuroprotective activities of stigmasterol against oxidative stress-induced neuronal cell death via sirtuin family. *Front. Nutr.* **2021**, *8*, 648995. [CrossRef]

60. Sun, J.; Li, X.; Liu, J.; Pan, X.; Zhao, Q. Stigmasterol exerts neuro-protective effect against ischemic/reperfusion injury through reduction of oxidative stress and inactivation of autophagy. *Neuropsychiatr. Dis. Treat.* **2019**, *15*, 2991. [CrossRef]
61. Burg, V.K.; Grimm, H.S.; Rothhaar, T.L.; Grösgen, S.; Hundsdörfer, B.; Haupenthal, V.J.; Zimmer, V.C.; Mett, J.; Weingärtner, O.; Laufs, U. Plant sterols the better cholesterol in Alzheimer's disease? A mechanistical study. *J. Neurosci.* **2013**, *33*, 16072–16087. [CrossRef]
62. Sun, Y.; Gao, L.; Hou, W.; Wu, J. β-Sitosterol alleviates inflammatory response via inhibiting the activation of ERK/p38 and NF-κB pathways in LPS-exposed BV2 cells. *BioMed Res. Int.* **2020**, *2020*, 7532306. [CrossRef]
63. Reddy, P.H. Role of mitochondria in neurodegenerative diseases: Mitochondria as a therapeutic target in Alzheimer's disease. *CNS Spectr.* **2009**, *14.S7*, 8–13. [CrossRef] [PubMed]
64. Bari, W.U.; Zahoor, M.; Zeb, A.; Khan, I.; Nazir, Y.; Khan, A.; Rehman, N.U.; Ullah, R.; Shahat, A.A.; Mahmood, H.M. Anticholinesterase, antioxidant potentials, and molecular docking studies of isolated bioactive compounds from *Grewia optiva*. *Int. J. Food Prop.* **2019**, *22*, 1386–1396. [CrossRef]
65. Koga, T.; Sakamoto, T.; Sakuradani, E.; Tai, A. Neurite Outgrowth-Promoting Activity of Compounds in PC12 Cells from Sunflower Seeds. *Molecules* **2020**, *25*, 4748. [CrossRef] [PubMed]

*nutrients*

MDPI

*Communication*

# *Pandanus amaryllifolius* Exhibits In Vitro Anti-Amyloidogenic Activity and Promotes Neuroprotective Effects in Amyloid-β-Induced SH-SY5Y Cells

**Mario A. Tan [1,*], Hayato Ishikawa [2] and Seong Soo A. An [3,*]**

1 College of Science and Research Center for the Natural and Applied Sciences, University of Santo Tomas, Manila 1015, Philippines
2 Graduate School of Pharmaceutical Sciences, Chiba University, 1-8-1 Inohana, Chuo-ku, Chiba 260-8675, Japan
3 Department of Bionano Technology, Bionano Research Institute, Gachon University, Seongnam-si 1342, Gyeonggi-do, Korea
* Correspondence: matan@ust.edu.ph (M.A.T.); seongaan@gachon.ac.kr (S.S.A.A.)

**Abstract:** Accumulation of amyloid-*beta* (Aβ) plaques leading to oxidative stress, mitochondrial damage, and cell death is one of the most accepted pathological hallmarks of Alzheimer's disease (AD). *Pandanus amaryllifolius*, commonly recognized as fragrant screw pine due to its characteristic smell, is widely distributed in Southeast Asia and is consumed as a food flavor. In search for potential anti-AD agents from terrestrial sources, *P. amaryllifolius* was explored for its in vitro anti-amyloidogenic and neuroprotective effects. Thioflavin T (ThT) assay and the high-throughput screening multimer detection system (MDS-HTS) assay were used to evaluate the extracts' potential to inhibit Aβ aggregations and oligomerizations, respectively. The crude alcoholic extract (CAE, 50 μg/mL) and crude base extract (CBE, 50 μg/mL) obstructed the Aβ aggregation. Interestingly, results revealed that only CBE inhibited the Aβ nucleation at 100 μg/mL. Both CAE and CBE also restored the cell viability, reduced the level of reactive oxygen species, and reversed the mitochondrial dysfunctions at 10 and 20 μg/mL extract concentrations in Aβ-insulted SY-SY5Y cells. In addition, the unprecedented isolation of nicotinamide from *P. amaryllifolius* CBE is a remarkable discovery as one of its potential bioactive constituents against AD. Hence, our results provided new insights into the promising potential of *P. amaryllifolius* extracts against AD and further exploration of other prospective bioactive constituents.

**Keywords:** Alzheimer's disease; amyloid-beta; neuroprotection; nicotinamide; *Pandanus amaryllifolius*; Thioflavin T; high-throughput screening multimer detection system (MDS-HTS) assay

**Citation:** Tan, M.A.; Ishikawa, H.; An, S.S.A. *Pandanus amaryllifolius* Exhibits In Vitro Anti-Amyloidogenic Activity and Promotes Neuroprotective Effects in Amyloid-β-Induced SH-SY5Y Cells. *Nutrients* **2022**, *14*, 3962. https://doi.org/10.3390/nu14193962

Academic Editors: Daniela Rigano, Justyna Godos and Paola Bontempo

Received: 12 August 2022
Accepted: 22 September 2022
Published: 24 September 2022



## 1. Introduction

Alzheimer's disease (AD) is a progressive neurodegenerative disorder commonly depicted by memory loss, mental dysfunctions, cognitive and learning impairments, and emotional instability [1]. With global prevalence as the seventh cause of death in 2020 and 2021, it is estimated that 44 million people worldwide are affected by AD [2]. With a complex plethora of etiologies, the excessive formation and deposition of amyloid-*beta* (Aβ) plaques leading to oxidative stress, mitochondrial dysfunctions, and eventual nerve cell death and tissue loss are the most accepted pathological hallmark of AD [3]. Currently, US FDA approved the acetylcholinesterase inhibitors donepezil galantamine, tacrine, and rivastigmine; the *N*-methyl-D-aspartate (NMDA) receptor antagonist memantine; and aducanumab, a monoclonal antibody targeting the aggregation of Aβ [4,5], to combat AD. However, these drugs only provide symptomatic relief of AD and come with adverse side effects [5,6]. Moreover, treatment of AD utilizing aducanumab would be very expensive at an annual price of USD 56,000 [7]. With the rapid growth of people worldwide with AD, the discovery of potential drugs and medicinal plants with minimal adverse effects, cost-effective, and may help to alleviate the progression of AD is warranted. Plants have been regarded as primary sources of diverse and pharmacologically important metabolites used in drug discovery research for the treatment of various diseases, including AD [8–11].

*Pandanus amaryllifolius*, commonly known as fragrant screw pine, is one of the 700 species of the Genus *Pandanus* widely distributed in tropical and sub-tropical environments such as the Southeast Asian countries, India, Taiwan, Papua New Guinea, and Sri Lanka [12]. It is utilized in Philippine traditional folk medicine to treat stomachache, high blood pressure, urinary tract infection, and kidney illness [13]. It is also used in cooking to impart color, flavor, and a distinct smell identified as 2-acetyl-1-pyrroline [14,15]. Phytochemical investigations on the leaf extract elaborated the identification of diverse alkaloid structures [16] and flavonoid and phenolic compounds [17,18].

In our continuing search for Philippine medicinal plants and their constituents with inhibitory effects against Aβ toxicity in vitro [19–23], we herein describe the neuroprotective effects and inhibition of Aβ aggregations and oligomerizations of *P. amaryllifolius* crude alcoholic (CAE) and crude base (CBE) extracts. The unprecedented isolation of nicotinamide from *P. amaryllifolius*, as a potential bioactive constituent of the crude base extract, is also reported.

## 2. Materials and Methods

### 2.1. Plant Material

Fresh, matured *P. amaryllifolius* leaves were collected in Santa Maria, Bulacan, Philippines, in April 2017. Leaves were authenticated at the Botany Division, National Museum of the Philippines (Control #17-04-533). Voucher specimens were deposited at the UST Herbarium, Research Center for Natural and Applied Sciences (USTH-3728).

### 2.2. Extraction of the Crude Extracts

Air-dried, ground *P. amaryllifolius* leaves (2 kg) were extracted with distilled MeOH five times and filtered. The combined filtrates were concentrated under reduced pressure yielding the crude alcoholic extract (CAE, 178 g). A portion of the CAE (170 g) was dissolved in 1 M HCl and partitioned with EtOAc thrice. The aqueous layer was basified to pH 9 using $Na_2CO_3$. The basified aqueous layer was extracted with 5% MeOH in $CHCl_3$ five times. The combined organic layer was dried with anhydrous $Na_2SO_4$ and concentrated in vacuo yielding the crude base extract (CBE, 3.3 g).

### 2.3. Thioflavin T (ThT) Assay

ThT assay was used to evaluate the inhibition of Aβ aggregation of the *P. amaryllifolius* extracts and phenol red (positive control) as previously described [21–23]. Briefly, Aβ42 (Aggresure™ AnaSpec, Fremont, CA, USA) in PBS was mixed with or without the *P. amaryllifolius* extracts or phenol red for 24 h at 37 °C. After the addition of ThT solution, the mixture was incubated for 15 min, and the fluorescence signal (Ex 450 nm; Em 510 nm) was measured. The percentage inhibition was calculated using the equation: $[(1 - I_{Fi}/I_{Fc}) \times 100\%]$, where $I_{Fi}$ (with inhibitor) and $I_{Fc}$ (without inhibitor) are the fluorescence signals after subtracting the background signal of the ThT solution.

### 2.4. High-Throughput Screening of Multimer Detection System (MDS-HTS) Assay

The ability of *P. amaryllifolius* extracts to inhibit the Aβ oligomerization was evaluated by the MDS-HTS assay (PeopleBio Inc., Sungnamsi, Gyeonggido, Korea), as previously described [24]. $A\beta_{42}$ (200 μg/mL), the crude extracts (100 μg/mL), and BPL-1 (100 μM, positive control) were dissolved in PBS and were utilized, following an enzyme-linked immunosorbent assay method [24–26]. Briefly, the 1000-times-diluted mixture of $A\beta_{42}$ and the extracts in PBST (100 μL) were added to the antibody-coated wells of a 96-well microtiter plate and incubated for 1 h at RT. Detection antibody with conjugated horseradish peroxidase (HRP, 100 μL) was employed to spot the Aβ oligomer attached to the capture antibody. After washing with washing buffer, a solution of 3,3′,5,5′-tetramethylbenzidine

(100 µL) was added and incubated at RT for 30 min. A stop solution (50 µL) was finally added, and the optical density was measured after 24 h. The significant difference ($p < 0.05$) against the negative control was determined.

### 2.5. Cell Culture

Neuroblastoma SH-SY5Y cells (ATCC CRL-2266) were obtained from the American Type Culture Collection (Manasas, VA, USA) and maintained in DMEM supplemented with 10% FBS and 1% kanamycin/penicillin at 37 °C and 5% $CO_2$. Cells at 80–90% confluency were used in the experiments.

### 2.6. Cell Cytotoxicity and Neuroprotection Assay

SH-SY5Y cells ($1 \times 10^4$ cells/well) were plated in 96-well plate and acclimatized for 24 h. Then, the cells were treated with 50, 20, 10, and 1 µg/mL *P. amaryllifolius* extracts for 24 h. After treatment, cells were washed with PBS, followed by the addition of 100 µL fresh media and incubation for 30 min. The cell viability was assessed following the ATP Luminescence (CellTiter-Glo®) (Promega, Madison, WI, USA) method as previously described [22,23]. The cell viability was expressed as % of the control cells (untreated).

For the neuroprotective experiment, SH-SY5Y cells ($1 \times 10^4$ cells/well) were subcultured in 96-well plate and incubated for 24 h. After incubation, cells were then treated with the 20, 10, and 1 µg/mL *P. amaryllifolius* extracts for 6 h. This was followed by Aβ (10 µM) treatment for 24 h. The % cell viability was determined in triplicate experiments using the ATP Luminescence (CellTiter-Glo®) method as previously described [22,23].

### 2.7. Determination of Intracellular Reactive Oxygen Species (ROS)

After 24 h incubation, SH-SY5Y ($1 \times 10^4$ cells/well) cells were pre-treated with 20, 10, and 1 µg/mL *P. amaryllifolius* extracts for 6 h and continued with 10 µM Aβ for 24 h. After treatment, cells were then incubated with 25 µM $H_2$DCFDA (Sigma Aldrich, St. Louis, MO, USA) for 2 h in 37 °C oven. Fluorescence intensity (Ex 495 nm, Em 520 nm) was measured, and the ROS level was calculated as a percentage of the control cells in triplicate experiments.

### 2.8. Mitochondrial Membrane Potential (ΔΨm) Assay

Following 24 h incubation, SH-SY5Y ($1 \times 10^4$ cells/well) cells were pre-treated with 20, 10, and 1 µg/mL *P. amaryllifolius* extracts for 6 h and incubated with 10 µM Aβ for 24 h. The process was followed by the addition of 1 µM TMRE staining solution (Abcam TMRE mitochondrial membrane kit) and incubation at 37 °C for 1 h. The fluorescence (Ex 549 nm, Em 575 nm) was measured, and the ΔΨm was calculated as a percentage of the untreated control (untreated) cells in triplicate experiments.

### 2.9. Isolation of Nicotinamide (1)

CBE (3 g) was initially separated by silica gel flash column chromatography using $CHCl_3$/MeOH mixtures in increasing polarity. The collected fractions were subjected to thin layer chromatography to obtain 4 pooled fractions, CBE1–CBE4. CBE 3 was partitioned by amino silica gel column chromatography using increasing polarity of EtOAc in hexane, neat EtOAC, 1:1 EtOAc/MeOH, and neat MeOH. The fraction eluting in neat EtOAc (10 mg) was subjected to MPLC using 1:1 $CHCl_3$/acetone as eluent to obtain nicotinamide as white crystals (1.8 mg). The schematic diagram of the isolation and identification of nicotinamide from *P. amaryllifolius* is presented in Figure 1. The $^1$H NMR, $^{13}$C NMR, and LCMS spectra of nicotinamide are found in the Supplementary Material.

Nicotinamide: white crystals; $^1$H NMR ($CDCl_3$, 400 MHz) δ 9.02 (1H, d, $J$ = 2.0 Hz), 8.77 (1H, dd, $J$ = 5.0, 1.8 Hz), 8.18 (1H, dt, $J$ = 8.0, 2.0 Hz), 7.43 (1H, dd, $J$ = 8.0, 5.0 Hz); $^{13}$C NMR ($CDCl_3$, 150 MHz) δ 167.1 (C=O), 152.8 (CH), 148.2 (CH), 135.5 (CH), 129.6 (C), 123.6 (CH).

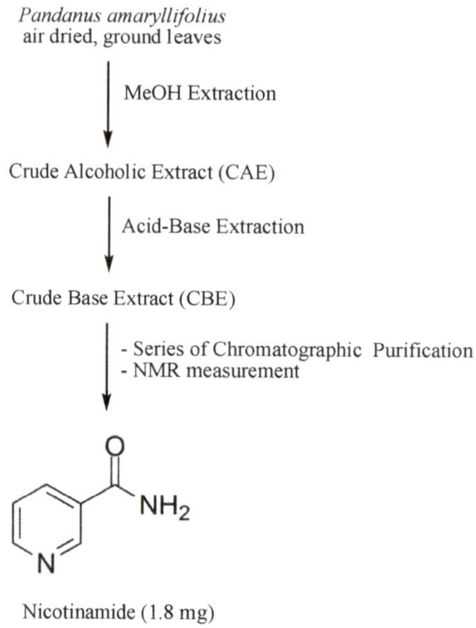


**Figure 1.** Schematic diagram on the isolation and identification of Nicotinamide.

### 2.10. General Considerations

$^{1}$H-NMR was recorded on an ECZ 400 FT-NMR spectrometer, and $^{13}$C-NMR spectra were recorded on an ECZ 600 FT-NMR spectrometer using $CDCl_3$ as solvent and TMS as internal standard. Thin layer chromatography was performed using Merck 60 F254 precoated silica gel plates (0.25 mm thickness). $UV_{254}$, followed by Dragendorff reagent, was used for visualization. Medium-pressure liquid chromatography (MPLC) was accomplished using a silica gel prepacked column CPS-HS-221-05. PerkinElmer Victor-3® multi-plate reader was used to measure the luminescence, fluorescence, or optical density in the biological assays.

### 2.11. Statistical Analysis

The quantitative data were reported as mean ± SD of at least three experiments. Statistical analysis was determined by one-way ANOVA (GraphPad Prism 5 software package, version 5.02, GraphPad Software Inc., San Diego, CA, USA), with the statistical significance considered at $p < 0.05$.

## 3. Results

### 3.1. Inhibitory Effect on the Aggregation and Oligomerization of Amyloid-Beta

The anti-amyloidogenic potential of *P. amaryllifolius* CAE and CBE fractions were elucidated using the ThT and MDS-HTS assays (Table 1). ThT binds to the Aβ aggregates resulting in an increase in the fluorescence signal. With the presence of potential inhibitors, the formation of Aβ fibrils and aggregates is prevented, thus, resulting in a decrease in the fluorescence signal [27]. As shown in Table 1, both extracts inhibited the Aβ aggregation. However, CBE revealed a stronger inhibitory effect at 89.53% in comparison to the positive control (70.06%) and CAE (73.67%). These results were also manifested in the MDS-HTS assay for screening Aβ breakers, which was an ELISA-based technique utilizing a capturing antibody against Aβ and a detecting antibody (horseradish peroxidase antibody, HRP), which allowed the selective detection of Aβ oligomers [26,27]. The inhibitors would break up the Aβ aggregates and reduce the concentrations of the captured oligomers, thus, resulting in a reduction of signal and preventing Aβ oligomerization [24–26]. The alkaloidal extract (CBE) showed a significant reduction in Aβ oligomer formation against the negative control. On the contrary, a significant difference was also observed with the

CAE against the negative control but may further promote oligomerization as indicated in the increase in optical density.

**Table 1.** Measurement of the Inhibition of Amyloid-β Aggregation using the Thioflavin T (ThT) Assay and the Inhibition of Amyloid-β Oligomerization using the Multimer Detection System (MDS) Assay.

| | ThT Assay [a] (% Inhibition) [c] | MDS Assay [b] (Optical Density) [c] |
|---|---|---|
| Crude Alcoholic Extract | 73.67 ± 3.54 | 2.559 ± 0.19 ^ |
| Crude Base Extract | 89.53 ± 5.21 * | 2.027 ± 0.080 ^ |
| Phenol Red (Positive Control) | 70.06 ± 2.87 | |
| BPL-1 (Positive Control) | | 1.509 ± 0.18 ^ |
| Negative Control | | 2.203 ± 0.068 |

[a] The extracts and positive control were measured at 50 μg/mL and 50 μM, respectively. [b] The extracts and positive control were measured at 100 μg/mL and 100 μM, respectively. [c] The % inhibition and optical density are expressed as mean ± SD of triplicate experiments. * Significant difference at $p < 0.05$ versus the positive control (phenol red). ^ Significant difference at $p < 0.05$ versus the negative control.

### 3.2. Cell Cytotoxicity and Neuroprotective effects of *P. amaryllifolius* Extracts

Prior to the neuroprotective experiments, SH-SY5Y cell viability of the crude extracts at 1, 10, 20, and 50 μg/mL concentrations were monitored (Figure 2). Results indicated that >95% cell viability was exhibited when the cells were treated with 1–20 μg/mL extract concentrations. However, 75.56 ± 2.97% for CAE and 80.64 ± 1.78% for cell viability were observed at 50 μg/mL. These cell viabilities were significantly different from the control (untreated) cells at $p < 0.05$. Hence, succeeding experiments utilized 1, 10, and 20 μg/mL concentrations.

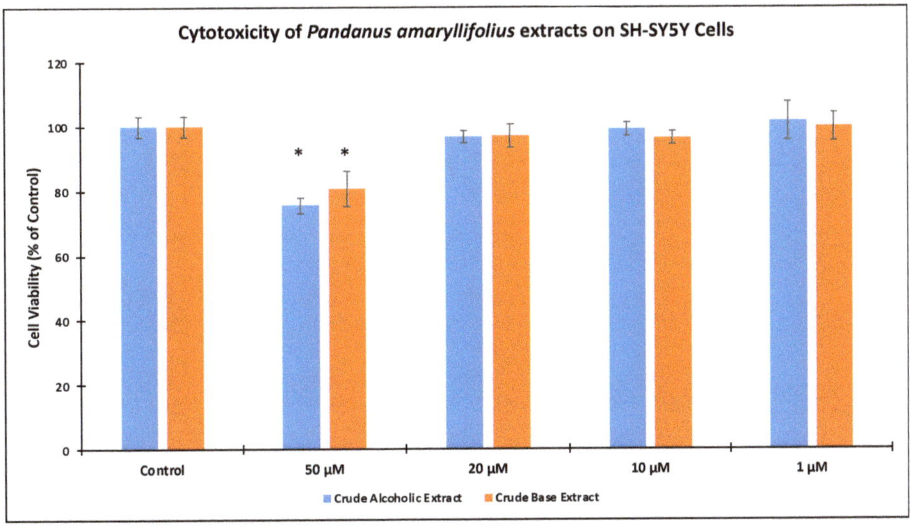


**Figure 2.** Cell cytotoxicity of *Pandanus amaryllifolius* Crude Alcoholic Extract (CAE) and Crude Base Extract (CBE). SH-SY5Y cells were treated with 1–50 μg/mL extract concentration and the cell viability were evaluated by ATP Luminescence assay. "Control" refers to untreated cells. The (*) showed significant difference with the control cells at $p < 0.05$.

In the Aβ-insulted SH-SY5Y cytotoxicity assay (Figure 3), 49.05 ± 2.86% cell viability was observed following the treatment of 10 μM Aβ. Pre-treatment of the extracts at 1–20 μg/mL followed by 10 μM Aβ significantly ($p < 0.05$) restored the cell viability. Both extracts at 20 μg/mL showed the highest % cell viabilities at 86.02 ± 4.76% (CAE) and 80.65 ± 4.81% (CBE).

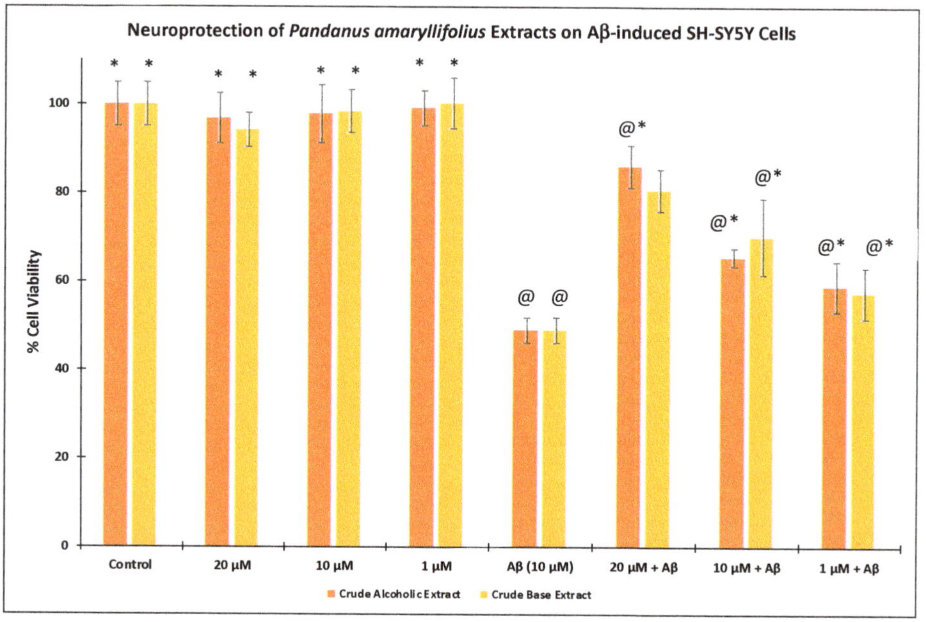


**Figure 3.** Effect of *Pandanus amaryllifolius* extracts on neuroprotective activity assay. Neuroblastoma SH-SY5Y cells were treated with the extracts for 6 h, followed by Aβ (10 μM) for 24 h. The % cell viability represents mean ± SD (triplicate experiments) and computed based on the control (untreated, 100%). @—statistically significant ($p < 0.05$) vs. control; *—statistically significant ($p < 0.05$) vs. Aβ-treated cells.

### 3.3. *Level of Intracellular Reactive Oxygen Species (ROS)*

The amount of intracellular ROS production when SH-SY5Y cells were induced with 10 μM Aβ was determined using 2′,7′-dichlorodihydrofluorescein diacetate ($H_2$DCFDA) reagent. As shown in Figure 4, the level of ROS was comparable with the control when the cells were treated with only the crude extracts (blue bars). Upon incubation only with 10 μM Aβ, 130.52 ± 2.63% ROS level was generated. Upon pre-treating the SH-SY5Y cells with the crude extracts for 6 h at 10 and 20 μg/mL, a significant reduction ($p < 0.05$) in the ROS level was observed when compared to the SH-SY5Y cells treated only with Aβ. CAE exhibited 120.31 ± 3.77% (20 μg/mL) and 123.56 ± 2.04% (10 μg/mL) reductions, while CBE showed 122.63 ± 3.06% (20 μg/mL) and 124.52 ± 1.05% (10 μg/mL) ROS levels. Both extracts did not efficiently lower the ROS level at 1 μg/mL.

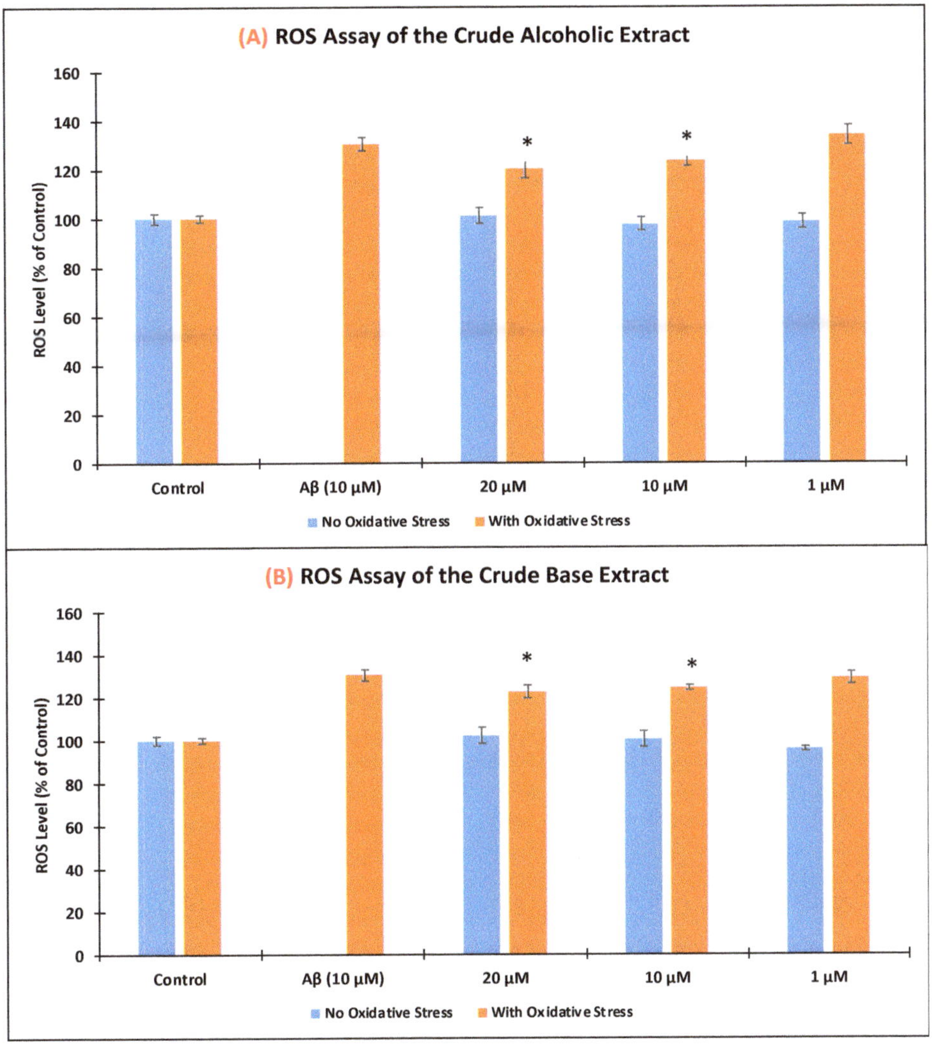


**Figure 4.** Effects of *Pandanus amaryllifolius* (**A**) crude alcoholic extract (**A**,**B**) crude base extract on intracellular ROS accumulation using the 2′,7′-dichlorodihydrofluorescein diacetate ($H_2$DCFDA) reagent. Neuroblastoma SH-SY5Y cells were incubated with the extracts for 6 h, followed by 10 μM Aβ for 24 h. "No Oxidative Stress" indicates non-Aβ treated cells. Intracellular ROS level (% of the control cells) was computed as mean ± SD of triplicate experiments. The (*) represents statistical difference ($p < 0.05$) of the ROS level with the Aβ-treated alone cells.

### 3.4. Effect of the *P. amaryllifolius* Extracts on the Mitochondrial Membrane Potential (Δψm)

As a result of Aβ induction in the neuroblastoma cells, an increase in ROS level was observed, which may result in mitochondrial dysfunctions. Hence, the capability of *P. amaryllifolius* extracts to restore the MMP was disclosed in Figure 5. Upon treatment with 10 Aβ only, SH-SY5Y cells gave 64.62 ± 3.21% Δψm. Pre-treatment with the extracts for 6 h prior to incubation with Aβ significantly ($p < 0.05$) restored the Δψm at 20 μg/mL (86.54 ± 4.61% for CAE and 84.55 ± 2.08% for CBE) and 10 μg/mL (74.43 ± 3.77% for CAE and 78.31 ± 1.06% for CBE). None of the extracts gave a substantial Δψm result at 1 μg/mL.

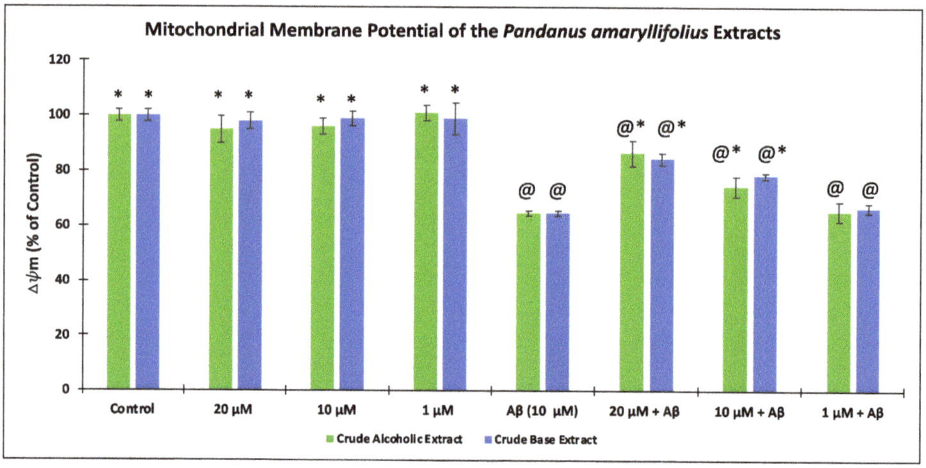


**Figure 5.** Assessment of the mitochondrial membrane potential (ΔΨm) of *Pandanus amaryllifolius* extracts using the tetramethylrhodamine methyl ester (TMRE) assay. SH-SY5Y cells were pre-treated with the extracts for 6 h, followed by Aβ (10 μM) for 24 h. The ΔΨm (% of the control cells) was expressed as the mean ± SD of triplicate measurements. @—statistically significant ($p < 0.05$) vs. control; *—statistically significant ($p < 0.05$) vs. Aβ-treated cells.

### *3.5. Isolation and Identification of Nicotinamide from the Crude Base Extract*

Fractionation of *P. amaryllifolius* CBE led to the isolation of white crystals elucidated as nicotinamide by spectroscopic analyses. This is the first report on the isolation of nicotinamide, a pyridine-containing alkaloid, from a *Pandanus* species. Nicotinamide was previously identified to influence a flower-inducing activity in *Lemna gibba* G3 [28] and was elaborated to be a bioactive repellent against the blue mussel *Mytilus edulis* in *Mallotus japonicus* [29]. Nicotinamide was also identified in functional food plants such as dried peas, red lentils, and chickpeas by the HPLC method using a standard sample [30].

Although nicotinamide was not subjected to bioassay experiments in this study, foregoing in vitro and in vivo studies, have described its potential effects against AD. Nicotinamide restored cognitive deficits in vivo using the 3x-Tg AD mice model by selectively reducing Thr231 in a Tau protein associated with microtubule depolymerization and increasing the acetylated α-tubulin [31]. In vitro cell models showed that nicotinamide prevented mitochondrial and autophagy dysfunctions and reduced neuronal susceptibility to Aβ toxicity and oxidative and metabolic inducers [32]. In vivo results also indicated that 3xTgAD mice treated with nicotinamide for 8 months showed an improved cognitive function and a reduction in the Aβ and hyperphosphorylated tau proteins in the hippocampus and cerebral cortex [32]. Nicotinamide-treated transgenic 5xFAD mice resulted in the attenuation of deficits in spine density derived from primary hippocampal neurons of the mice models [33]. Furthermore, nicotinamide is subjected (12 July 2017–30 August 2022) to a Phase 2 clinical trial to test its effects in adults with mild cognitive impairment or mild Alzheimer's disease [34,35].

## 4. Discussion

As part of our interest in exploring plant resources with anti-amyloidogenic activity against AD, we have investigated the crude alcoholic and crude alkaloidal extracts of *P. amaryllifolius*. *P. amaryllifolius* is an important medicinal plant in Philippine traditional folk medicine. Its ethnomedicinal uses were scientifically validated in pharmacological studies including anti-inflammatory [36], antioxidant [17], antiproliferative [17], hepatoprotective [37], antidiabetic [38], antimicrobial [39], and antiviral [40] properties. The anti-cholinesterase and β-secretase activities of a commercially available tea infusion of

*P. amaryllifolius* (Pandan herbal tea) from Thailand have been reported [41]. Thus far, the anti-amyloidogenic and neuroprotective effects on Aβ toxicity of *P. amaryllifolius* are still unknown. Recently, we described the neuroprotective activity of *P. clementis* [22] and the identification of phytosterols as possible bioactive agents [42]. *P. amaryllifolius* is also an excellent source of alkaloids, collectively known as *Pandanus* alkaloids, yet biological studies on these *Pandanus* alkaloids, including the crude alkaloids extract, are limited [16,39]. Hence, this study also contributes to the limited pharmacological information specifically focused on the alkaloidal fraction of *P. amaryllifolius*. Natural products, particularly alkaloids, have been the primary source of new scaffolds for the treatment of neurodegenerative diseases, including AD. Interestingly, the current drugs memantine, tacrine, rivastigmine, donepezil, and galantamine approved by the US FDA for AD are all alkaloids.

In the present work, we were able to disclose the alkaloidal extract of *P. amaryllifolius* to effectively inhibit the aggregation and oligomerization of Aβ. As a major contributor leading to AD, these Aβ aggregates lead to Aβ plaques causing various oxidative stressors, neuroinflammation, mitochondrial deterioration, and death. Hence, the neuroprotective effects of *P. amaryllifolius* extracts on Aβ-insulted neuronal cells were also assessed. As seen in Figure 3, *P. amaryllifolius* extracts rescued the cell damage caused by Aβ toxicity. This may be explained by reducing the levels of ROS and increasing the mitochondrial membrane potential. To further explore the possible bioactive scaffold, the alkaloid fraction was subjected to chromatographic purification. Unprecedentedly, we were able to isolate nicotinamide, a pyridine-type compound that possesses an array of in vitro and in vivo activities against AD [31–33]. Our results also corroborated the previous report [41] on *P. amaryllifolius* as a potential medicinal plant against AD. Since nicotinamide was isolated as a minor metabolite comprising only 0.06% of the crude alkaloid fraction, we may speculate that the *Pandanus* alkaloids may have synergistic inhibitory and neuroprotective effects and other potential bioactive *Pandanus* alkaloids are present in the extract. Hence, comprehensive isolation of known and possibly new *Pandanus* alkaloids is being undertaken to assess their in vitro activities against Aβ aggregation and toxicity. Moreover, inhibition of Aβ aggregation and oligomerization was evaluated using protein-based assays. From a future perspective, Aβ will transfect neuronal cells to exploit the anti-aggregation and anti-oligomerization of plant extracts and pure compounds in live cells. Furthermore, validation of the neuroprotective effects of the alkaloids on other neuronal cells and exploring other AD-related mechanistic assays are also warranted.

**Supplementary Materials:** The following supporting information can be downloaded at: https://www.mdpi.com/article/10.3390/nu14193962/s1, NMR and MS spectra of nicotinamide.

**Author Contributions:** Conceptualization—M.A.T. and S.S.A.A.; Methodology—M.A.T.; Analysis of results—M.A.T., H.I. and S.S.A.A.; Writing of manuscript—M.A.T.; Review and Editing—M.A.T., H.I. and S.S.A.A.; Project administration and funding acquisition—H.I. and S.S.A.A. All authors have read and agreed to the published version of the manuscript.

**Funding:** This study was supported by the Basic Science Research Program through the National Research Foundation of Korea (NRF), funded by the Ministry of Education (NRF-2020R1A2B5B01002463 and 2021R1A6A1A03038996) and the Japan Society for the Promotion of Science.

**Institutional Review Board Statement:** Not applicable.

**Informed Consent Statement:** Not applicable.

**Data Availability Statement:** Not applicable.

**Acknowledgments:** M.A.T. thanks the Japan Society for the Promotion of Science for the Invitational Research Fellowship at the Graduate School of Pharmaceutical Sciences, Chiba University. Crisfel del Mundo is gratefully acknowledged for the plant collection and crude alcoholic extraction. The authors are also grateful to Kyuhwan Shim and PeopleBio Inc. for the MDS-HTS assay.

**Conflicts of Interest:** The authors declare no conflict of interest.

## References


1. DeTure, M.A.; Dickson, D.W. The neuropathological diagnosis of Alzheimer's disease. *Mol. Neurodegener.* **2019**, *14*, 32. [CrossRef] [PubMed]
2. Alzheimer's Association Report. 2022 Alzheimer's disease facts and figures. *Alzheimers Dement.* **2022**, *18*, 700–789. [CrossRef] [PubMed]
3. Hampel, H.; Hardy, J.; Blennow, K.; Chen, C.; Perry, G.; Kim, S.H.; Villemagne, V.L.; Aisen, P.; Vendruscolo, M.; Iwatsubo, T.; et al. The amyloid-β pathway in Alzheimer's disease. *Mol. Psychiatry* **2021**, *26*, 5481–5503. [CrossRef] [PubMed]
4. Huang, J.-M.; Huang, F.-I.; Yang, C.-R. Moscatilin ameliorates tau phosphorylation and cognitive deficits in Alzheimer's disease models. *J. Nat. Prod* **2019**, *82*, 1979–1988. [CrossRef] [PubMed]
5. Woloshin, H.; Kesselheim, A.S. What to know about the Alzheimer drug Aducanumab (Aduhelm). *JAMA Intern. Med.* **2022**, *182*, 892. [CrossRef] [PubMed]
6. Ruangritchankul, S.; Chantharit, P.; Srisuma, S.; Gray, L.C. Adverse drug reactions of acetylcholinesterase inhibitors in older people living with dementia: A comprehensive literature review. *Ther. Clin. Risk Manag.* **2021**, *17*, 927–949. [CrossRef]
7. Sinha, P.; Barocas, J.A. Cost effectiveness of aducanumab to prevent Alzheimer's disease progression at current list price. *Alzheimers Dement.* **2022**, *8*, e12256. [CrossRef]
8. John, O.O.; Amarachi, I.S.; Chinazom, A.P.; Adaeze, E.; Kale, M.B.; Umare, M.D.; Upaganlawar, A.B. Phytotherapy: A promising approach for the treatment of Alzheimer's disease. *Pharmacol. Res. Mod. Chin. Med.* **2022**, *2*, 100030. [CrossRef]
9. Noori, T.; Dehpour, A.R.; Sureda, A.; Sobarzo-Sanchez, E.; Shirooie, S. Role of natural products for the treatment of Alzheimer's disease. *Eur. J. Pharmacol.* **2021**, *898*, 173974. [CrossRef]
10. Roy, A. Role of medicinal plants against Alzheimer's disease. *Int. J. Complement. Altern. Med.* **2018**, *11*, 205–208. [CrossRef]
11. Yadav, S. Potential role of medicinal plants against Alzheimer's disease. *Int. J. Complement. Altern. Med.* **2021**, *14*, 138–140. [CrossRef]
12. Wakte, K.V.; Nadaf, A.B.; Thengane, R.J.; Jawali, N. *Pandanus amaryllifolius* Roxb. cultivated as spice in coastal regions of India. *Genet. Resour. Crop Evol.* **2009**, *56*, 735–740. [CrossRef]
13. Ordas, J.A.D.; Nonato, M.G.; Moran, C.B. Ethnobotanical uses of Pandanaceae species in selected rural communities in the Philippines. *Econ. Bot.* **2021**, *74*, 411–428. [CrossRef]
14. Surojanametakul, V.; Boonbumrung, S.; Tungtrakul, P.; Varanyanoid, W.; Themtakul, K.; Yoshihashi, T. Encapsulation of natural flavor from *Pandanus amaryllifolius* Roxb. in rice starch aggregates. *Food Sci. Technol. Res.* **2019**, *25*, 577–585. [CrossRef]
15. Routray, W.; Rayaguru, K. Chemical constituents and post-harvest prospects of *Pandanus amaryllifolius* leaves: A review. *Food Rev. Int.* **2010**, *26*, 230–245. [CrossRef]
16. Tan, M.A.; Takayama, H. Recent progress in the chemistry of *Pandanus* alkaloids. In *The Alkaloids*; Knölker, H.J., Ed.; Academic Press: London, UK, 2019; Volume 82, pp. 1–28. [CrossRef]
17. Ghasemzadeh, A.; Jaafar, H.Z.E. Profiling of phenolic compounds and their antioxidant and anticancer activities in pandan (*Pandanus amaryllifolius* Roxb.) extracts from different locations of Malaysia. *BMC Complement. Altern. Med.* **2013**, *13*, 341. [CrossRef] [PubMed]
18. Suzuki, R.; Kan, S.; Sugita, Y.; Shirataki, Y. *p*-Coumaroyl malate derivatives of the *Pandanus amaryllifolius* leaf and their isomerization. *Chem. Pharm. Bull.* **2017**, *65*, 1191–1194. [CrossRef]
19. Tan, M.A.; An, S.S.A. Neuroprotective potential of the oxindole alkaloids isomitraphylline and mitraphylline in human neuroblastoma SH-SY5Y cells. *3 Biotech* **2020**, *10*, 517. [CrossRef]
20. Tan, M.A.; An, S.S.A. Potential therapeutic agents from Philippine medicinal plants against Alzheimer's disease. *Alzheimers Dement.* **2020**, *16*, e038909. [CrossRef]
21. Tan, M.A.; Gonzalez, S.J.B.; Alejandro, G.J.D.; An, S.S.A. Neuroprotective effects of vomifoliol, isolated from *Tarenna obtusifolia* Merr. (Rubiaceae), against amyoid-$beta_{1-42}$-treated neuroblastoma SH-SY5Y cells. *3 Biotech* **2020**, *10*, 424. [CrossRef]
22. Tan, M.A.; Tan, B.L.U.; Nonato, M.G.; An, S.S.A. Neuroprotective effects on amyloid-*beta* induced cytotoxicity of *Pandanus clementis* Merr. *3 Biotech* **2021**, *11*, 330. [CrossRef] [PubMed]
23. Tan, M.A.; Zacharova, E.; An, S.S.A. Diaportheone A analogues instigate a neuroprotective effect by protecting neuroblastoma SH-SH5Y cells from oxidative stress. *Biology* **2021**, *10*, 199. [CrossRef] [PubMed]
24. An, S.S.A.; Shim, K.H.; Kang, S.; Kim, Y.K.; Subedi, L.; Cho, H.; Hong, S.M.; Tan, M.A.; Jeon, R.; Chang, K.A.; et al. The potential anti-amyloidogenic candidate, SPA1413, for Alzheimer's disease. *Br. J. Pharmacol.* **2022**, *179*, 1033–1048. [CrossRef] [PubMed]
25. An, S.S.A.; Lee, B.-S.; Yu, J.S.; Lim, K.; Kim, G.J.; Lee, R.; Kim, S.; Kang, S.; Park, Y.H.; Wang, M.J.; et al. Dynamic changes of oligomeric amyloid β levels in plasma induced by spiked synthetic Aβ42. *Alzheimer's Res. Ther.* **2017**, *9*, 86. [CrossRef]
26. Wang, M.J.; Yi, S.; Han, J.-Y.; Park, S.Y.; Jang, J.-W.; Chun, I.K.; Kim, S.E.; Lee, B.S.; Kim, G.J.; Yu, J.S.; et al. Oligomeric forms of amyloid- β protein in plasma as a potential blood-based biomarker for Alzheimer's disease. *Alzheimer's Res. Ther.* **2017**, *9*, 98. [CrossRef] [PubMed]
27. Biancalana, M.; Koide, S. Molecular mechanism of Thioflavin-T binding to amyloid fibrils. *Biochim. Biophys. Acta* **2010**, *1804*, 1405–1412. [CrossRef]
28. Fujioka, S.; Sakurai, A.; Yamaguchi, I.; Murofushi, N.; Takahashi, N.; Kaihara, S.; Takimoto, A. Isolation and identification of L-pipecolic acid and nicotinamide as flower-inducing substances in *Lemna*. *Plant Cell Physiol.* **1987**, *28*, 995–1003. [CrossRef]

29. Yamashita, N.; Sakata, K.; Ina, H.; Ina, K. Isolation of nicotinamide from *Mallotus* leaves as an attaching repellent against the blue mussel, *Mytilus edulis*. *Agric. Biol. Chem.* **1989**, *53*, 3351–3352. [CrossRef]
30. Çatak, J. Determination of niacin profiles in some animal and plant based foods by high performance liquid chromatography: Association with healthy nutrition. *J. Anim. Sci. Technol.* **2019**, *61*, 138–146. [CrossRef]
31. Green, K.N.; Steffan, J.S.; Martinez-Coria, H.; Sun, X.; Schreiber, S.S.; Thompson, L.M.; LaFerla, F.M. Nicotinamide restores cognition in Alzheimer's disease transgenic mice via a mechanism involving sirtuin inhibition and selective reduction of Thr231-phosphotau. *J. Neurosci.* **2008**, *28*, 11500–11510. [CrossRef]
32. Liu, D.; Pitta, M.; Jiang, H.; Lee, J.H.; Zhang, G.; Chen, X.; Kawamoto, E.M.; Mattson, M.P. Nicotinamide forestalls pathology and cognitive decline in Alzheimer mice: Evidence for improved neuronal bioenergentics and autophagy procession. *Neurobiol. Aging* **2013**, *34*, 1564–1580. [CrossRef] [PubMed]
33. Kim, H.; Kim, B.; Kim, H.S.; Cho, J.Y. Nicotinamide attenuates the decrease in dendritic spine density in hippocampal primary neurons from 5xFAD mice, an Alzheimer's disease animal model. *Mol. Brain* **2020**, *13*, 17. [CrossRef] [PubMed]
34. Alzheimers.gov. Nicotinamide as an Early Alzheimer's Treatment (NEAT). 2022. Available online: https://www.alzheimers.gov/clinical-trials/nicotinamide-early-alzheimers-treatment-neat (accessed on 26 July 2022).
35. ClinicalTrials.gov. Nicotinamide as an Early Alzheimer's Disease Treatment (NEAT), ClinicalTrials.gov Identifier: NCT03061474. 2021. Available online: https://clinicaltrials.gov/ct2/show/record/NCT03061474 (accessed on 27 July 2022).
36. Reshidan, N.H.; Muid, S.A.; Mamikutty, N. The effects of *Pandanus amaryllifolius* (Roxb.) leaf water extracts on fructose-induced metabolic syndrome rat model. *BMC Complement. Med. Ther.* **2019**, *19*, 232. [CrossRef] [PubMed]
37. Thanebal, S.A.; Vun-Sang, S.; Iqbal, M. Hepatoprotective effects of *Pandanus amaryllifolius* against carbon tetrachloride ($CCl_4$) induced toxicity: A biochemical and hispathological study. *Arab. J. Chem.* **2021**, *14*, 103390. [CrossRef]
38. Chiabchalard, A.; Nooron, N. Antihyperglycemic effects of *Pandanus amaryllifolius* Roxb. leaf extract. *Pharmacogn. Mag.* **2015**, *11*, 117–122. [CrossRef] [PubMed]
39. Laluces, H.M.; Nakayama, A.; Nonato, M.G.; dela Cruz, T.E.; Tan, M.A. Antimicrobial alkaloids from the leaves of *Pandanus amaryllifolius*. *J. Appl. Pharm. Sci.* **2015**, *5*, 151–153. [CrossRef]
40. Ooi, L.; Sun, S.; Ooi, V. Purification and characterization of a new antiviral protein from the leaves of *Pandanus amaryllifolius* (Pandanaceae). *Int. J. Biochem. Cell Biol.* **2004**, *36*, 1440–1446. [CrossRef]
41. Suttisansanee, U.; Kunkeaw, T.; Thatsanasuwan, N.; Tonglim, J.; Temviriyanakul, P. The investigation on cholinesterase and BACE1 inhibitory activities in various tea infusions. *Walailak J. Sci. Technol.* **2019**, *16*, 165–174. [CrossRef]
42. Sharma, N.; Tan, M.A.; An, S.S.A. Phytosterols: Potential metabolic modulators in neurodegenerative diseases. *Int. J. Mol. Sci.* **2021**, *22*, 12255. [CrossRef]

*nutrients*

MDPI

*Article*

# Evaluation of the Anti-Shigellosis Activity of Dietary Isothiocyanates in *Galleria mellonella* Larvae

**Dariusz Nowicki, Klaudyna Krause, Monika Karczewska and Agnieszka Szalewska-Pałasz ***

Department of Bacterial Molecular Genetics, Faculty of Biology, University of Gdansk, Wita Stwosza 59, 80-308 Gdansk, Poland; dariusz.nowicki@ug.edu.pl (D.N.); klaudyna.krause@phdstud.ug.edu.pl (K.K.); monika.szalkowska@phdstud.ug.edu.pl (M.K.)
* Correspondence: agnieszka.szalewska-palasz@ug.edu.pl; Tel.: +48-58-523-6026

**Abstract:** Cruciferous vegetables, widely present in daily diets, are a rich source of organosulfur compounds with proven health benefits, especially chemopreventive or antioxidative effects. Isothiocyanate derivatives (ITCs) exhibit a broad spectrum of biological and pharmacological activity and recently, their antibacterial properties have been of particular importance. Here, we have focused on the anti-shigellosis activity of sulforaphane (SFN) and phenethyl ITC (PEITC). The genus *Shigella* causes gastroenteritis in humans, which constitutes a threat to public health. Production of a potent Stx toxin by *S. dysenteriae* type 1 results not only in more severe symptoms but also in serious sequela, including the hemolytic uremic syndrome. Here, we present evidence that two aliphatic and aromatic ITCs derivatives, SFN and PEITC, have an effective antibacterial potency against *S. dysenteriae*, also negatively regulating the *stx* gene expression. The molecular mechanism of this effect involves induction of the global stress-induced stringent response. ITCs also inhibit bacterial virulence against the Vero and HeLa cells. We present evidence for the therapeutic effect of sulforaphane and phenethyl ITC against a *S. dysenteriae* infection in the *Galleria mellonella* larvae model. Thus, our results indicate that isothiocyanates can be effectively used to combat dangerous bacterial infections.

**Keywords:** *Shigella dysenteriae*; isothiocyanates; sulforaphane; phenethyl isothiocyanate; stringent response; *Galleria mellonella*

**Citation:** Nowicki, D.; Krause, K.; Karczewska, M.; Szalewska-Pałasz, A. Evaluation of the Anti-Shigellosis Activity of Dietary Isothiocyanates in *Galleria mellonella* Larvae. *Nutrients* **2021**, *13*, 3967. https://doi.org/10.3390/nu13113967

Academic Editors: Daniela Rigano and Paola Bontempo

Received: 19 October 2021
Accepted: 5 November 2021
Published: 7 November 2021



## 1. Introduction

Natural products and their derivatives are an important source of bioactive compounds and represent more than one-third of all FDA-approved new molecules to date [1]. Among natural products, isothiocyanates (ITCs) are one of the major secondary metabolites that occur naturally and ubiquitously in plants. The dietary sources of ITCs are mainly vegetables, sprouts and seeds from the Brassicaceae family. These volatile substances are formed as a result of the substrate hydrolysis of glucosinolates, by the enzymatic activity of myrosinase in plant tissue for the protection against pathogens and animal-related damages [2]. Currently, it is well established that ITCs exhibit diverse biological activities, such as chemopreventive, anti-inflammatory and anticancer effects, and their role in combating bacterial infection has recently attracted increasing attention, making this class of molecules a potential source of new compounds for industry and medicine [3–7].

The *Shigella* genus is an etiological factor of bacterial diarrhea and dysentery worldwide, especially in the developing countries [8]. *Shigella* is considered a frequent bacterial causative agent in the diarrhoea of travellers visiting low-income areas. Apart from foreign travel, risk factors for shigellosis in industrialized countries include children who are <5 years or subjects in day-care centres. A dose as low as 10–100 bacteria is sufficient to cause an infection [9]. This high risk of infection ensures that the total number of shigellosis episodes remains constant, despite reduction in shigellosis-related deaths in the healthcare system in the past two decades [9,10]. The endemic outbreaks of shigellosis emerge even more often in developed countries these days [11]. Shigellosis can be caused by four main Shigella species: *S. sonnei*, *S. flexneri*, *S. boydii* and *S. dysenteriae* [8].

In addition to various virulence factors harbored by the representatives of this genus, serotype 1 of *S. dysenteriae* can produce a potent toxin that inhibits protein synthesis. The first report on the production of this toxin, called the Shiga toxin, dates from the late nineteenth century (1898), when Kiyoshi Shiga described cases of bloody diarrhoea [12]. Dysentery outbreaks caused by *S. dysenteriae* type 1 occurred several times in the 20th century, causing a serious threat to public health. Infections with *S. dysentieriae* result in a variety of symptoms, ranging from mild diarrhoea to hemorrhagic colitis, and are due to bacteria invading the mucosal surface of the intestines [13]. In some patients, the effect of toxin production during *S. dysenteriae* infection can lead to serious complications and sequela, including neurological problems and the hemolytic uremic syndrome caused by invasion of the vascular endothelium and subsequent lesions [14,15]. The Shiga toxin, responsible for the virulence of *S. dysenteriae* type 1, is encoded on the bacterial chromosome, although presence of bacteriophage-derived genes has been reported. However, the toxin production is not related to prophage excision and induction, contrary to enterohemorrhagic *E. coli* strains [16,17]. Due to extensive knowledge of *Shigella* virulence and genetics, representatives of this genus make good models for bacterial invasive pathogenesis [18].

As mentioned above, the Shiga toxins can be also harbored by *Escherichia coli* strains, resulting in their toxicity. In *E. coli*, the production of Shiga toxin is dependent on the induction of the lambdoid prophage, as the *stx* genes are under the control of this phage's late promoter. As the induction is mediated by the SOS response, DNA damage and other stresses, such as free radicals, can lead to phage induction and toxin release, which means that the use of standard antibiotics is discouraged in the course of EHEC infection. In our previous work, we showed that plant secondary metabolites, isothiocyanates (ITCs), effectively inhibit the growth of enterohaemorrhagic *E. coli* (EHEC) strains [19,20]. We showed that the induction of phage lytic development and the Stx toxin production was also inhibited by the action of ITC [19,20]. Moreover, the toxicity of bacterial lysates treated with ITC against human and simian cell cultures was significantly reduced [19,20]. The molecular mechanism underlying the antibacterial effect of phytochemicals involves the global stress response, the stringent response. Its alarmones, ppGpp and pppGpp, accumulate during various stress conditions and nutrient limitation, leading to remodeling of the gene expression pattern by direct and indirect action on the transcription machinery [21].

The stringent response is widespread among bacteria and both *E. coli* and *S. dysenteriae*, belonging to the Enterobacteriaceae family, can synthesize (p)ppGpp via two synthetases, RelA and SpoT. However, previously we presented evidence that the antibacterial effect of ITC is not always related to the stringent response induction, e.g., in *Bacillus subtilis*, and could be based on the disruption of the integrity of the cellular membrane integrity [22]. Nevertheless, we showed that ITC could inhibit the growth of several bacterial pathogens, including *Klebsiella pneumoniae* and *Staphylococcus aureus*. Recently, we also presented evidence for the antibacterial action of ITCs against another toxin-producing bacterium, *Vibrio cholerae* [23].

Developing a new possible line of infection treatment is important, especially nowadays, when multidrug-resistant bacterial strains emerge and when designing new antibiotics is a long process. In this work, we ask whether the growth of a human pathogen bearing the Stx encoding genes, *S. dysenteriae*, is affected by two representative ITC, sulforaphane (SFN) and phenethyl ITC (PEITC). We show that these two compounds can effectively inhibit bacterial growth, toxin gene expression and toxicity against eukaryotic cells. Furthermore, we presented the SFN efficient antishigellosis effect in the surrogate *Galleria mellonella* larvae model based on the enhanced phagocytosis of bacterial cells.

## 2. Materials and Methods

### 2.1. Bacterial Strains and Growth Conditions

The following strains were used in this study: *S. dysenteriae* SZ2855/97 Japan, *S. dysenteriae* Bonn/76, *S. dysenteriae* 156-36/88 Osaka, kindly provided by Dr. Anselm Lehmacher, *S. flexneri* ATCC 12022, *S. sonnei* ATCC 25931. Bacteria were typically grown in Lysogeny Broth (Lennox, Sigma-Aldrich, Darmstadt, Germany) at 37 °C with orbital shaking at 160 rpm.

### 2.2. Bacterial Growth Inhibition

To assess the minimum inhibitory and bactericidal concentrations (MIC/MBC) of the tested compounds, the twofold broth microdilution method was used according to that BSAC standard, with the inoculum of $5 \times 10^5$ cells/well in 96 well plates [22]. The concentration of ITC ranged from 10–0.0125 mM. The results were obtained using the EnSpire (Perkin Elmer Singapore Pte. Ltd, Singapore) instrument after 20 h incubation at 37 °C. MBC was assessed in the same procedure but after 20 h samples were serially diluted in PBS and spotted (10 µL) on solid agar LB plates. Then, a concentration of ITC that kills 99.9% of cells was determined after over-night incubation at 37 °C. The kinetics of growth inhibition were measured spectrophotometrically ($A_{600}$) in the presence of relevant concentrations of ITCs.

### 2.3. Assessment of (p)ppGpp Accumulation in Bacteria

Cellular levels of the stringent response alarmones, i.e., ppGpp and pppGpp, were measured basically as described previously [24], with modifications. The overnight bacterial culture was grown in the MOPS (4-morpholinepropanesulfonic acid) minimal medium and then was diluted in the same medium, but with a low phosphate concentration (0.4 mM) and then cultivated to $A_{600}$ of 0.2. Then, the cultures were diluted in a 1:10 ratio in the same medium, but with addition of [$^{32}$P]orthophosphoric acid (150 µCi/mL) and grown for two generations. Next, at time zero, ITCs at 1 × MIC, or serine hydroxamate (SHX), at 1 mM was added. At specific times, samples were collected and lysed with formic acid (13 M) in three cycles of the freeze-thaw procedure. Following centrifugation, nucleotides were separated by thin-layer chromatography on PEI cellulose plates (Sigma-Aldrich, Darmstadt, Germany) in 1.5 M potassium phosphate buffer (pH 3.4). The chromatograms were analyzed using the Typhoon 9200 Phosphorimager (GE Healthcare, Uppsala, Sweden). QuantityOne (BioRad, Hercules, CA, USA) software was used for densitometry analysis.

### 2.4. RNA Isolation and RT-PCR Analysis

Bacteria were grown to $OD_{600} = 0.1$ (t = 0) in LB and treated with SFN doses equal to 1 × MIC, $\frac{1}{4}$ × MIC and 1/16 × MIC, or SHX (1 mM), mitomycin C (0.5 µg/mL); for negative control untreated culture was used. Samples for RNA purification were collected 180 min after treatment. Total RNA was extracted in duplicate from two biological replicate cultures of each strain using the RNeasy MINI kit (Qiagen GmbH, Hilden, Germany) according to the manufacturer's protocol. EvoScript Reverse transcriptase (Roche, Basel, Switzerland) and random primers were used to create cDNA from RNA samples. Quantitative real-time PCR (qRT-PCR) was performed using LightCycler 480 (Roche Diagnostics AG, Rotkreuz, Switzerland) and the following primers: stx1—CGATGTTACGGTTTGTTACTGTGACAGC; stx1R—AATGCCACGCTTCCCAGAATTG; recAF—CTGGAAATCTGTGACGCCCT; and recAR—ATCGCCTGGCTCATCATACG. All data were normalized to levels of *recA*, and analyzed using the comparative cycle threshold (CT) method. Target gene expression levels were compared by the relative quantification method $2^{-\Delta\Delta CT}$ [25].

### 2.5. Assessment of Toxicity of Bacterial Lysates on the HeLa and Vero Cells

The procedure was performed basically as described [19]. Human HeLa or simian Vero cells were grown in the DMEM medium (Gibco, Waltham, MA, USA), pH 7.4, supplemented with 10% heat-inactivated fetal bovine serum (FBS) and 1 × Antibiotic and

Antimycotic solution (all purchased from Sigma, Darmstadt, Germany) and incubated at 37 °C in a humidified 5% $CO_2$ atmosphere. To evaluate bacterial toxicity towards the HeLa or Vero cells, bacterial lysates were prepared. Overnight cultures of *S. dysenteriae* were grown in LB medium. After 100-fold dilution in fresh MOPS minimal medium (containing 0.5% glucose), the cultivation was continued until $A_{600}$ reached 0.1. Then, mitomycin C, ITC, and/or SHX were added to the cultures to various final concentrations and cultures were incubated with shaking for 30 min. To prepare cell-free lysates, cultures were centrifuged and lysates were filtered using PVDF syringe filters (0.22 mm; Roth). Lysates were then serially diluted in DMEM. Cytotoxicity was assessed using the 3-(4,5-dimethyl-2-thiazolyl)-2,5-diphenyl-2H-tetrazolium bromide (MTT) assay, performed as described previously [26]. HeLa or Vero cell lines were seeded at $1 \times 10^4$ cells/well in 96-well plates and incubated overnight. Then, the growth medium was substituted with fresh medium supplemented with the tested compounds (5 µL of lysates were added to 95 mL of DMEM). After 48 h of incubation, the medium was changed to the MTT solution (100 µL) (MTT was diluted in PBS to 1 mg/mL). Plates were incubated at 37 °C for 4 h, and then the medium was removed. The purple formazan product was dissolved in 100 µL dimethyl sulfoxide (DMSO) and quantified by measuring $A_{570}$. The data are presented as means (with SD) from at least three independent experiments, relative to nontreated cultures.

### 2.6. Study of Pathogenicity of S. dysenteriae by Employing the GALLERIA Mellonella Larvae

The larvae of *Galleria mellonella* (the greater wax moth), TruLarv, were obtained from BioSystems Technology (Credition, UK). Larvae were stored in the darkness at room temperature before experiments. The infection was conducted basically, as described previously [27]. Briefly, the healthy larvae of 250–300 mg weight with no melanization symptoms were selected and transferred to the Petri dish. The larvae ($n$ = 15) were surface sterilized with ethanol before each injection procedure. Insects were infected with the varying number (from $10^2$ to $10^8$ c.f.u./larva) of *S. dysenteriae* cells using a 100 µL Hamilton syringe into the last left foreleg with a blunt tip needle. The control insects were injected with PBS. Caterpillars were examined every 24 h (up to 96 h) for changes in behavior and melanization; the number of living larvae was assessed. For the treatment experiment, SFN (25 mg/kg) and/or azithromycin (6.25 mg/kg) as a standard antibiotic used against gastrointestinal infection (the dosage corresponds to human treatment) were injected 1 h after infection. Three independent experiments were performed.

### 2.7. In Vivo Phagocytosis Assay

The heat-killed and FITC labelled *S. dysenteriae* SZ2855/97 Japan bacteria were prepared as described [28]. Briefly, the $10^9$/mL cells were heated to 70 °C for 1 h, washed twice and incubated in 0.5% carbonate buffer (pH 9.5) containing FITC (0.15 mg/mL) for 30 min at 37 °C. Larvae ($n$ = 10) were treated with SFN (25 mg/kg) or PBS as control and after 30 min were challenged with 10 µL of labelled bacteria injected into the larvae's last left foreleg. 3 h after infection, 5 larvae from each group were bled into Eppendorf tubes with heparin and trypan blue (1% in PBS). Samples (5 µL) were placed on the coverslips and visualized under a Leica DMI4000B microscope fitted with a DFC365FX camera (Leica Microsystems, Wetzlar, Germany). The phagocytosis rates were assessed from at least 5 independent fields (100 hemocytes) for each sample.

## 3. Results and Discussion

### 3.1. ITC Inhibit S. dysenteriae Growth and SHIGA Toxin Production

A desirable antibacterial compound must be safe for therapy, should inhibit bacterial growth and development and also preferably inhibit virulence factor secretion. Importantly, to avoid unnecessary burden for the patient's organism, the lowest possible effective concentration should be used. Thus, a first step to assess the antimicrobial potential of a given compound is determination of the minimum inhibitory concentration (MIC). Three different isolates of *S. dysenteriae* were used for the assessment of the ITC effect. For this work, we chose SFN and phenethyl PEITC as these ITCs have been quite well studied for their effects in humans. Moreover, PEITC was shown to be a potent antibacterial agent against *E. coli* and *S. aureus* as reported in [29], and also against EHEC strains [19]. The antibacterial effect of SFN and other aliphatic ITC was also reported [22,30] and furthermore, as part of the daily diet SFN is known for its chemopreventive and anticancer effect [31,32]. The determination of MIC (Table 1) indicates that both ITCs exert their effect against all tested *S. dysenteriae* isolates, as well as against *S. sonnei* and *S. flexneri*.

**Table 1.** Susceptibility of *Shigella* species to PEITC and SFN.

| Strain | MIC (MBC) [mM] | |
|---|---|---|
| | **PEITC** | **SFN** |
| *S. dysenteriae* SZ2855/97 Japan | 0.50 (1.00) | 0.25 (0.50) |
| *S. dysenteriae* Bonn/76 | 0.25 (0.25) | 0.06 (0.125) |
| *S. dysenteriae* 156-36/88 Osaka | 0.50 (0.50) | 0.25 (0.50) |
| *S. sonnei* | 1.00 (2.00) | 2.00 (2.00) |
| *S. flexneri* | 4.00 (ND [1]) | 4.00 (ND) |

[1] not determined.

The minimum inhibitory concentrations are relatively low for S. *dysenteriae* strains (in the range between 11–44 and 41–81 µg/mL, for SFN and PEITC, respectively), which enhances the possibility of employing these ITC in the therapy of *Shigella* infection. Notably, in comparison to the MIC for EHEC strains [19,20], the growth of *S. dysenteriae* representatives was impaired at the same or lower range of compounds. Nevertheless, the other two *Shigella* representatives, *S. sonnei* and *S. flexneri*, were less sensitive to the ITCs treatment with their MIC values greater than 100 µg/mL, and therefore we decided to exclude them from further analysis. Next, we assessed growth of *S. dysenteriae* in the presence of varying concentrations of the ITCs tested, and found that even at concentrations of $\frac{1}{4}$ MIC, the ITCs were able to significantly inhibit growth ($p$ <0.001) of bacterial cultures in a rich liquid medium (Figure 1). Interestingly, the PEITC antibacterial effect does not appear to be dose-dependent, as complete growth inhibition was observed for all concentrations tested and for all strains. We observed *S. dysenteriae* Boon/76 was the most sensitive to SFN. Thus, we can conclude that both SFN and PEITC act as effective antibacterial agents against *Shigella* strains, providing a promising course of further studies.

### *3.2. ITCs Reduce the Toxicity of S. dysenteriae*

The most harmful effect during *S. dysenteriae* infection is caused by production of the Shiga toxin, and thus not only growth inhibition but also toxin synthesis downregulation would be expected from a potential antimicrobial compound. Since the eukaryotic cell lines: human HeLa and simian Vero are sensitive to Shiga toxin [33], they can serve as good indicators for bacterial virulence. Thus, we tested viability of these cells when exposed to the ITC-treated or untreated bacterial lysates. We observed that the treatment of *S. dysenteriae* strains with PEITC and SFN at the concentrations of $\frac{1}{4}$ MIC (which effectively impaired bacterial growth) resulted in a statistically significant decrease in toxicity of bacterial lysates, as tested by the MTT assay (Figure 2). Alleviation of bacterial toxicity was similar for both ITCs tested. The results are in concert with those previously reported by us for the EHEC strain, where PEITC-treated cultures showed a notably lower virulence against the Vero and HeLa cells [19].

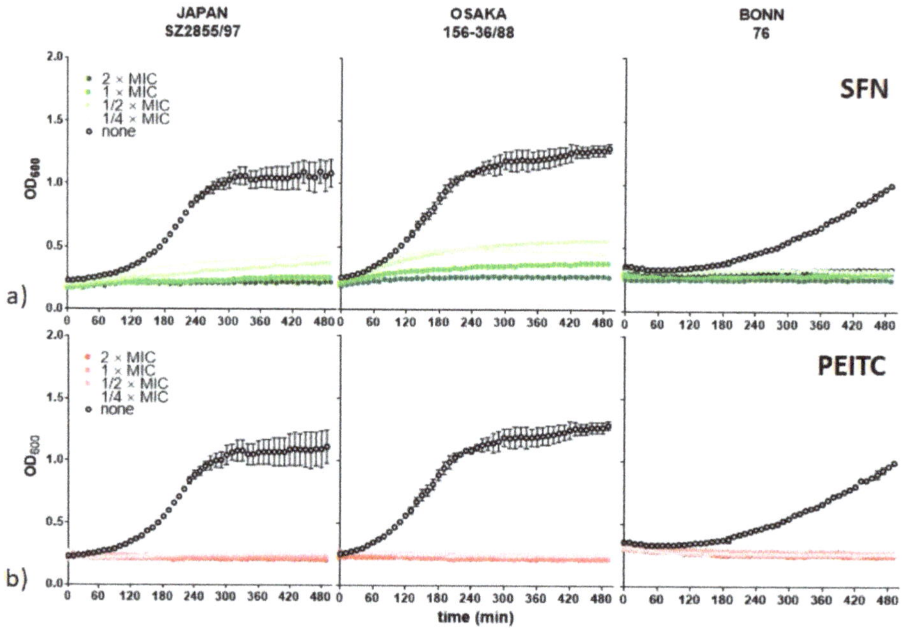


**Figure 1.** SFN and PEITC affect bacterial growth. *S. dysenteriae* isolates (indicated above graphs) were cultured in the presence of (**a**) SFN and (**b**) PEITC at indicated concentrations related to the MIC determined previously (Table 1).

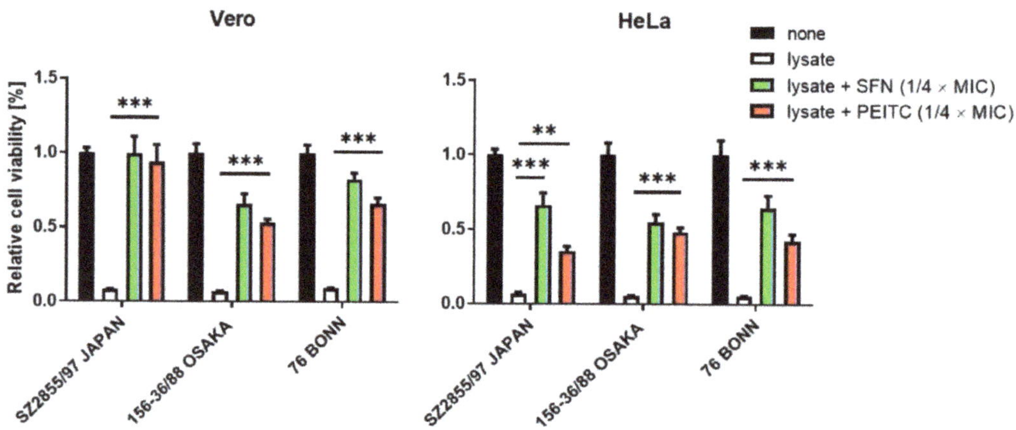


**Figure 2.** ITC reduces the toxicity of *S. dysenteriae* lysates. The effect of SFN on viability of HeLa and Vero cells treated with *S. dysenteriae* lysate was assessed after 74 h in the absence or presence of SFN at concentrations indicated. Untreated culture was used as a negative control (black bar) and a culture treated with bacterial lysate was used as a positive control (white bar). SFN (green bar) and PEITC (red bar) treated groups were assessed against positive control using Student's *t*-test. Statistical significance is marked with asterisks: *** $p < 0.001$; ** $p < 0.01$.

Here, we show that downregulation of Shiga toxin production by ITC (as shown in Figure 2) has a direct effect on increased cell viability. Moreover, there are no defects in cell morphology when tested eukaryotic cells are exposed to the lysates of bacterial cells treated with ITC, contrary to cells exposed to untreated bacterial lysates. It is worth mentioning that, as we showed previously [19] and as was already reported by others [31], when PEITC and SFN are used alone at the concentrations employed for this study, they do not have a detrimental effect on the cell lines' viability.

### *3.3. Induction of the Stringent Response Underlines the Antibacterial Effect of ITC on S. dysenteriae*

The observed growth inhibition of *Shigella* strains in the presence of ITC led us to search for the mechanisms underlying this process. As we know from our previous studies, in *E. coli* ITCs exerted its effects via induction of the stringent response. Therefore, we evaluated cellular levels of the stringent response alarmones, ppGpp and pppGpp, in the *S. dysenteriae* SZ2855/97 Japan strain treated with PEITC, SFN, or serine hydroxamate, the amino acid analog that serves as a control of the stringent response induction. We also compared the observed nucleotide levels visualized by thin-layer chromatography with an untreated control (Figure 3a,b). The strong signal of both tetra- and pentaphosphate is visible in the SHX control, indicating that bacteria react to amino acid starvation by alarmone accumulation. Interestingly, both SFN and PEITC, induced the stress alarmone accumulation to an extent comparable to that induced by SHX. Apparently, similar to many bacterial strains tested so far [22], the basis of the effect of ITC-mediated growth inhibition is induction of the global stress response. The stringent response, as a general stress mechanism, is widely conserved in the Enterobacteriaceae family and *Shigella sp.* genus. The *relA/spoT* genes display high sequence similarity to their *E. coli* homologues [34]. Accumulation of (p)ppGpp causes reprogramming of bacterial gene expression, shutting down such energy-consuming processes as ribosomal component synthesis, and, in general, expression of numerous household genes, while expression of genes involved in the stress response and those recognized by alternative sigma factors is activated. Such a global alteration in bacterial cellular metabolism results in growth inhibition, and importantly, leads to the downregulation of the Shiga toxin production (directly or indirectly). Another mode of action of ITCs can also involve destabilization of cellular membrane integrity, as we observed in our recent work [22] and as was suggested previously [29]. However, this phenomenon in *B. subtilis* was not triggered by the stringent response induction. Thus, it can be concluded from the data presented here that the major mechanism underlying growth inhibition of *S. dysenteriae* involves the alarmone accumulation. It was also suggested that ppGpp and its cofactor DksA play a certain role in *Shigella* virulence [35,36], and interestingly, concentrations of the alarmones during the onset of the stringent response induced by ITC significantly exceed those observed during normal growth. Thus, the effect of (p)ppGpp accumulation mediated by ITCs is growth inhibition, which provides a promising course for additional investigation of ITCs application.

Since the pathogenicity of *Shigella* strains is strictly related to the synthesis of Stx toxin, we tested expression of the *stx* gene in *S. dysenteriae* SZ2855/97 Japan by qRT-PCR after the SFN/PEITC treatment (Figure 3c). We assessed the effect of both compounds at $1 \times$ MIC and $\frac{1}{4}$ MIC, and found that they effectively downregulated the level of *stx* gene expression. Namely, the fold change observed for SFN treatment was $0.16 \pm 0.01$, $0.18 \pm 0.04$, while for PEITC it was $0.22 \pm 0.11$, $0.28 \pm 0.03$ for $1 \times$ MIC and $\frac{1}{4}$ MIC, respectively. However, even lower concentrations of SFN/PEITC ($1/16 \times$ MIC) also had an inhibiting effect ($0.47 \pm 0.09$ and $0.83 \pm 0.79$, fold level change for SFN and PEITC, respectively), although it was not as pronounced. On the other hand, mitomycin C, the SOS triggering factor, caused a significant increase in the *stx* gene transcription (Figure 3c). The latter effect is caused by the general stress exerted by mitomycin C, as it was reported that addition of this drug had caused an increase in production and secretion of Shiga toxins through the outer membrane vesicles [37]. Although the Shiga toxin production by *S. dysenteriae* was reported not to be directly related to the prophage induction, the presence of genes belonging to a defective prophage in the vicinity of *stx* genes may have an influence on the toxin gene expression. Furthermore, the *stx* encoding region within this genus is remarkably conserved [38]. Our results indicate that ITCs do not only inhibit the growth of *Shigella* strains, but more importantly, reduce expression of the Shiga toxin genes.

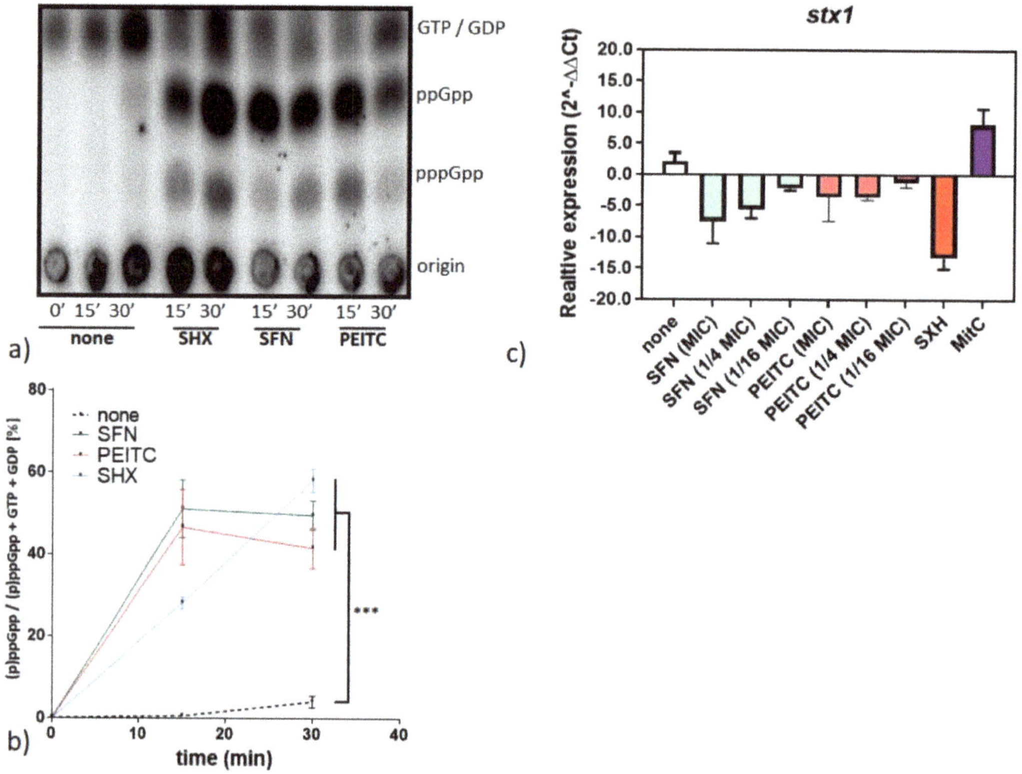


**Figure 3.** SFN and PEITC impair expression of the Shiga toxin genes and induce stringent response in *S. dysenterae*. (**a**) Synthesis of the stringent response alarmones, ppGpp and pppGpp, was assessed by culturing bacteria with the addition of [$^{32}$P]orthophosphoric acid (150 µCi/mL) followed by cell lysis and nucleotide separation by thin-layer chromatography. ITC or SHX were added at time zero. Samples were withdrawn at 15 and 30 min after the compound addition. The positions of ppGpp and pppGpp are indicated at the right side of the panel. (**b**) Relative cellular level of (p)ppGpp was assessed densitometricaly, where treated samples were compared to the nontreated negative control (at the 30 min endpoint) using Student's *t*-test. Statistical significance is marked with asterisks: *** $p < 0.001$; (**c**) Relative expression of *stx*1 in *S. dysenteriae* SZ2855/97 Japan was determined by qRT-PCR and is presented using $2^{-\Delta\Delta CT}$ quantification normalized to the *recA* gene expression level.

### 3.4. SFN Reduces the Toxicity of *S. dysenteriae In Vivo*

Our cell lines experiments described above provided further proof for potential therapeutic use of ITC in treating *Shigella* infections. However, more important are studies performed in living organisms, not only in isolated cell cultures. Thus, we decided to employ an animal model. The *Shigella* infections have been studied in many models, including primates, rodents and piglets [18]; nevertheless, the use of these animals is expensive, requires a specific facility and raises significant ethical issues.

Thus, a more convenient model was introduced, namely, the larvae of *Galleria mellonella*, the greater wax moth. This surrogate model is widely used for novel antibiotics' efficacy studies, pathogen virulence and resistance screening, and its use in monitoring *Shigella* infections has been described in detail [27]. Briefly, these insects are easy to maintain and handle, their size allows for injection of pathogenic bacteria and simple-to-monitor steps of infection were described. Furthermore, the insect immune system, although evolutionarily distant, presents significant similarities to vertebrates, which makes it useful to make conclusions about general response to pathogens. In particular, specific larvae

**Informed Consent Statement:** Not applicable.

**Data Availability Statement:** Not applicable.

**Acknowledgments:** We kindly thank Anselm Lehmacher for providing bacterial isolates. We are grateful to Anna Herman-Antosiewicz for discussions and support and Katarzyna Potrykus for the critical reading of the manuscript.

**Conflicts of Interest:** The authors declare no conflict of interest.

## References


1. Patridge, E.; Gareiss, P.; Kinch, M.S.; Hoyer, D. An analysis of FDA-approved drugs: Natural products and their derivatives. *Drug Discov. Today* **2016**, *21*, 204–207. [CrossRef]
2. Shirakawa, M.; Hara-Nishimura, I. Specialized Vacuoles of Myrosin Cells: Chemical Defense Strategy in Brassicales Plants. *Plant Cell Physiol.* **2018**, *59*, 1309–1316. [CrossRef]
3. Singh, S.V.; Singh, K. Cancer chemoprevention with dietary isothiocyanates mature for clinical translational research. *Carcinogenesis* **2012**, *33*, 1833–1842. [CrossRef] [PubMed]
4. Kurepina, N.; Kreiswirth, B.N.; Mustaev, A. Growth-inhibitory activity of natural and synthetic isothiocyanates against representative human microbial pathogens. *J. Appl. Microbiol.* **2013**, *115*, 943–954. [CrossRef] [PubMed]
5. Mazarakis, N.; Snibson, K.; Licciardi, P.V.; Karagiannis, T.C. The potential use of L-sulforaphane for the treatment of chronic inflammatory diseases: A review of the clinical evidence. *Clin. Nutr.* **2020**, *39*, 664–675. [CrossRef]
6. Romeo, L.; Iori, R.; Rollin, P.; Bramanti, P.; Mazzon, E. Isothiocyanates: An Overview of Their Antimicrobial Activity against Human Infections. *Molecules* **2018**, *23*, 624. [CrossRef]
7. Dufour, V.; Stahl, M.; Baysse, C. The antibacterial properties of isothiocyanates. *Microbiology* **2015**, *161*, 229–243. [CrossRef]
8. Kotloff, K.L.; Riddle, M.S.; Platts-Mills, J.A.; Pavlinac, P.; Zaidi, A.K.M. Shigellosis. *Lancet* **2018**, *391*, 801–812. [CrossRef]
9. Ekdahl, K.; Andersson, Y. The epidemiology of travel-associated shigellosis-regional risks, seasonality and serogroups. *J. Infect.* **2005**, *51*, 222–229. [CrossRef]
10. Toro, C.; Arroyo, A.; Sarria, A.; Iglesias, N.; Enríquez, A.; Baquero, M.; De Guevara, C.L. Shigellosis in subjects with traveler's diarrhea versus domestically acquired diarrhea: Implications for antimicrobial therapy and human immunodeficiency virus surveillance. *Am. J. Trop. Med. Hyg.* **2015**, *93*, 491–496. [CrossRef] [PubMed]
11. Baker, S.; The, H.C. Recent insights into Shigella. *Curr. Opin. Infect. Dis.* **2018**, *31*, 449–454. [CrossRef]
12. Trofa, A.F.; Ueno-Olsen, H.; Oiwa, R.; Yoshikawa, M. Dr. Kiyoshi Shiga: Discoverer of the dysentery bacillus. *Clin. Infect. Dis. Off. Publ. Infect. Dis. Soc. Am.* **1999**, *29*, 1303–1306. [CrossRef]
13. O'Loughlin, E.V.; Robins-Browne, R.M. Effect of Shiga toxin and Shiga-like toxins on eukaryotic cells. *Microbes Infect.* **2001**, *3*, 493–507. [CrossRef]
14. Bergan, J.; Dyve Lingelem, A.B.; Simm, R.; Skotland, T.; Sandvig, K. Shiga toxins. *Toxicon* **2012**, *60*, 1085–1107. [CrossRef] [PubMed]
15. Cherla, R.P.; Lee, S.-Y.; Tesh, V.L. Shiga toxins and apoptosis. *FEMS Microbiol. Lett.* **2003**, *228*, 159–166. [CrossRef]
16. Hale, T.L. Genetic basis of virulence in *Shigella* species. *Microbiol. Rev.* **1991**, *55*, 206–224. [CrossRef] [PubMed]
17. Schmidt, H. Shiga-toxin-converting bacteriophages. *Res. Microbiol.* **2001**, *152*, 687–695. [CrossRef]
18. Lampel, K.A.; Formal, S.B.; Maurelli, A.T. A Brief History of *Shigella*. *EcoSal Plus* **2018**, *8*. [CrossRef] [PubMed]
19. Nowicki, D.; Maciąg-Dorszyńska, M.; Kobiela, W.; Herman-Antosiewicz, A.; Węgrzyn, A.; Szalewska-Pałasz, A.; Węgrzyn, G. Phenethyl isothiocyanate inhibits shiga toxin production in enterohemorrhagic Escherichia coli by stringent response induction. *Antimicrob. Agents Chemother.* **2014**, *58*, 2304–2315. [CrossRef]
20. Nowicki, D.; Rodzik, O.; Herman-Antosiewicz, A.; Szalewska-Pałasz, A. Isothiocyanates as effective agents against enterohemorrhagic Escherichia coli: Insight to the mode of action. *Sci. Rep.* **2016**, *6*, 22263. [CrossRef]
21. Potrykus, K.; Cashel, M. (p)ppGpp: Still magical? *Annu. Rev. Microbiol.* **2008**, *62*, 35–51. [CrossRef] [PubMed]
22. Nowicki, D.; Maciąg-Dorszyńska, M.; Bogucka, K.; Szalewska-Pałasz, A.; Herman-Antosiewicz, A. Various modes of action of dietary phytochemicals, sulforaphane and phenethyl isothiocyanate, on pathogenic bacteria. *Sci. Rep.* **2019**, *9*, 13677. [CrossRef]
23. Krause, K.; Pyrczak-felczykowska, A.; Karczewska, M.; Narajczyk, M.; Herman-antosiewicz, A.; Szalewska-pałasz, A.; Nowicki, D. Dietary isothiocyanates, sulforaphane and 2-phenethyl isothiocyanate, effectively impair vibrio cholerae virulence. *Int. J. Mol. Sci.* **2021**, *22*, 10187. [CrossRef]
24. Mechold, U.; Potrykus, K.; Murphy, H.; Murakami, K.S.; Cashel, M. Differential regulation by ppGpp versus pppGpp in Escherichia coli. *Nucleic Acids Res.* **2013**, *41*, 6175–6189. [CrossRef] [PubMed]
25. Nowicki, D.; Bloch, S.; Nejman-Faleńczyk, B.; Szalewska-Pałasz, A.; Węgrzyn, A.; Węgrzyn, G. Defects in RNA polyadenylation impair both lysogenization by and lytic development of Shiga toxin-converting bacteriophages. *J. Gen. Virol.* **2015**, *96*, 1957–1968. [CrossRef] [PubMed]
26. Rasooly, R.; Do, P.M. Shiga toxin Stx2 is heat-stable and not inactivated by pasteurization. *Int. J. Food Microbiol.* **2010**, *136*, 290–294. [CrossRef]

27. Barnoy, S.; Gancz, H.; Zhu, Y.; Honnold, C.L.; Zurawski, D.V.; Venkatesan, M.M. The Galleria mellonella larvae as an in vivo model for evaluation of *Shigella* virulence. *Gut Microbes* **2017**, *8*, 335–350. [CrossRef] [PubMed]
28. Wu, G.; Liu, Y.; Ding, Y.; Yi, Y. Ultrastructural and functional characterization of circulating hemocytes from Galleria mellonella larva: Cell types and their role in the innate immunity. *Tissue Cell* **2016**, *48*, 297–304. [CrossRef]
29. Abreu, A.C.; Borges, A.; Simões, L.C.; Saavedra, M.J.; Simões, M. Antibacterial activity of phenyl isothiocyanate on Escherichia coli and Staphylococcus aureus. *Med. Chem.* **2013**, *9*, 756–761. [CrossRef]
30. Nowicki, D.; Krause, K.; Szamborska, P.; Żukowska, A.; Cech, G.M.; Szalewska-Pałasz, A. Induction of the Stringent Response Underlies the Antimicrobial Action of Aliphatic Isothiocyanates. *Front. Microbiol.* **2020**, *11*, 591802. [CrossRef]
31. Pawlik, A.; Wiczk, A.; Kaczyńska, A.; Antosiewicz, J.; Herman-Antosiewicz, A. Sulforaphane inhibits growth of phenotypically different breast cancer cells. *Eur. J. Nutr.* **2013**, *52*, 1949–1958. [CrossRef] [PubMed]
32. Wiczk, A.; Hofman, D.; Konopa, G.; Herman-Antosiewicz, A. Sulforaphane, a cruciferous vegetable-derived isothiocyanate, inhibits protein synthesis in human prostate cancer cells. *Biochim. Biophys. Acta* **2012**, *1823*, 1295–1305. [CrossRef] [PubMed]
33. Bouzari, S.; Oloomi, M.; Azadmanesh, K. Study on induction of apoptosis on HeLa and Vero cells by recombinant shiga toxin and its subunits. *Cytotechnology* **2009**, *60*, 105. [CrossRef]
34. Dorman, M.J.; Dorman, C.J. Regulatory Hierarchies Controlling Virulence Gene Expression in *Shigella flexneri* and *Vibrio cholerae*. *Front. Microbiol.* **2018**, *9*, 2686. [CrossRef]
35. Trastoy, R.; Manso, T.; Fernández-García, L.; Blasco, L.; Ambroa, A.; Pérez Del Molino, M.L.; Bou, G.; García-Contreras, R.; Wood, T.K.; Tomás, M. Mechanisms of Bacterial Tolerance and Persistence in the Gastrointestinal and Respiratory Environments. *Clin. Microbiol. Rev.* **2018**, *31*. [CrossRef] [PubMed]
36. Sharma, A.K.; Payne, S.M. Induction of expression of hfq by DksA is essential for *Shigella flexneri* virulence. *Mol. Microbiol.* **2006**, *62*, 469–479. [CrossRef]
37. Dutta, S.; Iida, K.; Takade, A.; Meno, Y.; Nair, G.B.; Yoshida, S. Release of Shiga toxin by membrane vesicles in *Shigella dysenteriae* serotype 1 strains and in vitro effects of antimicrobials on toxin production and release. *Microbiol. Immunol.* **2004**, *48*, 965–969. [CrossRef] [PubMed]
38. Sváb, D.; Falgenhauer, L.; Horváth, B.; Maróti, G.; Falgenhauer, J.; Chakraborty, T.; Tóth, I. Genome Analysis of a Historical *Shigella dysenteriae* Serotype 1 Strain Carrying a Conserved Stx Prophage Region. *Front. Microbiol.* **2020**, *11*, 614793. [CrossRef]
39. Ribeiro, C.; Brehélin, M. Insect haemocytes: What type of cell is that? *J. Insect Physiol.* **2006**, *52*, 417–429. [CrossRef] [PubMed]
40. Lavine, M.D.; Strand, M.R. Insect hemocytes and their role in immunity. *Insect Biochem. Mol. Biol.* **2002**, *32*, 1295–1309. [CrossRef]
41. Lapointe, J.F.; Dunphy, G.B.; Mandato, C.A. Hemocyte-hemocyte adhesion and nodulation reactions of the greater wax moth, Galleria mellonella are influenced by cholera toxin and its B-subunit. *Results Immunol.* **2012**, *2*, 54–65. [CrossRef] [PubMed]
42. Sowa-Jasiłek, A.; Zdybicka-Barabas, A.; Stączek, S.; Wydrych, J.; Mak, P.; Jakubowicz, T.; Cytryńska, M. Studies on the role of insect hemolymph polypeptides: *Galleria mellonella* anionic peptide 2 and lysozyme. *Peptides* **2014**, *53*, 194–201. [CrossRef] [PubMed]
43. Shih, Y.-L.; Wu, L.-Y.; Lee, C.-H.; Chen, Y.-L.; Hsueh, S.-C.; Lu, H.-F.; Liao, N.-C.; Chung, J.-G. Sulforaphane promotes immune responses in a WEHI-3-induced leukemia mouse model through enhanced phagocytosis of macrophages and natural killer cell activities in vivo. *Mol. Med. Rep.* **2016**, *13*, 4023–4029. [CrossRef]
44. Wu, M.; Gibbons, J.G.; DeLoid, G.M.; Bedugnis, A.S.; Thimmulappa, R.K.; Biswal, S.; Kobzik, L. Immunomodulators targeting MARCO expression improve resistance to postinfluenza bacterial pneumonia. *Am. J. Physiol. Lung Cell. Mol. Physiol.* **2017**, *313*, L138–L153. [CrossRef] [PubMed]

nutrients MDPI

Brief Report

# A Water Extract from *Chlorella sorokiniana* Cell Walls Stimulates Growth of Bone Marrow Cells and Splenocytes

**Susumu Ishiguro [1], Mary Roth [2], Ruth Welti [2], Mayme Loyd [1], Ravindra Thakkar [1], Morgan Phillips [1], Nicole Robben [1], Deepa Upreti [1], Ayaka Nakashima [3], Kengo Suzuki [3], Jeffrey Comer [1] and Masaaki Tamura [1,*]**

[1] Department of Anatomy & Physiology, Kansas State University College of Veterinary Medicine, Manhattan, KS 66506, USA; isusumu@vet.k-state.edu (S.I.); loydmayme@gmail.com (M.L.); ravithakkar@vet.k-state.edu (R.T.); mphillips30@ksu.edu (M.P.); ndrobben@vet.k-state.edu (N.R.); deepa07@vet.k-state.edu (D.U.); jeffcomer@ksu.edu (J.C.)

[2] Division of Biology, Kansas Lipidomics Research Center, Kansas State University, Manhattan, KS 66506, USA; mrroth@ksu.edu (M.R.); welti@ksu.edu (R.W.)

[3] Euglena Co., Ltd., Minato-ku, Tokyo 108-0014, Japan; nakashima@euglena.jp (A.N.); suzuki@euglena.jp (K.S.)

* Correspondence: mtamura@vet.k-state.edu; Tel.: +1-(785)-532-4825; Fax: +1-(785)-532-4557

**Abstract:** A water extract derived from the isolated cell walls of *Chlorella sorokiniana* (*C. sorokiniana*, Chlorella water extract, CWE) was analyzed for the presence of lipopolysaccharide (LPS)-related material via the Limulus amebocyte lysate (LAL) assay and evaluated for its growth stimulation effect on the bone marrow cells and splenocytes in vitro cell cultures. The extract contained low levels of LPS-related material, and a mass spectrum suggested that the extract contained many components, including a low level of a lipid A precursor, a compound known as lipid X, which is known to elicit a positive response in the LAL assay. Treatment with the CWE dose- and time-dependently stimulated the growth of mouse bone marrow cells (BMCs) and splenocytes (SPLs). Treatment with the CWE also increased specific BMC subpopulations, including antigen-presenting cells ($CD19^+$ B cells, $33D1^+$ dendritic cells and $CD68^+$ macrophages), and $CD4^+$ and $CD8^+$ T cells, but decreased the number of $LY6G^+$ granulocytes. Treatment with the CWE also increased cytokine mRNA associated with T cell activation, including TNFα, IFNγ, and granzyme B in human lymphoblasts. The present study indicates that the cell wall fraction of *C. sorokiniana* contains an LPS-like material and suggests a candidate source for the bioactivity that stimulates growth of both innate and adaptive immune cells.

**Keywords:** lipopolysaccharide-like substance; *Chlorella sorokiniana* cell wall extract; growth stimulation; bone marrow cells; splenocytes

**Citation:** Ishiguro, S.; Roth, M.; Welti, R.; Loyd, M.; Thakkar, R.; Phillips, M.; Robben, N.; Upreti, D.; Nakashima, A.; Suzuki, K.; et al. A Water Extract from *Chlorella sorokiniana* Cell Walls Stimulates Growth of Bone Marrow Cells and Splenocytes. *Nutrients* **2022**, *14*, 2901. https://doi.org/10.3390/nu14142901

Academic Editors: Daniela Rigano and Paola Bontempo

Received: 27 February 2022
Accepted: 13 July 2022
Published: 15 July 2022



## 1. Introduction

*Chlorella* is a genus of unicellular green algae found in fresh water and seawater [1]. Because this alga contains a variety of nutrients such as amino acids, carbohydrates, vitamins, minerals, and dietary fiber, as well as bioactive components such as pigments, organic compounds, fatty acids, peptides etc., it is taken as a nutritional and functional dietary supplement throughout the world [2–5]. Enhancement of specific nutritional/functional properties can be attainable by managements of culture conditions of algae [6–9]. In addition, preclinical studies have shown that whole dried powder and/or water/organic solvent extracts of *Chlorella* have therapeutic effects against various cancers [10–17]. These effects are attributable to the stimulation of host anti-tumor immune responses [11,14,16]. However, the origin and nature of the bioactive component(s) is yet to be fully clarified.

Lipopolysaccharide (LPS) is a glycolipid found in the bacterial outer membrane of Gram-negative bacteria [18,19], which is recognized as endotoxin by innate immune systems via Toll-like receptor 4 (TLR4) and induce inflammatory responses [20]. LPS is composed of three parts: O-antigen, Core oligosaccharide, and lipid A; O-antigen structures are highly diverse, while the lipid A structure is conserved at the species level [18]. It was reported that LPS-like molecules are found in some organisms such as Gram-negative photosynthetic prokaryotes, cyanobacteria (also called as blue-green alga) [21]; promastigotes, *Leishmania* [22]; algae, *Prototheca* [23]; and land plants, *Arabidopsis thaliana* [24]. In addition, LPS from Gram-negative photosynthetic prokaryotes, *Rhodobacter*, acts as LPS antagonist which shows anti-oxidant and anti-inflammatory effects [25] as well as anti-cancer effects prevent lung metastasis [26]. High doses of LPS lead to excessive immune responses, called endotoxin shock, which is a cause of sepsis [27]. However, low doses of LPS stimulate functional activation and maturation of immune cells, therefore, it is used as an adjuvant of vaccines; monophosphoryl lipid A as an adjuvant for cervical cancer vaccines, also known as human papillomavirus (HPV) vaccines [28].

The Chlorella cell wall is a thick membrane composed of a large amount of insoluble polysaccharide, a relatively small amount of protein/glycoprotein, and unidentified materials [29,30]. It is also reported that the cell wall contains lipopolysaccharide-like immune-reactive substances, although their chemical structures have yet to be determined [31]. Since the Chlorella cell wall is unique in structure and composition and makes up a relatively large portion of the Chlorella body, it is of interest to study the immune response related activities of the water extract from the Chlorella cell wall.

Our recent study suggested that water extract of the cell wall fraction from *Chlorella sorokiniana* attenuated colon carcinoma growth in cell culture and mice by inducing apoptosis of cancer cells [32]. This extract also stimulated cytotoxicity of T cells in three-dimensional spheroid culture with colon carcinoma cells, however a bioactive component(s) in the extract is yet to be identified. Here, we report for the first time that water extract from the cell wall fraction of *Chlorella sorokiniana* (CWE), which stimulates the growth of bone marrow cells and splenocytes in vitro in cell culture, contains a low level of lipopolysaccharide (LPS)-like substance (LPS-LS) which includes a molecule similar to the lipid X found in *Arabidopsis thaliana* [24].

## 2. Materials and Methods

### 2.1. Preparation of CWE

Water extract from the *Chlorella sorokiniana* cell wall fraction (CWE) was initially prepared by a proprietary method developed by the Euglena Co. Ltd. (Minato-ku, Tokyo, Japan). Briefly, *Chlorella sorokiniana* collected from Ishigaki, Japan, was aseptically cultured in modified Hannan and Patouille medium [33] supplemented with glucose as a source of carbon. The cell walls spontaneously suspended in Chlorella culture media were separated from intact cell bodies of *Chlorella sorokiniana* by nozzle separator (Y-250H: Saito Separator Ltd., Ohta-ku, Tokyo, Japan). The resultant Chlorella cell wall fraction was washed with deionized water twice by centrifuging at 6800× *g* for 5 min and lyophilized.

To extract bioactive components from the Chlorella cell wall fraction, lyophilized cell walls were suspended in phosphate buffered saline (PBS) at a concentration of 40 mg/mL and incubated at 4 °C for 12 h and then at 37 °C for 30 min with periodic sonication for 30 s and vortex mixing. Insoluble materials composed of Chlorella cell walls were removed by two steps of centrifugation at 2300× *g* and 11,800× *g*, respectively, at room temperature for 10 min each. The resultant supernatant fraction was filtered through a 0.22 µm disk filter (Midwest Scientific, Valley Park, MO, USA) and stored at −20 °C until use. This fraction was designated the CWE partially purified from Chlorella cell walls and subjected to the experiments described below. The amount of LPS-LS in this preparation was determined by the Limulus amebocyte lysate (LAL) assay (Pierce Chromogenic Endotoxin Quantitation Kit, Thermo Fisher Scientific, Waltham, MA, USA).

### 2.2. Analysis of Chlorella CWE for Lipid A and Lipid X by Mass Spectrometry

Chlorella cell walls were subjected to the method described by Henderson et al. for the preparation of Lipid A [34]. Lipid A from *Escherichia coli* (*E. coli*) was purchased from Sigma–Aldrich (St. Louis, MO, USA). Lipids were extracted from *Arabidopsis thaliana* (*A. thaliana*) was extracted by the method described by Welti et al. [35] and further extracted using "solvent H" as carried out by Markham et al. [36], based on Toledo et al. [37]. All samples were dissolved in chloroform/methanol/300 mM ammonium acetate in water (35/66.5/3.5, *v*/*v*/*v*), ionized by electrospray and subject to fragmentation in a Waters Xevo TQ-S triple quadrupole mass spectrometer (Waters Corporation, Milford, MA, USA) by direct infusion.

### 2.3. Electron Microscopy

For both scanning electron microscopy (SEM) and transmission electron microscopy (TEM), the lyophilized Chlorella cell wall fraction was rehydrated with PBS, fixed with Trump's fixative (1% glutaraldehyde and 4% formaldehyde in a 0.1 M phosphate buffer at pH 7.4) overnight at 4 °C, post-fixed with 1% osmium tetroxide in a 0.2 M phosphate buffer for 1 h, and dehydrated with a series of graded ethanol solutions. For the SEM analysis, ethanol in the dehydrated cell wall sample solution was replaced with hexamethyldisilazane by centrifugation and the membrane sample was sputtered with palladium using Denton Vacuum Desk II sputter coater (Denton Vacuum, Moorestown, NJ, USA). Sputter-coated Chlorella cell walls were analyzed using Hitachi S-3500 N Scanning Electron Microscope (Hitachi Science Systems Ltd., Minato-ku, Tokyo, Japan) at an accelerating voltage of 10 kV. For the TEM investigation, the Chlorella cell walls dehydrated with ethanol were washed with acetone and embedded in Spurr resin, followed by polymerization of sample block in flat embedding molds. The sample was thin-sectioned at a thickness of 700–900 Å using a Leica UC7 ultramicrotome (Leica Biosystems Inc., Buffalo Grove, IL, USA) and placed on a 200-mesh copper TEM grid. Ultrathin sections were analyzed using a FEI Tecnai G2 Spirit BioTWIN transmission electron microscope (FEI Company, Hillsboro, OR, USA) at an accelerating voltage of 80 kV. Electron micrographs were taken with a Tecnai 12 (FEI Company, Hillsboro, OR, USA) microscope, equipped with a Gatan CCD camera (Gatan, Inc., Pleasanton, CA, USA).

### 2.4. Cell Culture

The Jurkat human lymphoblast cell line (TIB-152) was purchased from American Type Culture Collection (ATCC; Manassas, VA, USA). The Jurkat cells were cultured in RPMI 1640 (Mediatech, Inc., Manassas, VA, USA) supplemented with 10% *v*/*v* FBS (EQUITECH-BIO Inc.; Kerrville, TX, USA) and 1% *v*/*v* penicillin-streptomycin (Lonza Rockland, Inc.; Allendale, NJ, USA). Cell culture was carried out at 37 °C in a humidified air atmosphere containing 5% $CO_2$. The cell line was authenticated by short tandem repeat (STR) DNA profiling. Both the cells were maintained in low passages (<15) for this study.

### 2.5. Animals

Female Balb/c mice were obtained from Charles River Laboratories International, Inc. All mice were housed in a clean facility under controlled conditions of temperature (20–26 °C), with 30–70% relative humidity and light (12:12 h light–dark cycles) and acclimatized for 10 days. All mice were housed humanely according to university, state, and federal guidelines (AAALAC) in the AAALAC-accredited animal resource facilities of the Kansas State University College of Veterinary Medicine. The mice's condition was observed daily, and body weights were obtained every other day. All animal experiments adhered strictly to protocols approved by the Kansas State University Institutional Animal Care and Use Committee (Protocol # 4346) and Institutional Biosafety Committee (Registration # 1317).

### 2.6. Effect of CWE on the Growth of Bone Marrow Cells and Splenocytes in Cell Culture

The effect of CWE on the proliferation of bone marrow cells (BMCs) and splenocytes (SPLs) was evaluated by MTT (3-(4,5-dimethylthiazol-2-yl)-2,5-diphenyltetrazolium bromide) assay in a 96-well plate. The Balb/c mice were sacrificed by exposure to saturated $CO_2$ followed by cervical dislocation. BMCs were harvested from the bone marrow in the hind legs and cultured in RPMI 1640 supplemented with 10% *v*/*v* FBS, 1% *v*/*v* penicillin-streptomycin. SPLs were collected from the spleens and cultured in RPMI 1640 supplemented with 10% $v/v$ FBS, 1% $v/v$ penicillin-streptomycin, and 20 μM 2-mercaptoethanol. BMCs ($2 \times 10^5$ cells/well) and SPLs ($5 \times 10^5$ cells/well) were cultured in a 96-well plate. Cell proliferation was evaluated for its dose- (at 0.1, 1 and 10 μg/mL CWE) and time-dependencies (at 24, 48, 72 and 96 h of treatment with 0.1, 1 and 10 μg/mL CWE). PBS and 100 ng/mL authentic LPS (*Escherichia coli* 026:B6; Sigma–Aldrich, St. Louis, MO, USA) served as negative and positive controls, respectively.

### 2.7. Flow Cytometry Analysis of CWE-Induced Proliferation of BMCs In Vitro

BMCs ($5 \times 10^6$ cells) were seeded into a 6-well plate and incubated with 10 μg/mL CWE for 48 h. Cells were immunostained with anti-CD4 (H129.19; helper T cells), anti-CD8b (YTS156.7.7; cytotoxic T cells), anti-CD19 (6D5; B cells), anti-DC marker (33D1; dendritic cells), anti-LY6G (1A8; neutrophil) and anti-CD68 (FA-11; macrophage) antibodies. Antibodies which match the host species and the class with primary antibodies above were used for the isotype control. All antibodies were obtained from BioLegend (San Diego, CA, USA). The changes of cell populations were analyzed by flow cytometry (BD LSRFortessa X-20; BD Biosciences, San Jose, CA, USA) and analyzed by BD FACSDiva (BD Bioscience, San Jose, CA, USA). PBS and authentic LPS (100 ng/mL) served as negative and positive treatment controls, respectively.

### 2.8. Effect of CWE on the Growth of Jurkat Cells

Jurkat cells (1000 cells/well) were seeded in a 96-well plate and treated with CWE (1–100 μg/mL) and authentic LPS (0.1–10 μg/mL) after 24 h. Cell proliferation was evaluated by MTT assay at 48 h after the treatment.

### 2.9. Cytokine Expression in CWE-Treated Jurkat Cells

Gene expression of tumor necrosis factor alpha (TNFα), interferon gamma (IFNγ) and granzyme B (GZMB) in CWE-treated Jurkat cells were measured by reverse transcription quantitative polymerase chain reaction (RT-qPCR). Jurkat cells ($1 \times 10^5$ cells/well) were seeded into a 12-well plate. After 24 h, the cells were treated with 1 or 10 μg/mL of CWE, or 5, 10 or 100 ng/mL authentic LPS. Total RNA was purified at 24 and 48 h after treatment, using the reagent TRIzol (InvitroGen; Thermo Fisher Scientific, Waltham, MA, USA). One step RT-qPCR was performed using the iTaq Universal SYBR Green One-Step Kit (Bio-Rad; Hercules, CA, USA). The RT-qPCR was performed as follows: 45 cycles of 15 s at 95 °C, and 60 s at 60 °C. The results were quantified by the comparative Ct method [38]. Table 1 displays the sequences of the primers.

**Table 1.** Primers used for RT-qPCR.

| Primer | | Sequence | Size |
|---|---|---|---|
| Human TNFα | Forward (5′-3′)<br>Reverse (5′-3′) | GCCAGAATGCTGCAGGACTT<br>GGCCTAAGGTCCACTTGTGTCA | 63 bp |
| Human IFNγ | Forward (5′-3′)<br>Reverse (5′-3′) | AGGGAAGCGAAAAAGGAGTCA<br>GGACAACCATTACTGGGATGCT | 64 bp |
| Human IL-2 | Forward (5′-3′)<br>Reverse (5′-3′) | ATGAGACAGCAACCATTGTAGAATTT<br>CACTTAATTATCAAGTCAGTGTTGAGATGA | 87 bp |
| Human GZMB | Forward (5′-3′)<br>Reverse (5′-3′) | TGCAGGAAGATCGAAAGTGCG<br>GAGGCATGCCATTGTTTCGTC | 180 bp |
| 18S | Forward (5′-3′)<br>Reverse (5′-3′) | GAGGTTCGAAGACGATCAGA<br>TCGCTCCACCAACTAAGAAC | 315 bp |

### 2.10. Morphological Observation of CWE-Treated Jurkat Cells

To evaluate the morphological differentiation of immature T lymphoblasts, Jurkat cells were cultured three-dimensionally (3D) as described previously with slight modifications [39].

The cells were treated with CWE (10 μg/mL), or authentic LPS (10 ng/mL) on Day 1 (24 h after Jurkat cell seeding) and Day 4 (96 h after Jurkat cell seeding). The image of Jurkat cells was taken by an inverted microscope IX51 (Olympus America Inc., Center Valley, PA, USA) equipped with cellSens Dimension software (Olympus America Inc., Center Valley, PA, USA) at Day 7.

### *2.11. Statistical Analysis*

All values are expressed as the mean ± standard deviation of mean. For all in vitro experiments, statistical significance was assessed by an unpaired t-test or ANOVA followed by Tukey's test. All experiments were conducted with multiple sample determinations with several samples ($n$ = 3–5). Statistical significance was set at *, $p < 0.05$.

## **3. Results**

### *3.1. Morphological Analysis Revealed the Chlorella Cell Wall Fraction Is Composed of Only Cell Walls and Membranes*

The morphological analyses by both SEM and TEM clearly indicate that the washed cell wall fraction is composed of only Chlorella cell walls and membranes and no intact Chlorella cell bodies or bacteria contaminated this fraction (Figure 1).

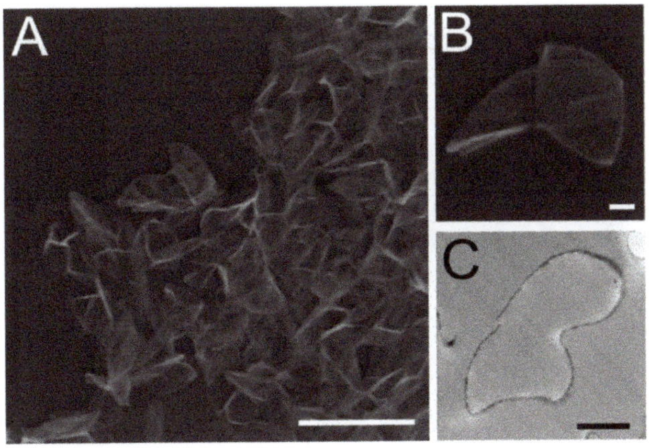


**Figure 1.** Scanning panels (**A**,**B**) and transmission panel (**C**) electron micrographs of washed Chlorella cell wall fraction. Scale bars in the panels (**A**–**C**) represent 10 μm, 1 μm, and 1 μm, respectively.

To clarify whether the bioactivity is attributable to an LPS-like substance in the final extract of the Chlorella cell wall fraction, the "LPS" levels in the final extract were determined using a commercially available Limulus amebocyte lysate (LAL) assay. This assay reacts positively with LPS and related compounds, including the LPS core lipid, lipid A, and its precursor, lipid X [40,41]. The results of the assay demonstrated that the final extract contained low levels of LPS-like material (average 0.200 ± 0.002 ng per 1 μg dry weight of preparations, $n$ = 3).

### *3.2. Mass Spectrometry Suggests the Presence of Lipid X, but Not Lipid A, in Chlorella Cell Wall Fraction*

Fragmentation of the $[M\text{-}H]^-$ ion of commercially obtained *E. coli* lipid A ($m/z$ 1796) in negative ion mode generated a spectrum with large pyrophosphate ($m/z$ 177 and 159) and phosphate ($m/z$ 97 and 79) peaks, as observed by Jones et al. [42]. Those fragments were not observed as products of $m/z$ 1796 in the Chlorella cell wall extract. The plant *A. thaliana*, which contains genes homologous to those in the *E. coli* lipid A biosynthetic pathway, does not contain observable lipid A, but does contain lipid X, 2,3-diacylglucosamine-1-phosphate, an intermediate in lipid A biosynthesis [24]. Thus, we considered whether the Chlorella cell wall might contain lipid X. Li et al. demonstrated that there are four characteristic fragments of the lipid X [M-H]$^-$ ion (*m/z* 710): phosphate (*m/z* 97 and 79), glucosamine-phosphate (*m/z* 240), and hydroxymyristoylglucosamine-phosphate (*m/z* 466) [24]. Negative-ion fragmentation of *m/z* 710 in both a wild-type *A. thaliana* extract and the Chlorella cell wall extract generated peaks at *m/z* 466, *m/z* 97, and *m/z* 79, among others that were likely derived from other compounds of the same nominal mass (Figure 2). These results suggest that a small amount of lipid X may be present in the Chlorella cell wall.

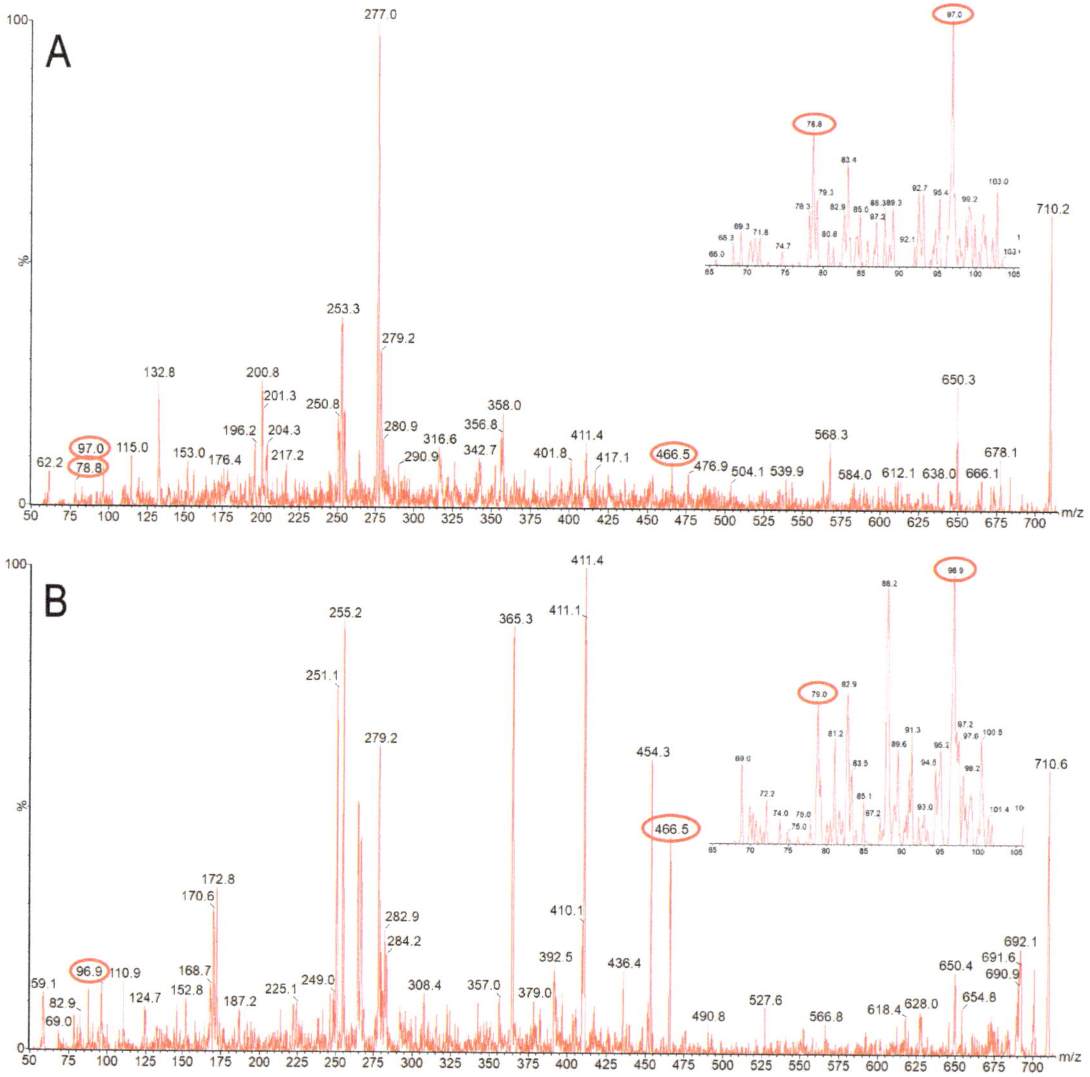


**Figure 2.** Product ion spectra of *m/z* 710.4, the *m/z* of the [M-H]$^-$ ion of lipid X. (**A**) *Arabidopsis thaliana* extract, (**B**) Chlorella cell wall fraction. Insets show the *m/z* range from 65 to 105. Fragments characteristic of lipid X are circled. Characteristic fragments include *m/z* 79, 97, 240, and 466, of which *m/z* 79, 97, and 466 were detected in both species.

### 3.3. CWE Treatment Stimulated the Growth of Bone Marrow Cells and Splenocytes

CWE treatment (0.1–10 μg/mL) significantly increased the growth of BMCs in a dose- and time-dependent manner (Figure 3A). However, the same dosage range of CWE treatment showed a highest growth stimulation of SPL at the lower dosage (0.1 μg/mL), but the higher dosages (1 and 10 μg/mL) showed less growth stimulation than that by the lower dosage (Figure 3B). The positive control, LPS derived from *Escherichia coli* 026:B6, also exhibited similar patterns with the CWE on BMCs. Comparison of CWE and authentic LPS demonstrated that their growth stimulation patterns are very similar. Hence, the aqueous extract of the Chlorella cell wall fraction contains LPS-like bioactivity, justifying our designation of this substance as an LPS-like substance. However, the functionality of the CWE is apparently different from that of the bacterial LPS; CWE can efficiently stimulate growth of hematopoietic precursor cells in bone marrow, but its effect on differentiated lymphocytes in spleen is limited, whereas the bacterial LPS can stimulate growth of hematopoietic precursor cells in bone marrow and differentiated lymphocytes in the spleen.

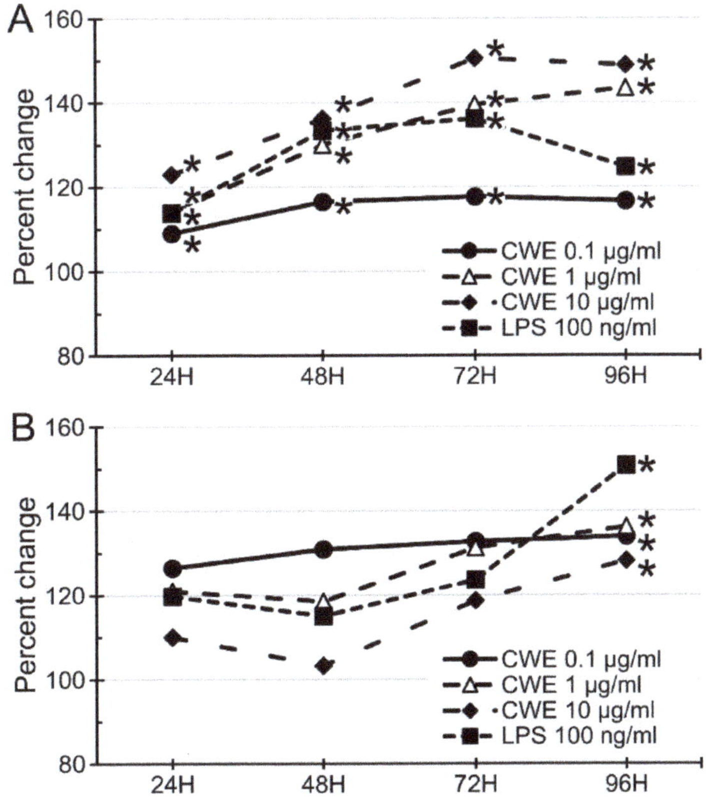


**Figure 3.** CWE dose- and time-dependently stimulated the growth of murine bone marrow cells (BMCs) and splenocytes (SPLs) in cell culture. CWE treatment dose-dependently (0.1–10 μg/mL) and time-dependently (24, 48 72 and 96 h) stimulated the growth of mouse BMCs (**A**) but not SPLs (**B**), 100 ng/mL LPS served as a positive control. The cell growth was evaluated by MTT (3-(4,5-dimethylthiazol-2-yl)-2,5-diphenyltetrazolium bromide) assay. The data was shown as percent change compared with phosphate buffered saline (PBS)-treated control. *, $p < 0.05$ compared to PBS-treated control ($n = 3$).

### 3.4. CWE Treatment Increased Lymphocyte and Antigen-Presenting Cell Populations but Decreased Neutrophil Population in Bone Marrow Cells

CWE treatment stimulated the growth of BMCs as a whole (Figure 3A). Detailed changes of immune cell populations in BMCs were analyzed by flow cytometry. As shown in Figure 4, treatment with 10 μg/mL CWE increased $CD4^+$ (56.1% increase vs. the PBS control group, $p < 0.05$), $CD8^+$ (78.2% increase, $p < 0.05$), $CD19^+$ (34.2% increase, $p < 0.05$), $CD68^+$ (77.6% increase, $p < 0.05$) and $33D1^+$ (41.4% increase, $p < 0.05$) cells, whereas the LY6G+ cell population was significantly decreased (86.2% decrease, $p < 0.05$) as compared with the PBS control group. Treatment with LPS (100 ng/mL) elicited a similar pattern of cell growth as that with CWE except $CD68^+$ macrophage population (LPS did not change this cell population, Figure 4). These results suggest that increases in cell populations caused by CWE treatment are primarily related to the adaptive immune system.

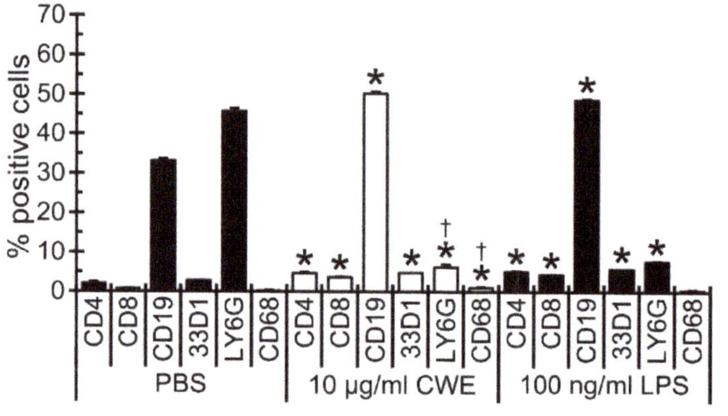


**Figure 4.** CWE treatment increased various immune cell populations in primary cultured mouse bone marrow cells (BMCs). Percentage of the population of BMCs treated with 10 μg/mL CWE and 100 ng/mL LPS for 48 h were evaluated by flow cytometry. These BMCs were labeled with various antibodies and relative quantities were analyzed by flow cytometry. PBS and LPS served as negative and positive controls, respectively. *, $p < 0.05$ compared to PBS control group; †, $p < 0.05$ compared to LPS-treated group ($n = 3$).

### 3.5. CWE Treatment Induced Expression of T Lymphocyte Activation-Associated Cytokines and Caused Morphological Differentiation of Lymphoblasts

As shown in Figure 4, CWE treatment increased the T lymphocyte population. In order to evaluate the effect of CWE for functional activation of T lymphocytes, the effect of CWE on cell proliferation and gene expression of TNFα, IFNγ, and granzyme B (GZMB) in Jurkat cells was measured by MTT assay and real-time PCR, respectively. Although CWE (1–100 μg/mL) only slightly altered the cell proliferation (a low concentration slightly increased, but a high concentration slightly decreased cell proliferation Figure 5A), LPS from *E. coli* significantly stimulated the cell proliferation in a dose-dependent manner. As shown in Figure 5B, the expression of TNFα, IFNγ, and GZMB was significantly upregulated at 24 h after CWE treatment and returned to same level as that of the PBS control group at 48 h. In contrast, LPS increased only TNFα and IFNγ expression, but not GZMB expression. In addition, when lymphoblasts were treated with CWE, they were enlarged and exhibited significant morphological differentiation (Figure 5C). The same kind of morphological differentiation was also observed in LPS-treated Jurkat cells (Figure 5C). These results suggest that CWE treatment stimulates activation of T lymphocytes. It also suggests that stimulatory activity of CWE for lymphoblast is different from that by an authentic LPS.

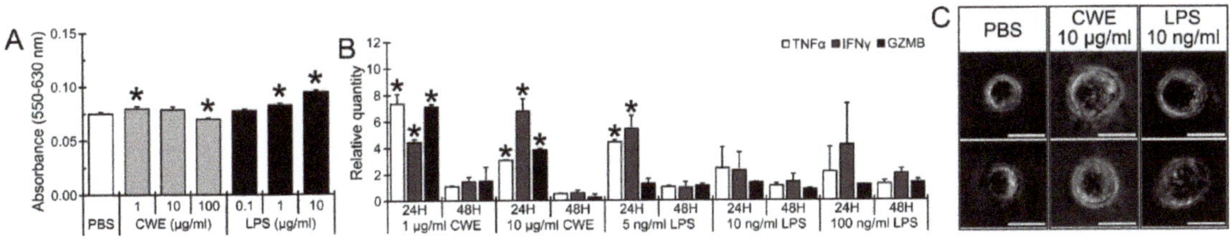

**Figure 5.** CWE treatment increased cytokine expressions and induced morphological differentiation in Jurkat cells. (**A**) Cell proliferation of Jurkat cells was evaluated by MTT assay in the presence or absence of CWE (1–100 μg/mL) or LPS (0.1–10 μg/mL) at 48 h after the treatment ($n = 3$). (**B**) T cell activation-associated cytokine expression in Jurkat cells treated with 1 or 10 μg/mL CWE, or 5, 10, or 100 ng/mL LPS at 24 and 48 h after treatment was measured by RT-qPCR. *, $p < 0.05$ compared to PBS control group ($n = 4$). (**C**) The Jurkat cells were grown in a U-shaped agar matrix (Day 0). The CWE (10 μg/mL) or LPS (10 ng/mL) was treated twice at Day 1 and Day 4. Typical pictures of CWE- or LPS-treated Jurkat cells in the agar matrix at Day 7. Scale bar, 20 μm.

## 4. Discussion

Chlorella has been used as a dietary supplement with high nutritional value worldwide, and a large number of publications indicate that Chlorella extracts may serve as sources of therapeutics against various diseases including cancers [14,43–46]. Practically all these studies have used a water or alcohol extract from whole Chlorella including crushed Chlorella powders. However, since whole Chlorella contains a large number of potentially bioactive substances including cytotoxic materials, it seems likely that most reported medicinal activities are due to a mixture of bioactive substances. On the other hand, specific roles for certain bioactive substances from Chlorella have been demonstrated. The polysaccharide of *C. pyrenoidosa* and *C. sorokiniana* stimulates innate immunity via TLR4 [47,48]. It has also been reported that a glycoprotein in *C. vulgaris* stimulates cytokine production in adherent splenocytes through the TLR2 signaling pathway in mice, suggesting that this glycoprotein stimulates innate immunity via TLR2 [49]. In addition, it should be noted that the cell wall of Chlorella contains lipopolysaccharide-like immunoreactivity [31,50]. Although it is not easy to pinpoint a single bioactive compound in natural products, it can be useful to determine the subcellular components from which the bioactive substances originate and their general chemical makeup. In an effort to identify the origin of the bioactivity and to reduce the number of bioactive substances in the extract, the present study focused on the cell wall fraction of *Chlorella sorokiniana* and its water extract, and on evaluating its immune modulatory abilities in in vitro.

Authenticity of the Chlorella cell wall was examined using both scanning and transmission electron microscopies. These morphological analyses clearly indicated that the materials used for the present study were composed entirely of cell walls and plasma membranes derived from Chlorella, and no evidence of intact cell body or bacterial contamination was found (Figure 1). The morphologies of the cell walls and associated membranes were identical to those of previously published paper by Northcote et al. [30].

Potential contamination by bacterial LPS was excluded by lack of detectable lipid A by mass spectrometry. On the other hand, the LAL assay, which responds to both lipid A, which is found in Gram-negative bacteria, but not in plants, and lipid X, which is found in both bacteria and plants, was present at low, but measurable levels (0.2 ng/μg dry extract). Mass spectrometry provided suggestive evidence for a low-level presence of the lipid A precursor, lipid X, but not lipid A, in the Chlorella cell wall fraction. As shown in Figure 2, in a parallel analysis of the plant *A. thaliana*, only lipid X was detected, while lipid A was not observed, as reported by Li et al. [24]. These data suggest that lipid X is potentially present in the Chlorella cell wall extract. However, whether lipid X is primarily responsive for CWE-dependent bioactivities awaits further studies, including complete purification and identification of its chemical nature.

The effect of CWE on the growth of murine bone marrow cells and splenocytes was evaluated in cell culture-based studies. Treatment with the CWE significantly increased the growth of immune cells in the bone marrow and spleen in a time- and dose-dependent manner (Figure 3). Clarification of the specific cell populations influenced by this treatment revealed that the CWE treatment increased the numbers of lymphocytes ($CD4^+$ and $CD8^+$ T cells, and $CD19^+$ B cells) and antigen-presenting cells ($33D1^+$ dendritic cells and $CD68^+$ macrophages) (Figure 4).

Furthermore, gene expression of T lymphocyte activation-associated cytokines was also found to be upregulated by CWE treatment in Jurkat cells (Figure 5B). CWE treatment also induced morphological differentiation in Jurkat cells in 3D culture (Figure 5C). However, gene expression profiles of T cell activation-associated cytokines in bacterial LPS-treated Jurkat cells were slightly different from those by CWE (Figure 5B). Although Jurkat cells are human lymphoblasts derived from an acute T cell leukemia patient, they have been used as a model cell for normal T lymphocytes. For example, Jurkat cells were used for the evaluation of IL-2-dependent granzyme B production, which is a marker of T lymphocyte activation [51,52]. Taken together, these results imply that CWE treatment appears to induce functional differentiation in T lymphocytes. Therefore, it appears that CWE is capable of T lymphocyte activation as well as leukocyte growth stimulation, which is very similar to bacteria cell wall-derived LPS [53,54]. In this regard, it should be noted that the cell wall of Chlorella contains lipopolysaccharide-like immunoreactivity [31,50]. In the present study, however, CWE slightly, but clearly, inhibited growth of Jurkat cells, whereas authentic LPS stimulated their growth in a parallel experiment (Figure 5A). Furthermore, analysis of BMC responses to CWE or LPS showed that $CD68^+$ macrophages respond only to CWE treatment (Figure 4). It is apparent that the action of CWE is different from LPS in stimulation of Jurkat and macrophages. These results suggest that the bioactive compound(s) in the Chlorella cell wall fraction is apparently distinct from bacterial LPS. However, determination of the detailed chemical nature of such bioactive compounds in the Chlorella cell wall fraction awaits future study. On the other hand, this CWE-induced direct differentiation and/or activation of T lymphocytes suggests that CWE is potentially a useful agent for cancer immune therapy, applicable to both primary and metastatic cancer. To the best of our knowledge, this is the first study to report that the extract from the Chlorella cell wall fraction stimulates the growth of immune cells in the spleen and bone marrow and induces the functional differentiation of T lymphocytes.

## 5. Conclusions

Analysis of CWE from the cell wall fraction of *Chlorella sorokiniana* indicated the presence of LPS-like immunoreactivity, and mass spectrometry suggested that a small amount of lipid X may be present in the Chlorella cell wall. The CWE significantly stimulates the growth of primary cultured mouse BMCs and SPLs. Treatment with the CWE also significantly increased the number of lymphocytes ($CD4^+$ and $CD8^+$ T cells, and $CD19^+$ B cells) and antigen-presenting cells ($33D1^+$ dendritic cells and $CD68^+$ macrophages) in primary cultured mouse BMCs. In a 3D culture, the CWE treatment caused morphological differentiation of lymphoblasts, i.e., Jurkat cells. These data show that CWE could be a useful agent for the stimulation of anti-tumor or anti-microbial immunity.

**Author Contributions:** Conceptualization, S.I., R.W. and M.T.; Methodology, S.I., R.W., M.R. and M.T.; Validation, S.I., M.R., R.W. and R.T.; Formal Analysis, S.I., M.R., R.W. and R.T.; Investigation, S.I., M.R., R.W., M.L., R.T., M.P., N.R. and D.U.; Resources, A.N. and K.S.; Data Curation, S.I., R.W. and M.T.; Writing—Original Draft Preparation, S.I., R.W., J.C. and M.T.; Writing—Review & Editing, S.I., R.W., J.C. and M.T.; Visualization, S.I., R.W., J.C. and M.T.; Supervision, R.W. and M.T.; Project Administration, R.W. and M.T.; Funding Acquisition, R.W. and M.T. All authors have read and agreed to the published version of the manuscript.

**Funding:** This work was supported in part by Kansas State University College of Veterinary Medicine Dean's funds [grant number 2015CVM-SMILE] (MT) and research funds from Euglena Co., Ltd. [grant number 2017EU-2] (MT). Instrument acquisition at the Kansas Lipidomics Research Center was supported by the National Science Foundation (including support from the Major Research Instrumentation program; most recent award DBI-1726527), K-IDeA Networks of Biomedical Research Excellence (INBRE) of National Institute of Health [P20GM103418], and Kansas State University.

**Institutional Review Board Statement:** All experiments were carried out under approvals by the Institutional Animal Care and Use Committee (Protocol # 4346) and Institutional Biosafety Committee (Registration # 1317) of the Kansas State University.

**Informed Consent Statement:** Not applicable.

**Data Availability Statement:** All data is available upon request to the corresponding author.

**Acknowledgments:** The authors thank Kaori Knights (Department of Diagnostic Medicine/Pathobiology, Kansas State University College of Veterinary Medicine) for his technical support on the flow cytometer. We are indebted to Yaeyama Shokusan Co. Ltd. (Okinawa, Japan) and Hideyuki Adachi (Euglena Co. Ltd.) for preparation of Chlorella cell wall.

**Conflicts of Interest:** The authors declare no conflict of interest.

## References

1. Darienko, T.; Rad-Menéndez, C.; Campbell, C.; Pröschold, T. Are there any true marine *Chlorella* species? Molecular phylogenetic assessment and ecology of marine Chlorella-like organisms, including a description of *Droopiella* gen. nov. *Syst. Biodivers.* **2019**, *17*, 811–829. [CrossRef] [PubMed]
2. de Morais, M.G.; Vaz, B.d.S.; de Morais, E.G.; Costa, J.A.V. Biologically Active Metabolites Synthesized by Microalgae. *BioMed Res. Int.* **2015**, *2015*, 835761. [CrossRef] [PubMed]
3. Kulkarni, S.; Nikolov, Z. Process for selective extraction of pigments and functional proteins from *Chlorella vulgaris*. *Algal Res.* **2018**, *35*, 185–193. [CrossRef]
4. Caporgno, M.P.; Mathys, A. Trends in Microalgae Incorporation into Innovative Food Products with Potential Health Benefits. *Front. Nutr.* **2018**, *5*, 58. [CrossRef] [PubMed]
5. Matos, J.; Cardoso, C.; Bandarra, N.M.; Afonso, C. Microalgae as healthy ingredients for functional food: A review. *Food Funct.* **2017**, *8*, 2672–2685. [CrossRef] [PubMed]
6. Ziganshina, E.E.; Bulynina, S.S.; Ziganshin, A.M. Comparison of the Photoautotrophic Growth Regimens of Chlorella sorokiniana in a Photobioreactor for Enhanced Biomass Productivity. *Biology* **2020**, *9*, 169. [CrossRef]
7. Chai, S.; Shi, J.; Huang, T.; Guo, Y.; Wei, J.; Guo, M.; Li, L.; Dou, S.; Liu, L.; Liu, G. Characterization of Chlorella sorokiniana growth properties in monosaccharide-supplemented batch culture. *PLoS ONE* **2018**, *13*, e0199873. [CrossRef]
8. Cecchin, M.; Benfatto, S.; Griggio, F.; Mori, A.; Cazzaniga, S.; Vitulo, N.; Delledonne, M.; Ballottari, M. Molecular basis of autotrophic vs mixotrophic growth in Chlorella sorokiniana. *Sci. Rep.* **2018**, *8*, 6465. [CrossRef]
9. Lizzul, A.M.; Lekuona-Amundarain, A.; Purton, S.; Campos, L.C. Characterization of Chlorella sorokiniana, UTEX 1230. *Biology* **2018**, *7*, 25. [CrossRef]
10. Justo, G.Z.; Silva, M.R.; Queiroz, M.L. Effects of the green algae *Chlorella vulgaris* on the response of the host hematopoietic system to intraperitoneal ehrlich ascites tumor transplantation in mice. *Immunopharmacol. Immunotoxicol.* **2001**, *23*, 119–132. [CrossRef]
11. Konishi, F.; Tanaka, K.; Himeno, K.; Taniguchi, K.; Nomoto, K. Antitumor effect induced by a hot water extract of *Chlorella vulgaris* (CE): Resistance to Meth-A tumor growth mediated by CE-induced polymorphonuclear leukocytes. *Cancer Immunol. Immunother. CII* **1985**, *19*, 73–78. [CrossRef] [PubMed]
12. Miyazawa, Y.; Murayama, T.; Ooya, N.; Wang, L.F.; Tung, Y.C.; Yamaguchi, N. Immunomodulation by a unicellular green algae (*Chlorella pyrenoidosa*) in tumor-bearing mice. *J. Ethnopharmacol.* **1988**, *24*, 135–146. [CrossRef]
13. Noda, K.; Ohno, N.; Tanaka, K.; Kamiya, N.; Okuda, M.; Yadomae, T.; Nomoto, K.; Shoyama, Y. A water-soluble antitumor glycoprotein from *Chlorella vulgaris*. *Planta Med.* **1996**, *62*, 423–426. [CrossRef]
14. Ramos, A.L.; Torello, C.O.; Queiroz, M.L. *Chlorella vulgaris* modulates immunomyelopoietic activity and enhances the resistance of tumor-bearing mice. *Nutr. Cancer* **2010**, *62*, 1170–1180. [CrossRef] [PubMed]
15. Tanaka, K.; Konishi, F.; Himeno, K.; Taniguchi, K.; Nomoto, K. Augmentation of antitumor resistance by a strain of unicellular green algae, *Chlorella vulgaris*. *Cancer Immunol. Immunother. CII* **1984**, *17*, 90–94. [CrossRef] [PubMed]
16. Tanaka, K.; Tomita, Y.; Tsuruta, M.; Konishi, F.; Okuda, M.; Himeno, K.; Nomoto, K. Oral administration of *Chlorella vulgaris* augments concomitant antitumor immunity. *Immunopharmacol. Immunotoxicol.* **1990**, *12*, 277–291. [CrossRef]
17. Tanaka, K.; Yamada, A.; Noda, K.; Hasegawa, T.; Okuda, M.; Shoyama, Y.; Nomoto, K. A novel glycoprotein obtained from *Chlorella vulgaris* strain CK22 shows antimetastatic immunopotentiation. *Cancer Immunol. Immunother. CII* **1998**, *45*, 313–320. [CrossRef]
18. Bertani, B.; Ruiz, N. Function and Biogenesis of Lipopolysaccharides. *EcoSal Plus* **2018**, *8*. [CrossRef]

19. Konovalova, A.; Kahne, D.E.; Silhavy, T.J. Outer Membrane Biogenesis. *Annu. Rev. Microbiol.* **2017**, *71*, 539–556. [CrossRef]
20. Rosadini, C.V.; Kagan, J.C. Early innate immune responses to bacterial LPS. *Curr. Opin. Immunol.* **2017**, *44*, 14–19. [CrossRef]
21. Durai, P.; Batool, M.; Choi, S. Structure and Effects of Cyanobacterial Lipopolysaccharides. *Mar. Drugs* **2015**, *13*, 4217–4230. [CrossRef] [PubMed]
22. Slutzky, G.M.; Londner, M.V.; Greenblatt, C.L. Lipid and lipopolysaccharide-like antigens of Leishmania promastigotes. *J. Protozool.* **1985**, *32*, 347–352. [CrossRef] [PubMed]
23. Bedick, J.C.; Shnyra, A.; Stanley, D.W.; Pardy, R.L. Innate immune reactions stimulated by a lipopolysaccharide-like component of the alga Prototheca (strain 289). *Die Nat.* **2001**, *88*, 482–485. [CrossRef] [PubMed]
24. Li, C.; Guan, Z.; Liu, D.; Raetz, C.R. Pathway for lipid A biosynthesis in Arabidopsis thaliana resembling that of *Escherichia coli*. *Proc. Natl. Acad. Sci. USA* **2011**, *108*, 11387–11392. [CrossRef]
25. Mehanna, E.T.; Ali, A.A.; El-Shaarawy, F.; Mesbah, N.M.; Abo-Elmatty, D.M.; Aborehab, N.M. Anti-Oxidant and Anti-Inflammatory Effects of Lipopolysaccharide from *Rhodobacter sphaeroides* against Ethanol-Induced Liver and Kidney Toxicity in Experimental Rats. *Molecules* **2021**, *26*, 7437. [CrossRef]
26. Murakami, K.; Kamimura, D.; Hasebe, R.; Uchida, M.; Abe, N.; Yamamoto, R.; Jiang, J.J.; Hidaka, Y.; Nakanishi, Y.; Fujita, S.; et al. *Rhodobacter azotoformans* LPS (RAP99-LPS) Is a TLR4 Agonist That Inhibits Lung Metastasis and Enhances TLR3-Mediated Chemokine Expression. *Front. Immunol.* **2021**, *12*, 675909. [CrossRef]
27. Van Amersfoort, E.S.; Van Berkel, T.J.; Kuiper, J. Receptors, mediators, and mechanisms involved in bacterial sepsis and septic shock. *Clin. Microbiol. Rev.* **2003**, *16*, 379–414. [CrossRef]
28. Baldridge, J.R.; Crane, R.T. Monophosphoryl lipid A (MPL) formulations for the next generation of vaccines. *Methods* **1999**, *19*, 103–107. [CrossRef]
29. Loos, E.; Meindl, D. Composition of the cell wall of *Chlorella fusca*. *Planta* **1982**, *156*, 270–273. [CrossRef]
30. Northcote, D.H.; Goulding, K.J.; Horne, R.W. The chemical composition and structure of the cell wall of *Chlorella pyrenoidosa*. *Biochem. J.* **1958**, *70*, 391–397. [CrossRef]
31. Armstrong, P.B.; Armstrong, M.T.; Pardy, R.L.; Child, A.; Wainwright, N. Immunohistochemical demonstration of a lipopolysaccharide in the cell wall of a eukaryote, the green alga, Chlorella. *Biol. Bull.* **2002**, *203*, 203–204. [CrossRef] [PubMed]
32. Ishiguro, S.; Robben, N.; Burghart, R.; Cote, P.; Greenway, S.; Thakkar, R.; Upreti, D.; Nakashima, A.; Suzuki, K.; Comer, J.; et al. Cell Wall Membrane Fraction of Chlorella sorokiniana Enhances Host Antitumor Immunity and Inhibits Colon Carcinoma Growth in Mice. *Integr. Cancer Ther.* **2020**, *19*, 1534735419900555. [CrossRef]
33. Hannan, P.J.; Patouillet, C. Gas Exchange with Mass Cultures of Algae. I. Effects of Light Intensity and Rate of Carbon Dioxide Input on Oxygen Production. *Appl. Microbiol.* **1963**, *11*, 446–449. [CrossRef] [PubMed]
34. Henderson, J.C.; O'Brien, J.P.; Brodbelt, J.S.; Trent, M.S. Isolation and chemical characterization of lipid A from gram-negative bacteria. *J. Vis. Exp. JoVE* **2013**, *79*, e50623. [CrossRef]
35. Welti, R.; Li, W.; Li, M.; Sang, Y.; Biesiada, H.; Zhou, H.E.; Rajashekar, C.B.; Williams, T.D.; Wang, X. Profiling membrane lipids in plant stress responses. Role of phospholipase D alpha in freezing-induced lipid changes in Arabidopsis. *J. Biol. Chem.* **2002**, *277*, 31994–32002. [CrossRef] [PubMed]
36. Markham, J.E.; Li, J.; Cahoon, E.B.; Jaworski, J.G. Separation and identification of major plant sphingolipid classes from leaves. *J. Biol. Chem.* **2006**, *281*, 22684–22694. [CrossRef]
37. Toledo, M.S.; Suzuki, E.; Straus, A.H.; Takahashi, H.K. Glycolipids from Paracoccidioides brasiliensis. Isolation of a galactofuranose-containing glycolipid reactive with sera of patients with paracoccidioidomycosis. *J. Med. Vet. Mycol. Bi-Mon. Publ. Int. Soc. Hum. Anim. Mycol.* **1995**, *33*, 247–251. [CrossRef]
38. Schmittgen, T.D.; Livak, K.J. Analyzing real-time PCR data by the comparative C(T) method. *Nat. Protoc.* **2008**, *3*, 1101–1108.
39. Ishiguro, S.; Uppalapati, D.; Goldsmith, Z.; Robertson, D.; Hodge, J.; Holt, H.; Nakashima, A.; Turner, K.; Tamura, M. Exopolysaccharides extracted from Parachlorella kessleri inhibit colon carcinoma growth in mice via stimulation of host antitumor immune responses. *PLoS ONE* **2017**, *12*, e0175064. [CrossRef]
40. Proctor, R.A.; Textor, J.A. Activation and inhibition of Limulus amebocyte lysate coagulation by chemically defined substructures of lipid A. *Infect. Immun.* **1985**, *49*, 286–290. [CrossRef]
41. Takayama, K.; Qureshi, N.; Raetz, C.R.; Ribi, E.; Peterson, J.; Cantrell, J.L.; Pearson, F.C.; Wiggins, J.; Johnson, A.G. Influence of fine structure of lipid A on Limulus amebocyte lysate clotting and toxic activities. *Infect. Immun.* **1984**, *45*, 350–355. [CrossRef]
42. Jones, J.W.; Shaffer, S.A.; Ernst, R.K.; Goodlett, D.R.; Turecek, F. Determination of pyrophosphorylated forms of lipid A in Gram-negative bacteria using a multivaried mass spectrometric approach. *Proc. Natl. Acad. Sci. USA* **2008**, *105*, 12742–12747. [CrossRef] [PubMed]
43. Ibusuki, K.; Minamishima, Y. Effect of *Chlorella vulgaris* extracts on murine cytomegalovirus infections. *Nat. Immun. Cell Growth Regul.* **1990**, *9*, 121–128. [PubMed]
44. Merchant, R.E.; Andre, C.A.; Sica, D.A. Nutritional supplementation with *Chlorella pyrenoidosa* for mild to moderate hypertension. *J. Med. Food* **2002**, *5*, 141–152. [CrossRef] [PubMed]
45. Okamoto, K.; Iizuka, Y.; Murakami, T.; Miyake, H.; Suzuki, T. Effects of chlorella alkali extract on blood pressure in SHR. *Jpn. Heart J.* **1978**, *19*, 622–623.
46. Sano, T.; Tanaka, Y. Effect of dried, powdered *Chlorella vulgaris* on experimental atherosclerosis and alimentary hypercholesterolemia in cholesterol-fed rabbits. *Artery* **1987**, *14*, 76–84.

47. Chou, N.T.; Cheng, C.F.; Wu, H.C.; Lai, C.P.; Lin, L.T.; Pan, I.H.; Ko, C.H. Chlorella sorokiniana-Induced Activation and Maturation of Human Monocyte-Derived Dendritic Cells through NF-kappaB and PI3K/MAPK Pathways. *Evid. Based Complement. Altern. Med. Ecam* **2012**, *2012*, 735396. [CrossRef]
48. Hsu, H.Y.; Jeyashoke, N.; Yeh, C.H.; Song, Y.J.; Hua, K.F.; Chao, L.K. Immunostimulatory bioactivity of algal polysaccharides from *Chlorella pyrenoidosa* activates macrophages via Toll-like receptor 4. *J. Agric. Food Chem.* **2010**, *58*, 927–936. [CrossRef]
49. Hasegawa, T.; Matsuguchi, T.; Noda, K.; Tanaka, K.; Kumamoto, S.; Shoyama, Y.; Yoshikai, Y. Toll-like receptor 2 is at least partly involved in the antitumor activity of glycoprotein from *Chlorella vulgaris*. *Int. Immunopharmacol.* **2002**, *2*, 579–589.
50. Armstrong, M.T.; Theg, S.M.; Braun, N.; Wainwright, N.; Pardy, R.L.; Armstrong, P.B. Histochemical evidence for lipid A (endotoxin) in eukaryote chloroplasts. *FASEB J. Off. Publ. Fed. Am. Soc. Exp. Biol.* **2006**, *20*, 2145–2146. [CrossRef]
51. Huang, C.; Bi, E.; Hu, Y.; Deng, W.; Tian, Z.; Dong, C.; Hu, Y.; Sun, B. A novel NF-kappaB binding site controls human granzyme B gene transcription. *J. Immunol.* **2006**, *176*, 4173–4181. [CrossRef]
52. Voskoboinik, I.; Whisstock, J.C.; Trapani, J.A. Perforin and granzymes: Function, dysfunction and human pathology. *Nat. Rev. Immunol.* **2015**, *15*, 388–400. [CrossRef] [PubMed]
53. Hornef, M.W.; Wick, M.J.; Rhen, M.; Normark, S. Bacterial strategies for overcoming host innate and adaptive immune responses. *Nat. Immunol.* **2002**, *3*, 1033–1040. [CrossRef] [PubMed]
54. Ulmer, A.J.; Flad, H.; Rietschel, T.; Mattern, T. Induction of proliferation and cytokine production in human T lymphocytes by lipopolysaccharide (LPS). *Toxicology* **2000**, *152*, 37–45. [CrossRef]

*nutrients*

Article

# Different Seasonal Collections of *Ficus carica* L. Leaves Diversely Modulate Lipid Metabolism and Adipogenesis in 3T3-L1 Adipocytes

Mariachiara Pucci [1,†], Manuela Mandrone [2,†], Ilaria Chiocchio [2], Eileen Mac Sweeney [1], Emanuela Tirelli [1], Daniela Uberti [1], Maurizio Memo [1], Ferruccio Poli [2], Andrea Mastinu [1,*,†] and Giulia Abate [1,†]

1 Department of Molecular and Translational Medicine, Division of Pharmacology, University of Brescia, 25123 Brescia, Italy; m.pucci003@unibs.it (M.P.); e.macsweeney@studenti.unibs.it (E.M.S.); e.tirelli004@unibs.it (E.T.); daniela.uberti@unibs.it (D.U.); maurizio.memo@unibs.it (M.M.); giulia.abate@unibs.it (G.A.)

2 Department of Pharmacy and Biotechnology (FaBiT), University of Bologna, Via Irnerio 42, 40126 Bologna, Italy; manuela.mandrone2@unibo.it (M.M.); ilaria.chiocchio2@unibo.it (I.C.); ferruccio.poli@unibo.it (F.P.)

* Correspondence: andrea.mastinu@unibs.it; Tel.: +39-030-371-7509

† These authors contributed equally to this work.

**Citation:** Pucci, M.; Mandrone, M.; Chiocchio, I.; Sweeney, E.M.; Tirelli, E.; Uberti, D.; Memo, M.; Poli, F.; Mastinu, A.; Abate, G. Different Seasonal Collections of *Ficus carica* L. Leaves Diversely Modulate Lipid Metabolism and Adipogenesis in 3T3-L1 Adipocytes. *Nutrients* **2022**, *14*, 2833. https://doi.org/10.3390/nu14142833

Academic Editor: Hayato Tada

Received: 28 May 2022
Accepted: 5 July 2022
Published: 10 July 2022



**Abstract:** Due to the high prevalence of obesity and type 2 diabetes, adipogenesis dysfunction and metabolic disorders are common features in the elderly population. Thus, the identification of novel compounds with anti-adipogenic and lipolytic effects is highly desirable to reduce diabetes complications. Plants represent an important source of bioactive compounds. To date, the antidiabetic potential of several traditional plants has been reported, among which *Ficus carica* L. is one of the most promising. Considering that plant metabolome changes in response to a number of factors including seasonality, the aim of this study was to evaluate whether *Ficus carica* leaves extracts collected in autumn (FCa) and spring (FCs) differently modulate lipid metabolism and adipogenesis in 3T3-L1 adipocytes. The $^1$H-NMR profile of the extracts showed that FCs have a higher content of caffeic acid derivatives, glucose, and sucrose than FCa. In contrast, FCa showed a higher concentration of malic acid and furanocoumarins, identified as psoralen and bergapten. In vitro testing showed that only FCa treatments were able to significantly decrease the lipid content (Ctrl vs. FCa 25 μg/mL, 50 μg/mL and 80 μg/mL; $p < 0.05$, $p < 0.01$ and $p < 0.001$, respectively). Furthermore, FCa treatments were able to downregulate the transcriptional pathway of adipogenesis and insulin sensitivity in 3T3-L1 adipocytes. In more detail, FCa 80 μg/mL significantly decreased the gene expression of PPARγ ($p < 0.05$), C/EBPα ($p < 0.05$), Leptin ($p < 0.0001$), adiponectin ($p < 0.05$) and GLUT4 ($p < 0.01$). In conclusion, this study further supports an in-depth investigation of *F. carica* leaves extracts as a promising source of active compounds useful for targeting obesity and diabetes.

**Keywords:** seasonality; *Ficus carica* L.; furanocoumarins; $^1$H-NMR profile; lipid metabolism; adipogenesis

## 1. Introduction

Metabolic syndrome and diabetes pose a significant health burden worldwide and their incidence was found to be highly increased due to recent and alarming changes in food habits and lifestyle [1]. Metabolic syndrome is described as a pre-diabetic condition that includes obesity, dyslipidemia, impaired fasting glucose and/or impaired glucose tolerance, and reduced insulin sensitivity [2]. Diabetes mellitus (DM) is instead a chronic condition characterized by abnormal glucose homeostasis, resulting in elevated plasma glucose levels [3]. When prolonged in life, chronic hyperglycemia is associated with a myriad of long-term DM-related clinical sequelae such as damage to pancreatic β cells and microvascular complications, including retinopathy, nephropathy, and neuropathy [3].

The International Diabetes Federation has estimated that 463 million adults lived with diabetes worldwide in 2019, with a projected increase to 700 million by 2045 if effective prevention or strict glycemic control methods are not be adopted in DM patients before disease progression [4].

Several risk factors are known to be involved in the development of diabetes, including genetic factors (i.e., family history), lifestyle, diet, age, ethnicity, high blood glucose during pregnancy, hypertension, dyslipidemia, and obesity [5]. Of note, the increasing literature highlights a contributing role of adipose tissue dysfunction and hyperproliferation in inducing metabolic syndrome and diabetes [6].

Adipose tissue is highly dynamic and is the largest organ in humans involved in lipid storage and mobilization on the basis of energy requirements. It consists of several types of cells, including mature adipocytes that can increase in number and size. Decreasing proliferation and adipogenesis at the early stage of adipocyte differentiation might represent potential targets for preventing or treating obesity and, in turn, diabetes [7].

Besides insulin supplementation, to reduce the cluster of inter-related events of hyperglycemia and hyperlipidemia, a multi-targeted approach controlling both glucose and lipid metabolism is needed. The currently available non-pharmacological options are exercise or dietary modification [8], while pharmacological options include treatment with anti-diabetic, and anti-hyperlipidemic drugs [3].

Unfortunately, these latter therapeutic treatments have undesirable side effects, leading to a growing interest in traditional medicine to treat metabolic disorders, focusing mainly on herbal medicines [9,10]. Noteworthy is the development of metformin, one of the main anti-hyperglycemic agents, which can be traced back to the traditional use of *Galega officinalis* to treat diabetes [11,12]. With specific regard to the treatment of diabetes, obesity, and blood glucose control, several traditional plants with anti-diabetic and anti-obesity activities have already been reported. Among them, *Ficus carica* L. is considered to be one of the most promising [13,14].

*Ficus carica* is the most popular species of the genus *Ficus* (Moraceae family). It is native to Southwest Asia but has been widely cultivated throughout the Mediterranean basin since ancient times. Overall, the genus *Ficus* is one of the most abundant and varied genera of Angiosperms, including more than 800 species widely distributed in the tropical and subtropical belt of Southeast Asia and South America [15,16]. Traditional systems of medicine such as Ayurveda, Unani, and Siddha use different organs of multiple *Ficus* species such as *F. benghalensis*, *F. religiosa*, *F. glumosa*, *F. deltoidea*, *F. racemosa* and *F. carica* to treat metabolic, respiratory, gastrointestinal, reproductive and infectious disorders [17,18]. In support of traditional use, the bioactivity of *F. carica* extracts was also the object of several studies demonstrating its numerous pharmacological properties, among which include antioxidant, anticancer, anti-inflammatory and, remarkably, antidiabetic and hypolipidemic properties, in both in vitro and in vivo models [19,20]. More specifically, the data in the literature identify *F. carica* leaf extracts as an optimal source of anti-diabetic, anti-hyperglycemic and anti-obesity compounds that can be used as food supplements in diabetes prevention [16,21].

The biological activities of *F. carica* are mainly associated with the presence of heterogeneous phytoconstituents such as phytosterols, anthocyanins, amino acids, organic acids, fatty acids, phenolics, furanocoumarins hydrocarbons and aliphatic alcohols, which have been found in the latex, leaves, fruits and roots [22]. Nevertheless, contradictory results are present, probably due not only to the lack of standardization but also to the high variability in the phytoconstituents. Both these parameters can be influenced by diverse factors such as different preparations of the plant extracts, plant harvesting seasonality, drought stress and soil conditions [23–25].

Therefore, the main objective of this study was to evaluate whether the effects of *F. carica* leaf extracts on glucose and lipid metabolism in fully differentiated 3T3-L1 adipocytes depended on different phytochemical profiles that, in turn, were potentially due to different harvesting seasons. *Ficus carica* leaves were collected in spring and autumn (before and after fructification), the differences in metabolite content were explored by $^1$H-NMR profiling and the activity of the extracts as key adipogenic gene modulators with anti-adipogenic and anti-obesity properties was tested in an *in vitro* model.

## 2. Materials and Methods

### 2.1. Plant Material and Extract Preparation

*Ficus carica* leaves were collected in November (FCa) and May (FCs) 2020 from a tree growing in the botanical garden of Bologna "Orto dei Semplici". Voucher specimen (BOLO0602009) were retained in the Herbarium of Alma Mater Studiorum-University of Bologna. The leaves, deprived of the stalk, were dried in stove for three days at 40 °C and consequentially grinded. For $^1$H-NMR profiling, 30 mg of plant material were extracted with 1 mL of mixture (1:1) of phosphate buffer (90 mM; pH 6.0) in $H_2O$-$d_2$ (containing 0.1% TMSP) and MeOH-$d_4$ by ultrasonication (TransSonic TP 690, Elma, Germany) for 20 min at 45 °C. After this procedure, samples were centrifuged for 10 min (17,000× *g*), then 700 μL of supernatant were transferred into NMR tubes. For bioactivity tests, 120 mg of plant material were extracted in 6 mL of MeOH:$H_2O$ (1:1), then sonicated for 30 min and centrifuged for 20 min. The supernatant was then divided into 3 × 2 mL test tubes and dried in a vacuum concentrator (Savant SpeedVac SPD210, Thermo Fisher Scientific, Waltham, MA, USA). Finally, it was resuspended in DMSO and then tested at different concentrations on cells.

### 2.2. Liquid-Liquid Partition of Ficus Carica Extracts

In order to elucidate the structure of the furanocoumarins visible in the $^1$H NMR profile of the extracts, FCa and FCs were subjected to a pre-purification procedure: 2 g of powder material were extracted in 100 mL of MeOH:$H_2O$ (1:1), then sonicated for 30 min and centrifuged for 15 min. The supernatant was dried in a rotary evaporator yielding 400 mg of extract, which was resuspended into 30 mL of water to undergo liquid-liquid partition with $CHCl_3$ and EtOAc used in sequence, repeating the procedure three times for each solvent. The three fractions were dried in a rotary evaporator and analyzed by $^1$H NMR. The chloroform fraction contained the diagnostic signals of the furanocoumarins; therefore, it was subjected to 2D NMR analysis and the two major compounds were identified as psoralen and bergapten

### 2.3. NMR Analysis

$^1$H NMR spectra, J-resolved (J-res), $^1$H-$^1$H homonuclear (COSY) and inverse detected $^1$H-$^{13}$C correlation experiments (HMBC, HSQC) were recorded at 25 °C on a Varian Inova (Milan, Italy) instrument (equipped with a reverse triple resonance probe). For $^1$H NMR profiling the instrument was operated at a $^1$H NMR frequency of 600.13 MHz, and $H_2O$-$d_2$ was used as the internal lock. Each $^1$H NMR spectrum consisted of 256 scans (corresponding to 16 min) with the relaxation delay (RD) of 2 s, acquisition time 0.707 s, and spectral width of 9595.8 Hz (corresponding to δ 16.0). A pre-saturation sequence (PRESAT) was used to suppress the residual water signal at δ 4.83 (power = −6 dB, pre-saturation delay 2 s). The spectra were manually phased and baseline corrected and calibrated to the internal standard trimethyl silyl propionic acid sodium salt (TMSP) at δ 0.0 using Mestrenova software (Mestrelab Research, Spain). The analysis of $^1$H-NMR profiles of extracts was performed based on an in-house library and in comparison with the literature [26–29].

### 2.4. Cells Culture and Treatment

3T3-L1 murine pre-adipocyte cells, purchased from Sigma Aldrich (Sigma-Aldrich, St. Louis, MO, USA), were cultured in a Dulbecco's Modified Eagle's Medium (DMEM) containing 10% FBS (Sigma Aldrich) at 37 °C and 5% $CO_2$ until ~80% cells confluence. Then, cells were seeded at a density of 1 × $10^4$ per well in 24-well plates. After 48 h, ~80% confluence was reached and cells were induced to differentiate (DAY 0) in MDI (Methylisobutylxanthine, Dexamethasone, Insulin) induction medium for 2 days. In particular, MDI induction medium is composed of 0.5 mM 3-isobutil-1-methylxantine, 1 μM dexamethasone and 10 μg/mL insulin in DMEM containing 10% FBS. At differentiation on DAY 3, the MDI medium was switched to a growth medium supplemented with 10 μg/mL insulin (DMEM containing 10% FBS and 10 μg/mL insulin). Medium was replaced every 2 days until fully differentiated adipocytes were obtained (DAY 11) [30,31]. Thus, at DAY 11, >90% of cells showed the characteristic adipocyte phenotype with accumulated lipid droplets. To test the effect of 48 h treatments with FC extracts, 3T3-L1 adipocytes were treated at DAY 9 with different dosages (25, 50, 80 and 100 μg/mL) of FCa and FCs and after 48 h (DAY 11), the MTT test was performed.

### 2.5. Cell Viability

The 3T3-L1 preadipocytes were seeded at a density of $1 \times 10^4$ cells per well in 96-well plates and incubated in culture medium. The cells were then treated with different dosages (25–100 μg/mL) of FCa and FCs. After 48 h, cells were incubated with 500 mg/mL of MTT (3-4,5-dimethylthiazol-2-yl)-2,5-diphenyltetrazolium bromide) for 1h at 37 °C. The supernatants were then removed and DMSO was added to each well. Plates were agitated to dissolve the formazan crystal products. The absorbance at 595 nm was measured using a Bio-Rad 3350 microplate reader (Bio Rad Laboratories, Richmond, CA, USA). The percentage of viable cells was calculated by defining the viability of untreated cells as 100%.

### 2.6. Red O Staining

3T3-L1 adipocytes were washed twice with PBS and fixed with 4% paraformaldehyde in PBS for 15 min at room temperature. After fixation, cells were washed two times with PBS and one time with 60% isopropanol. Then, cells were stained with filtered Oil-Red-O working solution (working solution: 60% Oil-Red-O stock solution and 40% distilled water) for 1 h at room temperature, washed 4 times with distilled water, and plates were dried and scanned for images. Oil-Red-O dye was then extracted using 100% isopropanol alcohol and measured spectrophotometrically at 490 nm.

### 2.7. Quantitative Real-Time PCR

Total RNA was extracted from $1 \times 10^4$ cells following TRIzol® reagent protocol (Invitrogen Corporation, Carlsbad, CA, USA). Then, 2μg of total RNA was retrotranscribed with M-MLV reverse transcriptase (Promega, Madison, Wisconsin, USA), using oligo-dT (15–18) as a primer in a final volume of 40μL reaction mix containing 1X RT buffer with $MgCl_2$ 5 mM, DDT 10 mM, oligo-dT 5 mM, dNTPs 1 mM, RNase inhibitor 1U/mL and Reverse Transcriptase (M-MLV Reverse Transcriptase, Invitrogen) 10 U/mL. The reaction was incubated at 70 °C for 10 min and then at 4 °C for 2 min, followed by 37 °C for 60 min as reported in [32]. SYBR Green-based Real-Time PCR was used to determine cDNA levels. PPAR-γ, C/EBP-α, Adiponectin, Leptin, GLUT4, FAS and β-actin primers reported in Table 1 were provided by Metabion (Metabion International AG, Planegg, Germany). β-actin was used as the endogenous reference. Quantitative RT-PCR was performed with the ViiA7 Real-Time PCR System (Applied Biosystems, Foster City, CA, USA) using the iQ™SYBR Green Supermix method (Bio-Rad Laboratories, Richmond, CA, USA) according to the manufacturer's instructions. Samples were run in triplicate in a 25 μL of reaction mix containing 12.5 μL x SYBR Green Master Mix (Bio-Rad), 12.5 pmol of each forward and reverse primer and 2 μL of diluted cDNA. The PCR mixtures were incubated at 95 °C for 10 min, followed by 40 cycles at 95 °C for 15 s and 60 °C for 60 s. A subsequent dissociation curve analysis verified the Ct for the target gene minus the mean of the $C_t$ for the internal control gene. The Ct represented the mean difference between the Ct of the test minus the $C_t$ of the calibrator. The N-fold differential expression in the target gene of the test compared with the calibrator (β-actin) was expressed as $2^{-\Delta\Delta Ct}$. Data analysis and graphics were performed using GraphPad Prism 9 software (GraphPad, San Diego, CA, USA).

Table 1. Primers used for q-PCR.

| Genes | Primer Sequences |
|---|---|
| Peroxisome Proliferator Activated Receptor Gamma (PPAR-γ) | f-5′-TCG CTG ATG CAC TGC CTA TG -3′; r-5′- GAG AGG TCC ACA GAG CTG ATT-3′ |
| CCAAT Enhancer Binding Protein Alpha (C/EBP-α) | f-5′-GTA ACC TTG TGC CTT GGA TAC T-3′; r-5′-GGA AGC AGG AAT CCT CCA AAT A-3′ |
| Leptin | f-5′-TCT TTC CGG AAC ATT TGG AG-3′; r-5′-TGT GAG ATC AAC CCT GGA CA-3′ |
| Adiponectin | f-5′-GAA GCC GCT TAT GTG TAT CGC-3′; r-5′-GAA TGG GTA CAT TGG GAA CAG T-3′ |
| Glucose Transporter type 4 (GLUT4) | f-5′-GAT TCT GCT GCC CTG TC-3′; r-5′-ATT GGA CGC TCT CTC TCC AA-3′ |
| Fatty Acid Synthase (FAS) | f-5′-AGA CCC GAA CTC CAA GTT ATT C-3′; r-5′-GCA GCT CCT TGT ATA CTT CTC C-3′ |
| Actin (β-actin) | f-5′-AGC CAT GTA CGT AGC CAT CC-3′ r-5′-CTC TCA GCT GTG GTG GTG AA-3′ |

### 2.8. Statistical Analysis

Statistical differences were determined by the analysis of variance (one-way ANOVA) followed, when significant, by an appropriate post hoc test; value of $p \leq 0.05$ was considered statistically significant. The results are reported as mean ± standard error mean (SEM) of at least three independent experiments.

## 3. Results

### 3.1. NMR Profiling and Compounds Identification

This work represents the first report of $^1$H-NMR profiling of *Ficus carica* leaves. As reported in Figure 1, the main metabolites identified in the samples were both primary such as: sucrose, α-glucose, β-glucose, malic acid, alanine and valine, and secondary such as: bergapten, psoralen and a caffeic acid derivative.

The structure of the furanocoumarins (bergapten and psoralen) was confirmed by further 2D NMR experiments (Tables 2 and 3 and Supplementary Figures S1–S4) performed on a pre-purified chloroform fraction obtained by liquid/liquid partition. The leaves collected in May (FCs) showed higher content of the caffeic acid derivative, glucose and sucrose compared to the sample collected in November (FCa). Conversely, the concentration of malic acid and furanocoumarins is higher in FCa than FCs. The quantity of psoralen was always greater than that of bergapten.

Table 2. NMR spectral references of psoralen.

| Position | Integrated Protons | $^1$H, δ, m, J (Hz) | $^{13}$C HSQC | HMBC Correlations | COSY Correlations |
|---|---|---|---|---|---|
| 2 | C | - | 161.59 | - | - |
| 3 | CH | 6.37, d, $J$ = 9.77 | 113.64 | 2,10 | 4 |
| 4 | CH | 8.03, d, $J$ = 9.77 | 145.06 | 2,5,9 | 3,8,5 |
| 5 | CH | 7.88, s | 120.28 | 4,7,8,9,3′ | 4,8 |
| 6 | C | - | 125.26 | - | - |
| 7 | C | - | 156.46 | - | - |
| 8 | CH | 7.52, s | 98.92 | 6,7,9,10 | 3′,4,2 |
| 9 | C | - | 151.94 | - | - |
| 10 | C | - | 115.48 | - | - |
| 2′ | CH | 7.86, d, $J$ = 2.33 | 147.23 | 6,3′ | 3′ |
| 3′ | CH | 6.95, dd, $J1$ = 2.33; $J2$ = 1.52 | 106.08 | 6,7,2′ | 8,2′ |

**Table 3.** NMR spectral references of bergapten.

| Position | Integrated Protons | $^1$H, δ, m, J (Hz) | $^{13}$C HSQC | HMBC Correlations | COSY Correlations |
|---|---|---|---|---|---|
| 2 | C | - | 161.81 | - | - |
| 3 | CH | 6.25, d, *J* = 9.77 | 111.52 | 10,2 | 4 |
| 4 | CH | 8.26, dd, *J*1 = 9.77; *J*2 = 1.27 | 139.84 | 2,9 | 3,8 |
| 5 | C | - | 149.82 | - | - |
| 6 | C | - | 112.75 | - | - |
| 7 | C | - | 158.62 | - | - |
| 8 | CH | 7.15, s | 92.76 | 6,7,9 | |
| 9 | C | - | 152.47 | - | - |
| 10 | C | - | 105.94 | - | - |
| 11 | CH3 | 4.30, s | 59.43 | 5 | - |
| 2′ | CH | 7.77, d, *J* = 2.29 | 145.17 | 6,7 | 3′ |
| 3′ | CH | 7.24, dd, *J*1 = 2.29; *J*2 = 1.27 | 104.89 | 6,7,2′ | 2′ |

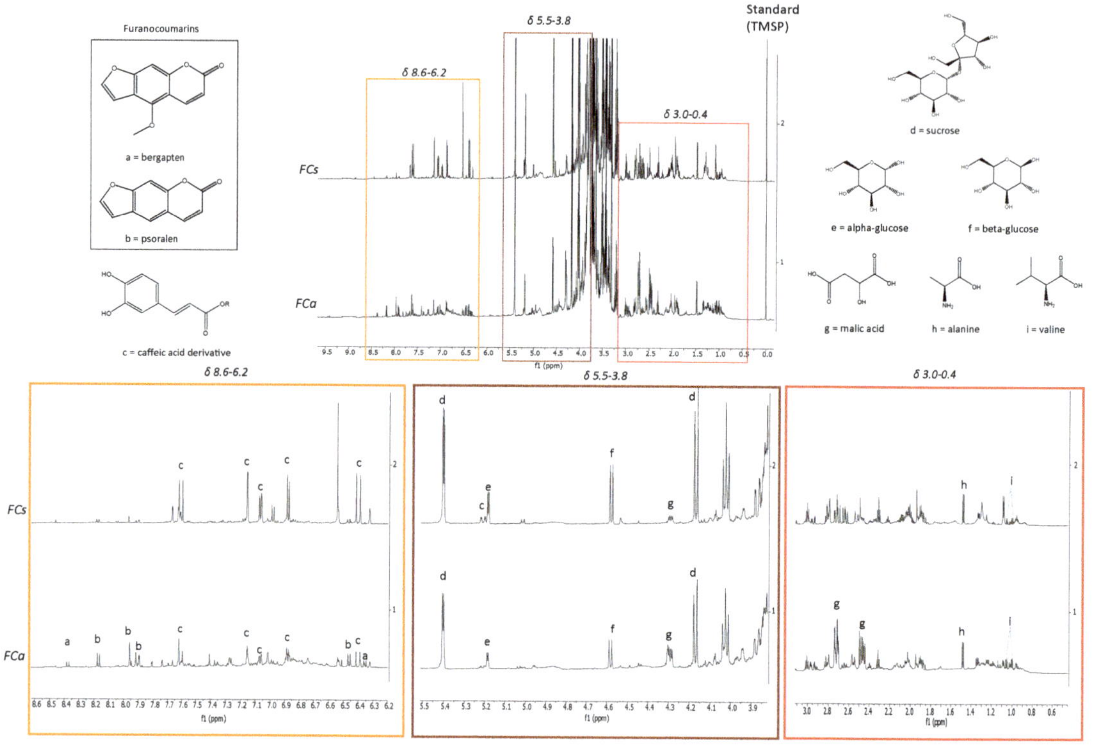


**Figure 1.** Comparison of $^1$H-NMR profiling of FCs and FCa: on top, entire spectrum; at the bottom, region from δ 8.6 to 6.2 (orange), region from δ 5.5 to 3.8 (brown), region from δ 3.0 to 0.4 (red) where a = bergapten, b = psoralen, c = caffeic acid derivative, d = sucrose, e = α-glucose, f = β-glucose, g = malic acid, h = alanine, i = valine.

### 3.2. *In Vitro Evaluation of Cell Viability and Biocompatibility of FCa and FCs Treatments*

From the phytochemical characterization of *Ficus carica* leaf extracts collected in autumn and spring, important differences in their composition were highlighted. Therefore, it was important to evaluate whether these variations in phytochemical composition consequently reflect different biological activities. Hence, at first, the possible effects of FC treatments on cell viability, biocompatibility and cell metabolic activity were assessed with MTT assay (Figure 2). MTT assay measures the activity of a mitochondrial enzyme and the colorimetric signal generated is an index of the number of viable cells and their metabolic activity [33].

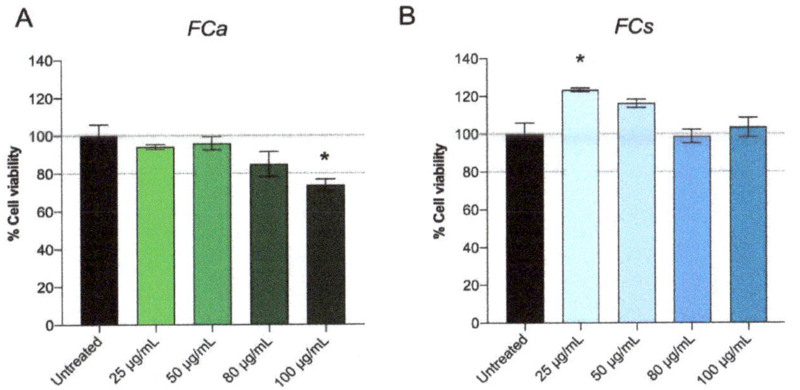


**Figure 2.** Effect of *Ficus carica* phytoextract collected (A) in autumn (FCa) and (B) in spring (FCs) on 3T3-L1 cell viability evaluated with MTT test. Data are shown as mean ± SEM; * $p < 0.05$ vs. control group (untreated).

Specifically, 3T3-L1 adipocytes were treated at different concentrations of FCa ranging from 25 µg/mL to 100 µg/mL for 48 h. MTT assay results showed that FCa treatment does not induce any cytotoxic effect on 3T3L-1 up to a concentration of 80 µg/mL, whereas dosages higher than 80 µg/mL resulted in cell suffering with a statistically significant decrease in cell viability ($p < 0.05$) (Figure 2A). On the contrary, FCs treatment showed good cell viability for all the concentrations tested (25, 50, 80 and 100 µg/mL) (Figure 2B) and interestingly, the lowest FCs dosage used (25 µg/mL) significantly increased cell metabolic activity ($p < 0.05$). Therefore, FC dosages at which no toxicity and good biocompatibility and cell viability was observed (25, 50 and 80 µg/mL) were selected for the following experiments.

### *3.3. FCa and FCs Differently Modulate Lipid Accumulation in Mature Adipocytes*

To further investigate the potential of *Ficus carica* phytoextracts in modulating lipid accumulation, 3T3-L1 preadipocyte were induced to differentiate by using MDI induction medium which contained pro-adipogenic factors such as insulin, dexamethasone, and isobutylmethyl xanthine. Different concentrations of FCa and FCs (25, 50, 80 µg/mL) were added to the medium at DAY 9 to observe the effects of 48 h treatments on mature adipocytes formation and lipid droplets accumulation. Thus, adipocytes treated and untreated with FC phytoextracts were stained by Oil Red-O at DAY 11.

Figure 3 shows that FCa treatments, but not FCs, were able to induce a dose-dependent decreasing trend of adipocytes stained with Oil Red-O that specifically binds to fatty acids and lipid depots [34]. In order to semi-quantitatively assess lipid accumulation in cultured adipocytes, Oil Red-O dye was extracted with isopropanol and then evaluated spectrophotometrically. As shown in Figure 3A, FCa treatment was able to significantly reduce lipid accumulation by 20% at the lowest dosages (25 µg/mL; $p < 0.05$), up to a reduction of nearly 40% at the highest dosage used (80 µg/mL; $p < 0.001$). On the contrary, no significant differences were observed for FCs treatment for any of the dosages used (Figure 3B).

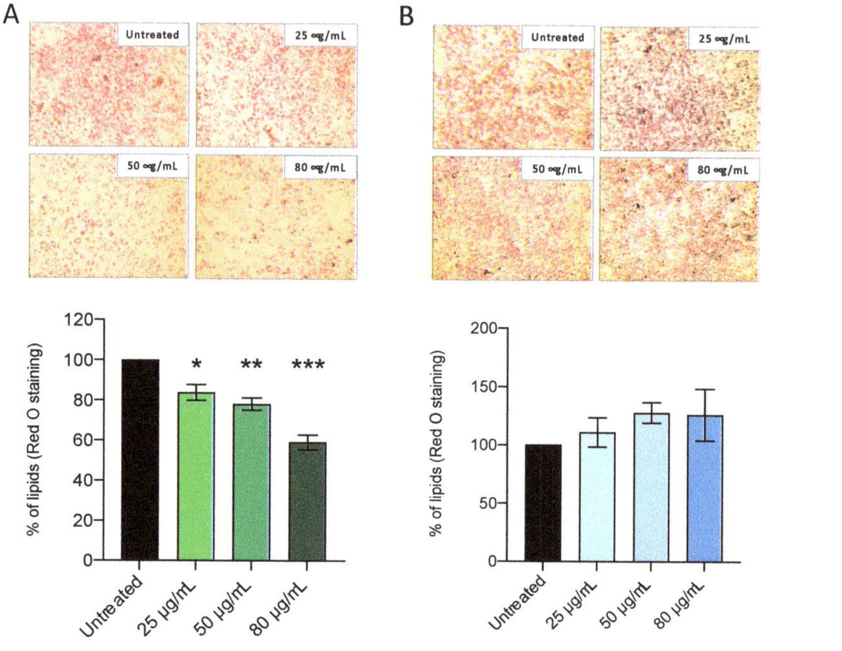


**Figure 3.** Effect of *Ficus carica* phytoextract collected (**A**) in autumn (FCa,) and (**B**) in spring (FCs) on 3T3-L1 lipid accumulation evaluated with Red O Staining. Data are shown as mean ± SEM; * $p< 0.05$; ** $p< 0.01$; *** $p< 0.001$ vs. control group (untreated).

### *3.4. FCa and FCs Modulate Genes Involved in Adipogenesis and Adipocytes Maturation*

Considering the results obtained from cell cytotoxicity and lipid accumulation, FC phytoextracts treatments at both 50 and 80 μg/mL were chosen to further investigate whether FCa and FCs were able to modulate genes involved in adipogenesis, adipocyte differentiation and lipid accumulation. In this regard, the gene expression of key adipogenesis activators such as Peroxisome proliferator-activated receptor γ (PPARγ) and CCAAT/enhancer-binding protein α (C/EBPα) were investigated [35]. They mutually induce the expression of each other and have been reported to cooperate in the activation of a few adipogenic marker genes as Leptin, Adiponectin and GLUT4 [36]. In particular, our results demonstrated that FCa significantly decreased the gene expression of PPARγ ($p < 0.05$, Figure 4A) at both dosages used.

Considering the gene expression of C/EBPα, only a reducing trend was observed for 50 μg/mL of FCa treatment but became significant when 80 μg/mL of FCa was used ($p < 0.01$, Figure 4A). Similarly, a decreasing trend in Adiponectin gene expression was found for the 50μg/mL dosage of FCa, which became significant when FCa 80 μg/mL was tested ($p < 0.05$). FCa treatment was also able to highly reduce the gene expression of Leptin and GLUT4 at both dosages used. In details, FCa at 50 μg/mL significantly reduced the gene expression of Leptin ($p < 0.01$) and GLUT4 ($p < 0.05$) and their reduction is even more evident at higher dosage ($p < 0.0001$ and $p < 0.01$, respectively). Interestingly, FCa phytoextract was unable for any of the concentrations used to modulate the gene expression of FAS, an adipogenic enzyme involved in the de novo lipogenesis.

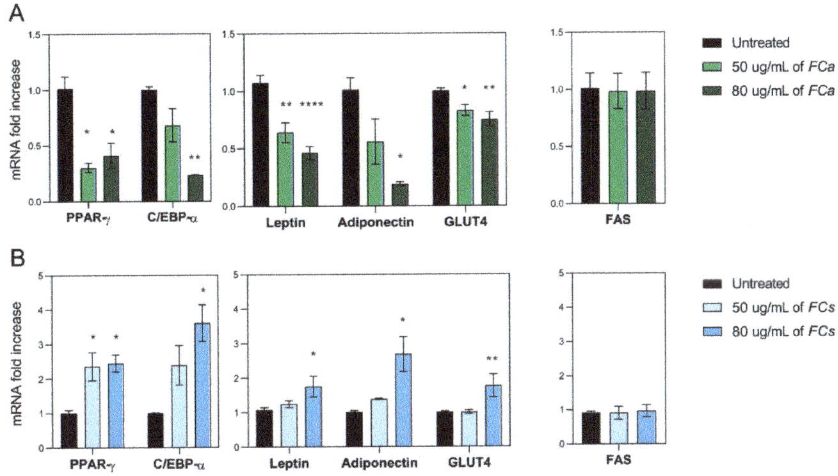


**Figure 4.** mRNA expressions of pro-adipogenic genes in 3T3-L1 adipocytes treated with *Ficus carica* phytoextract collected (**A**) in autumn (FCa,) and (**B**) in spring (FCs). mRNA expression of adipogenic transcription factors such as PPARγ and C/EBPα and of their related target genes as Leptin, Adiponectin and GLUT4. Additionally, the gene expression of the adipogenic enzyme FAS has been analyzed. Data are shown as mean ± SEM; * $p< 0.05$; ** $p< 0.01$; **** $p< 0.0001$ vs. control group (untreated).

Surprisingly, FCs phytoextracts showed an opposite effect when compared to FCa (Figure 4B). In fact, cells treated with FCs at 50 μg/mL showed a significant increase in PPARγ gene expression, while for C/EBPα, an increasing trend was found. These FCs effects were even more evident with increasing dosage of treatment. FCs at 80 μg/mL was able to induce a significant increase in the gene expression of PPARγ ($p < 0.05$) and C/EBPα ($p < 0.05$) and of their related target genes as Leptin ($p < 0.05$), Adiponectin ($p < 0.05$) and GLUT 4 ($p < 0.01$). Finally, not even FC phytoextract collected in spring were able to modulate FAS gene expression. Overall, these data suggest FCa phytoextract can exert an anti-adipogenic effect while FCs have shown an opposite pro-adipogenic-like effect on adipocyte differentiation and maturation.

## 4. Discussion

Targeting adipose tissue represents a valuable therapeutic strategy in obesity treatment and for diabetes prevention [37]. In fact, even though many therapies already exist for the treatment of adipose tissue-related disorders, nowadays nutritional interventions are recommended as first-line treatments and, among them, botanicals are considered promising coadjutants [18,38]. Despite the contradictory findings found in the literature [39], several studies have reported that *Ficus carica* L. extracts are endowed with potential health benefits against obesity-associated metabolic disorders and diabetes, but their mechanism of action is still debated [14,19] and the compounds responsible for the bioactivities have not yet been identified. Moreover, the plant metabolome is strongly affected by a number of biotic and abiotic factors, including seasonal variation [27]. The fluctuation in plant metabolome can be consequently reflected in their biological activities. In case of *F. carica* leaves, variations in its metabolites across the seasons were already reported. In particular, Innocenti et al. analyzed samples harvested from June to August, finding an increase in total coumarins content in *F. carica* leaves harvested in August [40]. Conversely, Marrelli et al. found that samples collected in June were endowed with a higher concentration of psoralen and bergapten than the ones collected in September [41].

This work was not designed to generically monitor the seasonal variation in *F. carica* metabolome, in fact, only two time points were chosen, therefore it is not possible here to draw general conclusions on seasonal variations. However, the $^1$H-NMR analysis suggested that coumarin content in the autumn extracts is higher than in the spring, while in the spring extracts there is a higher content of a caffeic acid derivative. The two samples not only showed differences in terms of phytochemical profile but also in their biological activity. In fact, in this study, only *F. carica* leaves harvested in autumn (FCa) were able to reduce lipid accumulation inside mature adipocytes. Moreover, FCa extract was able to modulate the expression of many genes involved in adipocyte differentiation, lipid accumulation and metabolism, whereas, notably, this occurs with opposite effects for *F. carica* leaves harvested in spring (FCs). In particular, in this work it has been demonstrated that FCa decreased the gene expression of PPAR-γ, the master regulator of adipogenesis, while FCs increased its expression. PPAR-γ plays a very important role, not only in adipocytes' differentiation but also in lipid and glucose homeostasis [42]. It is well documented that compounds able to reduce PPAR-γ expression have anti-obesity effects [43]. In mature adipocytes, the reduced gene expression of PPAR-γ may contribute to decreased lipid accumulation, as also confirmed by our results in Red O Staining.

FC extracts were also found able to modulate the gene expression of C/EBP-α, another key adipogenic gene that acts as a marker of adipocytes differentiation and remains elevated during cell life [42]. Interestingly, C/EBP-α has a role in promoting lipid accumulation, thus the inhibition of C/EBP-α can aid in reducing lipid content. Our data showed that C/EBP-α expression was oppositely and significantly modulated according to FC's seasonality, but only at the highest concentration (80 μg/mL) of both treatments. In detail, after 48 h treatment, a reduction in C/EBP-α gene expression was observed when FCa was used, while a decreasing tendency can also be noticed at the lower dosage. Differently, when leaves were harvested in spring, C/EBP-α gene expression was found to increase after 48 h of treatments.

Adipose tissue is also endowed with secretory properties as a huge number of proteins and other factors with biological activities can be secreted from adipocytes such as circulating hormones, pro-inflammatory and anti-inflammatory proteins [44]. Among these secreted factors, adipokines were found to play a role in the development of a metabolic syndrome and in adipose tissue expansion [45].

Among adipokines, we explored the potentiality of FC extracts in Adiponectin and Leptin gene expression modulation. Regarding Adiponectin, FCa caused a significant reduction in its expression only when the 80 μg/mL dosage of FCa was used, while at lower dosage (50 μg/mL) just a decreasing trend was observed. On the other thand, Adiponectin gene expression was increased after treating cells with FC harvested in spring at the highest dosage used. Adiponectin is also a pro-adipogenic marker [46], thus its reduction suggests an anti-adipogenic effect, while an opposite pro-adipogenic like effect can be attributed to its increased expression. These data supported that FC-dependent modulation of Adiponectin is upstream mediated by PPAR-γ modulation. In fact, it is well reported in the literature that adiponectin gene expression is under the control of PPAR-γ and C/EBP-α [46,47]. Of note, the seasonal opposite adipogenic effect of FC was also confirmed in the Leptin and GLUT4 gene expression data. Leptin is one of the adipokines responsible for regulating energy intake and expenditure [48]. The most well-known effect of Leptin is to regulate body weight, the feeding behavior of the animal, and energy balance [49], but it also has fundamental roles in tissue remodeling, inflammation, and glucose and lipid homeostasis [50,51]. GLUT-4 is the glucose transporter that is responsible for glucose uptake and whose function is disturbed in some conditions such as insulin resistance and type 2 diabetes [52]. During adipocytes differentiation, the gene expression of GLUT-4 favors glucose uptake and, in turn, glucose metabolism [53]. Thus, pharmacological drugs or natural compounds inhibiting both Leptin and GLUT-4 are highly desired for the treatment of obesity and diabetes. Interestingly, *F. carica* extracts did not modulate fatty acid synthase (FAS) levels, either when harvested in spring or in autumn. FAS is involved in the de novo lipogenesis [54] and, interestingly, it is not under the control of C/EBP-α-PPAR-γ axis [55,56], corroborating the hypothesis that PPAR-γ expression is a specific target gene of FC.

Overall, according to our results, FCa showed effective anti-adipogenic activities, by reducing lipid accumulation when 3T3-L1 mature adipocytes were treated for 48 h with FCa. These effects can be related to the FC phytochemicals content as demonstrated by the NMR spectra. In fact, most characterizing components, such as furanocoumarins were present only in FCa, which showed anti-obesity potentiality, while their absence was associated with reduced effectiveness: an almost opposite effect was observed. Furthermore, FCs NMR spectrum also revealed a higher presence of sucrose and glucose, two well-known energy sources involved in increasing the expression of pro-adipogenic regulators and fat storage [53,57], suggesting a highly relevant contributing role in exerting FCs effects on 3T3-L1 cells.

The data obtained in this work showed the promising activity of *F. carica* harvested in autumn as a potential source of anti-obesity and anti-adipogenic compounds, pointing out that this activity is strongly dependent on its phytochemical profile. These data lay the foundation for future investigations to fully explain seasonal variation in F. carica leaves metabolome, and how it could be reflected in the modulation of adipogenesis. Further biological experiments could also be conducted in vivo to investigate the possible adverse effects due to furanocoumarins content. In fact, different in vitro and in vivo studies have also demonstrated that furanocoumarins possess positive biological activities such as anti-inflammatory and anti-oxidative activities and bone health promoting effects [58,59]. However, they can also cause undesirable effects due to interactions with certain medications or by inducing phototoxicity [60,61].

In conclusion, this study further supports an in-depth investigation of *F. carica* leaves extract as a promising source of active compounds useful for targeting obesity and diabetes. However, whether the compounds contribute individually or accumulatively to the traditionally believed "medicinal" attributes of the plant is still unclear and under research.

**Supplementary Materials:** The following supporting information can be downloaded at: https://www.mdpi.com/article/10.3390/nu14142833/s1. Figure S1: HSQC spectrum of *Ficus carica* chloroform fraction where 1H-13C correlations of bergapten (blue) and psoralen (green) are highlighted. Figure S2: HMBC spectrum of *Ficus carica* chloroform fraction where 1H-13C correlations of bergapten are highlighted. Figure S3: HMBC spectrum of *Ficus carica* chloroform fraction where 1H-13C correlations of psoralen are highlighted. Figure S4: COSY spectrum of *Ficus carica* chloroform fraction where $^{1}H$-$^{1}H$ correlations of bergapten (blue) and psoralen (green) are highlighted.

**Author Contributions:** Conceptualization, M.P.; G.A.; A.M. and M.M. (Manuela Mandrone); investigation, M.P.; G.A.; I.C. and M.M. (Manuela Mandrone); formal analysis, M.P.; G.A. and A.M.; writing—original draft preparation, M.P.; G.A.; A.M. and M.M. (Manuela Mandrone); writing—review and editing, E.M.S.; E.T.; I.C.; D.U.; M.M. (Maurizio Memo) and F.P.; supervision, A.M.; project administration, G.A. All authors have read and agreed to the published version of the manuscript.

**Funding:** Giulia Abate benefits from funding FSE-REACT-EU, within the program PON "Research and Innovation" 2014–2020 (PON R&I), DM1062, Action IV.6 "Contratti di ricerca su tematiche Green".

**Data Availability Statement:** Data are available upon request to the corresponding author.

**Conflicts of Interest:** The authors declare no conflict of interest.

## References

1. Lin, X.; Xu, Y.; Pan, X.; Xu, J.; Ding, Y.; Sun, X.; Song, X.; Ren, Y.; Shan, P.-F. Global, regional, and national burden and trend of diabetes in 195 countries and territories: An analysis from 1990 to 2025. *Sci. Rep.* **2020**, *10*, 14790. [CrossRef] [PubMed]
2. Halpern, A.; Mancini, M.C.; Magalhães, M.E.C.; Fisberg, M.; Radominski, R.; Bertolami, M.C.; Bertolami, A.; de Melo, M.E.; Zanella, M.T.; Queiroz, M.S.; et al. Metabolic syndrome, dyslipidemia, hypertension and type 2 diabetes in youth: From diagnosis to treatment. *Diabetol. Metab. Syndr.* **2010**, *2*, 55. [CrossRef] [PubMed]
3. Campos, C. Chronic hyperglycemia and glucose toxicity: Pathology and clinical sequelae. *Postgrad. Med.* **2012**, *124*, 90–97. [CrossRef]

4. Saeedi, P.; Petersohn, I.; Salpea, P.; Malanda, B.; Karuranga, S.; Unwin, N.; Colagiuri, S.; Guariguata, L.; Motala, A.A.; Ogurtsova, K.; et al. Global and regional diabetes prevalence estimates for 2019 and projections for 2030 and 2045: Results from the International Diabetes Federation Diabetes Atlas. *Diabetes Res. Clin. Pract.* **2019**, *157*, 107843. [CrossRef]
5. Chehade, J.M.; Gladysz, M.; Mooradian, A.D. Dyslipidemia in type 2 diabetes: Prevalence, pathophysiology, and management. *Drugs* **2013**, *73*, 327–339. [CrossRef] [PubMed]
6. Burhans, M.S.; Hagman, D.K.; Kuzma, J.N.; Schmidt, K.A.; Kratz, M. Contribution of Adipose Tissue Inflammation to the Development of Type 2 Diabetes Mellitus. *Compr. Physiol.* **2018**, *9*, 1–58. [CrossRef] [PubMed]
7. Longo, M.; Zatterale, F.; Naderi, J.; Parrillo, L.; Formisano, P.; Raciti, G.A.; Beguinot, F.; Miele, C. Adipose Tissue Dysfunction as Determinant of Obesity-Associated Metabolic Complications. *Int. J. Mol. Sci.* **2019**, *20*, 2358. [CrossRef] [PubMed]
8. Raveendran, A.V.; Chacko, E.C.; Pappachan, J.M. Non-pharmacological Treatment Options in the Management of Diabetes Mellitus. *Eur. Endocrinol.* **2018**, *14*, 31–39. [CrossRef]
9. Kumar, S.; Mittal, A.; Babu, D.; Mittal, A. Herbal Medicines for Diabetes Management and its Secondary Complications. *Curr. Diabetes Rev.* **2021**, *17*, 437–456. [CrossRef]
10. Bilal, M.; Iqbal, M.S.; Shah, S.B.; Rasheed, T.; Iqbal, H.M.N. Diabetic Complications and Insight into Antidiabetic Potentialities of Ethno- Medicinal Plants: A Review. *Recent Pat. Inflamm. Allergy Drug Discov.* **2018**, *12*, 7–23. [CrossRef]
11. Bailey, C.J.; Day, C. Metformin: Its botanical background. *Pract. Diabetes Int.* **2004**, *21*, 115–117. [CrossRef]
12. Thomas, I.; Gregg, B. Metformin; a review of its history and future: From lilac to longevity. *Pediatr. Diabetes* **2017**, *18*, 10–16. [CrossRef]
13. Khan, V.; Najmi, A.K.; Akhtar, M.; Aqil, M.; Mujeeb, M.; Pillai, K.K. A pharmacological appraisal of medicinal plants with antidiabetic potential. *J. Pharm. Bioallied Sci.* **2012**, *4*, 27–42. [CrossRef]
14. Stephen Irudayaraj, S.; Christudas, S.; Antony, S.; Duraipandiyan, V.; Naif Abdullah, A.-D.; Ignacimuthu, S. Protective effects of *Ficus carica* leaves on glucose and lipids levels, carbohydrate metabolism enzymes and β-cells in type 2 diabetic rats. *Pharm. Biol.* **2017**, *55*, 1074–1081. [CrossRef] [PubMed]
15. Prasad, P.V.V.; Subhaktha, P.K.J.P.; Narayana, A.; Rao, M.M. Medico-historical study of "aśvattha" (sacred fig tree). *Bull. Indian Inst. Hist. Med. Hyderabad* **2006**, *36*, 1–20.
16. Mawa, S.; Husain, K.; Jantan, I. *Ficus carica* L. (Moraceae): Phytochemistry, Traditional Uses and Biological Activities. *Evid.-Based Complement. Altern. Med.* **2013**, *2013*, 974256. [CrossRef] [PubMed]
17. Kar, A.; Choudhary, B.K.; Bandyopadhyay, N.G. Comparative evaluation of hypoglycaemic activity of some Indian medicinal plants in alloxan diabetic rats. *J. Ethnopharmacol.* **2003**, *84*, 105–108. [CrossRef]
18. Salehi, B.; Ata, A.; Anil Kumar, N.V.; Sharopov, F.; Ramírez-Alarcón, K.; Ruiz-Ortega, A.; Abdulmajid Ayatollahi, S.; Tsouh Fokou, P.V.; Kobarfard, F.; Amiruddin Zakaria, Z.; et al. Antidiabetic Potential of Medicinal Plants and Their Active Components. *Biomolecules* **2019**, *9*, 551. [CrossRef]
19. Deepa, P.; Sowndhararajan, K.; Kim, S.; Park, S.J. A role of Ficus species in the management of diabetes mellitus: A review. *J. Ethnopharmacol.* **2018**, *215*, 210–232. [CrossRef]
20. Shahreen, S.; Banik, J.; Hafiz, A.; Rahman, S.; Zaman, A.T.; Shoyeb, M.A.; Chowdhury, M.H.; Rahmatullah, M. Antihyperglycemic activities of leaves of three edible fruit plants (*Averrhoa carambola*, *Ficus hispida* and *Syzygium samarangense*) of Bangladesh. *Afr. J. Tradit. Complement. Altern. Med. AJTCAM* **2012**, *9*, 287–291. [CrossRef]
21. Arafa, E.-S.A.; Hassan, W.; Murtaza, G.; Buabeid, M.A. *Ficus carica* and *Sizigium cumini* Regulate Glucose and Lipid Parameters in High-Fat Diet and Streptozocin-Induced Rats. *J. Diabetes Res.* **2020**, *2020*, 6745873. [CrossRef] [PubMed]
22. Li, Z.; Yang, Y.; Liu, M.; Zhang, C.; Shao, J.; Hou, X.; Tian, J.; Cui, Q. A comprehensive review on phytochemistry, bioactivities, toxicity studies, and clinical studies on *Ficus carica* Linn. leaves. *Biomed. Pharmacother.* **2021**, *137*, 111393. [CrossRef] [PubMed]
23. Isah, T. Stress and defense responses in plant secondary metabolites production. *Biol. Res.* **2019**, *52*, 39. [CrossRef] [PubMed]
24. Bayati, P.; Karimmojeni, H.; Razmjoo, J.; Pucci, M.; Abate, G.; Baldwin, T.C.; Mastinu, A. Physiological, Biochemical, and Agronomic Trait Responses of Nigella sativa Genotypes to Water Stress. *Horticulturae* **2022**, *8*, 193. [CrossRef]
25. Ashrafi, M.; Azimi-Moqadam, M.-R.; MohseniFard, E.; Shekari, F.; Jafary, H.; Moradi, P.; Pucci, M.; Abate, G.; Mastinu, A. Physiological and Molecular Aspects of Two Thymus Species Differently Sensitive to Drought Stress. *BioTech* **2022**, *11*, 8. [CrossRef]
26. Cappadone, C.; Mandrone, M.; Chiocchio, I.; Sanna, C.; Malucelli, E.; Bassi, V.; Picone, G.; Poli, F. Antitumor Potential and Phytochemical Profile of Plants from Sardinia (Italy), a Hotspot for Biodiversity in the Mediterranean Basin. *Plants* **2019**, *9*, 26. [CrossRef]
27. Mandrone, M.; Chiocchio, I.; Barbanti, L.; Tomasi, P.; Tacchini, M.; Poli, F. Metabolomic Study of Sorghum (*Sorghum bicolor*) to Interpret Plant Behavior under Variable Field Conditions in View of Smart Agriculture Applications. *J. Agric. Food Chem.* **2021**, *69*, 1132–1145. [CrossRef]
28. Mandrone, M.; Marincich, L.; Chiocchio, I.; Petroli, A.; Gođevac, D.; Maresca, I.; Poli, F. NMR-based metabolomics for frauds detection and quality control of oregano samples. *Food Control* **2021**, *127*, 108141. [CrossRef]
29. Chiocchio, I.; Prata, C.; Mandrone, M.; Ricciardiello, F.; Marrazzo, P.; Tomasi, P.; Angeloni, C.; Fiorentini, D.; Malaguti, M.; Poli, F.; et al. Leaves and Spiny Burs of Castanea Sativa from an Experimental Chestnut Grove: Metabolomic Analysis and Anti-Neuroinflammatory Activity. *Metabolites* **2020**, *10*, 408. [CrossRef]
30. Green, H.; Meuth, M. An established pre-adipose cell line and its differentiation in culture. *Cell* **1974**, *3*, 127–133. [CrossRef]

31. Zhao, X.; Hu, H.; Wang, C.; Bai, L.; Wang, Y.; Wang, W.; Wang, J. A comparison of methods for effective differentiation of the frozen-thawed 3T3-L1 cells. *Anal. Biochem.* **2019**, *568*, 57–64. [CrossRef] [PubMed]
32. Pucci, M.; Aria, F.; Premoli, M.; Maccarinelli, G.; Mastinu, A.; Bonini, S.; Memo, M.; Uberti, D.; Abate, G. Methylglyoxal affects cognitive behaviour and modulates RAGE and Presenilin-1 expression in hippocampus of aged mice. *Food Chem. Toxicol.* **2021**, *158*, 112608. [CrossRef] [PubMed]
33. Buranaamnuay, K. The MTT assay application to measure the viability of spermatozoa: A variety of the assay protocols. *Open Vet. J.* **2021**, *11*, 251–269. [CrossRef] [PubMed]
34. Daemen, S.; van Zandvoort, M.A.M.J.; Parekh, S.H.; Hesselink, M.K.C. Microscopy tools for the investigation of intracellular lipid storage and dynamics. *Mol. Metab.* **2016**, *5*, 153–163. [CrossRef]
35. Rosen, E.D.; Hsu, C.-H.; Wang, X.; Sakai, S.; Freeman, M.W.; Gonzalez, F.J.; Spiegelman, B.M. C/EBPalpha induces adipogenesis through PPARgamma: A unified pathway. *Genes Dev.* **2002**, *16*, 22–26. [CrossRef]
36. Lefterova, M.I.; Haakonsson, A.K.; Lazar, M.A.; Mandrup, S. PPARγ and the global map of adipogenesis and beyond. *Trends Endocrinol. Metab.* **2014**, *25*, 293–302. [CrossRef]
37. Kusminski, C.M.; Bickel, P.E.; Scherer, P.E. Targeting adipose tissue in the treatment of obesity-associated diabetes. *Nat. Rev. Drug Discov.* **2016**, *15*, 639–660. [CrossRef]
38. Durazzo, A.; Lucarini, M.; Santini, A. Plants and Diabetes: Description, Role, Comprehension and Exploitation. *Int. J. Mol. Sci.* **2021**, *22*, 3938. [CrossRef]
39. Olaokun, O.O.; McGaw, L.J.; Awouafack, M.D.; Eloff, J.N.; Naidoo, V. The potential role of GLUT4 transporters and insulin receptors in the hypoglycaemic activity of Ficus lutea acetone leaf extract. *BMC Complement. Altern. Med.* **2014**, *14*, 269. [CrossRef]
40. Innocenti, G.; Bettero, A.; Caporale, G. Determination of the coumarinic constituents of *Ficus carica* leaves by HPLC. *Farm. Sci.* **1982**, *37*, 475–485.
41. Marrelli, M.; Statti, G.A.; Tundis, R.; Menichini, F.; Conforti, F. Fatty acids, coumarins and polyphenolic compounds of *Ficus carica* L. cv. Dottato: Variation of bioactive compounds and biological activity of aerial parts. *Nat. Prod. Res.* **2014**, *28*, 271–274. [CrossRef] [PubMed]
42. Poulos, S.P.; Dodson, M.V.; Culver, M.F.; Hausman, G.J. The increasingly complex regulation of adipocyte differentiation. *Exp. Biol. Med.* **2016**, *241*, 449–456. [CrossRef] [PubMed]
43. Wang, L.; Waltenberger, B.; Pferschy-Wenzig, E.-M.; Blunder, M.; Liu, X.; Malainer, C.; Blazevic, T.; Schwaiger, S.; Rollinger, J.M.; Heiss, E.H.; et al. Natural product agonists of peroxisome proliferator-activated receptor gamma (PPARγ): A review. *Biochem. Pharmacol.* **2014**, *92*, 73–89. [CrossRef] [PubMed]
44. Richard, A.J.; White, U.; Elks, C.M.; Stephens, J.M. *Adipose Tissue: Physiology to Metabolic Dysfunction;* Feingold, K.R., Anawalt, B., Boyce, A., Chrousos, G., de Herder, W.W., Dhatariya, K., Dungan, K., Hershman, J.M., Hofland, J., Kalra, S., et al., Eds.; Endotext [Internet]; MDText.com, Inc.: South Dartmouth, MA, USA, 2000.
45. Zorena, K.; Jachimowicz-Duda, O.; Ślęzak, D.; Robakowska, M.; Mrugacz, M. Adipokines and Obesity. Potential Link to Metabolic Disorders and Chronic Complications. *Int. J. Mol. Sci.* **2020**, *21*, 3570. [CrossRef]
46. Fu, Y.; Luo, N.; Klein, R.L.; Garvey, W.T. Adiponectin promotes adipocyte differentiation, insulin sensitivity, and lipid accumulation. *J. Lipid Res.* **2005**, *46*, 1369–1379. [CrossRef]
47. Yang, W.; Yang, C.; Luo, J.; Wei, Y.; Wang, W.; Zhong, Y. Adiponectin promotes preadipocyte differentiation via the PPARγ pathway. *Mol. Med. Rep.* **2018**, *17*, 428–435. [CrossRef]
48. Landecho, M.F.; Tuero, C.; Valentí, V.; Bilbao, I.; de la Higuera, M.; Frühbeck, G. Relevance of Leptin and Other Adipokines in Obesity-Associated Cardiovascular Risk. *Nutrients* **2019**, *11*, 2664. [CrossRef]
49. Park, H.-K.; Ahima, R.S. Physiology of leptin: Energy homeostasis, neuroendocrine function and metabolism. *Metabolism* **2015**, *64*, 24–34. [CrossRef]
50. Ramos-Lobo, A.M.; Donato, J., Jr. The role of leptin in health and disease. *Temperature* **2017**, *4*, 258–291. [CrossRef]
51. Palhinha, L.; Liechocki, S.; Hottz, E.D.; da Pereira, J.A.S.; de Almeida, C.J.; Moraes-Vieira, P.M.M.; Bozza, P.T.; Maya-Monteiro, C.M. Leptin Induces Proadipogenic and Proinflammatory Signaling in Adipocytes. *Front. Endocrinol.* **2019**, *10*, 841. [CrossRef]
52. Govers, R. Molecular mechanisms of GLUT4 regulation in adipocytes. *Diabetes Metab.* **2014**, *40*, 400–410. [CrossRef] [PubMed]
53. Jackson, R.M.; Griesel, B.A.; Gurley, J.M.; Szweda, L.I.; Olson, A.L. Glucose availability controls adipogenesis in mouse 3T3-L1 adipocytes via up-regulation of nicotinamide metabolism. *J. Biol. Chem.* **2017**, *292*, 18556–18564. [CrossRef] [PubMed]
54. Wang, Y.; Jones Voy, B.; Urs, S.; Kim, S.; Soltani-Bejnood, M.; Quigley, N.; Heo, Y.-R.; Standridge, M.; Andersen, B.; Dhar, M.; et al. The human fatty acid synthase gene and de novo lipogenesis are coordinately regulated in human adipose tissue. *J. Nutr.* **2004**, *134*, 1032–1038. [CrossRef]
55. Collins, J.M.; Neville, M.J.; Pinnick, K.E.; Hodson, L.; Ruyter, B.; van Dijk, T.H.; Reijngoud, D.-J.; Fielding, M.D.; Frayn, K.N. De novo lipogenesis in the differentiating human adipocyte can provide all fatty acids necessary for maturation. *J. Lipid Res.* **2011**, *52*, 1683–1692. [CrossRef]
56. Ahmadian, M.; Suh, J.M.; Hah, N.; Liddle, C.; Atkins, A.R.; Downes, M.; Evans, R.M. PPARγ signaling and metabolism: The good, the bad and the future. *Nat. Med.* **2013**, *19*, 557–566. [CrossRef] [PubMed]
57. Krishna, M.S.; Revathy, V.M.; Jaleel, A. Adipocytes utilize sucrose as an energy source-Effect of different carbohydrates on adipocyte differentiation. *J. Cell. Physiol.* **2020**, *235*, 891–899. [CrossRef]

58. Mahendra, C.K.; Tan, L.T.H.; Lee, W.L.; Yap, W.H.; Pusparajah, P.; Low, L.E.; Tang, S.Y.; Chan, K.G.; Lee, L.H.; Goh, B.H. Angelicin-A Furocoumarin Compound with Vast Biological Potential. *Front. Pharmacol.* **2020**, *11*, 366. [CrossRef]
59. Liang, Y.; Xie, L.; Liu, K.; Cao, Y.; Dai, X.; Wang, X.; Lu, J.; Zhang, X.; Li, X. Bergapten: A review of its pharmacology, pharmacokinetics, and toxicity. *Phytother. Res.* **2021**, *35*, 6131–6147. [CrossRef]
60. Melough, M.M.; Cho, E.; Chun, O.K. Furocoumarins: A review of biochemical activities, dietary sources and intake, and potential health risks. *Food Chem. Toxicol.* **2018**, *113*, 99–107. [CrossRef]
61. Wang, X.; Lou, Y.-J.; Wang, M.-X.; Shi, Y.-W.; Xu, H.-X.; Kong, L.-D. Furocoumarins affect hepatic cytochrome P450 and renal organic ion transporters in mice. *Toxicol. Lett.* **2012**, *209*, 67–77. [CrossRef]

nutrients

MDPI

*Review*

# Health Benefits, Pharmacological Effects, Molecular Mechanisms, and Therapeutic Potential of α-Bisabolol

**Lujain Bader Eddin [1], Niraj Kumar Jha [2], Sameer N. Goyal [3], Yogeeta O. Agrawal [3], Sandeep B. Subramanya [4,5], Salim M. A. Bastaki [1] and Shreesh Ojha [1,5,*]**

1 Department of Pharmacology and Therapeutics, College of Medicine and Health Sciences, United Arab Emirates University, Al Ain P.O. Box 15551, United Arab Emirates; 201970113@uaeu.ac.ae (L.B.E.); sbastaki@uaeu.ac.ae (S.M.A.B.)
2 Department of Biotechnology, School of Engineering and Technology (SET), Sharda University, Greater Noida 201310, Uttar Pradesh, India; niraj.jha@sharda.ac.in
3 Shri Vile Parle Kelavani Mandal's Institute of Pharmacy, Dhule 424001, Maharashtra, India; sameer.goyal@svkm.ac.in (S.N.G.); yogeeta.goyal@svkm.ac.in (Y.O.A.)
4 Department of Physiology, College of Medicine and Health Sciences, United Arab Emirates University, Al Ain P.O. Box 15551, United Arab Emirates; sandeep.bs@uaeu.ac.ae
5 Zayed Bin Sultan Center for Health Sciences, United Arab Emirates University, Al Ain P.O. Box 15551, United Arab Emirates
* Correspondence: shreeshojha@uaeu.ac.ae

**Abstract:** α-Bisabolol is one of the important monocyclic sesquiterpenes, derived naturally from essential oils of many edible and ornamental plants. It was first obtained from *Matricaria chamomilla*, commonly known as chamomile or German chamomile. The available literature indicates that this plant along with other α-Bisabolol containing plants is popularly used in traditional medicine for potential health benefits and general wellbeing. Nutritional studies are indicative of the health benefits of α-Bisabolol. Numerous experimental studies demonstrated pharmacological properties of α-Bisabolol including anticancer, antinociceptive, neuroprotective, cardioprotective, and antimicrobial. This review aims to collectively present different pharmacological activities based on both in vitro and in vivo studies. In the present review using synoptic tables and figures, we comprehensively present that α-Bisabolol possesses therapeutic and protective activities, therefore, it can be used for potential health benefits based on pharmacological effects, underlying molecular mechanism, and favorable pharmaceutical properties. Based on the studies mostly performed on cell lines or animal models, it is evident that α-Bisabolol may be a promising nutraceutical and phytomedicine to target aberrant biological mechanisms which result in altered physiological processes and various ailments. Given the polypharmacological effects and pleiotropic properties, along with favorable pharmacokinetics, and dietary availability and safety, α-Bisabolol can be used as a dietary agent, nutraceutical or phytopharmaceutical agent or as an adjuvant with currently available modern medicines. The regulatory approval of this molecule for use as food additives, and in cosmetics and fragrance industry is also supportive of its human usage. Moreover, further studies are necessary to address pharmaceutical, pharmacological, and toxicological aspects before clinical or nutritional usage in humans. The biological actions and health benefits open opportunities for pharmaceutical development with pharmacological basis of its use in future therapeutics.

**Keywords:** α-Bisabolol; German chamomile tea; natural products; phytochemicals; sesquiterpene; pharmacology

**Citation:** Eddin, L.B.; Jha, N.K.; Goyal, S.N.; Agrawal, Y.O.; Subramanya, S.B.; Bastaki, S.M.A.; Ojha, S. Health Benefits, Pharmacological Effects, Molecular Mechanisms, and Therapeutic Potential of α-Bisabolol. *Nutrients* **2022**, *14*, 1370. https://doi.org/10.3390/nu14071370

Academic Editors: Daniela Rigano and Paola Bontempo

Received: 17 February 2022
Accepted: 15 March 2022
Published: 25 March 2022



## 1. Introduction

Sesquiterpenes, a subclass of terpenes has attracted a significant interest due to their wide range of biological properties that have been employed in pharmacological and therapeutic research applications. The high identification of sesquiterpenes compounds and their structural variability make this family more distinguishable compared with other plant sourced products. These phytochemicals occupy a large portion of the volatile fraction of aromatic plants. For example, α-bisabolene is a primary component in black pepper (*Piper nigrum*) and β-Caryophyllene in ylang ylang (*Cananga odorata*) [1].

Sesquiterpenes have been known for their wide range of biological functions including anti-infective, antioxidant, anti-inflammatory and anticancer activities and can be obtained from *Matricaria* genus. *Matricaria chamomilla* genus is an herbaceous plant, cultivated in many countries for commercial, pharmaceutical, and cosmeceutical purposes. It belongs to the family of flowering plants, *Asteraceae*. The flowers of *Matricaria chamomilla* have aromatic, coloring and flavoring properties which are implemented in several commercial products. The phytochemical composition of this genus varies and includes volatile sesquiterpene, flavonoid, polyacetylene, and coumarin. *Matricaria chamomilla* has been used anciently in folk medicine as a plant-based remedy for many ailments which survived until today [2]. The key and predominant constituent is α-Bisabolol, where it can be exuberantly found in the plant's flower buds.

α-Bisabolol also known as levomenol is a monocyclic sesquiterpene alcohol which was first identified and extracted from *Matricaria chamomilla* [3]. However, α-Bisabolol is also present abundantly in various medicinal plants essential oils. The percentage occurrence of α-Bisabolol in numerous plants are listed in Table 1. It exists in two configurations α and β. It's a clear colorless liquid; with fruity nutty aroma resembling coconut. α-Bisabolol possess favorable physicochemical properties and the characteristics are presented in Table 2. Most of the biological actions are attributed to the α form, on which extensive studies have been conducted. α-Bisabolol has been used as a skin conditioning agent where it is integrated in many cosmetic formulations due to its skin soothing effects, well documented dermal absorption and the absence of dermal irritations or photosensitivity following its application.

**Table 1.** Percentage occurrence of α-Bisabolol in different plants.

| Plants | Percentage Occurrence | Reference |
|---|---|---|
| *Stachys lavandulifolia* Vahl. (*Lamiaceae*) | 56.4 | [4] |
| *Vanillosmopsis arborea* Barker (*Asteraceae*) | 91.02 | [5] |
| *Nectandra megapotamicav* (Spreng.) Mez. (*Lauraceae*) | 93.7 | [6] |
| *Nepeta graciliflora* Benth. (*Lamiaceae*) | 8.97 | [7] |
| *Lantana achyranthifolia* Desf. (*Verbenaceae*) | 11.23 | [8] |
| *Hymenocrater yazdianus, Stachys obtusicrena Boiss*, and *Nepeta asterotricha* | 23.5% | [9] |
| *Tanacetum walteri* (*Anthemideae-Asteraceae*) | 6.3 | [10] |
| *Licaria, Nectrandra* and *Ocotea* Species (*Lauraceae*) | 59.7–93.7 | [11] |
| *Semenovia suffruticosa* | 13.3 | [12] |
| *Genus Matricaria* | 29–81 | [2] |
| *Algerian Eryngium tricuspidatum* L. | 32.6 | [13] |
| *Eupatorium adenophorum* | 9.53 | [14] |
| Candeia Tree (*Eremanthus erythropappus* (DC) McLeisch) | 66–91 | [15] |
| *Amomum biflorum* | 16.0 | [16] |
| *Acanthospermum hispidum* (*Asteraceae*) | 11.4 | [17] |
| *Helietta longifoliata* | 7.24 | [18] |
| *Betonica grandiflora* Willd. | 4.9 | [19] |
| *Rhaponticum acaule* DC | 4.8 | [20] |
| *Vismia macrophylla* | 14.9 | [21] |

**Table 1.** *Cont.*

| Plants | Percentage Occurrence | Reference |
|---|---|---|
| *Micromeria inodora* (Desf.) Benth. | 2.9 | [22] |
| *Ganoderma lucidum* and *Spongiporus leucomallellus* | 2 | [23] |
| *Laserpitium zernyi* | 30.9 | [24] |
| Araçá (*Psidium guineense* Sw.) | 6.5–18.1 | [25] |
| *Psidium myrtoides O. Berg (Myrtaceae)* | 5.3 | [26] |
| *Ferula hermonis* Boiss | 11.1 | [27] |
| *Plinia cerrocampanensis* | 42.8 | [28] |
| *Arnica longifolia*, *Aster hesperius*, and *Chrysothamnus nauseosus* | 8.2 | [29] |
| *Matricaria chamomilla* L. | 56.86 | [30] |
| *Teucrium polium* L. | 24.6 | [31] |
| *Angelica purpurascens* (Avé-Lall.) Gill | 22.93 | [32] |
| *Ferula asafoetida* | 9.75 | [33] |

**Table 2.** * Physicochemical properties of α-Bisabolol.

| | |
|---|---|
| PubChem CID | 1549992 |
| Molecular Formula | C15H26O |
| Synonyms | (+)-α-Bisabolol, D-α-Bisabolol, (2R)-6-Methyl-2-(4-methyl-3-cyclohexenyl)-5-heptene-2-ol, Dragosantol, Camilol, Hydagen B, (+)-6R,7R-α-Bisabolol, |
| Molecular Weight | 222.37 |
| XLogP3-AA | 3.8 |
| Hydrogen Bond Donor Count | 1 |
| Hydrogen Bond Acceptor Count | Rotatable Bond Count |
| Exact Mass | 222.198365449 |
| Monoisotopic Mass | 222.198365449 |
| Topological Polar Surface Area | 20.2 Å$^2$ |
| Heavy Atom Count | 16 |
| Formal Charge | 0 |
| Complexity | 284 |
| Isotope Atom Count | 0 |
| Defined Atom Stereocenter Count | 2 |
| Undefined Atom Stereocenter Count | 0 |
| Defined Bond Stereocenter Count | 0 |
| Undefined Bond Stereocenter Count | 0 |
| Covalently-Bonded Unit Count | 1 |
| Solubility | 1.688 mg/L @ 25 °C (est), Practically insoluble or insoluble in water, slightly soluble in ethanol |
| Density | 0.922–0.931 |
| LogP | 5.070 (est) |
| Refractive Index | 1.491–1.500 |
| Food additive class | Flavoring agent |

* Compiled from the PubChem substances.

The United States Food and Drug Administration (USFDA) has considered α-Bisabolol as a safe compound due to its low toxicity [34]. The effects of α-Bisabolol have been studied in different cell and animal models which indicated its promising beneficial actions. α-Bisabolol demonstrated neuro-, cardio- and nephro-protective, analgesic, anticancer as well as antimicrobial effects offering an intriguing line of evidence for its various therapeutic effects. It has also shown an appreciable chemopreventive activity rendering a great opportunity for further investigations and possible application in chemoprevention. It is thus noteworthy to state that more biological activities and biochemical modifications are likely to emanate from future studies targeting this sesquiterpene compound.

The present review comprehensively summarizes all available studies and aims to advance the understanding regarding the effectiveness of α-Bisabolol in treating different diseases. The broad range of pharmacological properties of α-Bisabolol offers the possibility of incorporating it into various medical formulations.

## 2. α-Bisabolol and Skin Disorders

Skin diseases are among the common illnesses affecting humans at all ages and imposing a huge burden on the global health system [35]. Skin is the body's largest organ, and it is the first line of defense against a large variety of stressors. Continuous stress can overwhelm defending dermal systems resulting in different cutaneous disorders. Phytochemicals have shown myriad of health effects in curtailing skin damage being used either as individual component or as amalgamated constituents = based formulations [36].

A randomized controlled study revealed that the topical application of a spray containing ozonated sunflower oil and α-Bisabolol achieved a complete healing for venous leg ulcers compared to the control group in which no complete healing was observed [37]. Moreover, the effectiveness of α-Bisabolol was evaluated against atopic dermatitis in a three-center assessor blinded trial on children. Results concluded that α-Bisabolol based cream mitigated eczematous flares compared to baseline showed by a significant decrease in the affected skin surface area, improvement in the total eczema severity score outcome as well as the single area eczema severity score which includes evaluating the infiltration, lichenification and scratching lesions [38].

α-Bisabolol was also found to enhance the effectiveness in attenuating pruritus and inflamed skin in patients with atopic dermatitis when combined with heparin in a topically applied formulation. Furthermore, an improvement was observed in the SCORing Atopic Dermatitis (SCORAD) values in patients who applied the combined therapy as compared with control and individualized therapy groups [39]. On the other hand, the integration of α-Bisabolol in a formulation intended for the treatment of melasma has shown an enhancement in the recovered melasmatic area measured by imaging techniques at the end of treatment duration compared with baseline. Obtained results reported a statistical significance in patients' satisfaction in terms of improved facial texture, skin oiliness, brightness, hydrated skin, and overall appearance following the treatment [40].

A clinical investigation reported that the use of topical formulation containing α-Bisabolol improved the inflammatory status and showed a reduction in the severity of eczema in treated children [41]. Studies that investigated the potential therapeutic effects of α-Bisabolol in different dermatological disorders are briefly stated in Table 3.

Wound healing is an essential recovery process that maintains skin integrity following a mechanical insult. It is a complex and highly regulated mechanism that consists of inflammatory, proliferation and remodeling phases working on restoring the barrier function of skin. The success of this process is dependent on the intricate interplay among different factors, any aberration which can hinder the normal healing and result in chronic discomfort [42].

**Table 3.** Therapeutic effects and demonstrated mechanisms of α-Bisabolol in skin disorders.

| Skin Disorders | | | | |
|---|---|---|---|---|
| **Compound** | **Dose/Route/Duration** | **Model** | **Major Mechanisms** | **Reference** |
| α-Bisabolol | 1% topical spray | Randomized controlled trial on chronic venous leg ulcer patients | ↓ wound and ulcer surface area ↑ healing speed | [37] |
| α-Bisabolol | 0.3 g/100 g cream twice daily for 8 weeks | Prospective, randomized, reference-controlled, double-blind, two-center and four-armed parallel group study on patients with atopic dermatitis | ↓ pruritis, inflammation ↑ healing | [39] |
| α-Bisabolol | 1% cream once-daily for 30 days | Single-center, single-arm, prospective, open-label study on patients with melasma | ↓ melasma area and severity index (MASI) score ↑ patient satisfaction | [40] |

Symbols indications: ↑; increase, ↓; decrease.

α-Bisabolol has been demonstrated to exert a cicatrizant effect on wounds with an $ED_{50}$ of 155 μg/mL. The cicatrization effect was evaluated on mice, where an incision was made, stitched and treatment was applied immediately. This effect is speculated to be attributed to the 1-methylcyclohexene and tertiary hydroxyl structural moieties. An unremarkable action of α-Bisabolol was observed on fibroblasts migration that can hasten the healing process. The study compared the efficacy of α-Bisabolol and the cytotoxicity on BALB/c 3T3 cells proliferation with that of taspine; a potent cicatrizant. It was found that although α-Bisabolol is a less potent compound, but it notably less cytotoxic than taspine [43]. Figure 1 represents the effects of α-Bisabolol on skin disorders and wound healing.

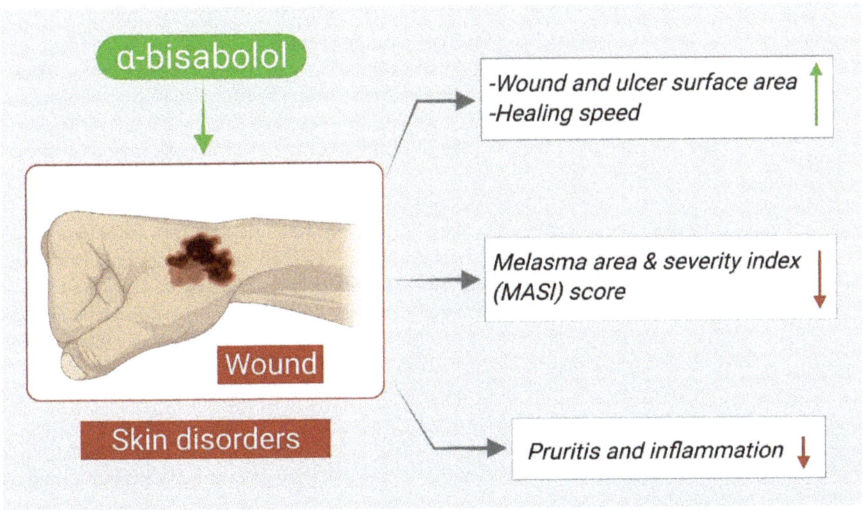


**Figure 1.** Effects of α-Bisabolol on skin disorders. ↑; increase, ↓; decrease.

## 3. α-Bisabolol and Neuroprotection

Neuroprotection refers to the integrated strategies that aim to preserve neuronal integrity and health against various insults including genetic mutations and environmental toxins to maintain neuronal function. There are many pathological mechanisms that can predispose the occurrence of neurodegenerative diseases; requiring neuroprotection. Neurodegenerative diseases are a subset of neurological disorders affecting primarily neurons [44].

Parkinson's disease (PD) is a progressive neurodegenerative disorder with neuropathological features of degraded dopaminergic neurons and accumulated α-synuclein depositions. Pharmacological treatment of PD is acquiring increased attention due to the challenges researchers facing during the quest of novel therapeutic approaches [45]. The neuroprotective effect of α-Bisabolol was first described on *Drosophila melanogaster* model of rotenone induced toxicity [46]. The study reported an improvement in locomotor activity, a reduction in the expression of thiol and a reinstate of the activity of mitochondrial complex-I. α-Bisabolol also increased the mRNA level of antioxidants proteins such as superoxide dismutase (SOD), catalase (CAT), and the keap1 gene product. A more recent study investigated the neuroprotective role of α-Bisabolol against rotenone induced neurodegeneration in a rat model of PD. Results showed a promising effective action of α-Bisabolol in abrogating neuronal degradation associated with PD in both Substantia Nigra and Striatum. α-Bisabolol attenuated oxidative insult by reducing malondialdehyde (MDA), restoring depleted glutathione (GSH) and improving SOD and CAT activity. It also attenuated neuroinflammation by reducing glial cells activation and subsequent release of proinflammatory cytokines (IL-1β, IL-6 and TNF-α) and mediators (iNOS and COX-2). Additionally, α-Bisabolol ameliorated apoptosis by reversing the downregulated Bcl-2, and the upregulated Bax, cleaved caspases-3 and 9 levels. It was also observed that α-Bisabolol restored mitochondrial function by preventing mitochondrial lipid peroxidation, cytochrome-C release and most importantly preserving Complex-I activity [47].

Another neurodegenerative disease is Alzheimer disease (AD), the pathogenesis of which is known to be attributed to the deposits of β-amyloid (Aβ) peptide affecting neuronal connectivity and synaptic function resulting in impaired cognitive function [48]. Therefore, the inhibition of Aβ aggregation has been considered the cornerstone in developing disease modifying agents for AD. Using $A\beta_{25-35}$ induced neurotoxicity in neuro2a cells and transgenic *Caenorhabditis elegans*, α-Bisabolol was evaluated for its efficacy in protecting against the neuronal insult. Cellular viability and morphology were preserved following α-Bisabolol treatment. As proposed by many hypotheses that reactive oxygen species (ROS) and reactive nitrogen species can trigger Aβ accumulation. Therefore, the level of ROS and nitrite were measured to explore the protective mechanisms of α-Bisabolol. The study concluded that α-Bisabolol safeguarded against the induced upsurge of ROS and nitrite. Lipid and protein peroxidation is considered the interconnected result of ROS accumulation. For further confirmation, finding showed that protein carbonyl content (PCC) was reduced in α-Bisabolol treated cells compared to untreated group. α-Bisabolol treatment also restored mitochondrial membrane potential (MMP) validating its antioxidant effect. Acetylcholinesterase (AChE) and butylcholinesterase are enzymes that control the degradation of Acetylcholine (Ach) and its subsequent abundance in synaptic cleft. Substantial evidence implicates that the inhibition of AChE forms a curative approach in AD patients. Therefore, the inhibitory activity of α-Bisabolol was evaluated in N2a cells and showed a significant reduction in AChE activity and an ability to avert Ach depletion. At the same time, α-Bisabolol protected cells from Aβ triggered apoptosis by reducing Bax and Caspase-3 and increasing Bcl-2 activity. Confocal laser scanning microscope also detected more viable cells with intact morphology than apoptotic cells in α-Bisabolol group. BACE1 (β-secretase βAPP-cleaving enzyme 1) is another enzyme that requires attention due to its participation in Aβ release.

Same study evaluated α-Bisabolol inhibitory activity on BACE1 and found a decrease in BACE1 activity following α-Bisabolol treatment. α-Bisabolol also preserved the viability of *Caenorhabditis elegans* by increasing their life span [49]. These findings were previously confirmed by another study evaluating α-Bisabolol effect on Aβ aggregation and survival of PC12 cells following exposure to Aβ due to the anti-aggregation property of α-Bisabolol against Aβ peptide. It also decreased apoptotic cell death and increased the cellular progenicity [50].

Recently, α-Bisabolol has been subjected for an additional assessment for acquiring corroborating evidence regarding its efficiency in combating AD associated pathology. α-Bisabolol was tested for its antioxidant, anti-aggregate, anti-AchE and anti-apoptotic activities. Neuro-2a cells treated with α-Bisabolol had better ROS and RNS scavenging potential, decreased MDA and protein carbonyl level, restored MMP loss and prevented induced apoptosis. Moreover, AChE, BuChE, β-secretase actions were decreased significantly in cells pretreated with α-Bisabolol [51]. On the other hand, a synthesized compound of α-Bisabolol; α-Bisabolol β-D-fucopyranoside (ABFP) was evaluated for its activity against β-amyloid toxicity in both in vivo and in silico experiments. The compound clearly illustrated a potent anti-AchE activity of 95.869% similar to the activity of donepezil, a standard drug. It had also a profound antioxidant potential demonstrated by both hydrogen peroxide and hydroxyl radical scavenging and a metal chelating activity. The compound also disaggregated $A\beta_{25-35}$ peptide and protected against its induced toxicity by increasing neuro2a cells viability [52]. In Figure 2, the mechanisms of α-Bisabolol in mediating neuroprotection are summarized.

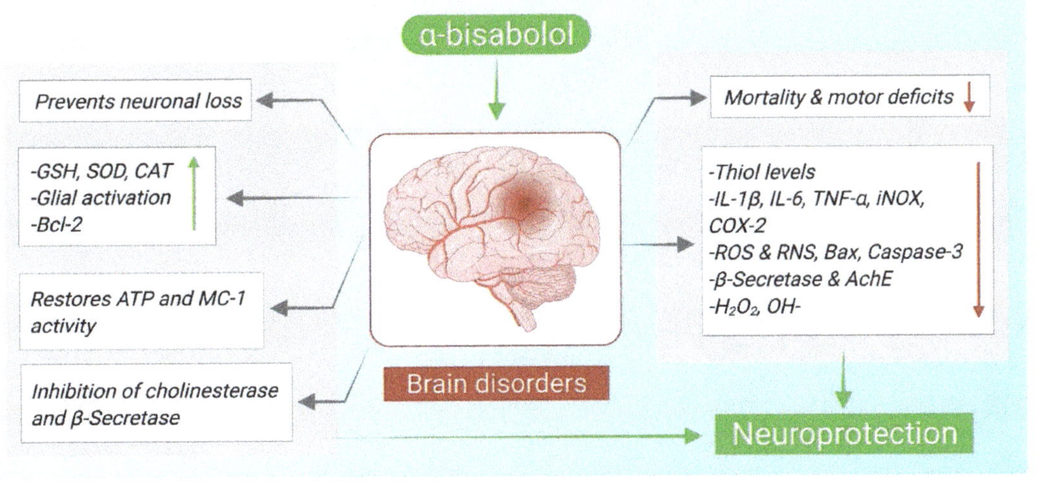


**Figure 2.** Neuroprotective mechanisms of α-Bisabolol against neurodegeneration. ↑; increase, ↓; decrease.

Cerebral ischemia is a neurological insult which can lead to irreversible damage or even death if not treated promptly and effectively. The principal in cerebral ischemia treatment is to restore blood flow to the deprived regions by reperfusion. This process can result in cerebral ischemia-reperfusion injury that requires neuroprotection due to the accompanying biochemical disturbances occurring in brain tissue [53]. In a model of permanent occlusion of the middle cerebral artery (pMCAO) inducing cerebral ischemia in mice, α-Bisabolol was evaluated for its neuroprotective effects. It was observed that there was a decrease in cerebral infarct area, a prevention of memory deficits and an improvement in motor performance in mice treated with α-Bisabolol following ischaemia. In addition, the cresyl-positive cells (viable cells) were increased by α-Bisabolol treatment compared with untreated mice. However, the counted Fluoro-Jade C positive cells (dead cells) were reduced in the group of mice treated with α-Bisabolol; indicating the inhibitory effect of α-Bisabolol on cellular degradation. At the same time, the induction of proinflammatory markers expression after pMCAO was suppressed by α-Bisabolol administration, where the expressions of myeloperoxidase (MPO), TNF-α, iNOS and the astroglial marker, GFAP were reduced in α-Bisabolol treated mice [54]. The neuroprotective effects and demonstrated pharmacological mechanisms of α-Bisabolol are presented in Table 4.

**Table 4.** Neuroprotective effects and underlying mechanisms of α-Bisabolol.

| | | Neuroprotective | | |
|---|---|---|---|---|
| **Compound** | **Dose/Route/Duration** | **Model** | **Major Mechanisms** | **Reference** |
| α-Bisabolol | 5, 25, and 250 μmol/L for 7 days | Rotenone (500 μmol/L) induced neurotoxicity in *Drosophila* | ↓ mortality and motor deficits, ↓ thiol level ↑ SOD, CAT and Keap1 | [46] |
| α-Bisabolol | 50 mg/kg i.p, 30 min before rotenone for 4 weeks | Rotenone (2.5 mg/kg) induced Parkinson's disease | ⇥ neuronal loss, ↓ MDA, ↑ GSH, SOD and CAT, ↓ glial activation, ↓ IL-1β, IL-6, TNF-α, iNOS and COX-2, ↑ Bcl-2, ↓ Bax, caspases-3, 9 and cytochrome-C, restored ATP and MC-I activity | [47] |
| α-Bisabolol | 5, 10 μg/mL for 2 h in N2a cells and 25, 50 and 100 μg/mL in elegans | $A\beta_{25-35}$ peptide (50 μM for 24 h) induced toxicity in N2a cells and *Caenorhabditis elegans* CL4176 and CL2006 | ⇥ cholinesterase and β-secretase, ↓ ROS and RNS ↓ Bax and caspase-3 ↓ ace-1, hsp-4 and Aβ genes | [49] |
| α-Bisabolol | 5 mg/mL for 2 h | $A\beta_{25-35}$ peptide (50 μM for 24 h) induced toxicity in PC12 cells | ↓ Aβ aggregation ↑ cell survival | [50] |
| α-Bisabolol | 5 and 10 μg/mL for 2 h | $A\beta_{25-35}$ peptide (50 μM for 24 h) induced toxicity in Neuro-2a cells | ↓ ROS and RNS, ↓ β-secretase and AchE activities, ↓ Bax, caspase3, and ↑ Bcl-2 | [51] |
| α-Bisabolol β-D-fucopyranoside | 10–50 μg/mL | $A\beta_{25-35}$ (100 μM for 24 h, 48 h, 96 h, 9 d) induced toxicity in Neuro 2a cells | Inhibited AChE, ↓ $H_2O_2$ and $OH^{\bullet}$, ↓ Aβ aggregation ↑ cell survival | [52] |
| α-Bisabolol | 50, 100 and 200 mg/kg/day, p.o | Permanent occlusion of the middle cerebral artery induced cerebral ischemia in mice | ↓ infarct size, ↑ motor performance, ↑ crossings and rearings | [54] |

Symbols indications: ↑; increase, ↓; decrease, ⇥; activity inhibition.

## 4. α-Bisabolol and Anticancer Effects

Cancer is characterized by cellular overgrowth with abnormal proliferation which can metastasize to surrounding tissues. Cancerous cells have dysregulated cell division cycle lacking the tight control required for regulated cell replication resulting in an altered mitotic activity. In addition to the uncontrolled growth, cancerous cells can easily evade apoptosis due to mutated genes responsible for apoptosis control [55]. Apoptosis plays an essential role in eliminating abundant cells and terminating continuous cell growth. Targeting apoptosis has been shown to be an effective strategy in treating cancer demonstrated by many anticancer drugs stimulating pro-apoptotic proteins and inhibiting anti-apoptotic ones [56]. The anticancer actions and underlying mechanisms of α-Bisabolol are illustrated in Figure 3.

In order to envisage the exact apoptotic mechanism of α-Bisabolol, a group of researchers hypothesized that α-Bisabolol can be embedded in the lipid rafts easily, which are known to be abundant in cancerous cells. At the same time, Bid; a member of the Bcl-2 pro-apoptotic family proteins, has been reported to be easily recruited in the lipid rafts by apoptosis-inducing agents, conferring it a role in the formation of a death-inducing signaling complex. With the aid of Gas Chromatography Mass Spectroscopy (GC-MS) which measured α-Bisabolol amount in the cell extracts, it is revealed that α-Bisabolol was adsorbed into flottillin rich structures known as the lipid rafts. Bid was also shown by western blotting to be present in the flottillin-enriched fraction, referring to α-Bisabolol induced movement of Bid to lipid rafts-rich membrane regions. The Surface Plasmon Resonance (SPR) analysis also indicated that α-Bisabolol interacts directly with Bid. Thus, it can be inferred that α-Bisabolol apoptotic action is mediated through its interaction with Bid protein [57].

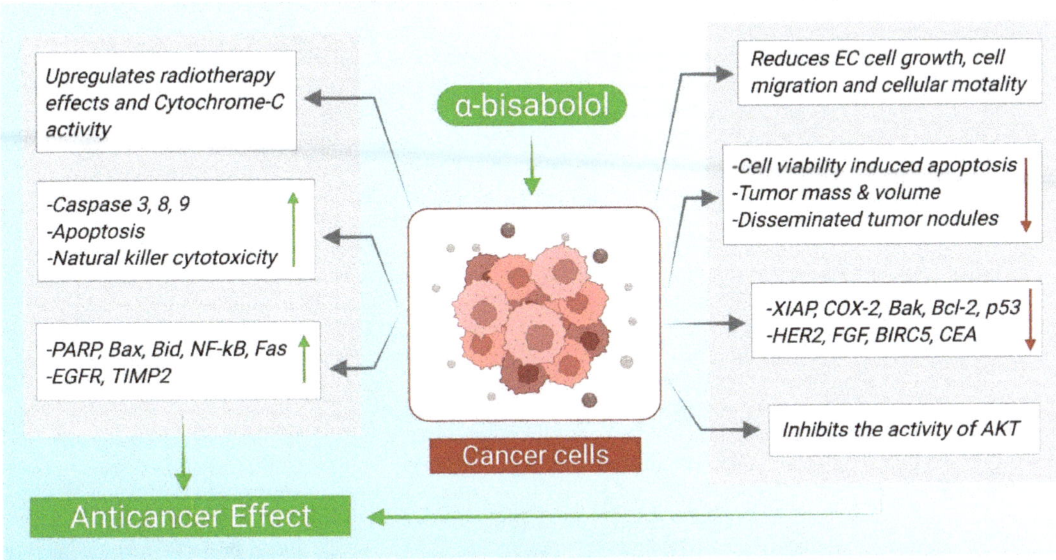


**Figure 3.** α-Bisabolol mediated anticancer actions and demonstrated mechanisms. ↑; increase, ↓; decrease.

The anticancer effects and mechanisms of α-Bisabolol in numerous types of cancer including, pancreatic, endometrial, breast and liver have been demonstrated in many experimental studies. Endometrial cancer is the main gynecological malignancy occurring in the inner lining of the uterus (endometrium) [58]. The anticancer activity of α-Bisabolol was investigated on endometrial cancer (EC) cell line on which α-Bisabolol exhibited an antitumor action by decreasing cellular viability, migration, and invasion ability of EC cells. α-Bisabolol was shown to promote apoptosis indicated by significantly increasing caspase-3 and correspondingly decreasing its direct inhibitor; X-linked inhibitor of apoptosis protein (XIAP). Cyclooxygenase-2 (COX-2) and Poly ADP-ribose polymerase (PARP) are considered as downstream mediators of caspase-3 effects. Therefore, their levels were further assessed revealing a reduced COX-2 and an increased cleaved PARP following α-Bisabolol treatment. There was also a synergistic improvement in the inhibitory outcome on EC cell growth when α-Bisabolol was combined with radiotherapy along with radiosensitivity enhancement [59].

The role of α-Bisabolol in glioblastoma, a tumor originating from glial cells of the central nervous system is well studied. Glioblastoma is considered among the most intimidating and aggressive types of tumors due to its poor prognosis and deleterious repercussions [60]. α-Bisabolol was evaluated for its cytotoxic effect against U138-MG human and C6 rat glioma cells. A3 is one of adenosine receptor subtypes on which is known as a cell death mediator in glioma. α-Bisabolol treated cells showed reduced viability and increased activity of ecto-5′-nucleotidase (ecto5′-NT/CD73); an enzymatic source for adenosine production. Since ecto-5′-NT/CD73 stimulation produces high levels of adenosine which might possibly be involved in inducing cell death through A3 adenosine receptor, the participation of this receptor was also investigated. Hence, C6 cells were pretreated with MRS1220, an A3 receptor antagonist. α-Bisabolol pretreatment significantly reversed the decrease in the number of the viable cells. An indication of the involvement of adenosinergic system represented by ecto-5′-nucleotidase/CD73 and $A_3$ receptor in the anti-proliferative action of α-Bisabolol on glioma cell lines [61]. Similar findings were reported by an earlier study which found that α-Bisabolol induced a cytotoxic effect on glioma cells with an inhibitory effect of 50% compared to untreated cells. The study also found that α-Bisabolol can mediate apoptosis by inducing cytochrome-C translocations from mitochondrial membrane [62].

Hepatocellular carcinoma is a malignancy affecting the liver with a growing tumor by increased number of cells resembling hepatocytes affecting patients with chronic liver diseases and cirrhosis [63]. The apoptosis inducing effect of α-Bisabolol was investigated on a model of human liver carcinoma cell line HepG2. It was observed that cleaved caspases 3, 8 and 9 were of a higher concentration in the α-Bisabolol treated cells as compared with untreated. In addition, α-Bisabolol decreased the level of mitochondrial cytochrome-C, increased the cytosolic cytochrome-C content along with an up-regulation in the pro-apoptotic proteins; Bax and Bid and a down-regulation of anti-apoptotic Bak and Bcl-2 proteins. The expression of p53 (a transcription factors whose products might lead to apoptosis), NF-κB and Fas was increased following α-Bisabolol treatment, indicating their function in mediating α-Bisabolol-induced apoptosis in the cancer cell line. At the same time, MTT assay showed the effectiveness of α-Bisabolol in inducing cytotoxicity in HepG2 cells [64].

Another study provided additional evidence for the anticancer potential of α-Bisabolol by examining it on non-small cell lung cancer (NSLC) cell line. Lung cancer is histologically subdivided into non-small cell lung cancer; a cancer type that is associated with major cases of cancer related mortality, and small cell lung cancer [65]. In efforts to elucidate the underlying mechanisms of α-Bisabolol anticancer effect, the study found that α-Bisabolol decreased p-PI3K and p-AKT activation; a signaling pathway, the activation of which is known to play a role in cancer pathogenesis. It was also demonstrated that α-Bisabolol suppresses the cellular proliferation at G2/M cell cycle phase. Bax was increased and Bcl-2 was decreased in α-Bisabolol-treated cells [66].

In pancreatic cancer, an aggressive adenocarcinoma with perineural and vascular growth rendering it resistant to surgical and conventional interventions [67], AKT phosphorylation was suppressed by α-Bisabolol when tested on pancreatic cancer cell line. It is speculated that AKT activation can suppress the activity of EGR1; a tumor suppressor molecule [68]. Consequently, the α-Bisabolol antitumor effect might be attributed to the activation of EGR1. The expression of EGR1 was highly elevated following the treatment with α-Bisabolol. TUNEL staining further confirmed the anti-apoptotic effect of α-Bisabolol evident by the high number of TUNEL-positive cells after pancreatic cancer cells being treated with α-Bisabolol [69].

Furthermore, α-Bisabolol was shown to inhibit the invasiveness of pancreatic cancer cell lines including KLM1, KP4 and Panc1. In looking into the key modulators of this effect, Polymerase chain reaction (PCR) analysis revealed an upregulated level in α-Bisabolol-treated cells for each of Kisspeptin-1 receptor (KISS1R), metastasis suppressor 1 (MTSS1) and tissue inhibitor of metalloproteinase 2 (TIMP2). To ensure the participation of KISS1R and its ligand KISS1 (metastasis suppressors) in the invasiveness inhibitory effect of α-Bisabolol, KISS1R was suppressed in the pancreatic cancer cells. This suppression was correlated with an attenuation of the inhibitory effect of α-Bisabolol on cancer cells migration [70].

Acute leukemia is a malignant disorder of blood constituting organs in a way where healthy bone marrow is replaced by undifferentiated cells [71]. Cells of acute leukemia were treated with α-Bisabolol and found sensitive to α-Bisabolol-induced growth inhibitory effect [72]. α-Bisabolol has been additionally scrutinized for its role against the carcinogenesis of breast cancer, a frequently diagnosed malignancy with the highest incidence in women. It was found that the treatment with α-Bisabolol reduced the incidence of tumor masses as shown by increased percent of tumor free mice and delayed appearance of tumors in HER2/neu transgenic mice. It is also reported that α-Bisabolol boosted the immunity response by T-cell subsets (CD4 and CD8) supplementation in treated mice. Gene expression profiling detected downregulated Spp1, Birc5 (carcinogenesis genes) and

Egfr, Fgf1 (angiogenesis genes), in addition to a significant downregulation in neu gene (Erbb2) following α-Bisabolol administration [73].

Moreover, α-Bisabolol derivative demonstrated a potent anti-proliferative effect on pancreatic cancer cells illustrated by the increased percentage of total apoptotic cells. α-Bisabolol also showed a growth inhibitory effect, but comparatively less that its derivative. The derivative also reduced the volume of xenograft tumor along with peritoneal metastasis, changed cell morphology with the loss of membrane asymmetry indicative for apoptosis induction, suppressed AKT expression and reduced the tumor makers CEA and CA19-9 [74].

More analogues were synthesized by glycosidation of α-Bisabolol which proved efficient cytotoxicity in vitro on human and rat glioma cells. α-Bisabolol α-L-rhamnopyranoside showed the most potent activity with IC50 reaching up to 64 M. Accordingly, it was concluded that the addition of a sugar moiety has amplified the cytotoxic effect of α-Bisabolol giving an emphasis on the importance of structural modifications [75]. In the same context, thiosemicarbazones; derived from α-Bisabolol were tested for their antitumor effect against human tumor cell lines. The synthesized derivatives were effective in inhibiting the growth of cells with the ketone analogue reported as the most potent [76]. The studies pertaining to the anticancer actions of α-Bisabolol are summarized in Table 5.

**Table 5.** Anticancer effects and demonstrated mechanisms of α-Bisabolol in the experimental models.

| | | **Anticancer Effects** | | |
|---|---|---|---|---|
| **Compound** | **Dose/Route/Duration** | **Model** | **Major Mechanisms** | **Reference** |
| α-Bisabolol | 0 to 32 μmol/L for 24 h | EC cell lines including RL95-2, ECC001 and ECC003 cells | (-) EC cells growth<br>↑ caspase-3, ↑ PARP<br>↓ XIAP, COX-2<br>↑ radiotherapy effect | [59] |
| α-Bisabolol | 35, 45 or 55 μM for C6 glioma cells and 55, 65 or 75 μM for U138-MG | U138-MG human and C6 rat glioma cell lines | ↓ cell viability<br>↑ ecto5′-NT/CD73 | [61] |
| α-Bisabolol | 100 and 250 μM | Human and rat glioma cell lines | ↓ cell viability<br>↑ Cytochrome-C | [62] |
| α-Bisabolol | 1 mM | Human prostate cancer cell line PC-3, human cervical carcinoma cell line Hela, human esophageal ECA-109, and human liver carcinoma cell line HepG2 | ↑ caspases 3, 8 and 9<br>↑ cytochrome-C<br>↑ Bax, Bid, ↓ Bak and Bcl-2, p53, ↑ NF-κB and Fas | [64] |
| α-Bisabolol | 0 t0 100 μM for 24 h (IC50 = 15 μM) | NSCLC cell line A549 | (-) migration of A549 cells, (-) PI3K/AKT,<br>↑ apoptotic cells<br>↑ Bax, ↓ Bcl-2,<br>triggers G2/M cell cycle arrest | [66] |
| α-Bisabolol | 0–250 μM<br>1000 mg/kg, (21–27 mg/mouse) once a week for 3 weeks | KLM1, Panc1, MIA Paca2 and KP4 human pancreatic cancer cell lines, BALB/c nude mice xenograft model inoculated with KLM1 and KP4 cells ($1 \times 10^7$ s.c.) in femoral area | ↓ cells viability,<br>↑ apoptosis, (-) AKT, ↑ EGR1,<br>↓ tumor volume and weight | [69] |
| α-Bisabolol | 1.56 μM | KLM1, KP4 and Panc1 human pancreatic cancer cell lines | (-) motility of cells<br>↑ KISS1R, MTSS1 and TIMP2 | [70] |

**Table 5.** *Cont.*

| Anticancer Effects | | | | |
|---|---|---|---|---|
| **Compound** | **Dose/Route/Duration** | **Model** | **Major Mechanisms** | **Reference** |
| α-Bisabolol | 0, 3, 15, 30, 60, 125, 250 μM | CML-T1primary human acute leukemia cell line | ↓ cells viability, induced apoptosis | [72] |
| α-Bisabolol | 300 μL intra-mammary injection (3.6 mg and 10 mg per mouse) | HER2/neu transgenic mice | ↓ tumor mass<br>↓ HER2/neu, Fgf and Birc5<br>↑ natural killer cytotoxicity | [73] |
| α-Bisabolol and its derivative | 62.5 μM and 125 μM, 1000 mg/kg | KLM1 and Panc1 human pancreatic cancer cell lines, BALB/c nude mice implanted with KLM1; cells ($1 \times 10^7$ cells/100 μL, s.c.) into femoral area | ↑ cell death<br>↓ volume of tumor<br>↓ CEA and CA19-9,<br>↓ disseminated tumorous nodules<br>⊣ AKT | [74] |
| α-Bisabolol β-D-fucopyranoside | IC50 > 100 μM | human lung carcinoma (A549), colon adeno-carcinoma (DLD-1), breast adeno-carcinoma (MCF-7), melanoma (SK-MEL-2), ovary teratocarcinoma (PA-1), prostate adeno-carcinoma (PC-3), pancreas adeno-carcinoma (PANC 05.04), glioma (U-251), glioblastoma (U-87) and murine glioma (GL-261) | ↑ α-Bisabolol cytotoxicity<br>↑ BBB penetration<br>↑ α-Bisabolol lipophilicity | [75] |
| α-Bisabolol-based thiosemicarbazones compounds | 0.25 to 250 mg/mL | Melanoma UACC-62, breast MCF-7, breast resistant NCI-ADR, lung NCI-460, leukemia K-562, ovarian OVCAR, prostate PCO-3, and colon HT29 cell lines | (-) cell growth | [76] |

Symbols indications: ↑; increase, ↓; decrease, (-); reduce, ⊣; activity inhibition.

## 5. α-Bisabolol and Antinociception

Nociception is a sensory mechanism of encoding noxious stimuli and perceiving them as pain. The process is transduced by nociceptors that can detect damaging sensory input and translate it as a painful stimulant [77]. α-Bisabolol has been speculated as a promising anti-nociceptive agent, where its analgesic effect and the corresponding implicated mechanisms have been subjected to an extensive investigation in different pain models. In an infraorbital nerve transection (IONX) induced nociception and trigeminal central sensitization in rats, α-Bisabolol administration was associated with a significant reversal of decreased threshold of mechanical sensitization in both ipsilateral and contralateral sides of the rat face as well as decreasing IONX-induced medullary dorsal horn (MDH) nociceptive neurons sensitization [78].

In a model of chronic pain, it is defined as a prolonged persistent pain that can be originated from inflammatory or neuropathic stimuli, α-Bisabolol was evaluated in reducing inflammatory and neuropathic associated pain induced by Freund's Complete Adjuvant (FCA) and partial lesion of the sciatic nerve (PLSN), respectively. When assessing mechanical hyperplasia, α-Bisabolol treated mice showed an increase in paw withdrawal threshold in hyperplasia induced by injected CFA and the partial sciatic nerve injury. α-Bisabolol also caused an anti-allodynic effect following acetone application on mice paw with lesioned nerve and it reduced gliosis evident by decreased immunostaining of ionized calcium-binding adapter protein (Iba-1) as well as decreased TNF-α production along with IL-10 release [79]. Moreover, the anti-nociceptive activity of α-Bisabolol was also assessed on formalin, capsaicin and glutamate induced orofacial pain in mice. α-Bisabolol administration resulted in decreased face-rubbing behavior in mice compared with control mice. This effect was suggested due to the observed inhibitory effect of α-Bisabolol on TNF-α production measured in pleural inflammatory exudate following carrageenan induced pleurisy [4].

In the same context, a study aimed to investigate the mechanism involved in α-Bisabolol attenuation of orofacial pain. The study found that the administration of HC-030031; a TRPA1 antagonist, enhanced α-Bisabolol efficacy in reducing orofacial pain induced by formalin, whereas the administration of naloxone (opioid antagonist), L-NAME (nitric oxide antagonist), and glibenclamide (ATP-sensitive $K^+$ channel ($K^+$ ATP blocker), did not alter the efficacy of α-Bisabolol. Furthermore, in cinnamaldehyde (a TRPA1 agonist) induced pain, HC-030031 did not result in an enhancement in anti-nociception of α-Bisabolol. Hence, it can be speculated the α-Bisabolol anti-nociceptive effect is due to TRPA1 receptor blockage [80]. Similar finding was reported by different study, where they found that the administration of ruthenium red (a non-competitive antagonist of TRPV1) led to additive anti-nociceptive effects with α-Bisabolol [81]. Figure 4 demonstrates the mechanisms by which α- Bisabolol exerts its anti-nociceptive effects.

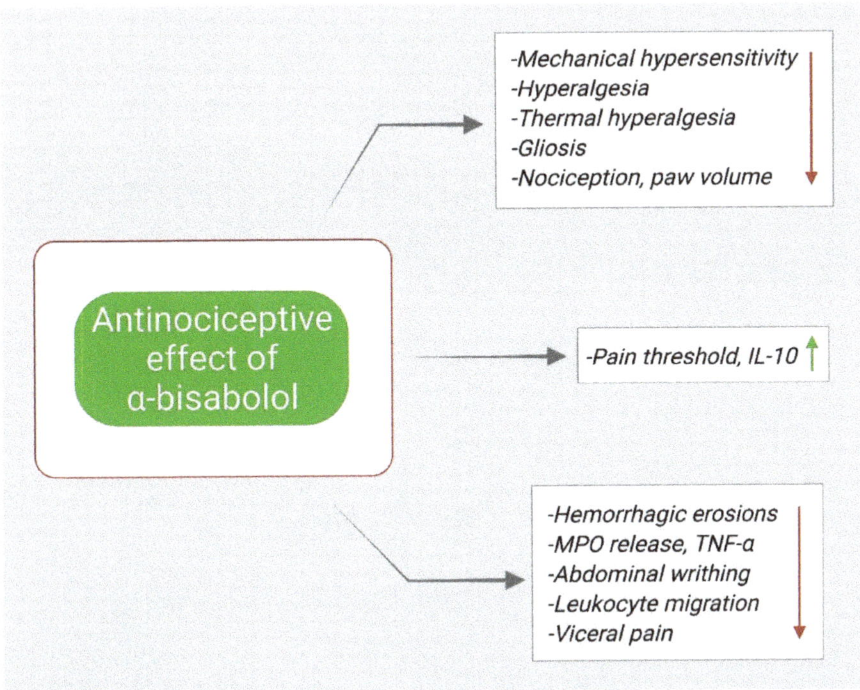


**Figure 4.** Antinociceptive mechanisms of α-Bisabolol. ↑; increase, ↓; decrease.

To elucidate the possible involved mechanism behind α-Bisabolol anti-nociception, a conducted study observed that α-Bisabolol inhibited nerve excitability of mice peripheral nervous system in a manner similar to lidocaine, suggesting a probable blockage of voltage dependent sodium channels [82]. In addition, it was observed that mice pretreated with α-Bisabolol displayed a reduction in paw-licking time following formalin injection and abdominal writhing induced by acetic acid. This is in consistent with a remarkable reduction in leukocytes migration, myeloperoxidase (MPO) activity, TNF-α in the peritoneal fluid of rats with peritonitis and decreasing edema volume induced by carrageenan, dextran, and 5-HT but not histamine [83]. In test models of visceral pain induced by cyclophosphamide and mustard oil, α-Bisabolol was evaluated and showed an ability to reduce pain related behavior in treated mice [84]. Similarly, corneal nociception was notably reduced, inferred from the decreased eye wipes by mice forepaws, on which α-Bisabolol containing ointment was applied [85]. Furthermore, it was reported that α-Bisabolol had a synergistic effect in reducing nociception, inflammation and gastric injury when co-administered with diclofenac [86].

Another aspect pertaining to α-Bisabolol efficiency enhancement was carried out to ensure more efficient anti-nociception effect. Nanocapsules containing α-Bisabolol were applied topically to mice eyes and provided a relief from triggered nociception [87]. Clinically, α-Bisabolol containing mouthwash was evaluated on thirty patients undergoing maxillofacial surgery in a randomized controlled trial. The study found similar efficiency of α-Bisabolol based mouthwash to that of chlorhexidine in reducing pain during brushing and the urge of rubbing or wiping sutures and lesions [88]. Collectively, this data (briefly summarized in Table 6) suggests that α-Bisabolol possess an anti-nociceptive effect that may led development of α-Bisabolol as a future analgesic agent to relieve pain of various origins.

**Table 6.** The anti-nociceptive actions of α-Bisabolol.

| | | **Antinociceptive Effects** | | |
|---|---|---|---|---|
| **Compound** | **Dose/Route/Duration** | **Model** | **Major Mechanisms** | **Reference** |
| α-Bisabolol | 200 mg/kg p.o | IONX-induced acute orofacial neuropathic pain in rats | ↓ mechanical hypersensitivity<br>↑ pain threshold | [78] |
| α-Bisabolol | 50 mg/kg, p.o | FCA (25 μL, i.p.) and PLSN induced pain in mice | ↓ mechanical and thermal hyperalgesia<br>↓ gliosis, ↑ IL-10, ↓ TNF-α | [79] |
| α-Bisabolol | 25 or 50 mg/kg, p.o 1 h before the local injection of inducing agents | Formalin (20 μL of 2% s.c.), capsaicin (20 μL of 2.5 μg, s.c.) or glutamate (40 μL of 25 mM, s.c.) induced orofacial nociception Carrageenan (100 μL of 1%*w/v* intrapleural) induced pleurisy in mice | ↓ orofacial pain<br>↓ TNF-α | [4] |
| α-Bisabolol | 30, 56, 100, and 180 mg/kg p.o. of α-Bisabolol alone α-Bisabolol -diclofenac (5.1, 10.3, 20.6, and 41.2 mg/kg) | Formalin (50 μL of 1%, s.c.) induced nociception<br>Carrageenan (100 μL of a 1%, s.c.)-induced inflammation in rats | ↓ nociception<br>↓ paw volume<br>↓ hemorrhagic erosion | [86] |
| α-Bisabolol | 25, 50, 100 and 200 mg/kg p.o | Carrageenan (20 μL 1%*w/v*, intraplantar injection), dextran (20 μL of 0.15%, *w*/*v*), histamine (200 μg/paw) or serotonin (200 μg/paw) induced inflammation, formalin (20 μL of 1%) induced nociception, acetic acid (0.1 mL/10 g of 0.6% solution)-induced abdominal writhing in rats | ↓ paw licking<br>↓ edema volume<br>↓ abdominal writhing<br>↓ leukocytes migration<br>↓ MPO release<br>↓ TNF-α | [83] |
| α-Bisabolol | 100, 200, or 400 mg/kg p.o., or 50, 100, or 200 mg/mL topical 60 min before injection | Formalin (20 μL of 1.5% s.c.), cinnamaldehyde (13.2 μg/lip) induced nociception in rodents | ↓ face rubbing<br>↓ head flinching | [80] |

**Table 6.** *Cont.*

| Antinociceptive Effects | | | | |
|---|---|---|---|---|
| **Compound** | **Dose/Route/Duration** | **Model** | **Major Mechanisms** | **Reference** |
| α-Bisabolol | 50, 100 or 200 mg/kg, p.o | Cyclophosphamide (400 mg/kg, i.p.), mustard oil (50 μL/animal intracolonic) induced visceral nociception in mice | ↓ visceral pain | [84] |
| α-Bisabolol | 50–200 mg/mL ointment | Hypertonic saline (20 μL of 5 M NaCl)-induced corneal nociception in mice | ↓ eye wiping | [85] |
| α-Bisabolol nanocapsules | 100 or 200 mg/mL | Hypertonic saline (20 μL of 5M NaCl)-induced corneal nociception in mice | ↓ eye wiping | [87] |
| α-Bisabolol | 50, 100 or 200 mg/kg, p.o | Acetic acid (0.6%, i.p.), Capsaicin (50 μL/animal, intracolonic), Formalin (10%, 10 μL/animal, intracolonic), (0.75%, 50 μL/animal, intracolonic) induced visceral nociception in mice | ↓ abdominal constrictions<br>↓ pain-related behavior | [81] |
| α-Bisabolol | 0.5, 1, 5 and 10 mM | Supramaximal stimulation consisted of 50–100 μs isolated rectangular voltage pulses applied on mice sciatic nerves | ↓ nerve excitability | [82] |
| α-Bisabolol | 1–0.5% mouthwash | postoperative complications of maxillofacial surgeries, a randomized, controlled, triple-blind clinical trial | ↓ pain during brushing<br>↓ lesion wiping | [88] |

Symbols indications: ↑; increase, ↓; decrease.

## 6. α-Bisabolol and Cardioprotection

Cardioprotection implies all interventions meant for preventing, ameliorating and repairing myocardial injuries following myocardial insults. It has been of great importance for research to develop pharmacological and mechanical interventions that induce cardioprotection. There are many injurers that necessitate the application of critical cardioprotection [89]. Myocardial infarction (MI) is an ischemic injury of the heart arises due to a blockage in arteries supplying heart with blood and oxygen leading to heart deprivation from these two nourishing elements [90].

To elucidate the underlying cellular perturbations in MI to develop effective therapeutic agents, α-Bisabolol was investigated for its protective properties in MI induced by isoproterenol in rats. The study has provided substantial evidence on the vital role α-Bisabolol has in ameliorating MI damage. α-Bisabolol was found to be capable of decreasing the level of serum lactate dehydrogenase (LDH), lipid peroxidation reflected by decreased amount of thiobarbituric acid reactive substances (TBARS) as well as myocardial infarct size. α-Bisabolol also exhibited an antioxidant effect manifested by elevating the activity of both SOD and CAT.

Lysosomal dysfunction was attenuated in α-Bisabolol receiving rats, where the level of lysosomal enzymes including β-glucuronidase, β-galactosidase, cathepsin-B and cathepsin-D were reduced as well as their leakage into cytosol was prevented. Furthermore, the study reported a decrease in TNF-α, IL-6, IL-1β levels and iNOS and COX-2 activities, concurrently with increased IL-10. The anti-inflammatory potential also included a NLRP3 inflammasome activation inhibition by suppressing TLR4 mediated NF-κB/MAPK pathway evidenced by measured (TLR4, p-NFκB-p65, p-IκBα), MAPK (p-ERK1/2, p-P38, p-JNK) and inflammasome complexes (NLRP3, ASC, pro-IL-1β, caspase-1-p20 and TXNIP) proteins in heart. Furthermore, an autophagy modulation was observed, where autophagic proteins (Beclin-1, LC3BI/II,) were increased, p-mTOR was decreased along with active formed autophagic vacuoles visualized under the electron microscope in α-Bisabolol treated rats compared with isoproterenol. The findings were further supported by the preservation of histological architecture of myocardium ascribed to the lessened inflammatory cells and reduced separation of cardiac muscle fibers [91].

The authors further investigated in a preceding study the effects of α-Bisabolol on mitochondrial dysfunction and apoptosis in MI induced by isoproterenol. The results proved that α-Bisabolol can protect against triggered mitochondrial dysfunction and apoptosis in MI. The mitochondrial lipid peroxidation was increased in isoproterenol injected rats shown by high levels of lipid hydroperoxide (LOOH) and TBARS, with decreased SOD, CAT activities and GSH level. However, rats co-treated with α-Bisabolol showed reduced LOOH and TBARS and increased SOD, CAT and GSH. α-Bisabolol also decreased $Ca^{2+}$ overload, mitochondrial swelling, and increased ATP production. The activities of Krebs cycle dehydrogenases (isocitrate dehydrogenase (ICDH), succinate dehydrogenase (SDH), malate dehydrogenase (MDH), α-ketoglutarate dehydrogenase (α-KGDH)) and inner mitochondrial electron transport chains (ETC) complexes (I, II, III and IV) were maintained by α-Bisabolol administration.

Furthermore, creatine kinase (CK) and LDH were increased and normal architecture of mitochondria with intact myofibrils were preserved by α-Bisabolol. Apoptosis induced by isoproterenol was inhibited by α-Bisabolol which downregulated Bax, P53, APAF-1, active caspase-3, active caspase-9, inhibited cytochrome-C release and upregulated the expression of Bcl-2 protein [92]. Figure 5 shows how α-Bisabolol protects the heart against myocardial insults.

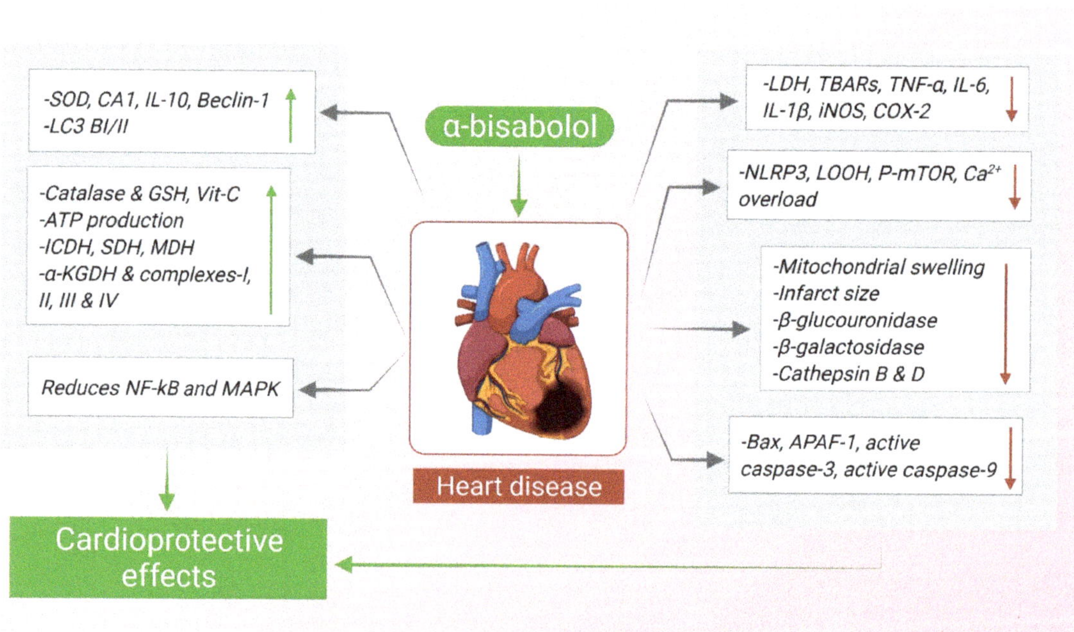


**Figure 5.** Schematic summary of cardioprotective actions of α-Bisabolol. ↑; increase, ↓; decrease.

α-Bisabolol is also able to reverse the increase in heart rate (HR), systolic (SBP), diastolic (DBP), and mean arterial pressure induced in MI and to maintain them within normal range [93]. An in silico based investigation by docking detected an ability of α-

Bisabolol to interact with receptors implicated in cardiovascular related health including α-glucosidase, angiotensin- converting enzymes, beta-2 adrenergic receptor, glucocorticoid, HMG-CoA reductase, insulin, mineralocorticoid, potassium channel receptors and peroxisome proliferator-activated receptor alpha, suggesting its likeliness to be a cardiovascular diseases modifying agent based on it its ability to interact on these receptors [94]. The cardioprotective effects and demonstrated mechanisms of α-Bisabolol are summarized in Table 7.

**Table 7.** Cardioprotective effects and mechanisms of α-Bisabolol.

| Cardioprotective Effects | | | | |
|---|---|---|---|---|
| **Compound** | **Dose/Route/Duration** | **Model** | **Major Mechanisms** | **Reference** |
| α-Bisabolol | 25 mg/kg, i.p for 10 days | Isoproterenol (85 mg/kg, s.c. for 2 days) induced myocardial infarction in rats | ↓ LDH, ↓ infarct size<br>↓ TBARS, ↑ SOD, CAT, ↓ β-glucuronidase, β-galactosidase, cathepsin-B &D, ↓ TNF-α, IL-6, IL-1β, iNOS and COX-2, ↑ IL-10, ↓ NLRP3, (-) NFκB/MAPK, ↑ Beclin-1, LC3BI/II, ↓ p-mTOR | [91] |
| α-Bisabolol | 25 mg/kg, i.p daily for 10 days | Isoproterenol (85 mg/kg, s.c. for 2 days) induced myocardial infarction in rats | ↑ CK and LDH, ↓ LOOH, TBARS, ↑ SOD, catalase and GSH, ↓ $Ca^{2+}$ overload ↓ mitochondrial swelling, ↑ ATP, ↑ ICDH, SDH, MDH, α-KGDH, and complexes I-IV, ↓ Bax, P53, APAF-1, active caspase-3 and 9, ↑ Bcl-2 | [92] |
| α-Bisabolol | 25 mg/kg, i.p daily for 10 days | Isoproterenol (85 mg/kg, s.c. for 2 days) induced myocardial infarction in rats | ↓ CK<br>↓ TBARS and LOOH<br>↑ GSH and vitamin-C<br>↓ HR, SBP and DBP | [93] |

Symbols indications: ↑; increase, ↓; decrease, (-); reduce.

## 7. α-Bisabolol and Antimicrobial Effects

Antimicrobials are drugs with an ability to inhibit or kill the pathogenic microorganisms that can cause infectious diseases which in turn can be a cause of morbidity and mortality [95]. The emergence of micro-organisms resistant to conventional antimicrobials with the concept of multiple drug resistance has urged scientists to look for alternative antimicrobial substances from various sources especially plant derived origin due to their abundance, biodiversity, and low side effects [96].

The antimicrobial activity of α-Bisabolol was evaluated against *Escherichia coli, Staphylococcus aureus, Candida albicans, Candida krusei, Candida tropicalis,* and multi-resistant bacterial strains, *Staphylococcus aureus* and *Escherichia coli.* Results showed that all strain were sensitive to α-Bisabolol which also exerted a synergism when applied with aminoglycosides [5]. In the same manner, α-Bisabolol demonstrated an antibacterial effect against *Staphylococcus aureus, Escherichia coli* and *Pseudomonas aeruginosa* as well as a synergism against *S. aureus,* when combined with the antibiotic norfloxacin and against *E. coli* when combined with gentamicin [97]. In digging into the antibacterial mechanisms associated with α-Bisabolol, α-Bisabolol was tested for a potential inhibition of TetK and NorA efflux pump responsible for pumping out antibiotics into extracellular space and thus rendering the bacteria resistant to the administered antibiotic. The combination of α-Bisabolol with tetracycline and norfloxacin resulted in potentiation of their action, illustrated by a reduction in their minimum inhibitory concentration (MIC) from 192 µg/mL to 128 µg/mL against SA IS-58 TetK pump expressing strain and from 256 µg/mL to 32 µg/mL against SA 1199 B NorA pump expressing strain, respectively [98].

Furthermore, α-Bisabolol exerted a concentration-dependent effect on decreasing the number of viable colonies of *Bacillus Solobacterium Moorei* associated with halitosis. The bacterial cell wall presented a concurrent enhanced permeability of antimicrobials such as tea tree oil after being exposed to α-Bisabolol, making it an effective agent in oral hygiene products for eliminating microbes from oral cavity and reducing halitosis [99]. *Staphylococcus aureus* and *Escherichia coli* were shown to be more sensitive to the effects of each of ciprofloxacin, clindamycin, erythromycin, gentamicin, tetracycline, and vancomycin when applied in concomitantly with α-Bisabolol [100]. This suggests the practical utility of α-Bisabolol in increasing the susceptibility of bacteria against antibiotics. The studies demonstrated anti-microbial potential of α-Bisabolol are presented in Table 8.

Fungal infections are considered predominant infectious diseases with a prominent resistance to commercially available synthetic antifungals which are associated with serious side effect such as hepatotoxicity caused by azole group [101]. Therefore, it is becoming of great importance to find compounds of plant origin with potent antifungal activity and safe to administer.

The antifungal activity of α-Bisabolol was evaluated on *Aspergillus fumigatus* species. Results demonstrated that α-Bisabolol is capable of inhibiting the growth of *A. fumigatus* as well as inhibiting the synthesis of ergosterol responsible for cellular membrane integrity. The expression of erg6 gene encoding for $\Delta^{24}$-sterol methyltransferase (24-SMT); an enzyme responsible for converting lanosterol to eburicol an integral contributor in ergosterol synthesis, was suppressed in *A. fumigatus* treated with α-Bisabolol. The activity of 24-SMT was also assayed for further assertion. The assay revealed a notable decrease in its functionality observable by the reduction in the transmethylation activity [102]. On nine different species of dermatophytes, α-Bisabolol along with other natural compounds were evaluated for their antidermatophytic activity. α-Bisabolol was superior to all other compounds in inhibiting the growth of fungal species, where it exerted over than 50% growth inhibition values on tested fungi. It is noteworthy that α-Bisabolol had an $IC_{50}$ value lower than that of fluconazole against some fungal species.

*Microsporum gypseum* which displayed a particular sensitivity for α-Bisabolol was used for carrying on further investigations. Resazurin Assay confirmed that α-Bisabolol is good at inhibiting spores' germination of *M. gypseum* with an increasing efficiency corresponds to an increase in the concentration. At the same time the transmission electron microscopy visualized the anomalies caused in *M. gypseum* organelles including irregular nucleus, vacuolization, and formation of multiple septa [103].

Additionally, α-Bisabolol exhibited a more potent growth inhibition for *Fusarium oxysporum* with low minimum inhibitory concentration (MIC) when compared with NaCl. The combination of both α-Bisabolol and NaCl resulted in an eightfold reduction in NaCl MIC, indicating a favorable synergism aimed to decrease administered antifungal agent [104]. On the other hand, α-Bisabolol was evaluated for an antiparasitic activity on *Trypanosoma cruzi*, where it was indicated that, α-Bisabolol induces cell death in *Trypanosoma cruzi* by triggering apoptosis, increasing ROS production, causing mitochondrial transmembrane potential dysfunction, and inhibiting tcGAPDH enzyme; detected by docking study which revealed that α-Bisabolol occupies the same site as chalepin (known inhibitor of tcGAPDH) does [105].

The biocide action of α-Bisabolol was also investigated on *Leishmania infantum promastigotes;* the main species causing Leishmaniases. α-Bisabolol marked a significant inhibitory effect at all concentrations tested with an inhibition of 100% for the highest one [106]. Another study also agreed that α-Bisabolol possess the highest efficiency in inhibiting *L. infantum* growth with low $IC_{50}$. Obtained results showed that α-Bisabolol-induced apoptosis, led to phosphatidylserine externalization, disrupted plasma membrane permeability, and decreased the potential of mitochondrial membrane as well as the ATP level [107]. A performed evaluation by light microscopy on the ultrastructural changes revealed that α-Bisabolol treated *Leishmania amazonensis promastigotes* and *amastigotes* showed mitochondrial swelling, numerous vacuoles formation, lipid inclusions of different sizes, condensed nucleus, and severely damaged cytoplasm [108].

The cytotoxic activity of α-Bisabolol against *Leishmania tropica promastigotes* was attributed to triggering ROS production, mitochondrial membrane depolarization, phosphatidilserine (PS) translocation, ultrastructural changes, and disrupted amastigotes cells [109] An in vivo study proved that α-Bisabolol is an effective orally administered compound in reducing the parasitic load in mice spleen and liver with no significant adverse effects on mammalian cells [110]. In a clinical and explanatory study, α-Bisabolol efficacy was assessed on canine leishmaniosis naturally infected dogs and compared with meglumine antimoniate as a reference drug. α-Bisabolol was effective in reducing the parasite load in bone marrow, lymph node and peripheral blood, increasing the release of IFNγ and inducing a Th1 cell mediated immunity. These effects were reproducible in most dogs received α-Bisabolol in comparison with the effect of the reference drug which were observed in only one dog out of six receiving it [111].

The topical application of α-Bisabolol in treating cutaneous leishmaniasis was evaluated on hamsters for an antileishmanial potential. The formulation containing α-Bisabolol was observed to be associated with a decrease in the lesion and footpad thickness and with up to 83% reduction in the parasite load in comparison with the reference group [112]. These findings are suggestive of the promising possession of antileishmanial properties of α-Bisabolol in treating visceral leishmaniosis.

**Table 8.** The antimicrobial actions of α-Bisabolol against different bacterial strains.

| Antimicrobial Actions | | | | |
|---|---|---|---|---|
| **Compound** | **Dose/Route/Duration** | **Model** | **Major Mechanisms** | **Reference** |
| α-Bisabolol | 4–512 μg/mL (for bacteria)<br>3–4096 μg/mL (for fungus) | *Staphylococcus aureus, Candida albicans, Candida krusei, Candida tropicalis* | Inhibited microbial growth | [5] |
| α-Bisabolol | 1024 μg/mL | *Staphylococcus aureus, Escherichia coli* and *Pseudomonas aeruginosa* | Inhibited microbial growth | [97] |
| α-Bisabolol | 1024 μg/mL | *Staphylococcus aureus* strains: pT181 carrying the TetK efflux pump protein that extrudes tetracycline; and the 1199B that presents resistance to norfloxacin by NorA pump expression | ↓ MIC for tetracycline and norfloxacin | [98] |
| α-Bisabolol | 0.1% | *Bacillus Solobacterium moorei* | ↓ colonies number<br>↑ effect of tea tree oil | [99] |
| α-Bisabolol | 0.5–2 mM | *Staphylococcus aureus* and *Escherichia coli* | ↑ effect of co-administered antibiotics | [100] |
| α-Bisabolol | 0.281–9 mM for 3 days | *Aspergillus fumigatus* Af239 | ↓ fungal growth<br>(-) 24-SMT, ↓ erg6 | [102] |
| α-Bisabolol | 5, 10, 20, 50, 100, 200 μg/mL | *Microsporum gypseum, Microsporum canis, Trichophyton violaceum, Nannizzia cajetani, Trichophyton mentagrophytes, Epidermophyton floccosum, Arthroderma gypseum, Trichophyton rubrum* and *Trichophyton tonsurans* | ↓ fungal growth<br>(-) spore germination<br>↑ morphological anomalies | [103] |
| α-Bisabolol | 1 μg/mL | *Fusarium oxysporum* | ↓ fungal growth<br>↓ MIC of NaCl | [104] |
| α-Bisabolol | 1000–31.25 μM | *Trypanosoma cruzi Y* infected LLC-MK2 cells | ↓ cell viability<br>↑ ROS, ↑ apoptosis | [105] |

**Table 8.** *Cont.*

| Antimicrobial Actions | | | | |
|---|---|---|---|---|
| **Compound** | **Dose/Route/Duration** | **Model** | **Major Mechanisms** | **Reference** |
| α-Bisabolol | 1000–6.25 μg/mL | *Leishmania infantum zymodeme 1* | ↓ parasite growth | [106] |
| α-Bisabolol | $IC_{50}$ = 9.5, 16.0 | *Promastigotes of Leishmania infantum* and *amazonensis* | ↓ parasite growth<br>↑ apoptosis, ↓ Δψm ↓ ATP,<br>↑ membrane permeabilization | [107] |
| α-Bisabolol | 1.86–60 μg/mL<br>($IC_{50}$ = 8.07 μg/mL) | MHOM/BR/76/Ma-76 *Leishmania amazonensis* strains | (-) parasite growth<br>↑ morphological changes | [108] |
| α-Bisabolol | 50, 200, and 1000 mg/kg p.o for 14 days | $10^7$ stationary-phase *promastigotes of Leishmania infantum* injected in mice | (-) parasite growth | [110] |
| α-Bisabolol | 25 and 100 μM<br>$IC_{50}$ = 25.2 μM | *Leishmania tropica promastigotes* | (-) parasite growth<br>↑ ROS, ↑ apoptosis<br>↑ ultrastructure changes<br>↑ PS externalization | [109] |
| α-Bisabolol | 30 mg/kg, p.o, once daily for 28 days | *Canine leishmaniosis* naturally infected dogs | ↓ parasite load<br>↓ antibody titers<br>↑ IFNγ,<br>↑ Th1/Th2 immunity | [111] |
| α-Bisabolol | 1%, 2.5%, 5% applied ointment, 200 mg/kg p.o. for 21 days | Inoculated 3 × $10^7$ parasites in the left hind footpad of hamsters | ↓ lesion thickness<br>↓ parasite load | [112] |

Symbols indications: ↑; increase, ↓; decrease, (-); reduce.

## 8. α-Bisabolol and Gastroprotection

The core function of stomach is to aid in food digestion which makes the stomach frequently exposed to various noxious factors such as hydrochloric acid, regurgitated bile acid and wide range of osmolality. The resilience of stomach in withstanding several insults is conferred by the physiological responses of gastric mucosa and the ability to self-repair induced gastric lining damages [113]. In ethanol induced gastric damage in the form of gastric hemorrhage, oral administration of α-Bisabolol was realized to attenuate gastric damage and to provide cytoprotection in stomach. Prostaglandins and nitric oxide are known for their important role in protecting gastric mucosal and maintaining gastric blood flow.

To investigate their participation in α-Bisabolol action; rats were injected with indomethacin (cyclooxygenase inhibitor) and L-NAME (nitric oxide synthase inhibitor). The results revealed an abrogation in α-Bisabolol following their administration, indicating the involvement of prostaglandins and nitric oxide in the protection effect of α-Bisabolol. Furthermore, the decrease in α-Bisabolol action following the administration of $K^+_{ATP}$ channel blocker; glibenclamide offered evidence for the possible role of $K^+_{ATP}$ channel activation in mediating α-Bisabolol action [114]. However, a different study reported the involvement of another mechanism which is based on reverting the depletion of GSH caused by ethanol, where α-Bisabolol restored non-protein sulfydryl's groups to which GSH belongs [115].

Furthermore, α-Bisabolol possesses antioxidant and anti-inflammatory properties which can protect the stomach. The antioxidant effect was demonstrated by causing an increase in SOD activity and a reduction in MDA levels. Whereas the anti-inflammatory effect was shown by the caused abolishment in MPO activity following ethanol injury leading to a decreased neutrophils influx into gastric mucosa [116]. Taken together, this demonstrates the anti-ulcerogenic activity of α-Bisabolol. The gastroprotective activity of α-Bisabolol is illustrated in Table 9.

**Table 9.** Effects of α-Bisabolol in preclinical models of gastrointestinal diseases.

| Dose/Route/Duration | Model | Major Mechanisms | Reference |
|---|---|---|---|
| 100 mg/kg p.o. | Ethanol (96%, 1 mL per animal) induced gastric damage in rats | ↓ gastric damage | [114] |
| 100 or 200 mg/kg p.o. | Ethanol (0.2 mL/animal p.o.) and Indomethacin (20 mg/kg p.o.) induced ulcer model in mice | ↓ gastric lesions<br>↑ GSH | [115] |
| 100 and 200 mg/kg, p.o. | Ethanol (0.2 mL) induced gastric lesion in mice | ↓ MDA, MPO, ↑ SOD,<br>↓ neutrophils influx | [116] |

Symbols indications: ↑; increase, ↓; decrease.

## 9. α-Bisabolol and Nephroprotection

Ischemia-Reperfusion injury (I/R) can result in a catastrophic tissue dysfunction and cellular loss due to sudden restored blood flow to ischemic organ. Reperfusion injury is a multifaceted injury that extensively destruct injured organ by inducing exacerbated oxidative and inflammatory response [117].

In a model of ischemia performed on rats' kidneys, α-Bisabolol was found to attenuate all induced biochemical alterations. For instance, α-Bisabolol abolished the increase in KIM-1 (kidney injury molecule-1), urinary osmolality, plasmatic creatinine, urea, uric acid, creatinine, proteinuria, and microalbuminuria. Additionally, α-Bisabolol protected against morphological changes including vacuolization, glomerular congestion and proteinaceous depositions in proximal tubules and external medulla [118].

The nephroprotection effect of α-Bisabolol was also tested in an in vitro model of I/R using HK2 cells. I/R caused significant cell death by both apoptosis and necrosis displayed by intense Annexin-V and (7AAD+/Anx+) staining. Cells treated with α-Bisabolol had a remarkable lower population of dead cells labelled with Annexin-V and (7AAD+/Anx+). α-Bisabolol reverted the increase in ROS and TBARS, enhanced GSH activity and most importantly induced NADPH production by inhibiting its oxidative degradation by NADPH oxidase, thus demonstrating an antioxidant defensive mechanism.

This was further validated by measuring NADPH oxidase gene (NOX4) expression, which was reduced after α-Bisabolol treatment [119]. The nephroprotective role of α-Bisabolol requires more extensive research to provide additional evidence on the effectiveness of α-Bisabolol and the mechanistic modulations against kidney injury. The nephroprotective effects of α-Bisabolol and the underlying mechanisms are summarized in Table 10.

**Table 10.** Nephroprotective effects of α-Bisabolol.

| | | | |
|---|---|---|---|
| 100 mg/kg p.o.<br>and 500, 250, 125, 62.5 and 31.25 μM | Clamping of the renal artery in the left kidney for 60 min. in rats and Ischemia/reperfusion model on tubular epithelial cells (LLC-$MK_2$) by anaerobic chamber method | ↓ creatinine, urea, uric acid<br>↓ urinary osmolality<br>↓ $FeNa^+$, $FeK^+$, $FeCl^-$<br>↓ microalbuminuria<br>↓ KIM-1, ↓ TBARS, ↑ GSH<br>↑ cell viability | [118] |
| 1000, 500, 250, 125, 62.5 and 31.25 μM | Ischemia/reperfusion model on human tubular kidney cells (HK2) by anaerobic chamber method | ↑ cell viability, ↓ apoptosis, ↓ TBARS, (-) NADPH oxidase ↑ GSH, ↓ NOX4, ↑ $\Delta\Psi m$, ↓ KIM-1 | [119] |

Symbols indications: ↑; increase, ↓; decrease, (-); reduce.

## 10. Anti-Inflammatory Effects of α-Bisabolol

Inflammation is a physiological response that includes an array of cellular and molecular reactions designed for promoting cellular growth and tissue restoration after an injury. Chronic inflammation can have far reaching medical consequences and it is reported to be existed as an underlying mediator of various illnesses such as diabetes, cancer, and autoimmune diseases [120]. The anti-inflammatory effect of α-Bisabolol is presented in Table 11.

**Table 11.** Anti-inflammatory effects of α-Bisabolol in different experimental models.

| Dose/Route/Duration | Model | Major Mechanisms | Reference |
|---|---|---|---|
| Cells treated with: 2.5, 5, 10 μM for 24 h Mice treated with 30 mg/kg/day p.o daily for 8 weeks | AGEs (50μg/mL for 2 h) induced OA in chondrocytes and Destabilization of the medial meniscus in mice | ↓ iNOS, COX-2, TNF-α, p65 PGE2, nitrite, IL-6, ↓ MMP13 ↑ collagen II, aggrecan ADAMT-S5, ↓ NF-κB, IκBα, ↓ pJNK, ↓ p-p38, ↑ chondrocytes and proteoglycans | [121] |
| 30, 50, or 100 mg/kg, p.o, 4 h before LPS | LPS (25 μg/25 μL intranasal) induced acute lung inflammation in mice | ↓ neutrophils, ↓ MPO, ↓ AHR, ↓ elastance, ↓ MIP-2 and KC ↓ alveolar wall thickening, inflammatory cell infiltration, alveolar hemorrhage, and lung tissue damage | [122] |
| 560, 860 and 1200 μM | LPS (10 μg/mL for 24 h) induced inflammation in human myometria biopsies | ↓ TNF-α, IL-1β | [123] |
| 50 mg/kg, p.o | Carrageenan (100 μL of 1% (*w*/*v*)) induced pleurisy in mice | ↓ TNF-α | [4] |
| In vitro: 0.5, 1, 3, 10, 30, or 90 μg/mL and in vivo: 50, 100, or 200 mg/kg p.o 1 h before surgery | Zymosan (1 mg/cavity, i.p.) induced neutrophils in peritoneal cavity of mice and Cecal ligation and puncture induced systemic infection | ↑ phagocytosis of neutrophils, ↓ leukocytes, ↓ NO, ↓ mortality ↓ colony forming units | [124] |
| 25 and 100 μM for 2h | LPS (500 ng/mL) induced inflammatory response in RAW264.7 macrophages cells | ↓ NO, PGE2, ↓ iNOS, COX-2, ↓ NF-jB, AP-1, ↓ pERK, p-p38 | [125] |

Symbols indications: ↑; increase, ↓; decrease.

In an in vitro model resembling osteoarthritis, the anti-inflammatory effect of α-Bisabolol was assessed using human chondrocytes. The model was developed by exposing chondrocytes to advanced glycation end products (AGE) that are known to accumulate with aging in joint cartilage and activate the expressed receptor of AGE (RAGE). The treatment with α-Bisabolol suppressed the AGE-induced inflammatory response by reducing the expression of iNOS, COX-2, TNF-α, prostaglandin E2 (PGE2), nitrite, and IL-6. It also attenuated the extracellular matrix (ECM) catabolism observable by the upregulation of collagen II and aggrecan with a concomitant downregulation of MMP13 and ADAMT-S5 levels. In investigating the signaling pathways, α-Bisabolol inhibited the activation of nuclear factor kappa B (NF-κB) as well as the nuclear translocation of P65. Furthermore, it suppressed the AGE-induced mitogen-activated protein kinase (MAPK) signaling by decreasing the phosphorylation of c-Jun N-terminal kinase (JNK) and p38 [121].

To improve the anti-inflammatory actions of α-Bisabolol, a drug delivery system was developed using lipid-core nanocapsules (LNC). α-Bisabolol loaded nanoparticles (α-bis-LNC) showed a protective effect on the LPS induced leukocytes influx where it decreased the inflammatory cell migration and neutrophils accumulation in the bronchoalveolar lavage fluid confirmed by the quantified level of myeloperoxidase (MPO) which showed an activity lower than that observed in the LPS challenged mice.

Moreover, α-bis-LNC reversed the LPS induced increase in cytokine levels which was evident by the reduced macrophage inflammatory protein-2 (MIP-2) and chemokine (KC) levels. Animals treated with α-bis-LNC exhibited a decrease in the phosphorylated forms of ERK, p38 and JNK along with an abolished airway hyperreactivity (AHR) [122]. Similarly, α-Bisabolol caused a decrease in LPS-induced TNFα and IL-1β production in human myometrium samples [123]. Consistent findings were reported regarding the efficiency of α-Bisabolol in attenuating inflammatory response by decreasing the release of TNF-α following an inflammatory trigger [4].

In a systemic model of infection induced by cecal ligation and puncture (CLP), α-Bisabolol administration significantly reduced the number of recruited leukocytes in mice peritoneal cavity and the release of nitrite in the peritoneal exudate and protected against mortality associated with the systemic infection [124]. Moreover, treating LPS- stimulated macrophages with α-Bisabolol caused a decrease in NO, PGE2, iNOS and COX-2 expression. The anti-inflammatory effect of α-Bisabolol was envisaged to be mediated by the inhibition of NF-κB and activator protein-1 (AP-1) signaling cascade, and it was further elucidated by α-Bisabolol-induced suppression of ERK and p38 phosphorylation [125]. The topical application of α-Bisabolol demonstrated a remarkable inhibition of pro-inflammatory cytokines in 12-O-tetradecanoyl-phorbol-13-acetate (TPA) induced skin inflammation in mice [126].

## 11. The Antioxidant Actions of α-Bisabolol

Oxidative stress occurs when the production of free radicals exceeds the antioxidant defense resulting in an imbalance between the production and the stabilizing agents. The excessively formed reactive oxygen species interact with biomolecules including proteins, lipids and DNA causing deleterious changes and cellular degeneration [127]. Phytochemicals have a major role as free radical scavengers which can depress the level of produced reactive oxygen species and contribute in treating as well as preventing the development of many chronic diseases [128].

α-Bisabolol was shown to exert an antioxidant effect evaluated by biochemical tests including luminol-amplified chemiluminescence (LACL) on induced human polymorphonuclear neutrophils burst by *Candida albicans* and soluble stimulants (N-formyl-methionyl-leucyl-phenylalanine, fMLP). The LACL inhibition was apparent after α-Bisabolol treatment. Test on cell-free systems using H2O2/HOCl(-) and 3-morpholino-sydnonimine (SIN-1), the hydrolysis of which leads to the formation of superoxide anions as a result of oxygen reduction, exhibited an oxidation scavenging action of α-Bisabolol by reducing the burden of produces free radicals [129].

Additionally, 2,2-diphenyl-1-picryl-hydrazyl-hydrate (DPPH) test revealed that α-Bisabolol has an antioxidant activity with a reported IC50 value of >10,000 μg/mL. The ABTS radical scavenging test illustrated the antiradical potential of α-Bisabolol with a similar IC50 value to that reported in DPPH rest [130]. These finding are summarized in Table 12.

**Table 12.** The antioxidant actions of α-Bisabolol.

| Dose/Route/Duration | Model | Major Mechanisms | Reference |
|---|---|---|---|
| 1.9 to 31 g/m | *Candida albicans* and *fMLP* induced Human polymorphonuclear neutrophils respiratory, Bursts and ROS production | ↑ LACL inhibition | [129] |
| 1000 μg/mL to 62.5 μg/mL | In vitro tests (DPPH and ABTS) | ↓ concentration of free radicals | [130] |

Symbols indications: ↑; increase, ↓; decrease.

## 12. Toxicity Assessment of α-Bisabolol

Phytochemicals are generally classified as safe compounds as they are present in plants and produced naturally. However, the risk of toxicity assessment requires full consideration to eliminate toxic reactions and ensure not exceeding the recommended dose [131]. There is a good number of studies that assessed the toxicity of α-Bisabolol based on different in vivo models, yet the clinical efficacy and safety remains scarce and to be determined. The acute toxicity of α-Bisabolol was reported in mice and rats with oral LD50 of 15.1 mL/kg and 14.9 mL/kg, respectively, where symptoms such as sedation, ataxia, dyspnea, apathy, and ruffled fur were observed when exceeding aforementioned doses. The sub-chronic toxicity was assessed on rats based on 28 days duration of repeated dose and No Observed Adverse Effects level (NOAEL) of 200 mg/kg/day was determined. This indicates that α-Bisabolol has good oral compatibility in mice and rats with incompatibility reaction occurring only at high doses. The intraperitoneal LD50 was determined to be 0.633 g/kg in mice with symptoms of trembling, staggering gait and hepatic adhesions and α-Bisabolol deposits in sacrificed animals. However, inhalational exposure to aerosolized α-Bisabolol caused no deaths or lesions in necropsy.

The dermal irritation of α-Bisabolol was tested on rabbits by applying occlusive patches containing α-Bisabolol on their back and flank. Slight erythema was developed after the initial 4 h but was subsided with time. In human repeat insult patch test (HRIPT), there was no sensitization induced by α-Bisabolol with a documented No Expected Sensitization Induction Level (NESIL) of 5500 μg/cm$^2$ [132]. Moreover, α-Bisabolol was found to be non-photoallergenic and no phototoxicity is reported post topical application of 5% and 0.5% α-Bisabolol solution on shaved guinea pigs exposed to UV radiation [133]. T

The mutagenicity of α-Bisabolol was evaluated in Salmonella/microsome assay which showed that α-Bisabolol did not increase the number of his$^+$ revertant bacteria colonies when compared with the control values [134]. Furthermore, the developmental and reproductive toxicological effects of α-Bisabolol were investigated on pregnant rats and a (NOAEL) was determined to be 980 mg/kg/day with no prenatal development effects. Whereas the dose of 1000 mg/kg was realized to increase the occurrence of resorption rate and to reduce the fetal number, body weight and feed intake as well [132].

In addition, α-Bisabolol was shown to protect the fertility of mice against daunorubicin, an inducer of free radicals. All mice treated with α-Bisabolol had their fertilization capacity maintained with sperm motility and concentration not affected. This reflects the chemoprotective potential of α-Bisabolol [135].

## 13. Conclusions

It is irrefutable that plants can provide a vast reservoir for biologically active natural compounds which can be integrated in novel drugs discovery. There is increased evidence for the use of α-Bisabolol as a potential therapeutic agent and nutraceutical. This is evident in the spurred growth in the in vitro and in vivo studies evaluating the bioefficacy of α-Bisabolol, where numerous preclinical studies have provided a solid basis for its efficacy against human diseases.

As discussed in this review, α-Bisabolol has shown therapeutic potential in several pathological conditions and possesses a wide range of biological activities putatively propitious for clinical applications such as the anticancer, antimicrobial, and anti-inflammatory. It also demonstrated a particular protective effect against various disorders affects different organ systems including nervous and cardiovascular. Conceivably, the underlying mediating mechanism for the exerted bio-efficiency of α-Bisabolol is anticipated via the modulation of signaling pathways and involved molecules. This review summarized the studies based on which, α-Bisabolol might be one of the potent plants derived agent for application as nutraceutical or pharmaceutical with pharmacological rationale and molecular mechanisms.

**Author Contributions:** S.O. has conceptualized, drafted the outline, and contributed to edit the manuscript. L.B.E. has collected the literature, interpreted the data, wrote the first draft, and revised the manuscript. N.K.J. conceptualized and draw the schemes. S.M.A.B., S.N.G., Y.O.A., S.B.S. read and edited the manuscript. All authors have read and agreed to the published version of the manuscript.

**Funding:** This research was funded by research grants from United Arab Emirates University, Al Ain, UAE. The APC has been funded by Office of Assistant Dean for Research and Graduate Studies, College of Medicine and Health Sciences, UAE University, Al Ain, UAE.

**Institutional Review Board Statement:** Not applicable.

**Informed Consent Statement:** Not applicable.

**Data Availability Statement:** This is a review article, and majority of the articles referred herein this manuscript has been suitably cited in the manuscript.

**Acknowledgments:** The authors sincerely acknowledge the research grant supports (Grant# 12R104 and 12R121) from United Arab Emirates University, P.O. Box: 15551, Al Ain, UAE.

**Conflicts of Interest:** The authors declare no conflict of interest.

## References

1. Da Silveira e Sá Rde, C.; Andrade, L.N.; de Sousa, D.P. Sesquiterpenes from essential oils and anti-inflammatory activity. *Nat. Prod. Commun.* **2015**, *10*, 1767–1774. [CrossRef] [PubMed]
2. Sharifi-Rad, M.; Nazaruk, J.; Polito, L.; Morais-Braga, M.F.B.; Rocha, J.E.; Coutinho, H.D.M.; Salehi, B.; Tabanelli, G.; Montanari, C.; Del Mar Contreras, M.; et al. Matricaria genus as a source of antimicrobial agents: From farm to pharmacy and food applications. *Microbiol. Res.* **2018**, *215*, 76–88. [CrossRef]
3. McKay, D.L.; Blumberg, J.B. A review of the bioactivity and potential health benefits of chamomile tea (*Matricaria recutita* L.). *Phytother. Res.* **2006**, *20*, 519–530. [CrossRef]
4. Barreto, R.S.S.; Quintans, J.S.S.; Amarante, R.K.L.; Nascimento, T.S.; Amarante, R.S.; Barreto, A.S.; Pereira, E.W.M.; Duarte, M.C.; Coutinho, H.D.M.; Menezes, I.R.A.; et al. Evidence for the involvement of TNF-α and IL-1β in the antinociceptive and anti-inflammatory activity of *Stachys lavandulifolia* Vahl. (Lamiaceae) essential oil and (-)-α-Bisabolol, its main compound, in mice. *J. Ethnopharmacol.* **2016**, *191*, 9–18. [CrossRef] [PubMed]
5. Rodrigues, F.F.G.; Colares, A.V.; Nonato, C.F.A.; Galvão-Rodrigues, F.F.; Mota, M.L.; Moraes Braga, M.F.B.; Costa, J. In vitro antimicrobial activity of the essential oil from Vanillosmopsis arborea Barker (Asteraceae) and its major constituent, α-Bisabolol. *Microb. Pathog.* **2018**, *125*, 144–149. [CrossRef]
6. Farias, K.S.; Kato, N.N.; Boaretto, A.G.; Weber, J.I.; Brust, F.R.; Alves, F.M.; Tasca, T.; Macedo, A.J.; Silva, D.B.; Carollo, C.A. Nectandra as a renewable source for (+)-α-Bisabolol, an antibiofilm and anti-Trichomonas vaginalis compound. *Fitoterapia* **2019**, *136*, 104179. [CrossRef] [PubMed]
7. Sharma, P.; Shah, G.C.; Sharma, R.; Dhyani, P. Chemical composition and antibacterial activity of essential oil of *Nepeta graciliflora* Benth. (Lamiaceae). *Nat. Prod. Res.* **2016**, *30*, 1332–1334. [CrossRef]
8. Hernández, T.; Canales, M.; Avila, J.G.; García, A.M.; Martínez, A.; Caballero, J.; de Vivar, A.R.; Lira, R. Composition and antibacterial activity of essential oil of *Lantana achyranthifolia* Desf. (Verbenaceae). *J. Ethnopharmacol.* **2005**, *96*, 551–554. [CrossRef]
9. Masoudi, S.; Rustaiyan, A.; Mohebat, R.; Mosslemin, M.H. Composition of the essential oils and antibacterial activities of *Hymenocrater yazdianus*, *Stachys obtusicrena* and *Nepeta asterotricha* three Labiatae herbs growing wild in Iran. *Nat. Prod. Commun.* **2012**, *7*, 117–120. [CrossRef]
10. Ghaderi, A.; Sonboli, A. Chemical composition and antimicrobial activity of the essential oil of Tanacetum walteri (Anthemideae-Asteraceae) from Iran. *Nat. Prod. Res.* **2019**, *33*, 1787–1790. [CrossRef]
11. Xavier, J.; Alves, N.S.F.; Setzer, W.N.; da Silva, J.K.R. Chemical diversity and biological activities of essential oils from Licaria, Nectrandra and Ocotea Species (Lauraceae) with occurrence in Brazilian biomes. *Biomolecules* **2020**, *10*, 869. [CrossRef]
12. Soltanian, S.; Mohamadi, N.; Rajaei, P.; Khodami, M.; Mohammadi, M. Phytochemical composition, and cytotoxic, antioxidant, and antibacterial activity of the essential oil and methanol extract of *Semenovia suffruticosa*. *Avicenna J. Phytomed.* **2019**, *9*, 143–152. [PubMed]
13. Merghache, D.; Boucherit-Otmani, Z.; Merghache, S.; Chikhi, I.; Selles, C.; Boucherit, K. Chemical composition, antibacterial, antifungal and antioxidant activities of Algerian *Eryngium tricuspidatum* L. essential oil. *Nat. Prod. Res.* **2014**, *28*, 795–807. [CrossRef]
14. Kurade, N.P.; Jaitak, V.; Kaul, V.K.; Sharma, O.P. Chemical composition and antibacterial activity of essential oils of *Lantana camara*, *Ageratum houstonianum* and *Eupatorium adenophorum*. *Pharm. Biol.* **2010**, *48*, 539–544. [CrossRef] [PubMed]
15. Alves Gomes Albertti, L.; Delatte, T.L.; Souza de Farias, K.; Galdi Boaretto, A.; Verstappen, F.; van Houwelingen, A.; Cankar, K.; Carollo, C.A.; Bouwmeester, H.J.; Beekwilder, J. Identification of the Bisabolol synthase in the endangered candeia tree (*Eremanthus erythropappus* (DC) McLeisch). *Front Plant Sci.* **2018**, *9*, 1340. [CrossRef] [PubMed]
16. Singtothong, C.; Gagnon, M.J.; Legault, J. Chemical composition and biological activity of the essential oil of *Amomum biflorum*. *Nat. Prod. Commun.* **2013**, *8*, 265–267. [CrossRef]
17. Alva, M.; Popich, S.; Borkosky, S.; Cartagena, E.; Bardón, A. Bioactivity of the essential oil of an argentine collection of *Acanthospermum hispidum* (Asteraceae). *Nat. Prod. Commun.* **2012**, *7*, 245–248. [CrossRef]
18. De Moura, N.F.; Simionatto, E.; Porto, C.; Hoelzel, S.C.; Dessoy, E.C.; Zanatta, N.; Morel, A.F. Quinoline alkaloids, coumarins and volatile constituents of *Helietta longifoliata*. *Planta Med.* **2002**, *68*, 631–634. [CrossRef]
19. Yousefi, M.; Gandomkar, S.; Habibi, Z. Essential oil from aerial parts of of *Betonica grandiflora* Willd. from Iran. *Nat. Prod. Res.* **2012**, *26*, 146–151. [CrossRef]
20. Boussaada, O.; Ammar, S.; Saidana, D.; Chriaa, J.; Chraif, I.; Daami, M.; Helal, A.N.; Mighri, Z. Chemical composition and antimicrobial activity of volatile components from capitula and aerial parts of Rhaponticum acaule DC growing wild in Tunisia. *Microbiol. Res.* **2008**, *163*, 87–95. [CrossRef]
21. Buitrago, A.; Rojas, J.; Rojas, L.; Velasco, J.; Morales, A.; Peñaloza, Y.; Díaz, C. Essential oil composition and antimicrobial activity of *Vismia macrophylla* leaves and fruits collected in Táchira-Venezuela. *Nat. Prod. Commun.* **2015**, *10*, 375–377. [CrossRef] [PubMed]

22. Benomari, F.Z.; Djabou, N.; Medbouhi, A.; Khadir, A.; Bendahou, M.; Selles, C.; Desjobert, J.M.; Costa, J.; Muselli, A. Chemical variability and biological activities of essential oils of micromeria inodora (Desf.) Benth. from Algeria. *Chem. Biodivers.* **2016**, *13*, 1559–1572. [CrossRef] [PubMed]
23. Campos Ziegenbein, F.; Hanssen, H.P.; König, W.A. Secondary metabolites from *Ganoderma lucidum* and *Spongiporus leucomallellus*. *Phytochemistry* **2006**, *67*, 202–211. [CrossRef]
24. Popović, V.; Petrović, S.; Pavlović, M.; Milenković, M.; Couladis, M.; Tzakou, O.; Duraki, S.; Niketić, M. Essential oil from the underground parts of *Laserpitium zernyi*: Potential source of alpha-Bisabolol and its antimicrobial activity. *Nat. Prod. Commun.* **2010**, *5*, 307–310. [CrossRef] [PubMed]
25. Figueiredo, P.L.B.; Silva, R.C.; da Silva, J.K.R.; Suemitsu, C.; Mourão, R.H.V.; Maia, J.G.S. Chemical variability in the essential oil of leaves of Araçá (*Psidium guineense* Sw.), with occurrence in the Amazon. *Chem. Cent. J.* **2018**, *12*, 52. [CrossRef]
26. Dias, A.L.B.; Batista, H.R.F.; Estevam, E.B.B.; Alves, C.C.F.; Forim, M.R.; Nicolella, H.D.; Furtado, R.A.; Tavares, D.C.; Silva, T.S.; Martins, C.H.G.; et al. Chemical composition and in vitro antibacterial and antiproliferative activities of the essential oil from the leaves of Psidium myrtoides O. Berg (Myrtaceae). *Nat. Prod. Res.* **2019**, *33*, 2566–2570. [CrossRef]
27. Al-Ja'fari, A.H.; Vila, R.; Freixa, B.; Tomi, F.; Casanova, J.; Costa, J.; Cañigueral, S. Composition and antifungal activity of the essential oil from the rhizome and roots of Ferula hermonis. *Phytochemistry* **2011**, *72*, 1406–1413. [CrossRef]
28. Vila, R.; Santana, A.I.; Pérez-Rosés, R.; Valderrama, A.; Castelli, M.V.; Mendonca, S.; Zacchino, S.; Gupta, M.P.; Cañigueral, S. Composition and biological activity of the essential oil from leaves of *Plinia cerrocampanensis*, a new source of alpha-Bisabolol. *Bioresour. Technol.* **2010**, *101*, 2510–2514. [CrossRef]
29. Tabanca, N.; Demirci, B.; Crockett, S.L.; Başer, K.H.; Wedge, D.E. Chemical composition and antifungal activity of *Arnica longifolia*, *Aster hesperius*, and *Chrysothamnus nauseosus* essential oils. *J. Agric. Food Chem.* **2007**, *55*, 8430–8435. [CrossRef]
30. Tolouee, M.; Alinezhad, S.; Saberi, R.; Eslamifar, A.; Zad, S.J.; Jaimand, K.; Taeb, J.; Rezaee, M.B.; Kawachi, M.; Shams-Ghahfarokhi, M.; et al. Effect of *Matricaria chamomilla* L. flower essential oil on the growth and ultrastructure of *Aspergillus niger* van Tieghem. *Int. J. Food Microbiol.* **2010**, *139*, 127–133. [CrossRef]
31. Sayyad, R.; Farahmandfar, R. Influence of *Teucrium polium* L. essential oil on the oxidative stability of canola oil during storage. *J. Food Sci. Technol.* **2017**, *54*, 3073–3081. [CrossRef] [PubMed]
32. Alkan Türkuçar, S.; Aktaş Karaçelik, A.; Karaköse, M. Phenolic compounds, essential oil composition, and antioxidant activity of *Angelica purpurascens* (Avé-Lall.) Gill. *Turk. J. Chem.* **2021**, *45*, 956–966. [CrossRef]
33. Niazmand, R.; Razavizadeh, B.M. Ferula asafoetida: Chemical composition, thermal behavior, antioxidant and antimicrobial activities of leaf and gum hydroalcoholic extracts. *J. Food Sci. Technol.* **2021**, *58*, 2148–2159. [CrossRef] [PubMed]
34. Kamatou, G.P.P.; Viljoen, A.M. A review of the application and pharmacological properties of α-bisabolol and α-bisabolol -rich oils. *J. Am. Oil Chem. Soc.* **2010**, *87*, 1–7. [CrossRef]
35. Hay, R.J.; Johns, N.E.; Williams, H.C.; Bolliger, I.W.; Dellavalle, R.P.; Margolis, D.J.; Marks, R.; Naldi, L.; Weinstock, M.A.; Wulf, S.K.; et al. The global burden of skin disease in 2010: An analysis of the prevalence and impact of skin conditions. *J. Invest. Derm.* **2014**, *134*, 1527–1534. [CrossRef]
36. Mintie, C.A.; Singh, C.K.; Ahmad, N. Whole fruit phytochemicals combating skin damage and carcinogenesis. *Transl. Oncol.* **2020**, *13*, 146–156. [CrossRef]
37. Solovăstru, L.G.; Stîncanu, A.; De Ascentii, A.; Capparé, G.; Mattana, P.; Vâţă, D. Randomized, controlled study of innovative spray formulation containing ozonated oil and α-Bisabolol in the topical treatment of chronic venous leg ulcers. *Adv. Ski. Wound Care* **2015**, *28*, 406–409. [CrossRef]
38. Licari, A.; Ruffinazzi, G.; M, D.E.F.; Castagnoli, R.; Marseglia, A.; Agostinis, F.; Puviani, M.; Milani, M.; Marseglia, G.L. A starch, glycyrretinic, zinc oxide and Bisabolol based cream in the treatment of chronic mild-to-moderate atopic dermatitis in children: A three-center, assessor blinded trial. *Minerva Pediatr.* **2017**, *69*, 470–475. [CrossRef]
39. Arenberger, P.; Arenbergerová, M.; Drozenová, H.; Hladíková, M.; Holcová, S. Effect of topical heparin and levomenol on atopic dermatitis: A randomized four-arm, placebo-controlled, double-blind clinical study. *J. Eur. Acad. Derm. Venereol.* **2011**, *25*, 688–694. [CrossRef]
40. Crocco, E.I.; Veasey, J.V.; Boin, M.F.; Lellis, R.F.; Alves, R.O. A novel cream formulation containing nicotinamide 4%, arbutin 3%, Bisabolol 1%, and retinaldehyde 0.05% for treatment of epidermal melasma. *Cutis* **2015**, *96*, 337–342.
41. Nemelka, O.; Bleidel, D.; Fabrizi, G.; Camplone, G.; Occella, C.; Marzatico, F.; Pecis, L.; Bocchietto, E. Experimental survey of a new topical anti-oxidant based on furfuryl palmitate in the treatment of child's and baby's dermatitis with eczema: Results from a multicenter clinical investigation. *Minerva Pediatr.* **2002**, *54*, 465–474. [PubMed]
42. Han, G.; Ceilley, R. Chronic wound healing: A review of current management and treatments. *Adv. Ther.* **2017**, *34*, 599–610. [CrossRef]
43. Villegas, L.F.; Marçalo, A.; Martin, J.; Fernández, I.D.; Maldonado, H.; Vaisberg, A.J.; Hammond, G.B. (+)-epi-Alpha-Bisabolol [correction of bisbolol] is the wound-healing principle of *Peperomia galioides*: Investigation of the in vivo wound-healing activity of related terpenoids. *J. Nat. Prod.* **2001**, *64*, 1357–1359. [CrossRef] [PubMed]
44. Dugger, B.N.; Dickson, D.W. Pathology of neurodegenerative diseases. *Cold Spring Harb. Perspect. Biol.* **2017**, *9*, a028035. [CrossRef] [PubMed]
45. Sarkar, S.; Raymick, J.; Imam, S. Neuroprotective and therapeutic strategies against Parkinson's Disease: Recent perspectives. *Int. J. Mol. Sci.* **2016**, *17*, 904. [CrossRef]

46. Leite, G.O.; Ecker, A.; Seeger, R.L.; Krum, B.N.; Lugokenski, T.H.; Fachinetto, R.; Sudati, J.H.; Barbosa, N.V.; Wagner, C. Protective effect of (-)-α-Bisabolol on rotenone-induced toxicity in *Drosophila melanogaster*. *Can. J. Physiol. Pharm.* **2018**, *96*, 359–365. [CrossRef]
47. Javed, H.; Meeran, M.F.N.; Azimullah, S.; Bader Eddin, L.; Dwivedi, V.D.; Jha, N.K.; Ojha, S. α-Bisabolol, a dietary bioactive phytochemical attenuates dopaminergic neurodegeneration through modulation of oxidative stress, neuroinflammation and apoptosis in rotenone-induced rat model of Parkinson's disease. *Biomolecules* **2020**, *10*, 1421. [CrossRef]
48. Murphy, M.P.; LeVine, H., 3rd. Alzheimer's disease and the amyloid-beta peptide. *J. Alzheimer's Dis.* **2010**, *19*, 311–323. [CrossRef]
49. Shanmuganathan, B.; Sathya, S.; Balasubramaniam, B.; Balamurugan, K.; Devi, K.P. Amyloid-β induced neuropathological actions are suppressed by *Padina gymnospora* (Phaeophyceae) and its active constituent α-Bisabolol in Neuro2a cells and transgenic *Caenorhabditis elegans* Alzheimer's model. *Nitric. Oxide* **2019**, *91*, 52–66. [CrossRef]
50. Shanmuganathan, B.; Suryanarayanan, V.; Sathya, S.; Narenkumar, M.; Singh, S.K.; Ruckmani, K.; Pandima Devi, K. Anti-amyloidogenic and anti-apoptotic effect of α-Bisabolol against Aβ induced neurotoxicity in PC12 cells. *Eur. J. Med. Chem.* **2018**, *143*, 1196–1207. [CrossRef]
51. Sathya, S.; Shanmuganathan, B.; Devi, K.P. Deciphering the anti-apoptotic potential of α-Bisabolol loaded solid lipid nanoparticles against Aβ induced neurotoxicity in Neuro-2a cells. *Colloids Surf. B Biointerfaces* **2020**, *190*, 110948. [CrossRef] [PubMed]
52. Jeyakumar, M.; Sathya, S.; Gandhi, S.; Tharra, P.; Suryanarayanan, V.; Singh, S.K.; Baire, B.; Pandima Devi, K. α-Bisabolol β-D-fucopyranoside as a potential modulator of β-amyloid peptide induced neurotoxicity: An in vitro &in silico study. *Bioorg. Chem.* **2019**, *88*, 102935. [CrossRef] [PubMed]
53. Nour, M.; Scalzo, F.; Liebeskind, D.S. Ischemia-reperfusion injury in stroke. *Interv. Neurol.* **2012**, *1*, 185–199. [CrossRef] [PubMed]
54. Fernandes, M.Y.D.; Carmo, M.; Fonteles, A.A.; Neves, J.C.S.; Silva, A.; Pereira, J.F.; Ferreira, E.O.; Lima, N.M.R.; Neves, K.R.T.; Andrade, G.M. (-)-α-Bisabolol prevents neuronal damage and memory deficits through reduction of proinflammatory markers induced by permanent focal cerebral ischemia in mice. *Eur. J. Pharm.* **2019**, *842*, 270–280. [CrossRef] [PubMed]
55. Ok, C.; Woda, B.; Kurian, E. *The Pathology of Cancer*; University of Massachusetts Medical School: Worcester, MA, USA, 2018.
56. Pfeffer, C.M.; Singh, A.T.K. Apoptosis: A target for anticancer therapy. *Int. J. Mol. Sci.* **2018**, *19*, 448. [CrossRef]
57. Darra, E.; Abdel-Azeim, S.; Manara, A.; Shoji, K.; Maréchal, J.-D.; Mariotto, S.; Cavalieri, E.; Perbellini, L.; Pizza, C.; Perahia, D.; et al. Insight into the apoptosis-inducing action of α-Bisabolol towards malignant tumor cells: Involvement of lipid rafts and Bid. *Arch. Biochem. Biophys.* **2008**, *476*, 113–123. [CrossRef]
58. Passarello, K.; Kurian, S.; Villanueva, V. Endometrial cancer: An overview of pathophysiology, management, and care. *Semin. Oncol. Nurs.* **2019**, *35*, 157–165. [CrossRef]
59. Fang, D.; Wang, H.; Li, M.; Wei, W. α-Bisabolol enhances radiotherapy-induced apoptosis in endometrial cancer cells by reducing the effect of XIAP on inhibiting caspase-3. *Biosci. Rep.* **2019**, *39*, BSR20190696. [CrossRef]
60. Omuro, A.; DeAngelis, L.M. Glioblastoma and other malignant gliomas: A clinical review. *JAMA* **2013**, *310*, 1842–1850. [CrossRef]
61. Mendes, F.B.; Bergamin, L.S.; Dos Santos Stuepp, C.; Braganhol, E.; Terroso, T.; Pohlmann, A.R.; Guterres, S.S.; Battastini, A.M. Alpha-bisabolol promotes glioma cell death by modulating the adenosinergic system. *Anticancer Res.* **2017**, *37*, 1819–1823. [CrossRef]
62. Cavalieri, E.; Mariotto, S.; Fabrizi, C.; de Prati, A.C.; Gottardo, R.; Leone, S.; Berra, L.V.; Lauro, G.M.; Ciampa, A.R.; Suzuki, H. alpha-Bisabolol, a nontoxic natural compound, strongly induces apoptosis in glioma cells. *Biochem. Biophys. Res. Commun.* **2004**, *315*, 589–594. [CrossRef] [PubMed]
63. Ibrahim, N.; Aboulthana, W.; Sahu, D.R. Hepatocellular carcinoma: Causes and prevention. *UK J. Pharm. Biosci.* **2018**, *6*, 48. [CrossRef]
64. Chen, W.; Hou, J.; Yin, Y.; Jang, J.; Zheng, Z.; Fan, H.; Zou, G. alpha-Bisabolol induces dose- and time-dependent apoptosis in HepG2 cells via a Fas- and mitochondrial-related pathway, involves p53 and NFkappaB. *Biochem. Pharm.* **2010**, *80*, 247–254. [CrossRef]
65. Akbulut, H.; İncedayı, S.; Atasoy, Ö. Non-small cell lung cancer and its treatment. *Demiroglu Sci. Univ. Florence Nightingale Transplant. J.* **2020**, *4*, 23–40. [CrossRef]
66. Wu, S.; Peng, L.; Sang, H.; Ping Li, Q.; Cheng, S. Anticancer effects of α-Bisabolol in human non-small cell lung carcinoma cells are mediated via apoptosis induction, cell cycle arrest, inhibition of cell migration and invasion and upregulation of P13K/AKT signalling pathway. *J. Buon* **2018**, *23*, 1407–1412. [PubMed]
67. Kleeff, J.; Korc, M.; Apte, M.; Vecchia, C.; Johnson, C.; Biankin, A.; Neale, R.; Tempero, M.; Tuveson, D.; Hruban, R.; et al. Pancreatic cancer. *Nat. Rev. Dis. Primers* **2016**, *2*, 16022. [CrossRef] [PubMed]
68. Yu, J.; Zhang, S.S.; Saito, K.; Williams, S.; Arimura, Y.; Ma, Y.; Ke, Y.; Baron, V.; Mercola, D.; Feng, G.-S.; et al. PTEN regulation by Akt-EGR1-ARF-PTEN axis. *EMBO J.* **2009**, *28*, 21–33. [CrossRef]
69. Seki, T.; Kokuryo, T.; Yokoyama, Y.; Suzuki, H.; Itatsu, K.; Nakagawa, A.; Mizutani, T.; Miyake, T.; Uno, M.; Yamauchi, K.; et al. Antitumor effects of α-Bisabolol against pancreatic cancer. *Cancer Sci.* **2011**, *102*, 2199–2205. [CrossRef]
70. Uno, M.; Kokuryo, T.; Yokoyama, Y.; Senga, T.; Nagino, M. α-bisabolol inhibits invasiveness and motility in pancreatic cancer through KISS1R activation. *Anticancer Res.* **2016**, *36*, 583–589.
71. Blundell, R. Acute Leukaemia. *Int. J. Mol. Med. Adv. Sci.* **2007**, *3*, 1380144.

72. Cavalieri, E.; Rigo, A.; Bonifacio, M.; Carcereri de Prati, A.; Guardalben, E.; Bergamini, C.; Fato, R.; Pizzolo, G.; Suzuki, H.; Vinante, F. Pro-apoptotic activity of α-Bisabolol in preclinical models of primary human acute leukemia cells. *J. Transl. Med.* **2011**, *9*, 45. [CrossRef] [PubMed]
73. Costarelli, L.; Malavolta, M.; Giacconi, R.; Cipriano, C.; Gasparini, N.; Tesei, S.; Pierpaoli, S.; Orlando, F.; Suzuki, H.; Perbellini, L.; et al. In vivo effect of alpha-Bisabolol, a nontoxic sesquiterpene alcohol, on the induction of spontaneous mammary tumors in HER-2/neu transgenic mice. *Oncol. Res.* **2010**, *18*, 409–418. [CrossRef] [PubMed]
74. Murata, Y.; Kokuryo, T.; Yokoyama, Y.; Yamaguchi, J.; Miwa, T.; Shibuya, M.; Yamamoto, Y.; Nagino, M. The anticancer effects of novel α-bisabolol derivatives against pancreatic cancer. *Anticancer Res.* **2017**, *37*, 589–598. [CrossRef]
75. Piochon, M.; Legault, J.; Gauthier, C.; Pichette, A. Synthesis and cytotoxicity evaluation of natural alpha-Bisabolol beta-D-fucopyranoside and analogues. *Phytochemistry* **2009**, *70*, 228–236. [CrossRef] [PubMed]
76. Da Silva, A.P.; Martini, M.V.; de Oliveira, C.M.; Cunha, S.; de Carvalho, J.E.; Ruiz, A.L.; da Silva, C.C. Antitumor activity of (-)-alpha-Bisabolol-based thiosemicarbazones against human tumor cell lines. *Eur. J. Med. Chem.* **2010**, *45*, 2987–2993. [CrossRef] [PubMed]
77. Smith, E. Advances in understanding nociception and neuropathic pain. *J. Neurol.* **2018**, *265*, 231–238. [CrossRef] [PubMed]
78. Melo, L.T.; Panchalingam, V.; Cherkas, P.; Campos, A.R.; Avivi-Arber, L.; Sessle, B.J. (-)-α-Bisabolol reduces nociception and trigeminal central sensitisation in acute orofacial neuropathic pain induced by infraorbital nerve injury. *Life Sci.* **2019**, *227*, 122–128. [CrossRef]
79. Fontinele, L.L.; Heimfarth, L.; Pereira, E.W.M.; Rezende, M.M.; Lima, N.T.; Barbosa Gomes de Carvalho, Y.M.; Afonso de Moura Pires, E.; Guimarães, A.G.; Bezerra Carvalho, M.T.; de Souza Siqueira Barreto, R.; et al. Anti-hyperalgesic effect of (-)-α-Bisabolol and (-)-α-Bisabolol / β-Cyclodextrin complex in a chronic inflammatory pain model is associated with reduced reactive gliosis and cytokine modulation. *Neurochem. Int.* **2019**, *131*, 104530. [CrossRef]
80. Melo, L.T.; Duailibe, M.A.; Pessoa, L.M.; da Costa, F.N.; Vieira-Neto, A.E.; de Vasconcellos Abdon, A.P.; Campos, A.R. (-)-α-Bisabolol reduces orofacial nociceptive behavior in rodents. *Naunyn. Schmiedebergs Arch. Pharm.* **2017**, *390*, 187–195. [CrossRef]
81. Leite Gde, O.; Fernandes, C.N.; de Menezes, I.R.; da Costa, J.G.; Campos, A.R. Attenuation of visceral nociception by α-Bisabolol in mice: Investigation of mechanisms. *Org. Med. Chem. Lett.* **2012**, *2*, 18. [CrossRef]
82. Alves Ade, M.; Gonçalves, J.C.; Cruz, J.S.; Araújo, D.A. Evaluation of the sesquiterpene (-)-alpha-Bisabolol as a novel peripheral nervous blocker. *Neurosci. Lett.* **2010**, *472*, 11–15. [CrossRef] [PubMed]
83. Rocha, N.F.; Rios, E.R.; Carvalho, A.M.; Cerqueira, G.S.; Lopes Ade, A.; Leal, L.K.; Dias, M.L.; de Sousa, D.P.; de Sousa, F.C. Anti-nociceptive and anti-inflammatory activities of (-)-α-Bisabolol in rodents. *Naunyn. Schmiedebergs Arch. Pharm.* **2011**, *384*, 525–533. [CrossRef] [PubMed]
84. Leite Gde, O.; Leite, L.H.; Sampaio Rde, S.; Araruna, M.K.; de Menezes, I.R.; da Costa, J.G.; Campos, A.R. (-)-α-Bisabolol attenuates visceral nociception and inflammation in mice. *Fitoterapia* **2011**, *82*, 208–211. [CrossRef]
85. Teixeira, G.F.; Costa, F.N.; Campos, A.R. Corneal antinociceptive effect of (-)-α-Bisabolol. *Pharm. Biol.* **2017**, *55*, 1089–1092. [CrossRef] [PubMed]
86. Ortiz, M.I.; Cariño-Cortés, R.; Ponce-Monter, H.A.; Castañeda-Hernández, G.; Chávez-Piña, A.E. Pharmacological interaction of α-Bisabolol and diclofenac on nociception, inflammation, and gastric integrity in rats. *Drug Dev. Res.* **2018**, *79*, 29–37. [CrossRef] [PubMed]
87. Teixeira, G.F.D.; Vieira-Neto, A.E.; da Costa, F.N.; ARA, E.S.; Campos, A.R. Antinociceptive effect of (-)-α-Bisabolol in nanocapsules. *Biomed. Pharm.* **2017**, *91*, 946–950. [CrossRef]
88. Amora-Silva, B.F.; Ribeiro, S.C.; Vieira, C.L.; Mendes, F.R.; Vieira-Neto, A.E.; Abdon, A.P.V.; Costa, F.N.; Campos, A.R. Clinical efficacy of new α-Bisabolol mouthwashes in postoperative complications of maxillofacial surgeries: A randomized, controlled, triple-blind clinical trial. *Clin. Oral Investig.* **2019**, *23*, 577–584. [CrossRef]
89. Heusch, G. Myocardial ischaemia-reperfusion injury and cardioprotection in perspective. *Nat. Rev. Cardiol.* **2020**, *17*, 773–789. [CrossRef]
90. Lu, L.; Liu, M.; Sun, R.; Zheng, Y.; Zhang, P. Myocardial infarction: Symptoms and treatments. *Cell Biochem. Biophys.* **2015**, *72*, 865–867. [CrossRef]
91. Nagoor Meeran, M.F.; Azimullah, S.; Laham, F.; Tariq, S.; Goyal, S.N.; Adeghate, E.; Ojha, S. α-Bisabolol protects against β-adrenergic agonist-induced myocardial infarction in rats by attenuating inflammation, lysosomal dysfunction, NLRP3 inflammasome activation and modulating autophagic flux. *Food Funct.* **2020**, *11*, 965–976. [CrossRef]
92. Nagoor Meeran, M.F.; Laham, F.; Azimullah, S.; Tariq, S.; Ojha, S. α-Bisabolol abrogates isoproterenol-induced myocardial infarction by inhibiting mitochondrial dysfunction and intrinsic pathway of apoptosis in rats. *Mol. Cell Biochem.* **2019**, *453*, 89–102. [CrossRef] [PubMed]
93. Meeran, M.F.N.; Laham, F.; Al-Taee, H.; Azimullah, S.; Ojha, S. Protective effects of α-Bisabolol on altered hemodynamics, lipid peroxidation, and nonenzymatic antioxidants in isoproterenol-induced myocardial infarction: In vivo and in vitro evidences. *J. Biochem. Mol. Toxicol.* **2018**, *32*, e22200. [CrossRef] [PubMed]
94. Hamedi, A.; Sakhteman, A.; Moheimani, S.M. An in silico approach towards investigation of possible effects of essential oils constituents on receptors involved in cardiovascular diseases (CVD) and associated risk factors (Diabetes Mellitus and Hyperlipidemia). *Cardiovasc. Hematol. Agents Med. Chem.* **2021**, *19*, 32–42. [CrossRef] [PubMed]

95. Amerikova, M.; Pencheva, I.; Maslarska, V.; Bozhanov, S.; Tachkov, K. Antimicrobial activity, mechanism of action, and methods for stabilisation of defensins as new therapeutic agents. *Biotechnol. Biotechnol. Equip.* **2019**, *33*, 671–682. [CrossRef]
96. Amenu, D. Antimicrobial activity of medicinal plant extracts and their synergistic effect on some selected pathogens. *Am. J. Ethnomed.* **2014**, *1*, 18–29.
97. Oliveira, F.S.; Freitas, T.S.; Cruz, R.P.D.; Costa, M.D.S.; Pereira, R.L.S.; Quintans-Júnior, L.J.; Andrade, T.A.; Menezes, P.D.P.; Sousa, B.M.H.; Nunes, P.S.; et al. Evaluation of the antibacterial and modulatory potential of α-Bisabolol, β-cyclodextrin and α-Bisabolol /β-cyclodextrin complex. *Biomed. Pharm.* **2017**, *92*, 1111–1118. [CrossRef]
98. Pereira da Cruz, R.; Sampaio de Freitas, T.; Socorro Costa, M.d.; Lucas Dos Santos, A.T.; Ferreira Campina, F.; Pereira, R.L.S.; Bezerra, J.W.A.; Quintans-Júnior, L.J.; De Souza Araújo, A.A.; Júnior, J.P.D.S.; et al. Effect of α-Bisabolol and its β-cyclodextrin complex as TetK and NorA efflux pump inhibitors in staphylococcus aureus strains. *Antibiotics* **2020**, *9*, 28. [CrossRef]
99. Forrer, M.; Kulik, E.M.; Filippi, A.; Waltimo, T. The antimicrobial activity of alpha-Bisabolol and tea tree oil against Solobacterium moorei, a Gram-positive bacterium associated with halitosis. *Arch. Oral. Biol.* **2013**, *58*, 10–16. [CrossRef]
100. Brehm-Stecher, B.F.; Johnson, E.A. Sensitization of *Staphylococcus aureus* and *Escherichia coli* to antibiotics by the sesquiterpenoids nerolidol, farnesol, Bisabolol, and apritone. *Antimicrob. Agents Chemother.* **2003**, *47*, 3357–3360. [CrossRef]
101. Elgharbawy, A.; Samsudin, N.; Hashim, Y.; Salleh, H.; Santhanam, J.; Ben belgacem, F. Phytochemicals with antifungal properties: Cure from nature. *Int. J. Eng. Computat.* **2020**, *16*, 323–345. [CrossRef]
102. Jahanshiri, Z.; Shams-Ghahfarokhi, M.; Asghari-Paskiabi, F.; Saghiri, R.; Razzaghi-Abyaneh, M. α-Bisabolol inhibits Aspergillus fumigatus Af239 growth via affecting microsomal Δ(24)-sterol methyltransferase as a crucial enzyme in ergosterol biosynthesis pathway. *World J. Microbiol. Biotechnol.* **2017**, *33*, 55. [CrossRef] [PubMed]
103. Romagnoli, C.; Baldisserotto, A.; Malisardi, G.; Vicentini, C.B.; Mares, D.; Andreotti, E.; Vertuani, S.; Manfredini, S. A multi-target approach toward the development of novel candidates for antidermatophytic activity: Ultrastructural evidence on α-bisabolol -treated microsporum gypseum. *Molecules* **2015**, *20*, 11765–11776. [CrossRef] [PubMed]
104. De Medeiros, C.A.C.; Pinto, Â.V.; de Oliveira, J.C.; Silva, G.S.; Arrua, J.M.M.; Lima, I.O.; Pereira, F.O. Evaluating the antifungal activity of α-bisabolol in association with NaCl on Fusarium oxysporum in maize grains. *Curr. Microbiol.* **2021**, *78*, 604–610. [CrossRef] [PubMed]
105. De Menezes, R.; Sampaio, T.L.; Lima, D.B.; Sousa, P.L.; de Azevedo, I.E.P.; Magalhães, E.P.; Tessarolo, L.D.; Marinho, M.M.; Dos Santos, R.P.; Martins, A.M.C. Antiparasitic effect of (-)-α-Bisabolol against Trypanosoma cruzi Y strain forms. *Diagn. Microbiol. Infect. Dis.* **2019**, *95*, 114860. [CrossRef] [PubMed]
106. Morales-Yuste, M.; Morillas-Márquez, F.; Martín-Sánchez, J.; Valero-López, A.; Navarro-Moll, M.C. Activity of (-)alpha-Bisabolol against Leishmania infantum promastigotes. *Phytomedicine* **2010**, *17*, 279–281. [CrossRef] [PubMed]
107. Hajaji, S.; Sifaoui, I.; López-Arencibia, A.; Reyes-Batlle, M.; Jiménez, I.A.; Bazzocchi, I.L.; Valladares, B.; Akkari, H.; Lorenzo-Morales, J.; Piñero, J.E. Leishmanicidal activity of α-Bisabolol from Tunisian chamomile essential oil. *Parasitol. Res.* **2018**, *117*, 2855–2867. [CrossRef]
108. Rottini, M.M.; Amaral, A.C.; Ferreira, J.L.; Silva, J.R.; Taniwaki, N.N.; Souza Cda, S.; d'Escoffier, L.N.; Almeida-Souza, F.; Hardoim Dde, J.; Gonçalves da Costa, S.C.; et al. In vitro evaluation of (-)α-Bisabolol as a promising agent against *Leishmania amazonensis*. *Exp. Parasitol.* **2015**, *148*, 66–72. [CrossRef]
109. Corpas-López, V.; Merino-Espinosa, G.; Díaz-Sáez, V.; Morillas-Márquez, F.; Navarro-Moll, M.C.; Martín-Sánchez, J. The sesquiterpene (-)-α-Bisabolol is active against the causative agents of Old World cutaneous leishmaniasis through the induction of mitochondrial-dependent apoptosis. *Apoptosis* **2016**, *21*, 1071–1081. [CrossRef]
110. Corpas-López, V.; Morillas-Márquez, F.; Navarro-Moll, M.C.; Merino-Espinosa, G.; Díaz-Sáez, V.; Martín-Sánchez, J. (-)-α-Bisabolol, a Promising oral compound for the treatment of visceral leishmaniasis. *J. Nat. Prod.* **2015**, *78*, 1202–1207. [CrossRef]
111. Corpas-López, V.; Merino-Espinosa, G.; Acedo-Sánchez, C.; Díaz-Sáez, V.; Navarro-Moll, M.C.; Morillas-Márquez, F.; Martín-Sánchez, J. Effectiveness of the sesquiterpene (-)-α-Bisabolol in dogs with naturally acquired canine leishmaniosis: An exploratory clinical trial. *Vet. Res. Commun.* **2018**, *42*, 121–130. [CrossRef]
112. Corpas-López, V.; Merino-Espinosa, G.; López-Viota, M.; Gijón-Robles, P.; Morillas-Mancilla, M.J.; López-Viota, J.; Díaz-Sáez, V.; Morillas-Márquez, F.; Navarro Moll, M.C.; Martín-Sánchez, J. Topical treatment of *Leishmania tropica* Infection using (-)-α-Bisabolol ointment in a hamster model: Effectiveness and safety assessment. *J. Nat. Prod.* **2016**, *79*, 2403–2407. [CrossRef] [PubMed]
113. Fornai, M.; Antonioli, L.; Colucci, R.; Tuccori, M.; Blandizzi, C. Pathophysiology of gastric ulcer development and healing: Molecular mechanisms and novel therapeutic options. In *Peptic Ulcer Disease*; IntechOpen: London, UK, 2011.
114. Bezerra, S.B.; Leal, L.K.; Nogueira, N.A.; Campos, A.R. Bisabolol-induced gastroprotection against acute gastric lesions: Role of prostaglandins, nitric oxide, and KATP+ channels. *J. Med. Food* **2009**, *12*, 1403–1406. [CrossRef] [PubMed]
115. Moura Rocha, N.F.; Venâncio, E.T.; Moura, B.A.; Gomes Silva, M.I.; Aquino Neto, M.R.; Vasconcelos Rios, E.R.; de Sousa, D.P.; Mendes Vasconcelos, S.M.; de França Fonteles, M.M.; de Sousa, F.C. Gastroprotection of (-)-alpha-Bisabolol on acute gastric mucosal lesions in mice: The possible involved pharmacological mechanisms. *Fundam. Clin. Pharm.* **2010**, *24*, 63–71. [CrossRef] [PubMed]
116. Rocha, N.F.; Oliveira, G.V.; Araújo, F.Y.; Rios, E.R.; Carvalho, A.M.; Vasconcelos, L.F.; Macêdo, D.S.; Soares, P.M.; Sousa, D.P.; Sousa, F.C. (-)-α-Bisabolol -induced gastroprotection is associated with reduction in lipid peroxidation, superoxide dismutase activity and neutrophil migration. *Eur. J. Pharm. Sci.* **2011**, *44*, 455–461. [CrossRef]

117. Soares, R.O.S.; Losada, D.M.; Jordani, M.C.; Évora, P.; Castro-e-Silva, O. Ischemia/Reperfusion injury revisited: An overview of the latest pharmacological strategies. *Int. J. Mol. Sci.* **2019**, *20*, 5034. [CrossRef]
118. Sampaio, T.L.; Menezes, R.R.; da Costa, M.F.; Meneses, G.C.; Arrieta, M.C.; Chaves Filho, A.J.; de Morais, G.B.; Libório, A.B.; Alves, R.S.; Evangelista, J.S.; et al. Nephroprotective effects of (-)-α-Bisabolol against ischemic-reperfusion acute kidney injury. *Phytomedicine* **2016**, *23*, 1843–1852. [CrossRef]
119. Sampaio, T.L.; Menezes, R.; Lima, D.B.; Costa Silva, R.A.; de Azevedo, I.E.P.; Magalhães, E.P.; Marinho, M.M.; Dos Santos, R.P.; Martins, A.M.C. Involvement of NADPH-oxidase enzyme in the nephroprotective effect of (-)-α-Bisabolol on HK2 cells exposed to ischemia - reoxygenation. *Eur. J. Pharmacol.* **2019**, *855*, 1–9. [CrossRef]
120. Roe, K. An inflammation classification system using cytokine parameters. *Scand. J. Immunol.* **2021**, *93*, e12970. [CrossRef]
121. Xu, C.; Sheng, S.; Dou, H.; Chen, J.; Zhou, K.; Lin, Y.; Yang, H. α-Bisabolol suppresses the inflammatory response and ECM catabolism in advanced glycation end products-treated chondrocytes and attenuates murine osteoarthritis. *Int. Immunopharmacol.* **2020**, *84*, 106530. [CrossRef]
122. D'Almeida, A.P.L.; Pacheco de Oliveira, M.T.; de Souza, É.T.; de Sá Coutinho, D.; Ciambarella, B.T.; Gomes, C.R.; Terroso, T.; Guterres, S.S.; Pohlmann, A.R.; Silva, P.M.; et al. α-Bisabolol-loaded lipid-core nanocapsules reduce lipopolysaccharide-induced pulmonary inflammation in mice. *Int. J. Nanomed.* **2017**, *12*, 4479–4491. [CrossRef]
123. Muñoz-Pérez, V.M.; Ortiz, M.I.; Ponce-Monter, H.A.; Monter-Pérez, V.; Barragán-Ramírez, G. Anti-inflammatory and utero-relaxant effect of α-Bisabolol on the pregnant human uterus. *Korean J. Physiol. Pharm.* **2018**, *22*, 391–398. [CrossRef]
124. Cavalcante, H.A.O.; Silva-Filho, S.E.; Wiirzler, L.A.M.; Cardia, G.F.E.; Uchida, N.S.; Silva-Comar, F.M.S.; Bersani-Amado, C.A.; Cuman, R.K.N. Effect of (-)-α-Bisabolol on the inflammatory response in systemic infection experimental model in C57BL/6 mice. *Inflammation* **2020**, *43*, 193–203. [CrossRef] [PubMed]
125. Kim, S.; Jung, E.; Kim, J.H.; Park, Y.H.; Lee, J.; Park, D. Inhibitory effects of (-)-α-Bisabolol on LPS-induced inflammatory response in RAW264.7 macrophages. *Food Chem. Toxicol.* **2011**, *49*, 2580–2585. [CrossRef] [PubMed]
126. Maurya, A.K.; Singh, M.; Dubey, V.; Srivastava, S.; Luqman, S.; Bawankule, D.U. α-(-)-Bisabolol reduces pro-inflammatory cytokine production and ameliorates skin inflammation. *Curr. Pharm. Biotechnol.* **2014**, *15*, 173–181. [CrossRef]
127. Jat, D.; Nahar, M. Oxidative stress and antioxidants: An overview. *IJARR Int. J. Adv. Res. Rev.* **2017**, *2*, 110–119.
128. Zhang, Y.-J.; Gan, R.-Y.; Li, S.; Zhou, Y.; Li, A.-N.; Xu, D.-P.; Li, H.-B. Antioxidant phytochemicals for the prevention and treatment of chronic diseases. *Molecules* **2015**, *20*, 21138–21156. [CrossRef] [PubMed]
129. Braga, P.C.; Dal Sasso, M.; Fonti, E.; Culici, M. Antioxidant activity of Bisabolol: Inhibitory effects on chemiluminescence of human neutrophil bursts and cell-free systems. *Pharmacology* **2009**, *83*, 110–115. [CrossRef]
130. Ren, G.; Xue, P.; Sun, X.; Zhao, G. Determination of the volatile and polyphenol constituents and the antimicrobial, antioxidant, and tyrosinase inhibitory activities of the bioactive compounds from the by-product of Rosa rugosa Thunb. var. plena Regal tea. *BMC Complement Altern. Med.* **2018**, *18*, 307. [CrossRef]
131. Thakur, M.; Singh, K.; Khedkar, R. 11—Phytochemicals: Extraction process, safety assessment, toxicological evaluations, and regulatory issues. In *Functional and Preservative Properties of Phytochemicals*; Prakash, B., Ed.; Academic Press: Cambridge, MA, USA, 2020; pp. 341–361.
132. Api, A.M.; Belsito, D.; Biserta, S.; Botelho, D.; Bruze, M.; Burton, G.A., Jr.; Buschmann, J.; Cancellieri, M.A.; Dagli, M.L.; Date, M.; et al. RIFM fragrance ingredient safety assessment, α-Bisabolol, CAS registry number 515-69-5. *Food Chem. Toxicol.* **2020**, *141*, 111238. [CrossRef]
133. Bhatia, S.P.; McGinty, D.; Letizia, C.S.; Api, A.M. Fragrance material review on alpha-Bisabolol. *Food Chem. Toxicol.* **2008**, *46*, S72–S76. [CrossRef]
134. Gomes-Carneiro, M.R.; Dias, D.M.; De-Oliveira, A.C.; Paumgartten, F.J. Evaluation of mutagenic and antimutagenic activities of alpha-Bisabolol in the Salmonella/microsome assay. *Mutat. Res.* **2005**, *585*, 105–112. [CrossRef] [PubMed]
135. Hernandez-Ceruelos, A.; Sánchez-Gutiérrez, M.; Mojica-Villegas, M.; Chamorro, G. Chemoprotection of fertility by chamomile essential oil over the toxic effect of. *Toxicol. Lett.* **2007**, *172*, S185–S186. [CrossRef]

*nutrients*

MDPI

Article

# *Dendropanax trifidus* Sap-Mediated Suppression of Obese Mouse Body Weight and the Metabolic Changes Related with Estrogen Receptor Alpha and AMPK-ACC Pathways in Muscle Cells

**Ahreum Lee [1,†], Eugene Koh [2,†], Dalnim Kim [1], Namkyu Lee [3], Soo Min Cho [4], Young Joo Lee [3], Ik-Hyun Cho [5] and Hyun-Jeong Yang [1,6,7,*]**

1. Korea Institute of Brain Science, Seoul 06022, Korea; dkfma5025@hanmail.net (A.L.); hoipig0326@naver.com (D.K.)
2. Temasek Life Sciences Laboratories, Singapore 117604, Singapore; eugene@tll.org.sg
3. Department of Integrated Bioscience and Biotechnology, College of Life Science, Sejong University, Seoul 05006, Korea; nam879@naver.com (N.L.); yjlee@sejong.ac.kr (Y.J.L.)
4. PharmCADD, R&D Center, Seoul 06180, Korea; chosm@pharmcadd.com
5. College of Korean Medicine, Kyung Hee University, Seoul 02447, Korea; ihcho@khu.ac.kr
6. Department of Integrative Health Care, University of Brain Education, Cheonan 31228, Korea
7. Department of Integrative Biosciences, University of Brain Education, Cheonan 31228, Korea

* Correspondence: yang@ube.ac.kr

† These authors contributed equally to this work.

**Citation:** Lee, A.; Koh, E.; Kim, D.; Lee, N.; Cho, S.M.; Lee, Y.J.; Cho, I.-H.; Yang, H.-J. *Dendropanax trifidus* Sap-Mediated Suppression of Obese Mouse Body Weight and the Metabolic Changes Related with Estrogen Receptor Alpha and AMPK-ACC Pathways in Muscle Cells. *Nutrients* **2022**, *14*, 1098. https://doi.org/10.3390/nu14051098

Academic Editors: Daniela Rigano and Paola Bontempo

Received: 17 February 2022
Accepted: 2 March 2022
Published: 5 March 2022



**Abstract:** *Dendropanax trifidus* (DT) is a medicinal herb native to East Asia, which has been used extensively for its therapeutic properties in traditional medicine. In this study, we examined the effects of DT sap on the regulation of body weight and muscle metabolism in mice. Obese model db/db mice were administered daily with DT sap or vehicle control over a 6-week period. The effects of DT sap on muscle metabolism were studied in C2C12 muscle cells, where glycolytic and mitochondrial respiration rates were monitored. As AMP-activated protein kinase (AMPK) is a master regulator of metabolism and plays an important function as an energy sensor in muscle tissue, signaling pathways related with AMPK were also examined. We found that DT sap inhibited body weight increase in db/db, db/+, and +/+ mice over a 6-week period, while DT sap-treated muscle cells showed increased muscle metabolism and also increased phosphorylation of AMPK and Acetyl-CoA Carboxylase (ACC). Finally, we found that DT sap, which is enriched in estrogen in our previous study, significantly activates estrogen alpha receptor in a concentration-dependent manner, which can drive the activation of AMPK signaling and may be related to the muscle metabolism and weight changes observed here.

**Keywords:** *Dendropanax trifidus*; body weight; C2C12; metabolism; glycolysis; mitochondrial respiration; AMPK; ACC; estrogen

## 1. Introduction

Decrease of energy expenditure can increase body weight [1,2]. In other words, a person with lower metabolic rate gains more weight under the same caloric intake. The skeletal muscle is the largest metabolic organ system in the body [3]. It comprises ~40% of body weight in non-obese individuals [4] and is responsible for 20~30% of total resting oxygen uptake [4,5]. Skeletal muscle metabolism is an important determinant for whole-body resting metabolic rate [6], and changes dramatically depending on exercise activity.

During exercise, skeletal muscle ATP consumption increases, followed by increase in intracellular AMP concentrations, resulting in increases in the ratios of AMP/ATP and ADP/ATP, and leading to the activation of AMP-activated protein kinase (AMPK) [7].

AMPK is a sensor for intracellular energy status which maintains energy storage by fine-tuning anabolic and catabolic pathways, and is especially important for skeletal muscle, which experiences rapid energy turnover [8]. In order to maintain intracellular energy homeostasis, AMPK regulates a broad range of intracellular downstream signaling molecules such as Acetyl-CoA Carboxylase (ACC) and 6-phosphofructo-2-kinase/fructose-2,6-biphosphatase3 (PFKFB3). ACC, which catalyzes the carboxylation of acetyl-CoA to malonyl-CoA [9], can be phosphorylated by AMPK, resulting in inhibition of its enzymatic activity [10]. ACC has been previously investigated as a potential target of anti-obesity drugs [11,12]. PFKFB3 is a glycolytic enzyme which plays roles in glycolysis, cell proliferation, and tumor growth, and its selective inhibition is potentially regarded as a therapeutic target [13]. Likewise, AMPK is regarded as a central regulator for energy homeostasis, thus compounds which activate AMPK become targets for drug development [14]. Intriguingly, like exercise, estrogen also increases AMPK phosphorylation [15,16]. Estrogen plays a critical role in skeletal muscle homeostasis and exercise capacity [17] and can improve exercise endurance and mitochondrial energy metabolism in mice [17].

*Dendropanax trifidus* (DT), phylogenetically similar to *Dendropanax morbiferus* (DM), is an evergreen shrub which is mainly distributed throughout eastern Asia (including Korea and Japan) and has been appreciated for its therapeutic properties in Korean traditional medicine. DM and DT are hard to distinguish due to their close genetical and morphological characters [18,19]. In this aspect, Lee and colleagues [20] claimed that DT and DM should be considered as conspecific under *D. trifidus.* DM leaf extract was found to inhibit adipogenesis in mouse 3T3-L1 cells [21], and reduced body weight in high fat diet-fed C57BL/6 mice [22]. (9Z,16S)-16-hydroxy-9, 17-octadecadiene-12, 14-diynoic acid (HOD), a component of DM leaf, reduced oleic acid-induced TG accumulation in HepG2 hepatocytes, contributing to the reduction of body weight [23]. Recently, in a 12-week RCT, participants administered with DT leaf extract exhibited significant reductions in HbA1c, insulin resistance level and systolic BP, compared to the participants with placebo, however, lipid parameters and body composition including body weight were not changed [24]. Compared to the studies with DM leaf extract, not much is known about the effects of DM or DT sap.

In this study, we aimed to understand the effects of DT sap on mouse body weight and muscle metabolism. We administered DT sap via oral injection over a 6-week period to obese (db/db), heterozygous (db/+), and control (+/+) mice and measured their body weight at weekly time points. We further examined the effects of DT sap on glycolysis and mitochondrial respiration parameters in C2C12 skeletal muscle cells. To study the underlying molecular mechanism, we examined whether the function of DT sap is mediated by estrogen receptor, as estrogen was one of the major components of DT sap [25]. Finally, we show here AMPK and its downstream signaling—including ACC and PFKFB3 which are significantly altered by DT sap treatment, implying that those cellular components might mediate glycolytic and metabolic changes within the cell, thus contributing to weight loss in mice treated with DT sap.

## 2. Materials and Methods

### 2.1. Mice

All experiments were performed in compliance with the relevant laws and institutional guidelines and were approved by the University of Brain Education's Animal Care and Use Committee (Approval number: 2018-AE-01Y-01). For obesity model mice, BKS.Cg-$Dock7^{m}$ +/+ $Lepr^{db}$/J lines, which are used for mouse model for type II diabetes and obesity, were purchased from ORIENT BIO Inc. (Seongnam, Korea). Homozygous mice (db/db) manifest morbid obesity. Ten-week old female and male mice with genotype of db/db, db/+, +/+ (3~10 mice per genotype per sex) were administered daily with 100 μL (per mouse) of vehicle or DT sap (0.2 mg/g for DT sap weight/mouse body weight) via oral gavage for six weeks. Body weights were measured every week (Figure 1).

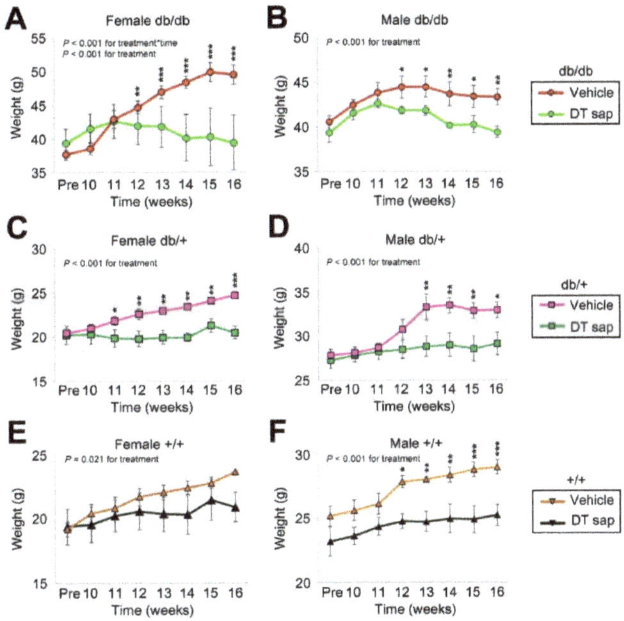


**Figure 1.** Body weight change by *Dendropanax trifidus* sap compared to vehicle administration in db/db, db/+, +/+ mice for 6 weeks. 0.2 mg/g DT sap or vehicle control (100 μL) was administered daily for 6 weeks to db/db (**A**,**B**), db/+ (**C**,**D**), and +/+ (**E**,**F**) mice (female: (**A**,**C**,**E**); male: (**B**,**D**,**F**)). Body weights were measured every week. Data was analyzed by two-way ANOVA. N = 3–10 mice per group. * $p < 0.05$, ** $p < 0.01$, *** $p < 0.001$. Bars indicate mean ± S.E.M.

### 2.2. Reagents

An estrogen receptor antagonist ICI 182,780 (Fulvestranat, 14409) and β-estradiol (E2758) was purchased from Sigma-Aldrich (Saint Louis, MO, USA). *Dendropanax trifidus* (Thunb.) Makino ex H.Hara from Simasi, Mieken, Japan was registered at Ibaraki Nature Museum (INM-2-212778). *Dendropanax trifidus* sap was ethanol-extracted, freeze-dried and stored at −20 °C. For experiments, the freeze-dried final product was dissolved in EtOH and diluted with water or medium [25].

### 2.3. Antibodies

Rabbit polyclonal or monoclonal antibodies against phospho-AMPKα (Thr172) (#2531), AMPK (#2532), phospho-HSL (#4137T), HSL (#4107T), phospho-ACC (Ser79) (#11818T), ACC (#3676T), ATGL (#2138S), phospho-ULK1(Ser555) (#5869T), and ULK1 (#8054T) were purchased from Cell Signaling. Rabbit polyclonal antibodies against PFKFB3 (#13763-1-AP) and b-Actin (#ab8227) were purchased from Proteintech (Rosemont, IL, USA) and Abcam (Cambridge, UK), respectively.

### 2.4. C2C12 Cell Culture

Murine myoblast cell line C2C12 cells were routinely cultured in growth medium containing 10% heat-inactivated fetal bovine serum (FBS)/1% penicillin streptomycin (PS)/Dulbecco's Modified Eagle's Medium (DMEM) without phenol red. Cells were cultured in chamber-slides at 37 °C in a humid atmosphere of 5% $CO_2$ in air. Cultures were passaged every 2 days with fresh medium. For differentiation, when cells reached confluence (at the 3rd day after seeding with a density of $1 \times 10^4$ cells/well for the 6-well plate), the medium was changed into 2% horse serum/1% PS/DMEM without phenol red. After three days of differentiation when cells were fused into myotubes, cells were treated with the indicated reagents for the indicated time duration and analyzed by Western Blot analysis.

### 2.5. Western Blot

Cells were lysed in cold RIPA buffer (WSE-7420, ATTO, DAWINBIO Inc, Hanam, Korea) on ice and collected for centrifugation at 15,000 RPM for 15 min. The resultant supernatant was quantified and diluted with sample buffer, boiled, subjected to SDS-PAGE, and transferred onto PVDF membranes. The membranes were blocked with EZBlock Chemi (AE-1475, ATTO, Tokyo, Japan) for 1 h at room temperature, incubated with the primary antibodies overnight at 4 °C, washed and further incubated with secondary antibody for 1 h at room temperature, and washed again. Protein signals were visualized by Super Signal West Pico PLUS Chemiluminescent Substrate (34580, Thermo Fisher Scientific, Waltham, MA, USA) and images were taken by Amersham Imager 600 (GE Healthcare, Chicago, IL, USA). Western blot images were quantified by Image J software (version 1.52p, NIH, Bethesda, USA).

### 2.6. Luciferase Assay and Cell Viability Test

For luciferase assay, the hERα-HeLa-9903 cell line expressing hERα was used. Cells were maintained in 10% FBS/DMEM. On the day of the experiment, cells were seeded in 10% DCC/FBS/DMEM on 96 well plates with a density of $1 \times 10^4$ cells/100 µL/well, and maintained at 37 °C in a 5% $CO_2$ incubator for 24 h. The cells were treated as indicated and further incubated for 48 h at 37 °C in a $CO_2$ incubator. The medium was removed and luciferase activity was measured by luminometer.

Cell viability check was used by Cell proliferation reagent (Thiazolyl Blue Tetrazolium Bromide) (Sigma, St Louis, MO, USA) for measurement at the absorption at 590 and 620 nm of wavelength by absorbance microplate reader (Bio-Tek, Winooski, VT, USA).

### 2.7. Glycolysis, Mitochondrial Stress Test, and Energy Map

C2C12 myoblast cells were seeded with a density of $0.5 \times 10^4$ cells/well on 96-well plates for the Agilent Seahorse XF assay, differentiated under the differentiation medium (2% horse serum/1% PS/DMEM) for three days, then treated with DT sap of the indicated concentrations (0, 62.5, 125, 250, 500 µg/mL) for 1 h or 24 h, and subjected to Glycolysis or Mitochondrial stress tests according to the manufacturer's instructions (Agilent Seahorse XF Glycolysis Stress Test Kit 103020-100, Agilent Seahorse XF Cell Mito Stress Test Kit 103015-100, Agilent Technologies, Santa Clara, CA, USA). Each treatment was performed in triplicate and independent experiments were performed in duplicate.

Agilent Seahorse XF technology provides a cell energy metabolism map (a cell energy phenotype profile, Figure 2A) by a simultaneous measurement of two major energy production pathways, i.e., mitochondrial respiration and glycolysis in live cells. Cells are distributed in the energy map according to the following attributes: Quiescent (not very energetic via either metabolic pathway), Energetic (utilizing both metabolic pathways), Aerobic (utilizing predominantly mitochondrial respiration), and Glycolytic (utilizing predominantly glycolysis).

For the Glycolysis Stress Test, 10 mM glucose, 1 µM oligomycin, and 50 mM 2-deoxy-glucose (2-DG) were applied according to the manufacturer's instructions (Figure 2B). Cells were incubated in the glycolysis stress test medium without glucose or pyruvate, then a saturating concentration of glucose was injected, inducing cells to catabolize it through the glycolytic pathway to pyruvate, producing ATP, NADH, water, and protons. The glucose-induced extrusion of protons into the surrounding medium rapidly increases extracellular acidification rate (ECAR), which is reported as the rate of glycolysis under basal conditions (Figure 2B, blue region). Glycolytic capacity is the maximum ECAR rate reached by a cell following the addition of oligomycin, which shuts down oxidative phosphorylation and drives the cell to reach its maximum glycolytic capacity (Figure 2B, green region). Glycolytic reserve indicates the capability of a cell to respond to an energetic demand and is calculated by the subtraction of the values of Glycolysis from Glycolytic Capacity (Figure 2B). Glycolytic reserve as a % is a calculation of (Glycolytic Capacity Rate)/(Glycolysis) x 100 (Figure 2B)).

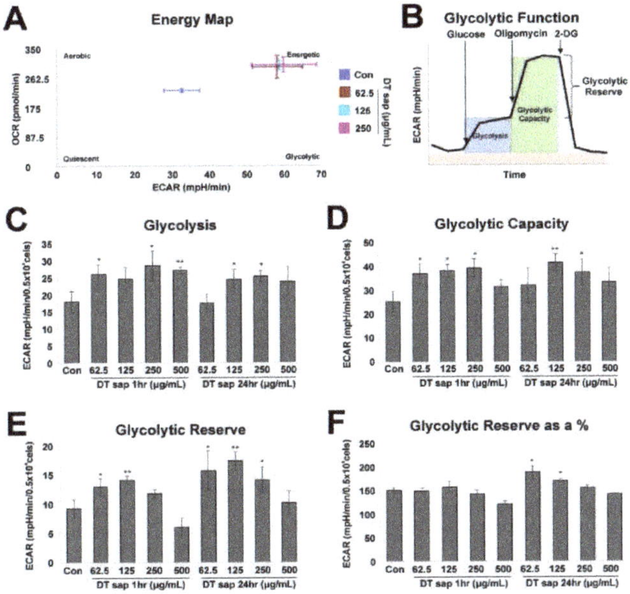

**Figure 2.** Glycolysis stress test for *Dendropanax trifidus* (DT) sap in C2C12 cells. (**A**) Energy phenotype profile of control cells (blue) and cells treated with DT sap of the indicated concentrations for 1 h (red, 62.5 μg/mL; light blue, 125 μg/mL; purple, 250 μg/mL). (**B–F**) Glycolysis Stress test for measuring glycolytic function in cells by direct measurement of the extracellular acidification rate (ECAR). (**B**) Agilent Seahorse XF Glycolysis Stress Test profile. Indicated compounds (i.e., 10 mM glucose, 1 μM oligomycin, and 50 mM 2-deoxy-glucose) were sequentially injected. Analysis details (i.e., Glycolysis, Glycolytic Capacity, and Glycolytic Reserve) are indicated. Image modified from Agilent Seahorse XF User Guide of the Test Kit. (**C–F**) Glycolysis stress test profiles are investigated by measuring ECAR from the cells of control or DT sap-treatment for 1 or 24 h. (**C**) Glycolysis. (**D**) Glycolytic capacity. (**E**) Glycolytic reserve. (**F**) Glycolytic reserve as a %. Con, Control; DT, *Dendropanax trifidus*; ECAR, extracellular acidification rate; OCR, oxygen consumption rate. N = 3. Student's *t*-test, comparison with the control: * $p < 0.05$; ** $p < 0.01$. Bars indicate mean ± S.D.

For the Mito Stress Test, 1 μM oligomycin, 1 μM FCCP, and 0.5 μM rotenone + 0.5 μM antimycin A were applied in order (Figure 3A). In the mito stress test, basal respiration, ATP production, maximal respiration, and spare respiratory capacity were measured. Basal respiration indicates energetic demand of the cell under baseline conditions. ATP production by mitochondria is measured by the reduced OCR upon the addition of ATP synthase inhibitor oligomycin. Maximal respiration is measured by the increased OCR upon the addition of the uncoupler FCCP, which induces rapid oxidation of substrates to stimulate the respiratory chain to operate at maximum capacity. Spare respiratory capacity is calculated by subtracting basal respiration from maximal respiration and indicates the cell's ability to respond to an energetic demand.

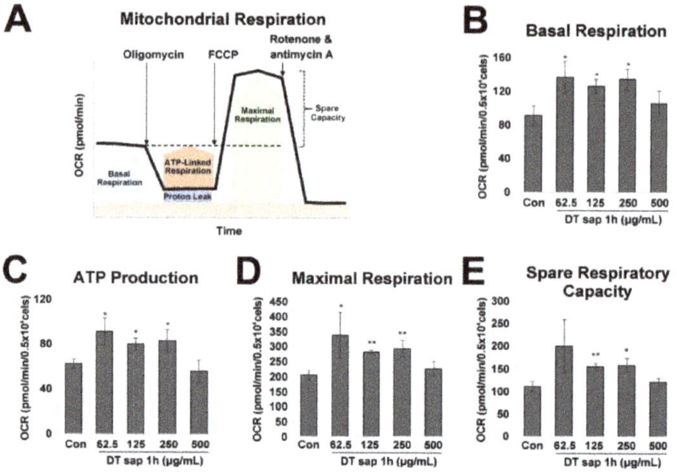


**Figure 3.** Mitochondrial stress test for *Dendropanax trifidus* (DT) sap in C2C12 muscle cells. (**A**) Agilent Seahorse XF Cell Mito Stress Test profile. Indicated compounds (i.e., 1 μM oligomycin, 1 μM FCCP, and 0.5 μM rotenone + 0.5 μM antimycin (**A**) were sequentially injected. Image modified from Agilent Seahorse XF User Guide of the Test Kit. (B–E) The oxygen consumption rate of C2C12 muscle cells was monitored at 1 h since the treatment of various concentrations of DT sap (0, 62.5, 123, 250, 500 μg/mL). (**B**) Basal respiration rate. (**C**) ATP production. (**D**) Maximal respiration. (**E**) Spare respiratory capacity. Con, Control; DT, *Dendropanax trifidus*; OCR, oxygen consumption rate. N = 3. Student's *t*-test, * $p < 0.05$, ** $p < 0.01$, comparison with vehicle control. Bars indicate mean ± S.D.

### *2.8. Statistics*

Two-way ANOVA and post hoc Holm–Sidak analysis were used for body weight analysis. Student's *t*-tests were used for other analyses.

## **3. Results**

### *3.1. Dendropanax trifidus Sap Suppresses Body Weight Increase in db/db, db/+, +/+ Mice of Both Sexes*

The homozygous (db/db) alleles of BKS.Cg-*Dock7*$^{m}$ +/+ *Lepr*$^{db}$/J lines, which have an obese phenotype, were used in this study. Female and male mice with genotypes of db/db, db/+, +/+ were administered daily with vehicle or DT sap over the course of 6 weeks and their body weights were measured weekly. Although we did not find a significant change in food or water consumption by DT sap administration (Supplementary Figure S1), DT sap-injected mice exhibited a suppression of body weight increase. This effect was clearly observed after 2–3 weeks of DTsap administration in all genotypes, and in both sexes, and was maintained until the end of the study period (Figure 1).

In detail, in female db/db, mice administered with vehicle control exhibited a steady rise in weight gain from week 9 to 15 which then plateaued till week 16, while mice injected with DT sap presented an increase only till week 11, and weight was stabilized thereafter (Figure 1A). Two-way ANOVA showed that there was a significant interaction in treatment x time ($p < 0.001$) and treatment ($p < 0.001$). In post hoc analysis, there were significant differences between the body weights of the vehicle group and the body weights of DT sap group throughout week 12~16 (post hoc Holm–Sidak, $p = 0.006$ at 12 week, $p < 0.001$ at 13~16 week, Figure 1A). In male db/db, body weights were also suppressed by the DT sap administration (Two-Way Anova, $p < 0.001$ for treatment, Figure 1B). The vehicle group exhibited a gradual increase in body weight till week 12 and was unchanged thereafter, while the DT sap group showed an increase till week 11 and a slight reduction since then, resulting in significant differences from week 12 to 16 (Holm–Sidak post hoc, $p < 0.05$ at 12, 13, 15 week, $p < 0.01$ at 14, 16 week, Figure 1B).

Suppression in body weight by DT sap was also observed in db/+ mice of both genders (two-way ANOVA, $p < 0.001$ for treatment, Figure 1C,D). Significant group differences in body weight were found from week 11 in female db/+ mice and from week 13 in male db/+ mice (Holm–Sidak post hoc, $p = 0.047$ at 11 week, $p < 0.01$ at 12–15 week, $p < 0.001$ at 16 week for female db/+, Figure 1C; $p < 0.01$ at 13, 14, 15 week, $p = 0.012$ at 16 week for male db/+, Figure 1D). In +/+ mice of both genders, body weight reduction was also observed (Two-Way Anova, for treatment: $p = 0.021$ for female +/+, Figure 1E; $p < 0.001$ for male +/+, Figure 1F). In female +/+ mice, no significant increase was found in post hoc analysis (Figure 1E), while there was a significant difference between groups from week 12 in male +/+ (Holm–Sidak post hoc, $p = 0.012$ at 12 week; $p < 0.01$ at 13, 14 week; $p < 0.001$ at 15, 16 week, Figure 1F). To summarize, compared to the vehicle control, DT sap administration exhibited a significant suppression against body weight increase in db/db, db/+, +/+ mice of both genders.

### 3.2. Glycolysis Stress Test in Skeletal Muscle Cells Treated with *Dendropanax trifidus* Sap

Muscle cells play the biggest roles in energy expenditure in the body [3,6]. If basal metabolic rate, which is variable between people due to genetic diversities [1,26,27], is low, the risk of body weight gain can rise [1,2]. We observed that DT sap administration suppressed the increase in body weight (Figure 1). In order to examine whether DT sap changes the energy expenditure of muscle cell, we investigated the changes in the glycolytic and mitochondrial function of C2C12 murine muscle cells treated with various concentrations of DTsap (Agilent Seahorse XF system, see Methods), (Figures 2 and 3). The Agilent Seahorse XF energy map revealed contrasting metabolic signatures between DT sap-treated cells and vehicle-treated cells (Figure 2A). Vehicle-treated cells (blue) were located in the middle of the energy map, while DT sap-treated cells (red, light blue, purple) were located near the 'Energetic' region which shows active utilization of both metabolic pathways (i.e., mitochondrial respiration as well as glycolysis), suggesting a higher energy expenditure in the cells treated with DT sap (Figure 2A).

The Agilent Seahorse XF Glycolysis Stress Test was performed according to the manufacturer's instructions. We observed that DT sap treatment significantly increased Glycolysis (Student's $t$-test, $p = 0.030, 0.027, 0.0079$ for 62.5, 250, 500 μg/mL DT sap 1 h treatment; $p = 0.0483, 0.0208$ for 125, 250 μg/mL DT sap 24 h treatment, N = 3, Figure 2C). DT sap treatment significantly improved the glycolytic capacity (Student's $t$-test, $p = 0.028, 0.012, 0.013$ for 62.5, 125, 250 μg/mL DT sap 1 hr treatment; $p = 0.0072, 0.040$ for 125, 250 μg/mL DT sap 24 h treatment, N = 3, Figure 2D). DT sap treatment exhibited significant increases in Glycolytic reserve (Student's $t$-test, $p = 0.040, 0.008$ for 62.5, 125 μg/mL DT sap 1 h treatment; $p = 0.041, 0.0032, 0.036$ for 62.5, 125, 250 μg/mL DT sap 24 h treatment, N = 3, Figure 2E). Glycolytic reserve was significantly increased after 24 h in for the 62.5 and 125 ug/mL concentrations (Student's $t$-test, $p = 0.019, 0.029$ for 62.5, 125 μg/mL DT sap 24 h treatment, N = 3, Figure 2F). In conclusion, DT sap increased glycolytic function in C2C12 myotubes.

### 3.3. Mitochondrial Stress Test in C2C12 Muscle Cells Treated with *Dendropanax trifidus* Sap

Next, to examine whether DT sap treatment changes mitochondrial respiration in the muscle, we performed Agilent Seahorse XF Mito Stress Test according to the manufacturer's instructions. OCR was directly measured and major parameters of mitochondrial function (basal respiration, ATP production, maximal respiration, and spare respiratory capacity) were calculated (Figure 3A).

DT sap-treated cells exhibited a higher basal respiration (Student's $t$-test, $p = 0.024, 0.012, 0.012$ for 62.5, 125, 250 μg/mL DT sap, N = 3, Figure 3B). ATP production was increased by DT sap treatment (Student's $t$-test, $p = 0.018, 0.011, 0.027$ for 62.5, 125, 250 μg/mL DT sap, N = 3, Figure 3C). Maximal respiration was also increased by DT sap treatment

(Student's *t*-test, $p = 0.043, 0.0014, 0.0084$ for 62.5, 125, 250 µg/mL DT sap, N = 3, Figure 3D). Spare respiratory capacity was significantly increased by DT sap treatment (Student's *t*-test, $p = 0.0049, 0.0131$ for 125, 250 µg/mL DT sap, N = 3, Figure 3E). The Mito Stress Test results showed that DT sap increases mitochondrial respiration in C2C12 muscle cells.

### 3.4. Estrogen Receptor Activation by *Dendropanax trifidus* Sap

In our previous study, DT sap was compared with Acer Saccharum (AS) sap using liquid chromatograph-tandem mass spectrometer, and we observed that estradiol was abundantly contained in DT sap compared to AS sap [25]. We were interested in knowing if estrogen may be an active component in DT sap-mediated signaling, thus we utilized human ERα-HeLa-9903 cell lines which contain a luciferase reporter gene at downstream of an estrogen responsive promoter element. Activation of this receptor by estrogen can then be studied by measuring luciferase activity through chemiluminescence detection. When the reporter cell line was treated with DT sap of various concentrations (0, 0.5, 2.5, 12.5, 25, 50 µg/mL) over 48 h, luciferase activity was observed to be concentration-dependent with saturation at 25 µg/mL (Student's *t*-test, $p = 0.0078, 0.0035, 0.00008, 0.00056, 0.0033$ for 0.5, 2.5, 12.5, 25, and 50 µg/mL, respectively, compared with negative control, N = 3, Figure 4A). This activity was significantly reduced by the estrogen receptor antagonist ICI, suggesting that its activation is via the estrogen receptor (Figure 4A). Estradiol (10 nM) alone was also used as a positive control, and showed a strong induction of luciferase activity. ICI alone did not induce luciferase activity. Cell viability was not damaged by any tested concentration of DT sap as well as other reagents (Figure 4B).

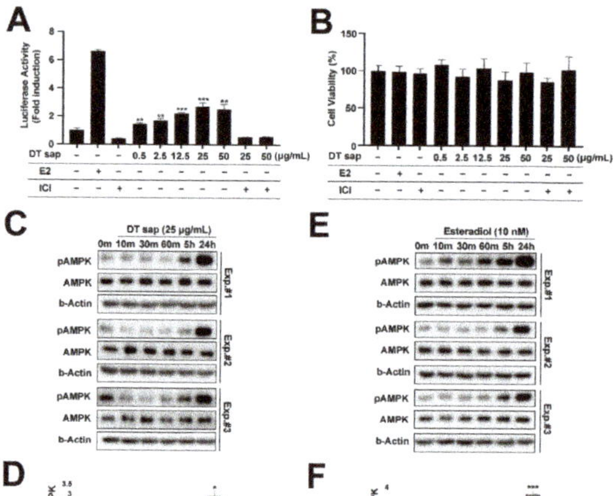

**Figure 4.** Estrogen receptor assay for *Dendropanax trifidus* (DT) sap and AMPK-mediated signaling. (**A**,**B**) "−" and "+" describe the absence or existence of the left-indicated reagent in the medium. (**A**) Estrogen-mediated luciferase reporter assay. hERα HeLa 9903 estrogen reporter cell line was used to measure estrogen activity in various concentrations (0, 0.5, 2.5, 12.5, 50 µg/mL) of DT sap. Estradiol (E2, 10 nM) was used as a positive control. Estrogen receptor antagonist ICI (1 µM) was used in a competition assay with 25 or 50 µg/mL of DT sap. N = 3. Student's *t*-test; ** $p < 0.01$, *** $p < 0.001$, comparison with negative control. (**B**) Cell viability assay. (**C**,**E**) Western blot images of C2C12 muscle cells treated with 25 µg/mL DT sap (**C**) or 10 nM estradiol (**E**) for the indicated time points and probed with antibodies against pAMPK, AMPK, and b-Actin. Three different experimental blots (Experiment #1, #2, #3) are shown. (**D**,**F**) Quantification of pAMPK/AMPK ratios of (**C**) or (**E**). N = 3. Student's *t*-test: * $p < 0.05$; *** $p < 0.001$, comparison with vehicle control. Bars indicate mean ± S.D. (**A**,**B**) or S.E.M. (**D**,**F**).

In C2C12 muscle cells, cellular metabolism including glycolysis and mitochondrial respiration was significantly increased by DT sap (Figures 2 and 3). AMPK signaling is an important regulator for muscle metabolism [8]. Therefore, to test whether the metabolism changes observed in C2C12 by DTsap (Figures 2 and 3) are related with AMPK signaling, we investigated AMPK signaling activation in C2C12 cells treated with DT sap over several time periods (0, 10, 30, 60 min, 5 h, 24 h) by western blot. Phosphorylation of AMPK was gradually increased, resulting in a significant increase at 24 h since the initial treatment (Student's $t$-test, $p = 0.020$ at 24 h since the treatment, N = 3 independent experiments, Figure 4C,D). Like DT sap treatment, estradiol treatment also activated AMPK in C2C12 muscle cells, and exhibited the maximum activity at 24 h during the investigated period (24 h) (Student's $t$-test, $p = 0.036$ for 0 vs. 5 h; $p = 0.00028$ for 0 vs. 24 h, N = 3, Figure 4E,F).

### 3.5. *AMPK/ACC Pathway Was Activated in C2C12 Muscle Cells by Dendropanax trifidus Sap Treatment*

AMPK is a master regulator of cellular metabolism, which is located upstream of various cellular signaling pathways including lipid metabolism, glucose metabolism, autophagy and mitochondrial homeostasis [28]. To investigate further the downstream effects of DT sap on AMPK signaling, we performed western blots against major downstream signaling molecules at 1 h or 24 h after the initial treatment of DT sap. The following molecules were examined: phospho-ACC, ACC, phospho-HSL, HSL, and ATGL for lipid metabolism; PFKFB3 for glucose metabolism; phospho-ULK1 and ULK1 for autophagy and mitochondrial homeostasis (Figure 5). While DT sap did not induce phosphorylation of AMPK at 1 h, it significantly induced phosphorylation of AMPK at 24 h in C2C12 cells (Figures 4C,D and 5A). Phospho-ACC/ACC was not altered at 1 h but significantly increased at 24 h after the treatment (Student's $t$-test, $p = 0.021$ at 24 h since the treatment, N = 3 independent cultures, Figure 5B,C). There were no significant changes in ACC/b-Actin, phospho-HSL/HSL, HSL/b-Actin, ATGL/b-Actin at both time points (Figure 5B,D–G). PFKFB3/b-Actin was not altered at 1 h but significantly decreased at 24 h after the treatment (Student's $t$-test, $p = 0.048$ at 24 h after the treatment, N = 3 independent cultures, Figure 5B,H). Moreover, phospho-ULK/ULK was significantly reduced at 1 h after the treatment but not after that (Student's $t$-test, $p = 0.043$ at 1 h after the treatment, N = 3 independent cultures, Figure 5I). There was a significantly transient increase in ULK1/b-Actin by DT sap at 1 h (Figure 5J), explaining the reduced pULK/1ULK1 by DT sap at 1 h. The Western blot analysis indicates DT sap affects the molecular signaling related with AMPK-ACC within a 24 h time window as well as potential other components in muscle cells.

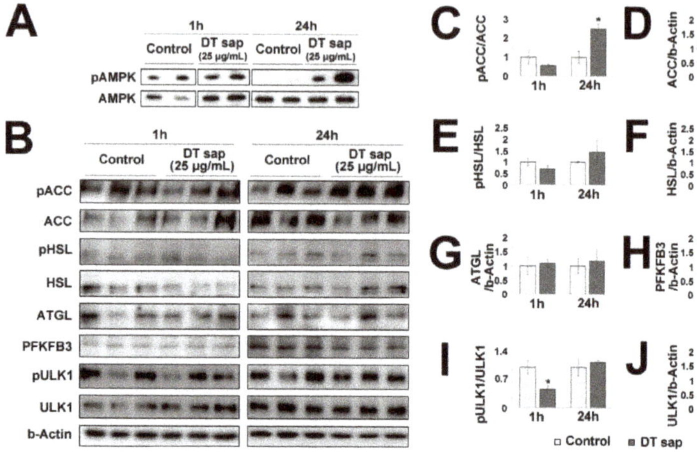

**Figure 5.** *Dendropanax trifidus* (DT) sap-induced signalings in C2C12 muscle cells. (**A**,**B**) Western blot images of C2C12 cells treated with vehicle control or 25 µg/mL DT for 1 h or 24 h and probed with pAMPK/AMPK antibodies (**A**) and AMPK-downstream signaling antibodies related with lipid metabolism (pACC, ACC, pHSL, HSL, ATGL); glucose metabolism (PFKFB3); autophagy and mitochondrial homeostasis (pULK1 and ULK1), and b-Actin (**B**). Three independent biological replicates are provided here. (**C**–**J**) Image analysis of western blot membranes of (**B**). Relative ratios for the indicated signals were calculated as follows: pACC/ACC (**C**); ACC/b-Actin (**D**); pHSL/HSL (**E**); HSL/b-Actin (**F**); ATGL/b-Actin (**G**); PFKFB3/b-Actin (**H**); pULK1/ULK1 (**I**); ULK1/b-Actin (**J**). 30 µg protein was loaded per lane. N = 3. Student's *t*-test: * $p < 0.05$, comparison with vehicle control. Bars indicate mean ± S.E.M.

## 4. Discussion

In this work, we studied the effects of DT sap on body weight in vivo and muscle metabolism in vitro. Daily oral administration of DT sap over six weeks suppressed the increase of body weight in db/db, db/+, +/+ mice of both genders. In C2C12 muscle cells, DT sap treatment increased glycolysis and mitochondrial respiration and activated the AMPK/ACC signaling pathway which is potentially mediated by estrogen receptor alpha.

The antibody against phospho-AMPKα (Thr172) which we used in this study detects the phosphorylation of threonine 172 in both endogenous AMPKα1 and α2 isoforms of the catalytic subunit but not the regulatory beta or gamma subunits. This phosphorylation is required for AMPK activation [29–31]. Once the AMPK complex is activated, it phosphorylates major targets, activating or suppressing them, resulting in rewiring of cellular metabolism including lipid metabolism, glucose metabolism, autophagy, and mitochondrial homeostasis [28]. As skeletal muscle is a major contributor to glucose and lipid metabolism, it plays an essential role in whole-body energy expenditure. In skeletal muscle, exercise activates AMPK, and AMPK affects exercise capacity by regulating mitochondrial content, contraction-stimulated glucose uptake, and fatty acid metabolism [32]. In previous investigations using obese mice, AMPK activation also increased exercise capacity [33]. Acute exercise increases AMPK phosphorylation in human skeletal muscle in an intensity-dependent manner [34,35]. In this work, the dramatic increase in AMPK phosphorylation by DT sap treatment in murine muscle cells (Figure 4C,D) suggests enhancement of muscle cell metabolism (exercise-like effects) by DT sap and a potential association with the DT sap induced suppression of body weight observed here (Figure 1).

ACC is a major enzyme for biosynthesis and oxidation of fatty acids [9], and is a potential target of anti-obesity drugs [11,12]. In rodents, ACC1 (ACCα, 265 kDa) is expressed in lipogenic tissues which forms lipids, while ACC2 (ACCβ, 280 kDa) is the main isoform in oxidative tissue such as skeletal muscle [9,36]. In humans, ACC2 is the predominant isoform for both lipogenic and oxidative tissues [9,36]. Phosphorylation at Ser79 of ACC by AMPK inhibits the enzymatic activity of ACC [10]. The antibody used in this study specifically detects phosphorylation of ACC at Ser79. We observed here that DT sap treatment increased phosphorylation of both AMPK and ACC in mouse C2C12 muscle cells (Figure 5). Thus, as AMPK phosphorylates ACC and ACC phosphorylation suppresses fatty acid biosynthesis, this process may contribute to suppression of body weight increase (Figure 1).

While overexpression of PFKFB3 in mouse skeletal muscle promotes glycolysis [37], its expression is downregulated during myogenic cell differentiation [38]. Moreover, during differentiation, metabolic pathways such as glycolysis are inhibited [39]. Here, while there was no change in PFKFB3 level after 1 h of DT sap treatment, there was a significant reduction in its expression after 24 h compared to the control (Figure 5H). However, as glycolysis and mitochondrial respiration rate are increased by DT sap treatment (Figures 2 and 3), the reduced expression of PFKFB3 by DT sap may indicate effects on myogenic cell differentiation rather than metabolism. ULK1 can be phosphorylated via both AMPK-dependent [40] and AMPK-independent pathways [41]. At 1 h of DT sap treatment, *p*-AMPK/AMPK was not altered (Figure 5A), while *p*-ULK1/ULK1 was significantly reduced due to increased ULK1 (Figure 5I,J), suggesting that enhanced *p*-ULK1/ULK1 here is via an AMPK-independent pathway.

In our previous report, we identified estrogen as one of the major components of DT sap [25]. In this study, we found that DT sap can signal through estrogen receptor alpha (Figure 4A). Estrogen receptor alpha is expressed in skeletal muscle and is critical for the regulation of metabolic homeostasis [42]. Its deletion in mouse increased adiposity caused by reductions in energy expenditure and exhibited glucose intolerance and insulin resistance [43–45]. In humans, women with the metabolic syndrome exhibit a reduced expression level of estrogen receptor alpha in muscle [46]. Therefore, at least partially, DT sap, which contains estrogen as a major component, might contribute to the control of body weight by modulating muscle metabolism via estrogen receptor alpha.

Taken together, our findings here allow us to hypothesize that DT sap increases energy expenditure in muscle cells, which may be via estrogen mediated AMPK signaling, thus contributing to the suppression of body weight increase. However, more studies will have to be performed to elucidate the molecular mechanisms by which this occurs. Possible approaches such as examining RNAi mediated regulation of the estrogen receptor under DT sap treatment, and also whether other compounds present in DT sap exhibit synergistic effects with estrogen on muscle cell energy expenditure and body weight reduction are expected for future study.

**Supplementary Materials:** The following supporting information can be downloaded at: https://www.mdpi.com/article/10.3390/nu14051098/s1. Figure S1: Average daily food or water consumption. Daily consumption of food (A) or water (B) was recorded and averaged after two weeks of daily oral administration of control vehicle or *Dendropanax trifidus* sap (0.2 mg/g body weight). N = 3 mice per group. Bars indicate mean ± S.E.M.

**Author Contributions:** Conceptualization, H.-J.Y.; methodology, A.L., N.L.; validation, D.K.; investigation, Y.J.L., H.-J.Y.; resources, I.-H.C.; data curation, S.M.C.; writing—original draft preparation, H.-J.Y.; writing—review and editing, E.K.; visualization, H.-J.Y.; supervision, H.-J.Y.; project administration, H.-J.Y.; funding acquisition, H.-J.Y. All authors have read and agreed to the published version of the manuscript.

**Funding:** This work was supported by the Basic Science Research Program through the National Research Foundation of Korea (NRF) funded by the Ministry of Education (2020R1I1A3A04038150).

**Institutional Review Board Statement:** The study was conducted according to the guidelines of the Declaration of Helsinki, and approved by the Institutional Animal Care and Use Committee of the University of Brain Education (Approval number: 2018-AE-01Y).

**Informed Consent Statement:** Not applicable.

**Data Availability Statement:** The data presented in this study are available on request from the corresponding author.

**Acknowledgments:** We appreciate Noriko Setou for her support and fruitful discussions for the study.

**Conflicts of Interest:** The authors declare no conflict of interest.

## References

1. Ravussin, E.; Lillioja, S.; Knowler, W.C.; Christin, L.; Freymond, D.; Abbott, W.G.; Boyce, V.; Howard, B.V.; Bogardus, C. Reduced rate of energy expenditure as a risk factor for body-weight gain. *N. Engl. J. Med.* **1988**, *318*, 467–472. [CrossRef] [PubMed]
2. Roberts, S.B.; Savage, J.; Coward, W.A.; Chew, B.; Lucas, A. Energy expenditure and intake in infants born to lean and overweight mothers. *N. Engl. J. Med.* **1988**, *318*, 461–466. [CrossRef] [PubMed]
3. Gheller, B.J.; Riddle, E.S.; Lem, M.R.; Thalacker-Mercer, A.E. Understanding Age-Related Changes in Skeletal Muscle Metabolism: Differences Between Females and Males. *Annu. Rev. Nutr.* **2016**, *36*, 129–156. [CrossRef] [PubMed]
4. Owen, E.; Reichard, G.A., Jr.; Boden, G.; Patel, M.S.; Trapp, V.E. Interrelationships among key tissues in the utilization of metabolic substrate. *Adv. Mod. Nutr.* **1978**, *2*, 517–550.
5. Wade, L.; Bishop, A.J.M. *Cardiac Output and Regional Blood Flow*; Blackwell Scientific Publications: Oxford, UK, 1962.
6. Zurlo, F.; Larson, K.; Bogardus, C.; Ravussin, E. Skeletal muscle metabolism is a major determinant of resting energy expenditure. *J. Clin. Investig.* **1990**, *86*, 1423–1427. [CrossRef]
7. Gowans, G.J.; Hardie, D.G. AMPK: A cellular energy sensor primarily regulated by AMP. *Biochem. Soc. Trans.* **2014**, *42*, 71–75. [CrossRef]
8. Kjobsted, R.; Hingst, J.R.; Fentz, J.; Foretz, M.; Sanz, M.N.; Pehmoller, C.; Shum, M.; Marette, A.; Mounier, R.; Treebak, J.T.; et al. AMPK in skeletal muscle function and metabolism. *FASEB J.* **2018**, *32*, 1741–1777. [CrossRef]
9. Castle, J.C.; Hara, Y.; Raymond, C.K.; Garrett-Engele, P.; Ohwaki, K.; Kan, Z.; Kusunoki, J.; Johnson, J.M. ACC2 is expressed at high levels in human white adipose and has an isoform with a novel N-terminus [corrected]. *PLoS ONE* **2009**, *4*, e4369. [CrossRef]
10. Ha, J.; Daniel, S.; Broyles, S.S.; Kim, K.H. Critical phosphorylation sites for acetyl-CoA carboxylase activity. *J. Biol. Chem.* **1994**, *269*, 22162–22168. [CrossRef]
11. Levert, K.L.; Waldrop, G.L.; Stephens, J.M. A biotin analog inhibits acetyl-CoA carboxylase activity and adipogenesis. *J. Biol. Chem.* **2002**, *277*, 16347–16350. [CrossRef]
12. Abu-Elheiga, L.; Matzuk, M.M.; Abo-Hashema, K.A.; Wakil, S.J. Continuous fatty acid oxidation and reduced fat storage in mice lacking acetyl-CoA carboxylase 2. *Science* **2001**, *291*, 2613–2616. [CrossRef] [PubMed]
13. Bartrons, R.; Rodriguez-Garcia, A.; Simon-Molas, H.; Castano, E.; Manzano, A.; Navarro-Sabate, A. The potential utility of PFKFB3 as a therapeutic target. *Expert Opin. Ther. Targets* **2018**, *22*, 659–674. [CrossRef] [PubMed]
14. Steinberg, G.R.; Carling, D. AMP-activated protein kinase: The current landscape for drug development. *Nat. Rev. Drug Discov.* **2019**, *18*, 527–551. [CrossRef] [PubMed]
15. Rogers, N.H.; Witczak, C.A.; Hirshman, M.F.; Goodyear, L.J.; Greenberg, A.S. Estradiol stimulates Akt, AMP-activated protein kinase (AMPK) and TBC1D1/4, but not glucose uptake in rat soleus. *Biochem. Biophys. Res. Commun.* **2009**, *382*, 646–650. [CrossRef]
16. D'Eon, T.M.; Rogers, N.H.; Stancheva, Z.S.; Greenberg, A.S. Estradiol and the estradiol metabolite, 2-hydroxyestradiol, activate AMP-activated protein kinase in C2C12 myotubes. *Obesity* **2008**, *16*, 1284–1288. [CrossRef]
17. Ikeda, K.; Horie-Inoue, K.; Inoue, S. Functions of estrogen and estrogen receptor signaling on skeletal muscle. *J. Steroid Biochem. Mol. Biol.* **2019**, *191*, 105375. [CrossRef]
18. Han, S.H.; Jung, Y.H.; Oh, M.H.; Ko, M.H.; Oh, Y.S.; Koh, S.C.; Kim, M.H.; Oh, M.Y. Phytogenetic relationships of the Dendropanax morbifera and D. trifidus based on PCR-RAPD. *Kor. J. Genet.* **1998**, *20*, 173–181.
19. Moon, M.O.; SIhm, B.S.; Chung, Y.C.; Kang, Y.J. Taxonomic Appraisal of Dendropanax morbifera Leveille and D. trifidus (Thunb. ex Murray) Makino based on Morphological Characters. *Korean J. Plant Taxon.* **1999**, *29*, 231–248. [CrossRef]
20. Lee, H.S.; Jeon, J.I.; Chang, C.S. Foliar flavonoids of two sections of genus dendropanax in China, Japan, and Korea. *For. Sci. Technol.* **2005**, *1*, 45–50.
21. Song, J.H.; Kang, H.B.; Kim, J.H.; Kwak, S.; Sung, G.J.; Park, S.H.; Jeong, J.H.; Kim, H.; Lee, J.; Jun, W.; et al. Antiobesity and Cholesterol-Lowering Effects of Dendropanax morbifera Water Extracts in Mouse 3T3-L1 Cells. *J. Med. Food* **2018**, *21*, 793–800. [CrossRef]
22. Song, J.H.; Kim, H.; Jeong, M.; Kong, M.J.; Choi, H.K.; Jun, W.; Kim, Y.; Choi, K.C. In Vivo Evaluation of Dendropanax morbifera Leaf Extract for Anti-Obesity and Cholesterol-Lowering Activity in Mice. *Nutrients* **2021**, *13*, 1424. [CrossRef] [PubMed]
23. Kang, M.J.; Kwon, E.B.; Ryu, H.W.; Lee, S.; Lee, J.W.; Kim, D.Y.; Lee, M.K.; Oh, S.R.; Lee, H.S.; Lee, S.U.; et al. Polyacetylene From Dendropanax morbifera Alleviates Diet-Induced Obesity and Hepatic Steatosis by Activating AMPK Signaling Pathway. *Front. Pharmacol.* **2018**, *9*, 537. [CrossRef] [PubMed]
24. Jun, J.E.; Hwang, Y.-C.; Ahn, K.J.; Chung, H.Y.; Choung, S.Y.; Jeong, I.-K. The efficacy and safety of Dendropanax morbifera leaf extract on the metabolic syndrome: A 12-week, placebo controlled, double blind, and randomized controlled trial. *Nutr. Res. Pract.* **2021**, *15*, e46. [CrossRef]

25. Lee, A.; Sugiura, Y.; Cho, I.H.; Setou, N.; Koh, E.; Song, G.J.; Lee, S.; Yang, H.J. In Vivo Hypoglycemic Effects, Potential Mechanisms and LC-MS/MS Analysis of *Dendropanax trifidus* Sap Extract. *Nutrients* **2021**, *13*, 4332. [CrossRef]
26. Bogardus, C.; Lillioja, S.; Ravussin, E.; Abbott, W.; Zawadzki, J.K.; Young, A.; Knowler, W.C.; Jacobowitz, R.; Moll, P.P. Familial dependence of the resting metabolic rate. *N. Engl. J. Med.* **1986**, *315*, 96–100. [CrossRef] [PubMed]
27. Bouchard, C.; Tremblay, A.; Nadeau, A.; Despres, J.P.; Theriault, G.; Boulay, M.R.; Lortie, G.; Leblanc, C.; Fournier, G. Genetic effect in resting and exercise metabolic rates. *Metabolism* **1989**, *38*, 364–370. [CrossRef]
28. Herzig, S.; Shaw, R.J. AMPK: Guardian of metabolism and mitochondrial homeostasis. *Nat. Rev. Mol. Cell Biol.* **2018**, *19*, 121–135. [CrossRef]
29. Hawley, S.A.; Davison, M.; Woods, A.; Davies, S.P.; Beri, R.K.; Carling, D.; Hardie, D.G. Characterization of the AMP-activated protein kinase kinase from rat liver and identification of threonine 172 as the major site at which it phosphorylates AMP-activated protein kinase. *J. Biol. Chem.* **1996**, *271*, 27879–27887. [CrossRef]
30. Lizcano, J.M.; Goransson, O.; Toth, R.; Deak, M.; Morrice, N.A.; Boudeau, J.; Hawley, S.A.; Udd, L.; Makela, T.P.; Hardie, D.G.; et al. LKB1 is a master kinase that activates 13 kinases of the AMPK subfamily, including MARK/PAR-1. *EMBO J.* **2004**, *23*, 833–843. [CrossRef]
31. Shaw, R.J.; Kosmatka, M.; Bardeesy, N.; Hurley, R.L.; Witters, L.A.; DePinho, R.A.; Cantley, L.C. The tumor suppressor LKB1 kinase directly activates AMP-activated kinase and regulates apoptosis in response to energy stress. *Proc. Natl. Acad. Sci. USA* **2004**, *101*, 3329–3335. [CrossRef]
32. O'Neill, H.M.; Holloway, G.P.; Steinberg, G.R. AMPK regulation of fatty acid metabolism and mitochondrial biogenesis: Implications for obesity. *Mol. Cell. Endocrinol.* **2013**, *366*, 135–151. [CrossRef]
33. Marcinko, K.; Bujak, A.L.; Lally, J.S.; Ford, R.J.; Wong, T.H.; Smith, B.K.; Kemp, B.E.; Jenkins, Y.; Li, W.; Kinsella, T.M.; et al. The AMPK activator R419 improves exercise capacity and skeletal muscle insulin sensitivity in obese mice. *Mol. Metab.* **2015**, *4*, 643–651. [CrossRef] [PubMed]
34. Wojtaszewski, J.F.; Nielsen, P.; Hansen, B.F.; Richter, E.A.; Kiens, B. Isoform-specific and exercise intensity-dependent activation of 5′-AMP-activated protein kinase in human skeletal muscle. *J. Physiol.* **2000**, *528 Pt 1*, 221–226. [CrossRef]
35. Egan, B.; Carson, B.P.; Garcia-Roves, P.M.; Chibalin, A.V.; Sarsfield, F.M.; Barron, N.; McCaffrey, N.; Moyna, N.M.; Zierath, J.R.; O'Gorman, D.J. Exercise intensity-dependent regulation of peroxisome proliferator-activated receptor coactivator-1 mRNA abundance is associated with differential activation of upstream signalling kinases in human skeletal muscle. *J. Physiol.* **2010**, *588*, 1779–1790. [CrossRef] [PubMed]
36. Kreuz, S.; Schoelch, C.; Thomas, L.; Rist, W.; Rippmann, J.F.; Neubauer, H. Acetyl-CoA carboxylases 1 and 2 show distinct expression patterns in rats and humans and alterations in obesity and diabetes. *Diabetes Metab. Res. Rev.* **2009**, *25*, 577–586. [CrossRef]
37. Xiang, C.; Zhang, Y.; Chen, Q.; Sun, A.; Peng, Y.; Zhang, G.; Zhou, D.; Xie, Y.; Hou, X.; Zheng, F.; et al. Increased glycolysis in skeletal muscle coordinates with adipose tissue in systemic metabolic homeostasis. *J. Cell. Mol. Med.* **2021**, *25*, 7840–7854. [CrossRef] [PubMed]
38. Riera, L.; Obach, M.; Navarro-Sabate, A.; Duran, J.; Perales, J.C.; Vinals, F.; Rosa, J.L.; Ventura, F.; Bartrons, R. Regulation of ubiquitous 6-phosphofructo-2-kinase by the ubiquitin-proteasome proteolytic pathway during myogenic C2C12 cell differentiation. *FEBS Lett.* **2003**, *550*, 23–29. [CrossRef]
39. McGraw, T.E.; Mittal, V. Metabolism regulates differentiation. *Nat. Chem. Biol.* **2010**, *6*, 176–177. [CrossRef]
40. Kim, J.; Kundu, M.; Viollet, B.; Guan, K.L. AMPK and mTOR regulate autophagy through direct phosphorylation of Ulk1. *Nat. Cell. Biol.* **2011**, *13*, 132–141. [CrossRef]
41. Kwon, K.Y.; Viollet, B.; Yoo, O.J. CCCP induces autophagy in an AMPK-independent manner. *Biochem. Biophys. Res. Commun.* **2011**, *416*, 343–348. [CrossRef]
42. Hevener, A.L.; Zhou, Z.; Moore, T.M.; Drew, B.G.; Ribas, V. The impact of ERalpha action on muscle metabolism and insulin sensitivity—Strong enough for a man, made for a woman. *Mol. Metab.* **2018**, *15*, 20–34. [CrossRef] [PubMed]
43. Heine, P.A.; Taylor, J.A.; Iwamoto, G.A.; Lubahn, D.B.; Cooke, P.S. Increased adipose tissue in male and female estrogen receptor-alpha knockout mice. *Proc. Natl. Acad. Sci. USA* **2000**, *97*, 12729–12734. [CrossRef] [PubMed]
44. Ribas, V.; Nguyen, M.T.; Henstridge, D.C.; Nguyen, A.K.; Beaven, S.W.; Watt, M.J.; Hevener, A.L. Impaired oxidative metabolism and inflammation are associated with insulin resistance in ERalpha-deficient mice. *Am. J. Physiol. Endocrinol. Metab.* **2010**, *298*, E304–E319. [CrossRef] [PubMed]
45. Bryzgalova, G.; Gao, H.; Ahren, B.; Zierath, J.R.; Galuska, D.; Steiler, T.L.; Dahlman-Wright, K.; Nilsson, S.; Gustafsson, J.A.; Efendic, S.; et al. Evidence that oestrogen receptor-alpha plays an important role in the regulation of glucose homeostasis in mice: Insulin sensitivity in the liver. *Diabetologia* **2006**, *49*, 588–597. [CrossRef]
46. Ribas, V.; Drew, B.G.; Zhou, Z.; Phun, J.; Kalajian, N.Y.; Soleymani, T.; Daraei, P.; Widjaja, K.; Wanagat, J.; de Aguiar Vallim, T.Q.; et al. Skeletal muscle action of estrogen receptor alpha is critical for the maintenance of mitochondrial function and metabolic homeostasis in females. *Sci. Transl. Med.* **2016**, *8*, 334ra354. [CrossRef]

*nutrients*

MDPI

*Article*

# Saponins from *Camellia sinensis* Seeds Stimulate GIP Secretion in Mice and STC-1 Cells via SGLT1 and TGR5

**Huanqing Zhu [1,†], Kaixi Wang [1,†], Shuna Chen [1], Jiaxin Kang [1], Na Guo [1], Hongbo Chen [2], Junsheng Liu [1], Yuanyuan Wu [1], Puming He [1], Youying Tu [1] and Bo Li [1,*]**

1 Department of Tea Science, Zhejiang University, 866 Yuhangtang Road, Hangzhou 310058, China
2 Department of Tea Science, Zhejiang Shuren University, 8 Shuren Road, Hangzhou 310000, China
* Correspondence: drlib@zju.edu.cn
† These authors contributed equally to this work.

**Abstract:** Glucose-dependent insulinotropic polypeptide (GIP) is one of the important incretins and possesses lots of physiological activities such as stimulating insulin secretion and maintaining glucose homeostasis. The pentacyclic triterpenoid saponins are the major active ingredients in tea (*Camellia sinensis*) seeds. This study aimed to investigate the effect of tea seed saponins on the GIP secretion and related mechanisms. Our data showed that the total tea seed saponins (TSS, 65 mg/kg BW) and theasaponin $E_1$ (TSE1, 2–4 μM) could increase the GIP mRNA and protein levels in mice and STC-1 cells. Phlorizin, the inhibitor of Sodium/glucose cotransporter 1 (SGLT1), reversed the TSE1-induced increase in $Ca^{2+}$ and GIP mRNA level. In addition, TSE1 upregulated the protein expression of Takeda G protein-coupled receptor 5 (TGR5), and TGR5 siRNA significantly decreased GIP expression in TSE1-treated STC-1 cells. Network pharmacology analysis revealed that six proteins and five signaling pathways were associated with SGLT1, TGR5 and GIP regulated by TSE1. Taken together, tea seed saponins could stimulate GIP expression via SGLT1 and TGR5, and were promising natural active ingredients for improving metabolism and related diseases.

**Keywords:** tea seed; Saponin; Theasaponin $E_1$; GIP; SGLT1; TGR5

**Citation:** Zhu, H.; Wang, K.; Chen, S.; Kang, J.; Guo, N.; Chen, H.; Liu, J.; Wu, Y.; He, P.; Tu, Y.; et al. Saponins from *Camellia sinensis* Seeds Stimulate GIP Secretion in Mice and STC-1 Cells via SGLT1 and TGR5. *Nutrients* **2022**, *14*, 3413. https://doi.org/10.3390/nu14163413

Academic Editor: Sonia de Pascual-Teresa

Received: 18 July 2022
Accepted: 17 August 2022
Published: 19 August 2022



## 1. Introduction

Incretins are a class of hormone peptides released by gut enteroendocrine cells, and they possess lots of physiological functions such as regulating appetite, blood glucose, gastrointestinal motility and lipid metabolism, etc. [1]. Glucose-dependent insulinotropic polypeptide (GIP) and glucagon-like peptide 1 (GLP-1) are two major incretins which respond to nutrients and control glucose homeostasis. Incretin-based pharmacotherapies for diabetes and obesity have received enormous attention. The GLP-1 analogues, GLP-1/GIP dual-agonists and dipeptidyl peptidase-4 (DPP-4) inhibitors that prolong GLP-1 half-life could offer effective treatment for diabetic patients [2,3].

GIP is a 42-amino-acid peptide secreted from K entero-endocrine cells in the duodenum, and its receptor (GIPR) expresses in various organs [4]. GIP can inhibit gastric emptying and motility, control thyroid morphogenesis upon nutrient ingestion, and regulate inflammation and adaptive thermogenesis via restraining myeloid-cell-derived S100 calcium-binding protein heterodimer S100A8/A9 [5–7]. In addition, GIP improves energy utilization, reduces inflammation, and exhibits neuroprotective activity in Parkinson's and Alzheimer 's disease models [8]. It protects Hippocampal HT-22 cells from glutamate-induced oxidative stress. The underlying molecular mechanisms is that GIP suppresses ferroptosis through activating the mitogen-activated protein kinase (MAPK) pathway [9].

So far, the molecular mechanisms of incretin secretion and related signaling are not very clear. Incretins can be induced via stimulating different receptors by small or macromolecular substances. For example, the sensor receptors for carbohydrates are sodium-dependent glucose transporter 1/3 (SGLT1/3), glucose transporter 2 (GLUT2) and type 1 taste receptor 2/3 (T1R2/T1R3). SGLT1 transfers two $Na^+$ molecules when transporting one glucose molecule, causing cell depolarization, opening of voltage-dependent calcium channel (VDCC) and enhancement of intracellular $Ca^{2+}$ concentration, which in turn promotes GLP-1 secretion. Cyclic adenosine monophosphate (cAMP) and $Ca^{2+}$ are considered second messengers for GLP-1 and GIP [10]. Lipid sensor receptors are mainly free fatty acid receptors (FFAR1-4, namely GPR40, GPR43, GPR41 and GPR120) and G protein-coupled receptor 119 (GPR 119). The known sensor receptors of proteins, polypeptides and amino acids are calcium-sensitive receptor (CaSR), T1R1/T1R3G, small peptide transporter (PepT1) and protein-coupled receptor family C group 6 subtype A (GPRC6A) receptor [11]. Cholic acid can stimulate the secretion of GLP-1 and Peptide YY (PYY) via Takeda G protein-coupled receptor 5 (TGR5) and Nuclear farnesoid X receptor (FXR), respectively [12,13]. Lipopolysaccharide induces GLP-1 secretion via Toll-like receptor 4 (TLR4) on intestinal cells, accompanied by an increase in intracellular $Ca^{2+}$ concentration [14].

Tea (*Camellia sinensis* L. (O) Kuntze) is famous worldwide as a popular beverage. In addition to the buds and leaves, tea seeds have received increasing attention due to their abundant active ingredients, including saponins, fatty acids, flavonoid glycosides and polysaccharides. Tea seed saponins are oleanane-type triterpene saponins, and have beenfound to possess various bioactivities, including regulation of gastrointestinal system, weight reduction, anti-microorganism, anti-inflammation, neuroprotection, anti-allergy properties, etc. [15]. Although tea seed saponins have been found to regulate gastric emptying and gastrointestinal transit, their effects on incretins are still not clear. In this study, the effects of total saponins from tea seeds (TSS) and one of the major saponin compound namely theasaponin $E_1$ (TSE1) on GIP secretion were evaluated in vitro and in vivo. Moreover, the underlying mechanisms and signaling networks were also investigated.

## 2. Materials and Methods

### 2.1. Chemicals and Regents

TSS and TSE1 (98% purity) were prepared as previously reported [16]. Briefly, hulled tea seeds were ground into powder and extracted with 70% methanol solution at 70 °C. Then, the crude extract was extracted sequentially with petroleum ether, ethyl acetate (EtOAc) and 1-butanol (n-BuOH), and the n-BuOH fraction was eluted with 30–70% ethanol on D101 macroporous resin to give TSS. TSE1 was isolated from TSS by a reversed-phase preparative HPLC system (GE ÄKTA purifier100, Uppsala, Sweden), and identified by UPLC-PDA-MS/MS, $^{13}C$- and $^{1}H$-NMR spectroscopy. Phlorizin (purity $\geq$ 98%) was obtained from Shanghai GE Biological Technology Co., Ltd. (Shanghai, China). Penicillin, streptomycin and Dulbecco's Modified Eagle's Medium (DMEM) were provided by Shanghai Guan & Dao Biological Engineering Co., Ltd. (Shanghai, China). Fetal bovine serum (FBS) was purchased from GBICO (Grand Island, NY, USA). The primary antibodies against TGR5 and β-actin and the HRP-linked secondary antibody (anti-mouse IgG) were obtained from Cell Signaling Technology, Inc. (Danvers, MA, USA).

### 2.2. Animal Experiment

Twenty-seven 4-week-old male ICR mice were provided by the Experimental animal center of Zhejiang University. The mice were housed under standard conditions (20–25 °C, 60–70% relative humidity, 12-h light/dark cycle) with free access to water and food. After a 2-week acclimation period, animals were randomly divided into 9 groups (3 mice per cage). Eight groups were administrated with TSS by gavage at a dose of 65 mg/kg BW, and sacrificed after 0.5, 1, 2, 3, 4, 6, 12 and 24 h. The control group received water only, and were sacrificed immediately after gavage. Small intestines were rinsed with 0.9% saline, frozen in liquid nitrogen and stored at −80 °C. All animal experiments were approved by the Experimental Animals Ethics Committee of Zhejiang University (protocol code 12531).

### 2.3. Cell Culture

The mouse intestinal endocrine STC-1 cells were provided by Shanghai Guan & Dao Biological Engineering Co., Ltd. (Shanghai, China). Cells were cultured in DMEM medium supplemented with 10% FBS at 37 °C, and maintained in a thermostatic cell incubator with 5% $CO_2$.

### 2.4. Cell Viability Assay

Cell growth was determined using a MTT cell proliferation and cytotoxicity detection kit (KeyGEN BioTECH, Jiangsu, China). Briefly, cells were seeded into 96-well plates, incubated overnight, and treated with TSE1 (0–10 μM) or phlorizin (0–30 μM) for 24 h. Subsequently, 50 μL of MTT solution (1×) was added and incubated for additional 4 h at 37 °C. Then, the supernatant in each well was removed, and 150 uL DMSO was added to dissolve the MTT-formazan crystals. The absorbance at 490 nm was determined with a microplate reader (BioTek, Shanghai, China).

### 2.5. $Ca^{2+}$ Measurement

The intracellular calcium level was determined using a BBcellProbe F03 Assay Kit (BestBio, Shanghai, China). Cells were incubated with calcium-indicating dye for 40 min at 37 °C, and monitored with excitation at 488–495 nm and emission at 516 nm by a microplate reader (BioTek, Shanghai, China).

### 2.6. Quantitative Real-Time PCR (qRT-PCR) Analysis

The total RNA of tissues and cells was extracted using an Eastep® Super Total RNA Extraction Kit (Promega, Wisconsin, WI, USA). Subsequently, the RNA was reversely transcribed to cDNA by a PrimeScript™ RT Reagent Kit with gDNA Eraser (TaKaRa, Kyoto, Japan) according to the operation manual. qRT-PCR was performed on a StepOnePlus™ Real-Time PCR System (Applied Biosystems, Foster City, CA, USA) with a TB Green® Premix Ex Taq™ (Tli RNaseH Plus) Kit (TaKaRa, Kyoto, Japan), and the comparative Ct method (ΔΔCt) was used for analyzing the relative mRNA expression. The primer pairs for GIP and GAPDH were as follows: 5′-GTGGCTTTGAAGACCTGCTC-3′ and 5′-TTGTTGTCGGATCTTGTCCA-3′ (GIP); 5′-GAAGGTGAAGGTCGGAGTC-3′ and 5′-GAAGATGGTGATGGGATTC-3′ (GAPDH).

### 2.7. Enzyme Linked Immunosorbent Assay (ELISA)

The GIP protein concentrations in the small intestine tissues and STC-1 cell culture supernatant were determined by a Mouse GIP (Gastric Inhibitory Polypeptide) ELISA Kit (Elabscience, Wuhan, China) following the operation manual. The plates were read by a microplate reader (BioTek, Shanghai, China) at 450 nm wavelength.

### 2.8. Western Blot Analysis

Cells were lysed with RIPA lysis buffer (Biosharp, Anhui, China) containing a protease inhibitor mixture (Biosharp, Anhui, China). The protein content was analyzed using a BCA Protein Assay Kit (Meilunbio, Dalian, China). Cell lysates were separated by sodium dodecyl sulfate–polyacrylamide gel electrophoresis (SDS-PAGE) and transferred to polyvinylidene difluoride (PVDF) membrane by a Mini-Protean 3 System (BioRad). Afterwards, the membrane was blocked with 5% defatted milk for 1 h, incubated with the primary antibody at 4 °C overnight, and then incubated with the secondary antibody for 2 h at the room temperature. Protein bands were visualized by a LAS-3000 Image Reader (Fujifilm, Tokyo, Japan) with a BeyoECL Plus kit (Beyotime Biotech Inc, Shanghai, China), and analyzed by the ImageJ 1.52v software (NIH, Bethesda, MD, USA).

### 2.9. Transfection with Small Interfering RNA (siRNA)

STC-1 cells were cultured in 6-well plates for 24 h, and transfected with control siRNA or TGR5 siRNA (Santa Cruz Biotechnology, Inc. TX, USA) using the HighGene transfection reagent (Abclonal, Wuhan, China). After a 6 h transfection period, cells were treated with TSE1 for 24 h and used for Western blot analysis.

### 2.10. Protein–Protein Interaction (PPI) Network Construction

The interaction between target proteins was illustrated by the STRING database and GeneMANIA database. The settings in the String system are as follows: Organism, *Homo sapiens*; Network type, full STRING network; Meaning of network edges, evidence; Active interaction sources, all selected; Minimum required interaction score, high confidence (0.700); Max number of interactors to show, select "no more than 10 interactors" in the first shell and "none" in the second shell. In the GeneMANIA system, all default settings were used.

### 2.11. Statistical Analysis

Data from three independent biological replicates were expressed as means ± standard deviation (SD). Mean comparison between two groups and multiple comparisons were assessed by Student's t-test and one-way analysis of variance (ANOVA) followed with Student–Newman–Keuls (SNK), respectively. $p < 0.05$ and $p < 0.01$ were considered statistically significant and statistically highly significant, respectively. All statistical analyses were performed using SPSS Statistics 26.0 (SPSS Inc., Chicago, IL, USA). Power analysis was performed by PASS 2021 (NCSS, LLC, Kaysville, UT, USA). One-Way Analysis of Variance Assuming Equal Variances (F-Tests) was used to assess the sample size.

## 3. Results

### 3.1. Effect of TSS on GIP Expression in the Small Intestine of Mice

Mice were administered with TSS by single gavage at the dose of 65mg/kg BW. As shown in Figure 1, the mRNA and protein levels of GIP in the small intestine of mice both significantly increased during the first several hours and then decreased ($p < 0.05$). The highest expression of GIP mRNA and protein occurred at 1 and 3 h, which were 6.8-fold and 2.7-fold higher than that of the control group, respectively. Numerous studies have demonstrated asynchrony in mRNA and protein expression, possibly due to differences in the timing and location of eukaryotic gene transcription and translation.

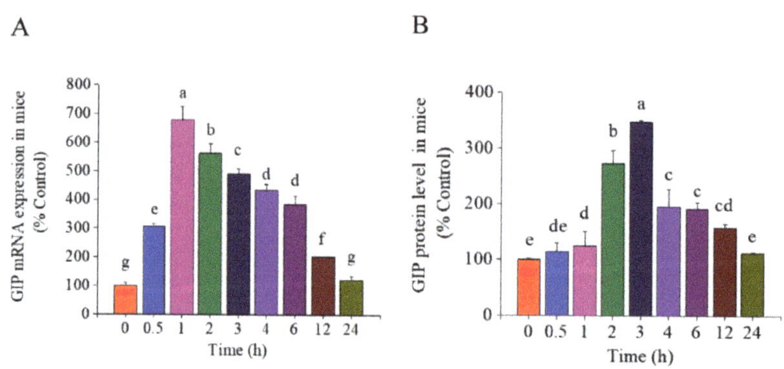


**Figure 1.** Effect of total tea seed saponins (TSS) on the GIP expression in the small intestine of mice. (**A**) The mRNA expression of GIP determined by RT-PCR. (**B**) The protein level of GIP determined by ELISA. Different letters indicate significant difference among groups ($p < 0.05$).

### 3.2. Effect of TSE1 on GIP Secretion in STC-1 Cells

To further evaluate the effect of tea seed saponins on intestinal GIP secretion, enteroendocrine STC-1 cells were treated with the individual saponin compound TSE1 for 24 h. Figure 2A shows that TSE1 at the concentrations of 0–10 μM did not inhibit the STC-1 cell proliferation ($p > 0.05$), indicating that TSE1was not toxic to cells within this dose range. The mRNA expression of GIP increased to 3.5 and 6.6 times that of the control at 2 and 4 μM of TSE1 ($p < 0.05$), and the protein level in the cell culture media was 1.3-fold higher than control at 4 μM ($p < 0.05$) (Figure 2B,C). These data suggest that TSE1 could enhance the GIP expression at relatively low concentrations.

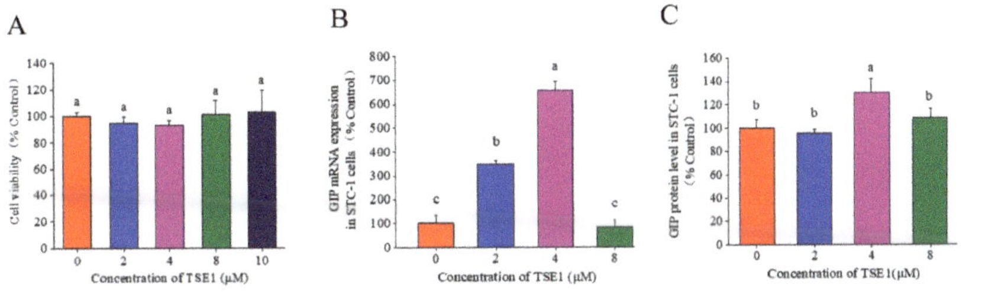


**Figure 2.** Effect of TSE1 on the GIP expression in STC-1 cells at 24 h. (**A**) Cell viability after 24 h treatment of TSE1 determined by MTT assay. (**B**) The mRNA expression of GIP determined by RT-PCR. (**C**) The protein level of GIP in the cell culture media determined by ELISA. Different letters indicate significant difference among groups ($p < 0.05$).

### 3.3. Role of SGLT 1 in TSE1-Induced GIP Expression

Phlorizin, a dihydrochalcone from the bark of pears, apples, cherries and other fruits, is a known inhibitor of SGLT1 [17]. In this work, phlorizin was used to clarify the effect of SGLT1 on TSE1-induced GIP secretion. Figure 3A shows that the cell viability did not change obviously at 10 and 25 μM of phlorizin, and slightly decreased by 17% at 50 μM ($p > 0.05$). Therefore, 25 μM of phlorizin was used in subsequent experiments. As shown in Figure 3B,C, phlorizin could significantly reverse the increase in intracellular $Ca^{2+}$ concentration ($p < 0.05$) and GIP mRNA level ($p < 0.01$) induced by TSE1 at 2 and 4 μM, indicating that SGLT1 played a key role in TSE1-induced GIP expression.

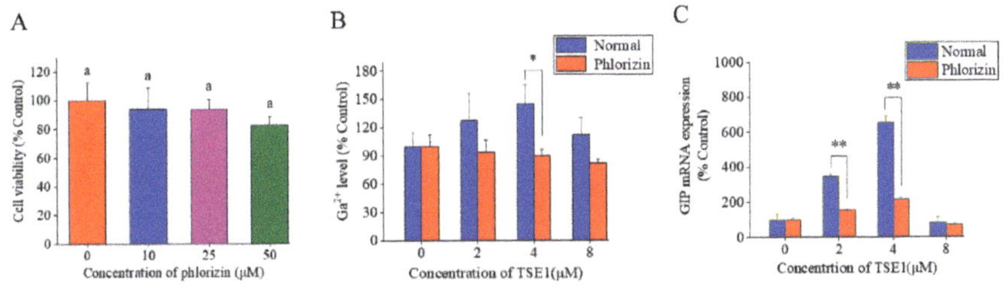


**Figure 3.** Effect of phlorizin on the $Ca^{2+}$ concentration and GIP expression in STC-1 cells at 24 h. (**A**) Cell viability after 24 h treatment of phlorizin determined by MTT assay. (**B**) The intracellular $Ca^{2+}$ concentration determined by BBcellProbe F03 assay. (**C**) The mRNA expression of GIP determined by RT-PCR. Different letters indicate significant difference among groups ($p < 0.05$). * $p < 0.05$, ** $p < 0.01$, compared with normal group.

### 3.4. Effect of TSE1 on TGR5 Expression in STC-1 Cells

Considering that TGR5 is a target of oleanolic acid, and contributes to the GLP-1 and -2 production [18,19], the effect of TSE1 on TGR5 expression was tested. Figure 4 shows that the protein level of TGR5 was significantly enhanced by TSE1 at 2 and 4 μM in a dose-dependent manner ($p < 0.05$), suggesting that TGR5 might be a target of tea seed saponins.

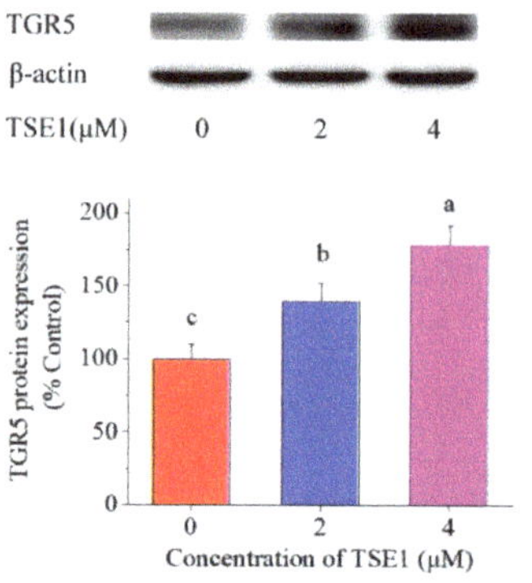


**Figure 4.** Effect of TSE1 on the TGR5 protein expression in STC-1 cells at 24 h. Protein levels were detected by Western blot, and analyzed by the ImageJ software. Different letters indicate significant difference among groups ($p < 0.05$).

### 3.5. Role of TGR5 in TSE1-Induced GIP Expression

In order to clarify whether TGR5 is involved in the regulation of GIP by TSE1, TGR5 was knocked down in STC-1 cells by siRNA approach. As shown in Figure 5A,B, TGR5 siRNA effectively reduced the protein expression of TGR5 compared with the controls, meanwhile, it significantly suppressed TSE1-induced upregulation of TGR5 ($p < 0.05$). Figure 5C shows that TSE1 did not enhance GIP mRNA expression after treatment with TGR5 siRNA, indicating that TSE1-induced GIP upregulation through activating TGR5 in the STC-1 cells.

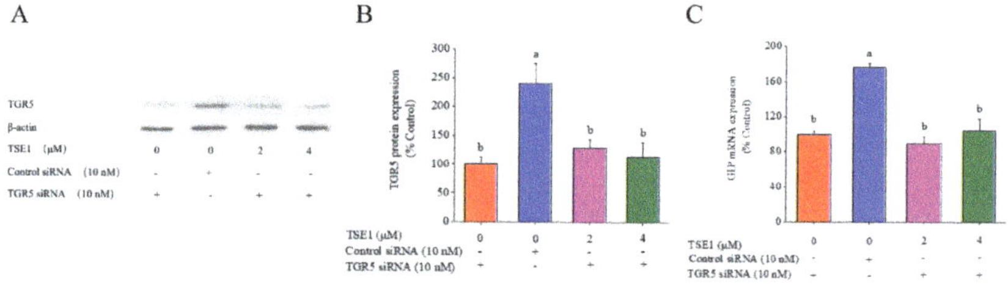


**Figure 5.** Role of TGR5 in TSE1-induced GIP expression in STC−1cells. (**A**) Effect of TGR5 siRNA on the TGR5 protein expression determined by Western blot. (**B**) Quantification of protein bands analyzed by the ImageJ software. (**C**) Effect of TGR5 siRNA on the GIP mRNA expression determined by RT-PCR. Different letters indicate significant difference among groups ($p < 0.05$).

### 3.6. PPI Network and KEGG Analysis

In order to further explore the regulatory network of TSE1-induced GIP expression, the three proteins, TGR5 (also called G-protein-coupled bile acid receptor 1, GPBAR1), SGLT1 (also called Solute carrier family 5 member 1, SLC5A1) and GIP, were uploaded to the STRING database and GeneMANIA database to construct a PPI network. As shown in Figure 6A,B, six common proteins, including insulin gene enhancer binding protein-1(ISL1), gastric inhibitory polypeptide receptor (GIPR), glucagon (GCG), G protein-coupled receptor 119 (GPR119), free fatty acid receptor 1 (FFAR1) and neuroendocrine convertase 1 (PCSK1), occurred in both databases. Figure 6A shows that GCG was a key node connecting GPBAR1 and SLC5A1 with GIP. Figure 6B shows that GIPR, GPR119 and FFAR1 were related with GPBAR1 and GIP, PCSK1 was associated with SLC5A1 and GIP, and GCG was connected with GPBAR1, SLC5A1 and GIP. These data provided a possible network of protein interactions that involved in the GIP regulation by TSE1. KEGG analysis demonstrated that the six common proteins genes together with GPBAR1 and

SLC5A1 were significantly enriched in five pathways including insulin secretion, cAMP signaling pathway, neuroactive ligand–receptor interaction, carbohydrate digestion and absorption, and mineral absorption ($p < 0.05$) (Figure 6C, Table S1), indicating that these signaling pathways are involved in TSE1-mediated upregulation of GIP expression and related biological functions.

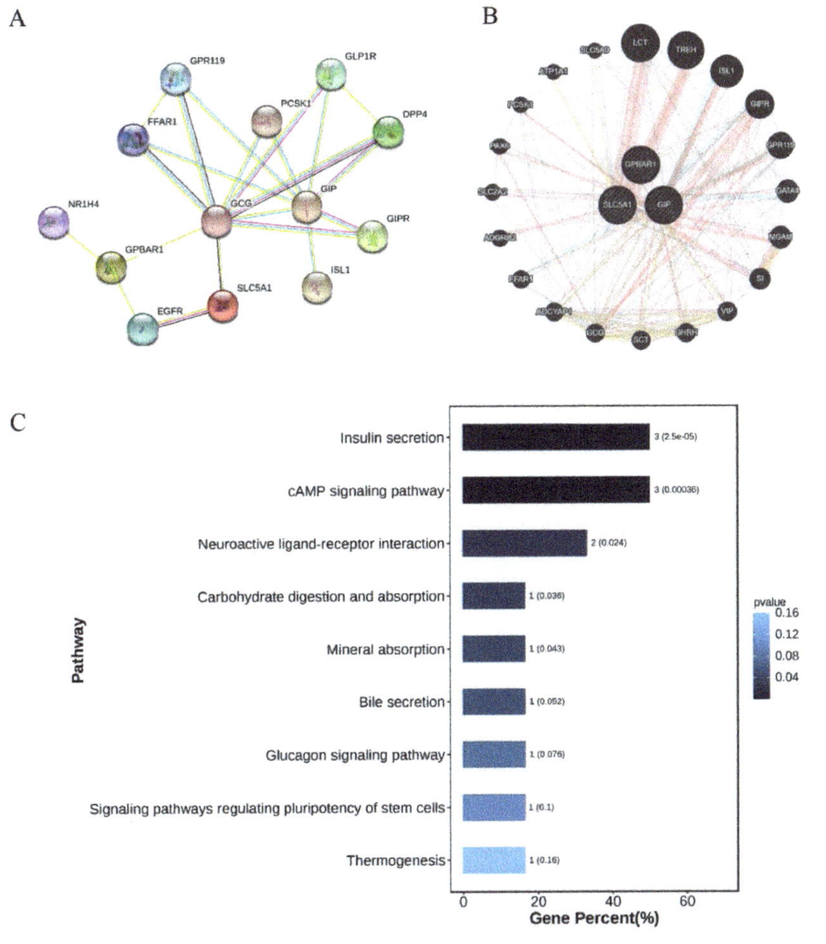


**Figure 6.** Protein–protein interaction (PPI) network of TGR5 (GPBAR1), SGLT1 (SLC5A1) and GIP. (**A**) The PPI network plotting constructed by the STRING database. (**B**) The PPI network plotting constructed by the GeneMANIA database. (**C**) KEGG enrichment analysis of target proteins.

## 4. Discussion

The intestine is not only the main site of digestion, but also the largest endocrine organ of human body. The incretin hormones have received increasing attention over the past few decades due to their modulatory effects on appetite, blood sugar, insulin secretion, lipid metabolism, gastrointestinal motility and immune function [20]. These bioactivities confer incretins as the potential therapeutic targets for the treatment of obesity, diabetes and cardiovascular disease [3,21].

Numerous studies have focused on GLP-1 and GIP receptor agonists and DPP-4 inhibitors for development of incretin-based antidiabetic drugs [22]. Recently, some nutrients and natural products such as L-tryptophan, lauric acid, α-linolenic acid and nobiletin were found to regulate glycaemia, gastric emptying, food intake and circadian rhythm via incretin stimulation [23–25]. Two saponins, ginsenosides and glycyrrhizic acid, have been reported to increase GLP-1 secretion, which may be related to their antidiabetic effects [13,26]. The triterpene saponins from tea seeds could modulate gastrointestinal system and blood lipids in vivo. However, little is known about their effects on incretin. In this work, it was demonstrated that TSS (65 mg/kg BW) and a major saponin compound, TSE1 (2 and 4 μM), could significantly increase the mRNA and protein levels of GIP in the mouse small intestine and STC-1 cells within 24 h (Figures 1 and 2). These results suggest that stimulation of GIP secretion might be an important mechanism by which tea seed saponins regulate intestinal function and metabolism.

SGLT1 (SLC5A1) is a key glucose transporter regulating glucose absorption in the gastrointestinal tract, and is considered a potential target for treating obesity and diabetes [27]. Several researches have demonstrated that SGLT1 participates in glucose- and high-fat-dependent GIP secretion in the normal, obese and diabetic states [28–30]. Activation of SGLT1 at the brush border of the small intestine leads to $Na^+$ influx, membrane depolarization, and opening of $Ca^{2+}$ voltage-gated channels, resulting in intracellular $Ca^{2+}$ accumulation and secretion of intestinal peptides by enteroendocrine cells [31]. SGLT1 inhibitors block glucose-mediated GLP-1 secretion, and reduce intracellular cAMP and $Ca^{2+}$ concentrations [32]. Our data showed that the SGLT1 inhibitor phlorizin (25 μM) could significantly reverse the TSE1-induced enhancement of intracellular $Ca^{2+}$ concentration and GIP mRNA expression (Figure 3), suggesting that TSE1 regulates GIP expression at least in part by triggering SGLT1 in the small intestine.

TGR5 (GPBAR-1) is highly expressed in intestine, brown adipose tissue, macrophages/ monocytes, spleen and gallbladder [33]. It regulates biliary homeostasis and related metabolism, prevents gastrointestinal and liver inflammation, and is considered a candidate target for improving obesity, dyslipidemia, type 2 diabetes and nonalcoholic fatty liver disease [34–36]. Pentacyclic triterpenoids such as betulinic acid and oleanolic acid exhibit TGR5 agonist capacity, possibly due to their structural similarity to bile acids [37,38]. Glycyrrhizic acid, a triterpenoid saponin, can increase GLP-1 secretion via activating TGR5 in intestinal NCI-H716 cells and type 1-like diabetic rats [13]. In this work, TSE1 was found to increase the TGR5 protein expression by approximately 30–80% in a dose-dependent manner (Figure 4). Silence of TGR5 obviously counteracted the elevated GIP mRNA expression in the STC-1 cells induced by TSE1 (Figure 5). Our data revealed for the first time that TGR5 was involved in the GIP regulation, which might be a common mechanism of oleanane-type saponins to stimulate GIP secretion.

Network pharmacology is an effective strategy for identifying drug components and their global mechanisms [39]. In the present study, the STRING and GeneMANIA databases were used to analyze the protein-protein relationship, and six key proteins, ISL1, GIPR, GPR119, GCG, FFAR1 and PCSK1, were found to be associated with SGLT1, TGR5 and GIP. KEGG enrichment analysis showed that these proteins were mainly involved in insulin secretion, cAMP signaling and neuroprotection (Figure 6). ISL1, a DNA-binding transcriptional activator, not only regulates the gene expression of insulin and glucagon to maintain the glucose homeostasis, but also plays a key role in cell proliferation, differentiation and tumorigenesis [40]. GIPR activation promotes insulin secretion through increasing cAMP and proinsulin gene transcription, and regulates pancreatic beta cell survival and proliferation independent of insulin [41]. GPR119 and FFAR1 are free fatty acid sensors that mediate the release of incretin hormones and insulin [42,43]. Glucagon from α-cells was found to be stimulated by GIP in hyperglycemia of Type 2 diabetics, but not in healthy people [44]. PCSK1, mainly expressed in neuronal and endocrine cells, could cleave many protein precursors involved in energy homeostasis [45]. Downregulation of PCSK1 suppresses the generation of GIP and GLP-1 in STC-1 cells [46]. These proteins are involved in tea seed saponin-induced GIP secretion and related physiological functions, and should be further studied as the targets of these saponins in the future.

## 5. Conclusions

In summary, total tea seed saponins and individual saponin compound TSE1 could stimulate GIP mRNA and protein expression in normal mice and STC-1 cells. Inhibition of SGLT1 by phlorizin or silence of TGR5 by siRNA both reverse the TSE1-induced up-regulation of GIP expression. Network pharmacology analysis indicated that ISL1, GIPR, GPR119, GCG, FFAR1 and PCSK1 were involved in the regulation of SGLT1, TGR5 and GIP by TSE1. These genes participate in insulin secretion, cAMP signaling, neuroprotection, carbohydrate digestion and absorption, and mineral absorption, suggesting that tea seed saponins possess these bioactivities and are worthy of further study as functional food ingredients or drugs for improving metabolic diseases.

**Supplementary Materials:** The following supporting information can be downloaded at: https://www.mdpi.com/article/10.3390/nu14163413/s1, Table S1: Five significantly enriched KEGG pathways and involved genes.

**Author Contributions:** Conceptualization, B.L.; methodology, H.Z. and K.W.; validation, S.C. and J.K.; formal analysis, H.Z. and N.G.; investigation, N.G. and H.Z.; resources, H.C.; writing—original draft preparation, H.Z.; writing—review and editing, B.L.; supervision, Y.T., P.H. and B.L.; project administration, B.L., Y.W. and J.L.; funding acquisition, B.L. All authors have read and agreed to the published version of the manuscript.

**Funding:** This research was funded by the Natural Science Foundation of Zhejiang Province, China (grant number LY21C200010 and LY15C200007), the Fundamental Research Funds for the Central Universities (grant number 2018QNA6024), and The APC was funded by the Science and Technology Innovation Platform Project of Fujian Provincial Department of Science and Technology (grant number 2018N2004).

**Institutional Review Board Statement:** The animal study protocol was approved by the Experimental Animals Ethics Committee of Zhejiang University (protocol code 12531, date of approval: 5 March 2018).

**Informed Consent Statement:** Not applicable.

**Data Availability Statement:** Not applicable.

**Conflicts of Interest:** The authors declare no conflict of interest.

## References

1. Yaribeygi, H.; Maleki, M.; Butler, A.E.; Jamialahmadi, T.; Sahebkar, A. The impact of incretin-based medications on lipid metabolism. *J. Diabetes. Res.* **2021**, *2021*, 1815178. [CrossRef]
2. Stemmer, K.; Finan, B.; DiMarchi, R.D.; Tschop, M.H.; Muller, T.D. Insights into incretin-based therapies for treatment of diabetic dyslipidemia. *Adv. Drug Deliv. Rev.* **2020**, *159*, 34–53. [CrossRef] [PubMed]
3. Tan, Q.M.; Akindehin, S.E.; Orsso, C.E.; Waldner, R.C.; DiMarchi, R.D.; Muller, T.D.; Haqq, A.M. Recent advances in incretin-based pharmacotherapies for the treatment of obesity and diabetes. *Front. Endocrinol.* **2022**, *13*, 838410. [CrossRef] [PubMed]
4. Uhlen, M.; Fagerberg, L.; Hallstrom, B.M.; Lindskog, C.; Oksvold, P.; Mardinoglu, A.; Sivertsson, A.; Kampf, C.; Sjostedt, E.; Asplund, A.; et al. Tissue-based map of the human proteome. *Science* **2015**, *347*, 1260419. [CrossRef] [PubMed]
5. Rhodes, R.S.S.; Singh, S.K.; Rajendran, V.M.; Walk, S.T.; Coon, S.D. Regulation of glucose insulinotropic peptide and intestinal glucose transporters in the diet-induced obese mouse. *J. Diabetes Res.* **2022**, *2022*, 5636499. [CrossRef]
6. Takagishi, M.; Aleogho, B.M.; Okumura, M.; Ushida, K.; Yamada, Y.; Seino, Y.; Fujimura, S.; Nakashima, K.; Shindo, A. Nutritional control of thyroid morphogenesis through gastrointestinal hormones. *Curr. Biol.* **2022**, *32*, 1485–1496.e4. [CrossRef] [PubMed]
7. Mantelmacher, F.D.; Zvibel, I.; Cohen, K.; Epshtein, A.; Pasmanik-Chor, M.; Vogl, T.; Kuperman, Y.; Weiss, S.; Drucker, D.J.; Varol, C.; et al. GIP regulates inflammation and body weight by restraining myeloid-cell-derived S100A8/A9. *Nat. Metab.* **2019**, *1*, 58–69. [CrossRef]
8. Zhang, Z.Q.; Holscher, C. GIP has neuroprotective effects in Alzheimer and Parkinson's disease models. *Peptides* **2020**, *125*, 170184. [CrossRef]
9. Ko, J.; Jang, S.; Kwon, W.; Kim, S.Y.; Jang, S.; Kim, E.; Ji, Y.R.; Park, S.; Kim, M.O.; Choi, S.K.; et al. Protective effect of GIP against monosodium glutamate-induced ferroptosis in mouse hippocampal HT-22 Cells through the MAPK signaling pathway. *Antioxidants* **2022**, *11*, 189. [CrossRef]
10. Bhatti, J.S.; Kumar, S.; Vijayan, M.; Bhatti, G.K.; Reddy, P.H. Therapeutic strategies for mitochondrial dysfunction and oxidative stress in age-related metabolic disorders. *J. Diabetes Res.* **2017**, *146*, 13–46.
11. Pais, R.; Gribble, F.M.; Reimann, F. Stimulation of incretin secreting cells. *Ther. Adv. Endocrinol. Metab.* **2016**, *7*, 24–42. [CrossRef]

12. Stepanov, V.; Stankov, K.; Mikov, M. The bile acid membrane receptor TGR5: A novel pharmacological target in metabolic, inflammatory and neoplastic disorders. *J. Recept. Signal Transduct.* **2013**, *33*, 213–223. [CrossRef]
13. Wang, L.Y.; Cheng, K.C.; Li, Y.; Niu, C.S.; Cheng, J.T.; Niu, H.S. Glycyrrhizic acid increases glucagon like peptide-1 secretion via TGR5 activation in type 1-like diabetic rats. *Biomed. Pharmacother.* **2017**, *95*, 599–604. [CrossRef] [PubMed]
14. Lebrun, L.J.; Lenaerts, K.; Kiers, D.; Pais de Barros, J.P.; Le Guern, N.; Plesnik, J.; Thomas, C.; Bourgeois, T.; Dejong, C.H.C.; Kox, M.; et al. Enteroendocrine L cells sense LPS after gut barrier injury to enhance GLP-1 secretion. *Cell Rep.* **2017**, *21*, 1160–1168. [CrossRef]
15. Guo, N.; Tong, T.T.; Ren, N.; Tu, Y.Y.; Li, B. Saponins from seeds of genus *Camellia*: Phytochemistry and bioactivity. *Phytochemistry* **2018**, *149*, 42–55. [CrossRef] [PubMed]
16. Wu, X.J.; Jia, L.Y.; Wu, J.F.; Liu, Y.W.; Kang, H.; Liu, X.B.; Li, P.; He, P.M.; Tu, Y.Y.; Li, B. Simultaneous determination and quantification of triterpene saponins from *Camellia sinensis* seeds using UPLC-PDA-QTOF-MS/MS. *Molecules* **2019**, *24*, 3794. [CrossRef] [PubMed]
17. Rieg, T.; Vallon, V. Development of SGLT1 and SGLT2 inhibitors. *Diabetologia* **2018**, *61*, 2079–2086. [CrossRef]
18. Hunt, J.E.; Billeschou, A.; Windelov, J.A.; Hartmann, B.; Ullmer, C.; Holst, J.J.; Kissow, H. Pharmacological activation of TGR5 promotes intestinal growth via a GLP-2-dependent pathway in mice. *Am. J. Physiol.-Gastroint. Liver Physiol.* **2020**, *318*, G980–G987. [CrossRef]
19. Zhai, H.N.; Li, Z.; Peng, M.A.; Huang, Z.Q.; Qin, T.F.; Chen, L.X.; Li, H.B.; Zhang, H.; Zhang, W.Z.; Xu, G.Y. Takeda G protein-coupled receptor 5-mechanistic target of rapamycin Complex 1 signaling contributes to the increment of glucagon-like peptide-1 production after Roux-en-Y gastric bypass. *Ebiomedicine* **2018**, *32*, 201–214. [CrossRef]
20. Campbell, J.E.; Drucker, D.J. Pharmacology, physiology, and mechanisms of incretin hormone action. *Cell Metab.* **2013**, *17*, 819–837. [CrossRef]
21. Ussher, J.R.; Greenwell, A.A.; Nguyen, M.A.; Mulvihill, E.E. Cardiovascular effects of incretin-based therapies: Integrating Mechanisms with cardiovascular outcome trials. *Diabetes* **2022**, *71*, 173–183. [CrossRef]
22. Radbakhsh, S.; Sathyapalan, T.; Banach, M.; Sahebkar, A. Incretins and microRNAs: Interactions and physiological relevance. *Pharmacol. Res.* **2020**, *153*, 104662. [CrossRef] [PubMed]
23. Hajishafiee, M.; McVeay, C.; Lange, K.; Rehfeld, J.F.; Horowitz, M.; Feinle-Bisset, C. Effects of intraduodenal infusion of lauric acid and L-tryptophan, alone and combined, on glucoregulatory hormones, gastric emptying and glycaemia in healthy men. *Metab.-Clin. Exp.* **2022**, *129*, 155140. [CrossRef] [PubMed]
24. Kamakura, R.; Raza, G.S.; Makila, E.; Riikonen, J.; Kovalainen, M.; Ueta, Y.; Lehto, V.P.; Salonen, J.; Herzig, K.H. Colonic delivery of alpha-linolenic acid by an advanced nutrient delivery system prolongs glucagon-like peptide-1 secretion and inhibits food intake in Mice. *Mol. Nutr. Food Res.* **2022**, *66*, e2100978. [CrossRef] [PubMed]
25. Martchenko, A.; Biancolin, A.D.; Martchenko, S.E.; Brubaker, P.L. Nobiletin ameliorates high fat-induced disruptions in rhythmic glucagon-like peptide-1 secretion. *Sci. Rep.* **2022**, *12*, 7271. [CrossRef]
26. Liu, C.; Zhang, M.; Hu, M.Y.; Guo, H.F.; Li, J.; Yu, Y.L.; Jin, S.; Wang, X.T.; Liu, L.; Liu, X.D. Increased glucagon-like peptide-1 secretion may be involved in antidiabetic effects of ginsenosides. *J. Endocrinol.* **2013**, *217*, 185–196. [CrossRef]
27. Song, P.; Onishi, A.; Koepsell, H.; Vallon, V. Sodium glucose cotransporter SGLT1 as a therapeutic target in diabetes mellitus. *Expert Opin. Ther. Targets* **2016**, *20*, 1109–1125. [CrossRef]
28. Gorboulev, V.; Schurmann, A.; Vallon, V.; Kipp, H.; Jaschke, A.; Klessen, D.; Friedrich, A.; Scherneck, S.; Rieg, T.; Cunard, R.; et al. $Na^+$-D-glucose cotransporter SGLT1 is pivotal for intestinal glucose-absorption and glucose-dependent incretin secretion. *Diabetes* **2012**, *61*, 187–196. [CrossRef]
29. Coon, S.D.; Rajendran, V.M.; Singh, S.K. Prolonged high-fat feeding decreases SGLT1 and GLUT2 activity and blunts their upregulation by glucose-dependent insulinotropic polypeptide (GIP) in obese mice. *Gastroenterology* **2014**, *146*, S493. [CrossRef]
30. Ogata, H.; Seino, Y.; Harada, N.; Iida, A.; Suzuki, K.; Izumoto, T.; Ishikawa, K.; Uenishi, E.; Ozaki, N.; Hayashi, Y.; et al. K-ATP channel as well as SGLT1 participates in GIP secretion in the diabetic state. *J. Endocrinol.* **2014**, *222*, 191–200. [CrossRef] [PubMed]
31. Mace, O.J.; Schindler, M.; Patel, S. The regulation of K- and L-cell activity by GLUT2 and the calcium-sensing receptor CasR in rat small intestine. *J. Physiol.* **2012**, *590*, 2917–2936. [CrossRef]
32. Zhou, L.; Wang, F.; Song, X.D.; Shi, M.; Liang, G.D.; Zhang, L.R.; Huang, F.; Jiang, G.R. 3-Deoxyglucosone reduces glucagon-like peptide-1 secretion at low glucose levels through down-regulation of SGLT1 expression in STC-1 cells. *Arch. Physiol. Biochem.* **2021**, *127*, 311–317. [CrossRef]
33. Cipriani, S.; Mencarelli, A.; Chini, M.G.; Distrutti, E.; Renga, B.; Bifulco, G.; Baldelli, F.; Donini, A.; Fiorucci, S. The bile acid receptor GPBAR-1 (TGR5) modulates integrity of intestinal barrier and immune response to experimental colitis. *PLoS ONE* **2011**, *6*, e25637. [CrossRef]
34. Chavez-Talavera, O.; Tailleux, A.; Lefebvre, P.; Staels, B. Bile acid control of metabolism and inflammation in obesity, type 2 diabetes, dyslipidemia, and nonalcoholic fatty liver disease. *Gastroenterology* **2017**, *152*, 1679–1694.e3. [CrossRef] [PubMed]
35. Merlen, G.; Bidault-Jourdainne, V.; Kahale, N.; Glenisson, M.; Ursic-Bedoya, J.; Doignon, I.; Garcin, I.; Humbert, L.; Rainteau, D.; Tordjmann, T. Hepatoprotective impact of the bile acid receptor TGR5. *Liver Int.* **2020**, *40*, 1005–1015. [CrossRef]
36. Garibay, D.; Zaborska, K.E.; Shanahan, M.; Zheng, Q.N.; Kelly, K.M.; Montrose, D.C.; Dannenberg, A.J.; Miller, A.D.; Sethupathy, P.; Cummings, B.P. TGR5 protects against colitis in mice, but vertical sleeve gastrectomy increases colitis severity. *Obes. Surg.* **2019**, *29*, 1593–1601. [CrossRef]

37. Maczewsky, J.; Kaiser, J.; Gresch, A.; Gerst, F.; Dufer, M.; Krippeit-Drews, P.; Drews, G. TGR5 activation promotes stimulus-secretion coupling of pancreatic -cells via a PKA-dependent pathway. *Diabetes* **2019**, *68*, 324–336. [CrossRef]
38. Lo, S.H.; Cheng, K.C.; Li, Y.X.; Chang, C.H.; Cheng, J.T.; Lee, K.S. Development of betulinic acid as an agonist of TGR5 receptor using a new in vitro assay. *Drug Des. Dev. Ther.* **2016**, *10*, 2669–2676.
39. Athanasios, A.; Charalampos, V.; Vasileios, T.; Ashraf, G.M. Protein-Protein interaction (PPI) network: Recent advances in drug discovery. *Curr. Drug Metab.* **2017**, *18*, 5–10. [CrossRef]
40. Guo, T.; Bai, Y.H.; Cheng, X.J.; Han, H.B.; Du, H.; Hu, Y.; Jia, S.Q.; Xing, X.F.; Ji, J.F. Insulin gene enhancer protein 1 mediates glycolysis and tumorigenesis of gastric cancer through regulating glucose transporter 4. *Cancer Commun.* **2021**, *41*, 258–272. [CrossRef] [PubMed]
41. Mayendraraj, A.; Rosenkilde, M.M.; Gasbjerg, L.S. GLP-1 and GIP receptor signaling in beta cells—A review of receptor interactions and co-stimulation. *Peptides* **2022**, *151*, 170749. [CrossRef] [PubMed]
42. Psichas, A.; Larraufie, P.F.; Goldspink, D.A.; Gribble, F.M.; Reimann, F. Chylomicrons stimulate incretin secretion in mouse and human cells. *Diabetologia* **2017**, *60*, 2475–2485. [CrossRef] [PubMed]
43. Cvijanovic, N.; Isaacs, N.J.; Rayner, C.K.; Feinle-Bisset, C.; Young, R.L.; Little, T.J. Lipid stimulation of fatty acid sensors in the human duodenum: Relationship with gastrointestinal hormones, BMI and diet. *Int. J. Obes.* **2017**, *41*, 233–239. [CrossRef] [PubMed]
44. El, K.; Campbell, J.E. The role of GIP in alpha-cells and glucagon secretion. *Peptides* **2020**, *125*, 170213. [CrossRef]
45. Aerts, L.; Terry, N.A.; Sainath, N.N.; Torres, C.; Martin, M.G.; Ramos-Molina, B.; Creemers, J.W. Novel homozygous inactivating mutation in the PCSK1 gene in an infant with congenital malabsorptive diarrhea. *Genes* **2021**, *12*, 710. [CrossRef]
46. Zhang, Y.F.; Huang, S.F.; Li, P.P.; Chen, Q.; Li, Y.Z.; Zhou, Y.Z.; Wang, L.T.; Kang, M.X.; Zhang, B.; Yang, B.; et al. Pancreatic cancer-derived exosomes suppress the production of GIP and GLP-1 from STC-1 cells in vitro by down-regulating the PCSK1/3. *Cancer Lett.* **2018**, *431*, 190–200. [CrossRef]

nutrients

MDPI

*Article*

# Olfactory Stimulation with Volatile Aroma Compounds of Basil (*Ocimum basilicum* L.) Essential Oil and Linalool Ameliorates White Fat Accumulation and Dyslipidemia in Chronically Stressed Rats

**Da-Som Kim [1,†], Seong-Jun Hong [2,†], Sojeong Yoon [2], Seong-Min Jo [2], Hyangyeon Jeong [2], Moon-Yeon Youn [1], Young-Jun Kim [3], Jae-Kyeom Kim [4] and Eui-Cheol Shin [1,2,5,*]**

1. Department of Food Science, Gyeongsang National University, Jinju 52725, Korea; kim94dasom@naver.com (D.-S.K.); ringspot@naver.com (M.-Y.Y.)
2. Department of GreenBio Science, Gyeongsang National University, Jinju 52725, Korea; 01028287383a@gmail.com (S.-J.H.); dbsthwjd0126@naver.com (S.Y.); jojo9875@naver.com (S.-M.J.); giddus9967@naver.com (H.J.)
3. Department of Food and Biotechnology, Korea University, Sejong 30019, Korea; yk46@korea.ac.kr
4. Department of Behavioral Health and Nutrition, University of Delaware, Newark, DE 19716, USA; jkkim@udel.edu
5. Division of Food Science and Technology, Agri-Food Bio Convergence Institute, Gyeongsang National University, Jinju 52725, Korea

* Correspondence: eshin@gnu.ac.kr; Tel.: +82-55-772-3271

† These authors contributed equally to this work.

**Citation:** Kim, D.-S.; Hong, S.-J.; Yoon, S.; Jo, S.-M.; Jeong, H.; Youn, M.-Y.; Kim, Y.-J.; Kim, J.-K.; Shin, E.-C. Olfactory Stimulation with Volatile Aroma Compounds of Basil (*Ocimum basilicum* L.) Essential Oil and Linalool Ameliorates White Fat Accumulation and Dyslipidemia in Chronically Stressed Rats. *Nutrients* **2022**, *14*, 1822. https://doi.org/10.3390/nu14091822

Academic Editors: Daniela Rigano and Paola Bontempo

Received: 8 April 2022
Accepted: 26 April 2022
Published: 27 April 2022



**Abstract:** We explored the physiological effects of inhaling basil essential oil (BEO) and/or linalool and identified odor-active aroma compounds in BEO using gas chromatography/mass spectrometry (GC–MS) and GC–olfactometry (GC–O). Linalool was identified as the major volatile compound in BEO. Three groups of rats were administered BEO and linalool via inhalation, while rats in the control group were not. Inhalation of BEO for 20 min only reduced the total weight gain (190.67 ± 2.52 g) and increased the forced swimming time (47.33 ± 14.84 s) compared with the control group (219.67 ± 2.08 g, 8.33 ± 5.13 s). Inhalation of BEO for 5 min (392 ± 21 beats/min) only reduced the pulse compared with the control group (420 ± 19 beats/min). Inhalation of linalool only reduced the weight of white adipose tissue (5.75 ± 0.61 g). The levels of stress-related hormones were not significantly different among the groups. The total cholesterol and triglyceride levels decreased after inhalation of BEO for 20 min (by more than −10% and −15%, respectively). Low-density lipoprotein cholesterol levels were lowered (by more than −10%) by the inhalation of BEO and linalool, regardless of the inhalation time. In particular, BEO inhalation for 20 min was associated with the lowest level of low-density lipoprotein cholesterol (53.94 ± 2.72 mg/dL). High-density lipoprotein cholesterol levels increased after inhalation of BEO (by more than +15%). The atherogenic index and cardiac risk factors were suppressed by BEO inhalation. Animals exposed to BEO and linalool had no significant differences in hepatotoxicity. These data suggest that the inhalation of BEO and linalool may ameliorate cardiovascular and lipid dysfunctions. These effects should be explored further for clinical applications.

**Keywords:** *Ocimum basilicum* L.; essential oil; volatile compounds; linalool; stress lipid metabolism

## 1. Introduction

Stress is classified as either acute or chronic and can influence the physiological regulation of hormones and inflammatory cytokine secretion through several pathways, involving psychological, social, physical, and chemical factors [1–3]. Chronic stress usually disturbs the autonomic nervous system (ANS), which maintains internal homeostasis responding to changes in the external environment and controls the metabolism of substances in the body. Furthermore, the ANS generally regulates the sympathetic and parasympathetic systems, and thus, chronic stress can interfere with the activation of the sympathetic and parasympathetic systems [4]. The deterioration of the ANS usually increases blood pressure, pulse, total cholesterol level, and low-density lipoprotein cholesterol (LDL) levels and decreases high-density lipoprotein cholesterol (HDL) levels. Accordingly, pathologies of the ANS can induce a deterioration in cardiovascular health, leading to hypertension and arteriosclerosis [4,5]. Therefore, researchers have attempted to improve the cholesterol levels and prevent the progression of cardiovascular diseases using natural products with physiological effects [6].

Basil (*Ocimum basilicum Licorice; O. basilicum* L.) is a member of the *Lamiaceae* family. The leaf and stem parts, are used as culinary ingredients and/or as medicinal herbs [7]. Additionally, basil contains a unique fragrance that has been used in the perfume industry. Furthermore, the intake of basil has beneficial effects on the cholesterol level; the intake of basil improves lipid metabolism in high-cholesterol-affected animal models [7]. In addition, orally administered linalool, one of the major compounds in basil, improves cholesterol levels and, when administered by inhalation, induces sedative and relaxing effects [6,8]. Generally, the volatile profiles of essential oil are affected by many factors, such as the geographical area of sampling [9], the variety of/accession to the plants [10], the harvest year [11], the harvest date [12], the extraction system [13], so on. Therefore, the major aroma compound (linalool) of basil is mainly affected by geographical conditions and the harvesting periods, and the concentration of linalool increases according to the flowering periods [14]. In addition, linalool concentration is also affected by the extraction method. In particular, the hydro-distillation extraction method (18.1%) yields a higher concentration of linalool than supercritical fluid extraction (12.6%) [15].

When fragrant products are inhaled, individual fragrance compounds bind to nasal olfactory receptors, and a signal is transmitted to the cerebrum. When a volatile compound is inhaled, it dissolves in the mucus of the nasal mucous membrane and moves to the olfactory epithelium. Subsequently, volatile compounds bind to the olfactory receptors of cilia. Olfactory receptors bind only to certain volatile compounds, and the generated electrical signal reaches the olfactory bulb in the frontal lobe via axons. Therefore, information on individual volatile compounds is delivered to the olfactory cortex and the cerebrum. Individual volatile compounds can be distinguished and recognized according to this signaling pathway [16]. Olfactory stimulation influences the central nervous system (CNS) and ANS activities; thus, olfactory stimulation can control the function of the sympathetic and parasympathetic nervous systems. The involved nerves generally influence energy and lipid metabolism; thus, food intake and cholesterol levels can be controlled by the sympathetic and parasympathetic nervous systems [5,17].

Improvements in lipid metabolism in vivo by the intake of basil and sedative and anti-stress effects of linalool contained in basil have been reported [6–8]. However, the ameliorating effects of inhaling volatile compounds (present in basil) on dyslipidemia caused by chronic stress have not been elucidated. Accordingly, this study observed changes in lipid parameters, stress hormone levels, pulse, body weight, and food intake after inhalation of basil essential oil (BEO) in chronically stressed rats. Furthermore, changes in metabolic parameters following linalool inhalation were observed.

## 2. Materials and Methods

### 2.1. Essential Oil

The basil used in this study was cultivated in Austria in 2017, and the essential oil was extracted by the distillation method using the leaves (100%). The BEO used in this study was a commercial product, purchased from the Aroma Care Solution (Helga-Stolz GmbH Co., Grafenwoerth, Austria) and stored at 4 °C in a dark place until experiments were performed. The grade of this product was for aroma therapy. The experiments in the present study were conducted in 2018–2020.

### 2.2. Odor-Active Aroma Compounds

Odor-active aroma compounds (OAACs) in BEO were collected using solid-phase microextraction (SPME) fibers (Supelco Co., Bellafonte, PA, USA), i.e., fibers coated with 100 μm of polydimethylsiloxane (1 cm in size). BEO (1 g) was placed in a glass vial tightly sealed with an aluminum cap. The OAACs were collected in the headspace while heating the sample to 50 °C. The SPME fibers were injected into the injector of a gas chromatography–mass spectrometry selective detector (GC–MS; Agilent 7890A & 5975C, Agilent Technologies, Santa Clara, CA, USA) at 220 °C, and the analysis was performed after desorption for 10 min. The column was HP-5MS (30 m (length) × 0.25 mm (inner diameter), 0.25 μm (film thickness)), and helium carrier gas was used at 1 mL/min, with a split ratio of 1:10. The initial oven temperature was set at 40 °C for 5 min, increased to 200 °C at a rate of 5 °C/min, and maintained for 10 min. An inlet temperature of 220 °C was set in the splitless mode. OAACs, separated by a total ionization chromatogram, were identified using the National Institute of Standards and Technology (NIST) mass spectral library (NIST version 12). Pentadecane (0.005 μg) was used as an internal standard. According to the peak area and concentration of the internal standard, the concentrations of the OAACs in BEO were expressed in μg/mL. To explore the odor-active characteristics of BEO, the volatile profiles were separated by the GC column and assessed using a GC–olfactometry port (GC-O) (ODP 3, Gerstel Co., Linthicum, MD, USA). Odor-active intensity was divided into four levels, with higher levels representing stronger odor-active intensity, as described previously [5].

### 2.3. Animal Care and Experimental Design

This study was approved by the Animal Experimental Ethics Committee (Animal protocol #: IACUC-4). Forty-five male Sprague–Dawley rats (4 weeks old) were obtained from Coretec Co., (Busan, Korea). The rats were acclimated to a normal diet for a week and randomly classified into four groups. After classification, chronic stress was applied to all groups for five weeks in total. Chronic mild stress was applied in the first week. Chronic mild stress (CMS) is a complex stress that includes food deprivation, restricted access to food, water deprivation, roommate separation, overnight illumination, and tilting the cage by 45°. From the second week, the rats were exposed to chronic stress with distilled water (DW) inhalation for 5 min/day in the control group (CON; $n$ = 6), chronic stress with linalool inhalation for 5 min/day in the positive control group (POS; $n$ = 6), chronic stress with BEO inhalation for 5 min/day in the third group (5 MIN; $n$ = 6), and chronic stress with BEO inhalation for 20 min/day in the fourth group (20 MIN; $n$ = 6) (Figure 1). Linalool and volatile compounds in BEO flowed at a rate of 8 mL/h, achieved by using a humidifier (Aroma diffuser humidifier; Cactus Co., Shanghai, China).

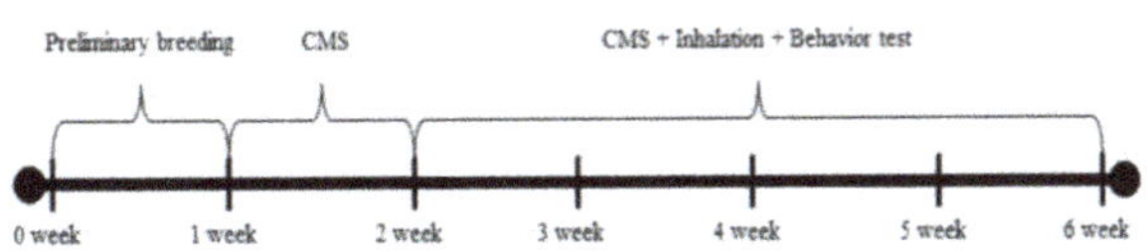


**Figure 1.** The plan of the animal study showing a week of preliminary breeding, a week of chronic mild stress (CMS), and four weeks of CMS + inhalation + behavior testing.

Food intake and body weight were measured once weekly. The rats were fasted for 16 h before dissection. Blood was collected from the heart using syringes containing 20 mg of ethylenediaminetetraacetic acid. The collected blood samples were kept for 30 min, then centrifuged at 1000 G to separate the serum. Finally, organs and tissues (the liver, kidneys, heart, white adipose tissue, and brown adipose tissue) were extracted and weighed. In addition, the organs and tissues were stored in a −80 °C freezer [5].

### 2.4. Forced Swimming Test

The forced swimming test was performed weekly with a standard behavioral despair test. Water (25 °C) was placed in a chamber (40 × 25 × 26.5 cm) at a height of 16 cm. During each experiment, the animals were placed in the chamber and allowed to swim (mobility). Immobility was assessed after swimming. Immobility was defined as when the animals stood upright and floated without movement, exposing only the head [18].

### 2.5. Pulse

The animal's pulse was measured using the tail-cuff method with BP-2000 (Visitech Systems Co., Apex, NC, USA). Eight measurements were taken, excluding the highest and lowest values and deviations. Finally, three measured values were expressed as the average and standard deviation (SD) [5].

### 2.6. Stress Hormones

Cortisol (450 nm) in the serum was analyzed using an ELISA kit (YH ELISA Kit, Shanghai Yehua Biological Technology Co., Shanghai, China), and serotonin (450 nm) in brain tissue was analyzed using another ELISA kit (Serotonin ELISA Kit, Bio Vision Co., Milpitas, CA, USA) by absorbance measurement according to the manufacturer's instructions [19,20].

### 2.7. Analyses of Serum Biomarkers and Hepatotoxicity

Total cholesterol (500 nm), HDL (500 nm), triglyceride (TG) (550 nm), and hepatotoxicity, including aspartate transaminase (AST) (505 nm) and alanine transaminase (ALT) (505 nm), were analyzed using a commercial kit (Asan Reagents, Asan Pharm Co., Seoul, Korea) by absorbance measurement according to the manufacturer's instructions [5].

### 2.8. Statistical Analysis

Experiments were performed in triplicate, and the results are presented as the average and SD. Non-parametric comparison was used to compare paired groups using the Friedman test with chi-square distribution. Differences were considered statistically significant at $p$-values less than 0.05 (SAS Institute Inc., Cary, NC, USA).

## 3. Results and Discussion

### 3.1. Odor-Active Aroma Compounds

Odor-active aroma profiles were detected using GC–MS and GC–O (Table 1 and Figure 2). A total of 17 aroma compounds were detected in BEO. In particular, four OAACs were identified, including linalool, linalool oxide, menthane, and carvone. Linalool elicits basil essential oil odor activation, and linalool oxide elicits the activation of grass and herb odors. Menthane also elicits herb odor and menthol activation. In addition, carvone elicits lemon odor activation.

Linalool had the highest concentration of OAAC in BEO (Table 1). Linalool is a common and major terpenoid, containing most herbal essential oils and has been identified as a forest-like odor using GC–O [5,21]. In addition, linalool can control the lipid metabolism in vivo [21]. Linalool oxide has shown anxiolytic-like effects in mouse anxiety models via inhalation [22]. Menthene is a hydrocarbon with colorless characteristics and an herb odor [23]. D-Carvone has anti-inflammatory and anti-microbial effects, and this volatile compound was identified as OAAC in essential oils by GC–O [5,24]. In general, the genus *Ocimum* includes approximately 150 species distributed worldwide, and different volatile profiles are characteristic of *Ocimum* species, as reported by a previous study. Importantly, high concentrations of linalool were detected in O. *basilicum* L. (25.6%) and O. *sanctum* L. (21.9%) but not in O. *gratissimum* L. (0.1%) and O. *kilimandscharicum* L. (1.4%) [25].

**Table 1.** Aroma profiles and odor-active aroma compounds in basil essential oil identified using GC–MS and GC–O.

| Compounds | Retention Time(min) | Retention Index | Content (μg/mL) | Odor Intensity | Odor Description |
|---|---|---|---|---|---|
| Alcohols (5) | | | | | |
| 1,8-Cineole | 16.29 | 1061 | 38.91 | | |
| Linalool oxide | 17.58 | 1100 | 7.33 | 3 | Herb |
| Linalool | 18.37 | 1128 | 107.65 | 4 | Sweet, Fruit |
| Fenchyl alcohol | 18.85 | 1145 | 32.11 | | |
| *trans*-Anethole | 23.02 | 1291 | 14.39 | | |
| Hydrocarbons (10) | | | | | |
| Camphene | 13.62 | 978 | 5.02 | | |
| $\beta$-Myrcene | 14.98 | 1019 | 15.72 | | |
| $\gamma$-Terpinene | 15.42 | 1034 | 11.24 | | |
| $\alpha$-Terpinene | 15.80 | 1046 | 6.22 | | |
| $\beta$-Cymene | 16.07 | 1054 | 7.77 | | |
| D-Limonene | 16.20 | 1058 | 34.64 | | |
| Ocimene | 16.44 | 1066 | 13.43 | | |
| Menthene | 20.82 | 1212 | 17.33 | 4 | Herb, Basil, Xylitol |
| 2-Hydroxy phenyl butane | 24.15 | 1335 | 22.29 | | |
| 1-Methoxy ethyl benzene | 24.34 | 1342 | 14.65 | | |
| Ketones (2) | | | | | |
| Menth-4-en-3-one | 22.82 | 1284 | 45.40 | | |
| D-Carvone | 22.89 | 1287 | 6.44 | 1 | Lemon |

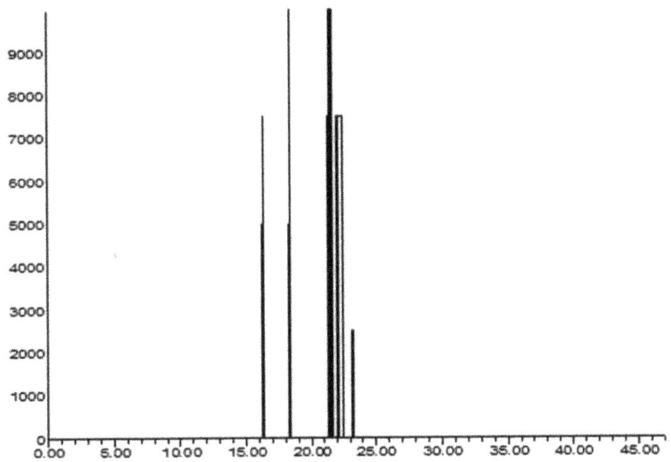


**Figure 2.** Representative aromagram of odor-active aroma compounds (OAACs) in basil (*Ocimum basilicum* L.) essential oil identified by GC–MS and GC–O test.

*3.2. Total Food Intake and Total Weight Gain*

Total food intake in the BEO-inhaled and linalool-inhaled group was much lower compared to that in the control group ($p > 0.05$) (Table 2); however, there were no significant differences among all groups. In the case of total weight gain, the 20 min BEO-inhaled group showed significant less weight gain compared to the control group ($p < 0.05$); however, the 5 min BEO-inhaled group and the linalool-inhaled group did not show any significant differences compared to the control group.

**Table 2.** Total food intake and total weight gain during the animal experiment. CON: chronic stress-exposed control group; POS: linalool (positive control) inhalation by chronic stress-exposed rats; 5 MIN: BEO inhalation for 5 min by chronic stress-exposed rats; 20 MIN: BEO inhalation for 20 min by chronic stress-exposed rats.

| | Total Food Intake (g) | Total Weight Gain (g) |
|---|---|---|
| CON | 499.51 ± 10.60 [a1] | 219.67 ± 2.08 [a] |
| POS | 483.37 ± 9.73 [a] | 204.67 ± 12.34 [ab] |
| 5 MIN | 467.21 ± 24.37 [a] | 201.67 ± 11.15 [ab] |
| 20 MIN | 479.90 ± 11.72 [a] | 190.67 ± 2.52 [b] |

Data are given as mean ± SD values from experiments performed in triplicate. [1] Mean values with different letters within the same row are significantly different according to the non-parametric Friedman test, followed by Dunn's test ($p < 0.05$).

Basil can modulate body weight, and linalool plays an important role as a ligand of peroxisome proliferator-activated receptor α (PPARα) [7,21]. PPARα can modulate fatty acid uptake and fatty acid oxidation and inhibit the occurrence of obesity. Linalool is commonly used for medicinal functions [21]. This study showed that 20 min of BEO inhalation suppressed total weight gain. In contrast, linalool inhalation did not result in decreased body weight. Previous research reported that BEO induced a decrease in body weight [26] and reduced the average body weight [27]. A linalool-containing essential oil has anti-obesity effects, including decreasing body weight and/or promoting lipolysis [5,28]. In addition, Baek et al. reported that linalool inhibits body weight gain [6]. However, linalool inhalation only suppressed the average body weight gain in this study. Therefore, the reduction in body weight could be due to the complex effects of the aroma components of BEO, rather than the sole effect of linalool.

### 3.3. Forced Swimming Test

Changes in swimming records were measured during the study period (Table 3). During the initial period, no significant differences were observed between the groups. In the final period, the control group had the lowest swimming time among all groups ($p < 0.05$), while the other groups showed increasing swimming time. The BEO-inhaled groups showed increased swimming time in an inhalation time-dependent manner. When comparing the BEO- and the linalool-inhaled groups, the 20 min BEO-inhaled group showed a significant increase in swimming time compared to the control group ($p < 0.05$).

**Table 3.** Forced swimming test during the initial and final periods. CON: chronic stress-exposed control group; POS: linalool (positive control) inhalation with chronic stress-exposed rats; 5 MIN: BEO inhalation for 5 min by chronic stress-exposed rats; 20 MIN: BEO inhalation for 20 min by chronic stress-exposed rats.

| | Initial (s) | Final (s) |
|---|---|---|
| CON | 71.67 ± 22.19 [a1] | 8.33 ± 5.13 [b] |
| POS | 44.67 ± 3.51 [a] | 16.67 ± 16.97 [b] |
| 5-MIN | 63.00 ± 6.93 [a] | 18.67 ± 9.87 [ab] |
| 20-MIN | 72.66 ± 10.26 [a] | 47.33 ± 14.84 [a] |

Data are given as mean ± SD values from experiments performed in triplicate. [1] Mean values with different letters within the same row are significantly different according to the non-parametric Friedman test, followed by Dunn's test ($p < 0.05$).

Chronic stress can cause oxidative stress, and animals exposed to oxidative stress have an increased immobility period during forced swimming tests [29,30]. In addition, the period of immobility in rats is decreased by reducing oxidative stress [30]. The results of this study also identified differences between stress-exposed rats and stress-relieved rats following the inhalation of BEO and linalool. A previous study indicated that linalool inhalation upregulated plasma biomarkers and gene expression in rat models of stress [31], while another study indicated that BEO ameliorated oxidative stress in rats [32]. In addition, BEO significantly increased the ambulatory activity via the stimulation of the CNS, and this BEO is considered a potent CNS regulator [33].

### 3.4. Pulse

During the initial period, no significant differences in pulse were observed between the groups (Table 4). In contrast, during the final measurement, inhalation of BEO for 5 min attenuated the pulse rate compared to that in the control group ($p < 0.05$). Inhalation of BEO for 20 min and of linalool only showed a tendency to decrease the pulse, and changes were not significant ($p > 0.05$).

**Table 4.** Pulse assessment using the tail-cuff method in rats.

| Pulse (beats/min) | Initial | Final |
|---|---|---|
| CON | 406 ± 13 [a1] | 420 ± 19 [a] |
| POS | 407 ± 17 [a] | 416 ± 14 [a,b] |
| 5 MIN | 402 ± 14 [a] | 351 ± 19 [b] |
| 20 MIN | 414 ± 3 [a] | 392 ± 21 [a,b] |

Data are given as mean ± SD values from experiments performed in triplicate. [1] Mean values with different letters within the same row are significantly different according to the non-parametric Friedman test, followed by Dunn's test ($p < 0.05$).

The pulse is controlled by the ANS, which includes sympathetic and parasympathetic nerves, and a reduced pulse is associated with decreased sympathetic and increased parasympathetic nerve activity [4]. In this study, inhalation of BEO and linalool resulted in significantly and/or relatively decreased pulse rates. Previous studies indicated that linalool and linalool-containing essential oils attenuated renal sympathetic nerve activity and enhanced parasympathetic nerve activity by olfactory stimulation, and linalool-containing essential oil inhalation decreased the pulse rates in rats [28,33]. In addition, inhalation of linalool has a sedative effect in animal models [8], and BEO inhalation also induced a sedative effect by decreasing the arousal response measured on the basis of electroencephalographic activity [34].

### 3.5. Organ Weights

The liver, kidney, heart, white adipose tissue (WAT), and brown adipose tissue (BAT) were weighed (Table 5). There were no significant differences in liver weights among the groups. The kidney weights in the 20 min BEO-inhaled group was lower than that of the control group ($p < 0.05$). Meanwhile, inhalation of BEO for 5 min and of linalool induced no significant decrease in liver weight. In terms of heart weight, there were no significant differences between the groups. The control group had the highest WAT weight among all groups. The BEO-inhaled groups showed a decreasing tendency in WAT weights; however, these changes were not significantly different. The linalool-inhaled group showed a decrease in WAT weight compared with the control group ($p < 0.05$). BAT weights were measured in all groups, and there were no significant differences.

WAT is related to oxidative stress, and increased WAT and oxidative stress can increase the metabolic risk [29]. Accumulation of WAT generally increases cardiovascular disorders, being associated with increased levels of TC, LDL, and TG, as well as decreased levels of HDL [17]. This study showed that inhalation of BEO and linalool decreased WAT weight. Linalool inhalation significantly decreased WAT weight compared with the control group ($p < 0.05$). Therefore, the reduction in WAT appeared to occur in a linalool concentration-dependent manner. A previous study reported that linalool reduced WAT weight in mice [6], while another study found that linalool induced lipolysis by upregulating PPAR$\alpha$ activity, fatty acid oxidation, and energy metabolism [21]. In addition, linalool treatment significantly reduced lipid accumulation in 3T3-L1 cells [35]. Moreover, research has found decreased fat accumulation following linalool odor stimulation [36].

**Table 5.** Changes in rat organ weights.

| | Liver (g) | Kidney (g) | Heart (g) | WAT (g) | BAT (g) |
|---|---|---|---|---|---|
| CON | 11.74 ± 2.87 [a1] | 2.09 ± 0.03 [a] | 1.38 ± 0.03 [a] | 6.86 ± 0.40 [a] | 0.62 ± 0.06 [a] |
| POS | 8.83 ± 0.19 [a] | 1.97 ± 0.02 [a,b] | 1.33 ± 0.11 [a] | 5.75 ± 0.61 [b] | 0.71 ± 0.10 [a] |
| 5 MIN | 9.00 ± 0.28 [a] | 2.11 ± 0.05 [a] | 1.30 ± 0.29 [a] | 6.46 ± 0.29 [a,b] | 0.68 ± 0.05 [a] |
| 20 MIN | 9.96 ± 0.29 [a] | 1.86 ± 0.03 [b] | 1.38 ± 0.08 [a] | 6.15 ± 0.16 [a,b] | 0.59 ± 0.03 [a] |

Data are given as mean ± SD values from experiments performed in triplicate. [1] Mean values with different letters within the same row are significantly different according to the non-parametric Friedman test, followed by Dunn's test ($p < 0.05$).

### 3.6. Stress Hormones

Stress hormones, including cortisol and serotonin, were measured using ELISA kits. Cortisol levels in the control group were the highest among all groups (Table 6). Inhalation of BEO and linalool decreased the cortisol levels; however, these decreases were not statistically significant. The levels of serotonin in the control group were the lowest among all groups. Inhalation of BEO and linalool induced an increase of serotonin; however, these increases showed no significant differences.

**Table 6.** Changes in stress hormones in rats.

| | Cortisol (ng/mL) | Serotonin (ng/mL) |
|---|---|---|
| CON | 25.34 ± 1.12 [a1] | 8.25 ± 0.90 [a] |
| POS | 20.99 ± 8.96 [a] | 9.72 ± 0.38 [a] |
| 5 MIN | 24.77 ± 3.14 [a] | 9.09 ± 1.01 [a] |
| 20 MIN | 24.16 ± 4.24 [a] | 9.34 ± 0.86 [a] |

Data are given as mean ± SD values from experiments performed in triplicate. [1] Mean values with different letters within the same row are significantly different according to the non-parametric Friedman test, followed by Dunn's test ($p < 0.05$).

### 3.7. Serum Biomarkers and Hepatotoxicity Indicators

Serum biomarkers were measured using a commercial kit. TC in the 20 min BEO-inhaled group was lower than in the control group ($p < 0.05$) (Table 7). However, there were no significant differences when comparing the control, 5 min BEO-inhaled, and linalool-inhaled groups. In the case of HDL levels, the control and linalool-inhaled groups had the lowest levels compared with the BEO-inhaled groups, regardless of the BEO inhalation time ($p < 0.05$). Thus, BEO inhalation upregulated the HDL levels. The control group had the highest LDL levels among all groups ($p < 0.05$). BEO inhalation ameliorated the levels of LDL, and linalool ameliorated the LDL levels compared to the control group ($p < 0.05$). The TG level in the control group was relatively higher than in the other groups ($p > 0.05$). BEO and linalool inhalation were associated with a decreasing tendency of TG levels. Inhalation of BEO for 20 min showed decreased the TG levels compared to the control group ($p < 0.05$). Meanwhile, linalool was associated with a decreasing trend in TG levels compared to the control group; however, there were no significant changes between the control and the linalool-inhaled group. Regarding the atherogenic index (AI) and cardiac risk factors (CRF) in the control group, inhalation of BEO ameliorated the AI and CRF indices in a time-dependent manner ($p < 0.05$). Inhalation of linalool induced no significant effects on AI or CRF. Inhalation of BEO and linalool significantly decreased ($p < 0.05$) the levels of LHR when compared with the control group. In particular, BEO inhalation for 20 min showed the lowest LHR among all the groups ($p < 0.05$). Hepatotoxicity indicators, including AST and ALT, were measured using a commercial kit. When comparing all groups, the AST and ALT levels did not show any significant differences (Table 8). In addition, the AST/ALT ratio was not significantly different among the groups.

**Table 7.** Effects of basil essential oil and linalool inhalation on lipid metabolism in chronically stressed rats. CON: chronic stress-exposed control group; POS: linalool (positive control) inhalation by chronic stress-exposed rats; 5 MIN: BEO inhalation for 5 min by chronic stress-exposed rats; 20 MIN: BEO inhalation for 20 min by chronic stress-exposed rats.

| | TC (mg/dL) | HDL (mg/dL) | LDL (mg/dL) | TG (mg/dL) | AI | CRF | LHR (mg/dL) |
|---|---|---|---|---|---|---|---|
| CON | 131.23 ± 6.29 [a1] | 46.79 ± 2.83 [b] | 84.43 ± 3.47 [a] | 46.78 ± 5.33 [a] | 1.81 ± 0.26 [a] | 2.81 ± 0.26 [a] | 180.71 ± 9.05 [a] |
| POS | 121.62 ± 4.43 [ab] | 47.17 ± 2.88 [b] | 73.78 ± 1.51 [b] | 40.81 ± 3.10 [ab] | 1.58 ± 0.07 [ab] | 2.58 ± 0.07 [ab] | 156.84 ± 11.28 [b] |
| 5 MIN | 131.10 ± 3.28 [a] | 55.34 ± 1.60 [a] | 75.76 ± 1.97 [b] | 43.32 ± 1.54 [a] | 1.37 ± 0.05 [b] | 2.37 ± 0.05 [b] | 136.91 ± 1.76 [b] |
| 20 MIN | 110.30 ± 0.59 [b] | 56.46 ± 2.70 [a] | 53.94 ± 2.72 [c] | 34.36 ± 0.17 [b] | 0.97 ± 0.10 [c] | 1.97 ± 0.10 [c] | 95.83 ± 9.36 [c] |

Data are given as mean ± SD values from experiments performed in triplicate. [1] Mean values with different letters within the same row are significantly different according to the non-parametric Friedman test, followed by Dunn's test ($p < 0.05$).

**Table 8.** Effects of basil essential oil and linalool inhalation on hepatotoxicity in chronically stressed rats. CON: chronic stress-exposed control group; POS: linalool (positive control) inhalation with chronic stress-exposed rats; 5 MIN: BEO inhalation for 5 min b chronic stress-exposed rats; 20 MIN: BEO inhalation for 20 min by chronic stress-exposed rats.

| | AST (Karmen/mL) | ALT (Karmen/mL) | AST/ALT |
|---|---|---|---|
| CON | 153.16 ± 1.82 [a1] | 25.24 ± 0.07 [a] | 6.07 ± 0.09 [a] |
| POS | 155.65 ± 2.67 [a] | 24.84 ± 0.58 [a] | 6.27 ± 0.23 [a] |
| 5 MIN | 151.94 ± 1.43 [a] | 24.05 ± 0.55 [a] | 6.18 ± 0.03 [a] |
| 20 MIN | 151.34 ± 1.89 [a] | 24.58 ± 0.12 [a] | 6.29 ± 0.14 [a] |

Data are given as mean ± SD values from experiments performed in triplicate. [1] Mean values with different letters within the same row are significantly different according to the non-parametric Friedman test, followed by Dunn's test ($p < 0.05$).

Oxidative stress influences the lipid metabolism [37] and usually increases the prevalence of atherosclerosis by increasing reactive oxygen species, nitric oxygen, and oxidized-LDL production and decreasing the levels of antioxidants [38]. LDL is generally associated with the weight of WAT, and increased levels of LDL can promote cardiovascular diseases such as atherosclerosis and dyslipidemia [5]. In contrast, HDL is associated with anti-inflammatory indicators and the presence of antioxidants [5]. Accordingly, HDL plays an important role in cardiovascular health and can reduce the prevalence rate of cardiovascular diseases [39,40]. The results of this study showed that linalool and BEO inhalation decreased the LDL levels in chronically stressed rats. In addition, BEO inhalation ameliorated the levels of HDL in chronically stressed rats, regardless of its concentration. Therefore, BEO and linalool inhalation upregulated the LDL and HDL levels in this study. In particular, the AI and CRF were reduced by BEO inhalation regardless of its concentration (Table 7). A recent study reported that the administration of BEO decreased the levels of TC, LDL, and TG [7]. Similarly, the administration of purple basil essential oil improved hyperlipidemia, lowering triglyceride and total cholesterol levels, similar to our results [31]. Additionally, linalool has cholesterol-lowering, antioxidant, and anti-inflammatory activities [41]. Previous research showed that linalool decreased the LDL levels [6] and activated hepatic PPAR$\alpha$ [21]. Therefore, linalool ameliorated dyslipidemia by lowering LDL activity [21]. BEO, a linalool-containing essential oil, has been reported to reduce hyperlipidemia and oxidative stress in rats [31]. Therefore, linalool inhalation played an important role in improving the LDL levels in this study.

The AST and ALT levels are associated with hepatic damage. The AST/ALT ratio is an indicator of liver function impairment [42]. In this study, there were no significant differences in the levels of AST and ALT and in the AST/ALT ratios among all groups. Therefore, inhalation of BEO and linalool had no adverse effects on hepatic and liver function.

## 4. Conclusions

In conclusion, these findings suggest that BEO and linalool inhalation suppresses stress responses, including dyslipidemia. Nevertheless, these findings are limited to animal models of chronic stress. Therefore, further research should be performed to investigate the effects of the inhalation of BEO and linalool in clinical trials. Furthermore, these findings can be of interest in the industry field and suggest the use of odor-active aroma compounds in BEO and linalool to suppress stress, without the intake and/or oral administration of health-promoting compounds.

**Author Contributions:** Formal Analysis and Writing-Original Draft Preparation, D.-S.K. and S.-J.H.; Formal Analysis and Writing-Original Draft Preparation, S.Y.; Data Curation, S.-M.J. and H.J.; Writing-Review & Editing, M.-Y.Y., Y.-J.K. and J.-K.K.; Supervision, Project Administration, and Funding Acquisition, E.-C.S. All authors have read and agreed to the published version of the manuscript.

**Funding:** This study was funded by the Basic Science Research Program, through the National Research Foundation of Korea (NRF) funded by the Ministry of Education (NRF-2018R1D1A1B07045431).

**Institutional Review Board Statement:** Not applicable.

**Informed Consent Statement:** Not applicable.

**Data Availability Statement:** The data presented in this study are available.

**Conflicts of Interest:** The authors declare no conflict of interest.

## References

1. Ochoa-Amaya, J.E.; Malucelli, B.E.; Cruz-Casallas, P.E.; Nasello, A.G.; Felicio, L.F.; Carvalho-Freitas, M.I.R. Acute and chronic stress and the inflammatory response in hyperprolactinemic rats. *Neuroimmunomodulation* **2010**, *17*, 386–395. [CrossRef] [PubMed]
2. Piskunov, A.; Stepanichev, M.; Tishkina, A.; Novikova, M.; Levshina, I.; Gulyaeva, N. Chronic combined stress induces selective and long-lasting inflammatory response evoked by changes in corticosterone accumulation and signaling in rat hippocampus. *Metab. Brain Dis.* **2016**, *31*, 445–454. [CrossRef] [PubMed]
3. Pacák, K.; Palkovits, M. Stressor specificity of central neuroendocrine responses: Implications for stress-related disorders. *Endocr. Rev.* **2001**, *22*, 502–548. [CrossRef] [PubMed]
4. Steptoe, A.; Kivimäki, M. Stress and cardiovascular disease. *Nat. Rev. Cariol.* **2012**, *9*, 360–370. [CrossRef]
5. Hong, S.J.; Cho, J.; Boo, C.G.; Youn, M.Y.; Pan, J.H.; Kim, J.K.; Shin, E.C. Inhalation of patchouli (*Pogostemon Cablin* Benth.) essential oil improved metabolic parameters in obesity-induced sprague dawley rats. *Nutrients* **2020**, *12*, 2077. [CrossRef]
6. Baek, H.; Kim, S.; Lee, I.; Kang, S.; Yoo, J.; Yoon, W.; Kim, Y.; Kim, H.; Kim, J. Anti-obesity and anti-lipidemic effects of linalool in high-fat diet-induced obese mice. *J. Biomed. Res.* **2012**, *13*, 229–235. [CrossRef]
7. Harnafi, H.; Aziz, M.; Amrani, S. Sweet basil (*Ocimum basilicum* L.) improves lipid metabolism in hypercholesterolemic rats. *e-SPEN Eur. J. Clin. Nutr.* **2009**, *4*, e181–e186. [CrossRef]
8. de Moura Linck, V.; da Silva, A.L.; Figueiró, M.; Piato, A.L.; Herrmann, A.P.; Birck, F.D.; Caramão, E.B.; Nunes, D.S.; Moreno, P.R.H.; Elisabetsky, E. Inhaled linalool-induced sedation in mice. *Phytomedicine* **2009**, *16*, 303–307. [CrossRef]
9. Şanli, A.; Karadoğan, T. Geographical impact on essential oil composition of endemic *Kundmannia anatolica* Hub.-Mor.(Apiaceae). *Afr. J. Tradit. Complement. Altern. Med.* **2017**, *14*, 131–137. [CrossRef]
10. Maria, G.A.; Riccardo, N. Citrus bergamia, Risso: The peel, the juice and the seed oil of the bergamot fruit of *Reggio Calabria* (South Italy). *Emir. J. Food Agric.* **2020**, *32*, 522–532. [CrossRef]
11. Gioffrè, G.; Ursino, D.; Labate, M.L.C.; Giuffrè, A.M. The peel essential oil composition of bergamot fruit (*Citrus bergamia*, Risso) of Reggio Calabria (Italy): A review. *Emir. J. Food Agric.* **2020**, *32*, 835–845. [CrossRef]
12. Inan, M.; Kirpik, M.; Kaya, D.A.; Kirici, S. Effect of harvest time on essential oil composition of *Thymbra spicata* L. growing in flora of Adıyaman. *Adv. Environ. Biol.* **2011**, *5*, 356–358.
13. Chatterjee, S.; Gupta, S.; Variyar, P.S. Comparison of essential oils obtained from different extraction techniques as an aid in identifying aroma significant compounds of nutmeg (*Myristica fragrans*). *Nat. Prod. Commun.* **2015**, *10*, 1443–1446. [CrossRef]
14. Türkmen, M. The effect of different Phenological periods and harvest times on the essential oil ratio and components of basil genotypes. *J. Essent. Oil-Bear. Plants.* **2021**, *24*, 94–109. [CrossRef]
15. Coelho, J.; Veiga, J.; Karmali, A.; Nicolai, M.; Pinto Reis, C.; Nobre, B.; Palavra, A. Supercritical CO2 extracts and volatile oil of basil (*Ocimum basilicum* L.) comparison with conventional methods. *Separations* **2018**, *5*, 21. [CrossRef]
16. Firestein, S. How the olfactory system makes sense of scents. *Nature* **2001**, *413*, 211–218. [CrossRef]
17. Riera, C.E.; Tsaousidou, E.; Halloran, J.; Follett, P.; Hahn, O.; Pereira, M.M.; Ruud, L.E.; Alber, J.; Tharp, K.; Anderson, C.M.; et al. The sense of smell impacts metabolic health and obesity. *Cell Metab.* **2017**, *26*, 198–211. [CrossRef]

18. da Silva Haeser, A.; Sitta, A.; Barschak, A.G.; Deon, M.; Barden, A.T.; Schmitt, G.O.; Landgraff, S.; Gomez, R.; Barros, H.M.T.; Vargas, C.R. Oxidative stress parameters in diabetic rats submitted to forced swimming test: The clonazepam effect. *Brain Res.* **2007**, *1154*, 137–143. [CrossRef]
19. Atescelik, M.; Yilmaz, M.; Korkmaz, S.; Goktekin, M.C.; Gurger, M.; Ilhan, N. The relationship between ghrelin and copeptin levels, and anxiety and depression levels in suicide attempts. *Clin. Psychopharmacol. Neurosci.* **2017**, *15*, 256. [CrossRef]
20. Dora, M.F.; Taha, N.M.; Lebda, M.A.; Hashem, A.E.; Elfeky, M.S.; El-Sayed, Y.S.; Jaouni, S.A.; El-Far, A.H. Quercetin attenuates brain oxidative alterations induced by iron oxide nanoparticles in rats. *Int. J. Mol. Sci.* **2021**, *22*, 3829. [CrossRef]
21. Jun, H.; Lee, J.H.; Kim, J.; Jia, Y.; Kim, K.H.; Hwang, K.Y.; Yun, E.J.; Do, K.; Lee, S. Linalool is a PPARα ligand that reduces plasma TG levels and rewires the hepatic transcriptome and plasma metabolome. *J. Lipid Res.* **2014**, *55*, 1098–1110. [CrossRef] [PubMed]
22. Souto-Maior, F.N.; de Carvalho, F.L.; de Morais, L.C.S.L.; Netto, S.M.; de Sousa, D.P.; de Almeida, R.N. Anxiolytic-like effects of inhaled linalool oxide in experimental mouse anxiety models. *Pharmacol. Biochem. Behav.* **2011**, *100*, 259–263. [CrossRef] [PubMed]
23. Eggersdorfer, M. Terpenes. In *Ullmann's Encyclopedia of Industrial Chemistry*; Wiley-VCH Verlag GmbH & Co. KGaA: Weinheim, Germany, 2000; pp. 29–45. [CrossRef]
24. Da Fonsêca, D.V.; da Silva Maia Bezerra Filho, C.; Lima, T.C.; de Almeida, R.N.; de Sousa, D.P. Anticonvulsant essential oils and their relationship with oxidative stress in epilepsy. *Biomolecules* **2019**, *9*, 835. [CrossRef] [PubMed]
25. Padalia, R.C.; Verma, R.S. Comparative volatile oil composition of four *Ocimum* species from northern India. *Nat. Prod. Res.* **2011**, *25*, 569–575. [CrossRef] [PubMed]
26. Riyazi, S.R.; Ebrahimnezhad, Y.; Hosseini, S.A.; Meimandipour, A.; Ghorbani, A. Comparison of the effects of basil (*Ocimum basilicum*) essential oil, avilamycin and protexin on broiler performance, blood biochemistry and carcass characteristics. *Arch. Anim. Breed.* **2015**, *58*, 425–432. [CrossRef]
27. Sentari, M.; Harahap, U.; Sapiie, T.W.A.; Ritarwan, K. Blood cortisol level and blood serotonin level in depression mice with basil leaf essential oil treatment. *Open Access Maced. J. Med. Sci.* **2019**, *7*, 2652. [CrossRef]
28. Shen, J.; Niijima, A.; Tanida, M.; Horii, Y.; Maeda, K.; Nagai, K. Olfactory stimulation with scent of lavender oil affects autonomic nerves, lipolysis and appetite in rats. *Neurosci. Lett.* **2005**, *383*, 188–193. [CrossRef]
29. Aschbacher, K.; Kornfeld, S.; Picard, M.; Puterman, E.; Havel, P.J.; Stanhope, K.; Lustig, R.H.; Epel, E. Chronic stress increases vulnerability to diet-related abdominal fat, oxidative stress, and metabolic risk. *Psychoneuroendocrinology* **2014**, *46*, 14–22. [CrossRef]
30. Nakamura, A.; Fujiwara, S.; Matsumoto, I.; Abe, K. Stress repression in restrained rats by (R)-(−)-linalool inhalation and gene expression profiling of their whole blood cells. *J. Agric. Food Chem.* **2009**, *57*, 5480–5485. [CrossRef]
31. Gökçe, Y.; Kanmaz, H.; Er, B.; Sahin, K.; Hayaloglu, A.A. Influence of purple basil (*Ocimum basilicum* L.) extract and essential oil on hyperlipidemia and oxidative stress in rats fed high-cholesterol diet. *Food Biosci.* **2021**, *43*, 101228. [CrossRef]
32. Umezu, T. Evaluation of central nervous system acting effects of plant-derived essential oils using ambulatory activity in mice. *J. Pharm. Pharmacol.* **2013**, *4*, 29704. [CrossRef]
33. Tanida, M.; Niijima, A.; Shen, J.; Nakamura, T.; Nagai, K. Olfactory stimulation with scent of lavender oil affects autonomic neurotransmission and blood pressure in rats. *Neurosci. Lett.* **2006**, *398*, 155–160. [CrossRef] [PubMed]
34. Ch, M.A.; Naz, S.B.; Sharif, A.; Akram, M.; Saeed, M.A. Biological and pharmacological properties of the sweet basil (*Ocimum basilicum*). *J. Pharm. Res. Int.* **2015**, *7*, 330–339. [CrossRef]
35. Cheng, B.H.; Sheen, L.Y.; Chang, S.T. Hypolipidemic effects of S-(+)-linalool and essential oil from *Cinnamomum osmophloeum* ct. linalool leaves in mice. *J. Tradit. Complement. Med.* **2018**, *8*, 46–52. [CrossRef] [PubMed]
36. Hirano, N.; Sakamoto, K. Linalool odor stimulation improves heat stress tolerance and decreases fat accumulation in nematodes. *Biosci. Biotechnol. Biochem.* **2019**, *83*, 148–154. [CrossRef] [PubMed]
37. Manting, L.; Haihong, Z.; Jing, L.; Shaodong, C.; Yihua, L. The model of rat lipid metabolism disorder induced by chronic stress accompanying high-fat-diet. *Lipids Health Dis.* **2011**, *10*, 1–4. [CrossRef]
38. Kattoor, A.J.; Pothineni, N.V.K.; Palagiri, D.; Mehta, J.L. Oxidative stress in atherosclerosis. *Curr. Atheroscler. Rep.* **2017**, *19*, 42. [CrossRef]
39. Zhao, H.L.; Cho, K.H.; Ha, Y.W.; Jeong, T.S.; Lee, W.S.; Kim, Y.S. Cholesterol-lowering effect of platycodin D in hypercholesterolemic ICR mice. *Eur. J. Pharmacol.* **2006**, *537*, 166–173. [CrossRef]
40. Myers, M.G.; Cowley, M.A.; Münzberg, H. Mechanisms of leptin action and leptin resistance. *Annu. Rev. Physiol.* **2008**, *70*, 537–556. [CrossRef]
41. Aprotosoaie, A.C.; Hăncianu, M.; Costache, I.I.; Miron, A. Linalool: A review on a key odorant molecule with valuable biological properties. *Flavour Fragr. J.* **2014**, *29*, 193–219. [CrossRef]
42. Ahn, H.; Go, G.W. Pinus densiflora bark extract (*PineXol*) decreases adiposity in mice by down-regulation of hepatic de novo lipogenesis and adipogenesis in white adipose tissue. *J. Microbiol. Biotechnol.* **2017**, *27*, 660–667. [CrossRef] [PubMed]

*nutrients*

*Article*

# Effect of Acute Dietary Nitrate Supplementation on the Venous Vascular Response to Static Exercise in Healthy Young Adults

**Anna Oue [1,*], Yasuhiro Iimura [2], Akiho Shinagawa [2], Yuichi Miyakoshi [1] and Masako Ota [1]**

1 Faculty of Food and Nutritional Sciences, Toyo University, Gunma 374-0193, Japan
2 Graduate School of Food and Nutritional Sciences, Toyo University, Gunma 374-0193, Japan
* Correspondence: oue@toyo.jp; Tel.: +81-276-82-9145; Fax: +81-276-82-9033

**Abstract:** The purpose of this study was to test the hypothesis that acute intake of inorganic nitrate ($NO_3^-$) via supplementation would attenuate the venoconstriction and pressor response to exercise. Sixteen healthy young adults were assigned in a randomized crossover design to receive beetroot juice (BRJ) or an $NO_3^-$-depleted control beverage (prune juice: CON). Two hours after consuming the allocated beverage, participants rested in the supine position. Following the baseline period of 4 min, static handgrip exercise of the left hand was performed at 30% of the maximal voluntary contraction for 2 min. Mean arterial pressure (MAP) and heart rate (HR) were measured. Changes in venous volume in the right forearm and right calf were also measured using venous occlusion plethysmography while cuffs on the upper arm and thigh were inflated constantly to 30–40 mmHg. The plasma $NO_3^-$ concentration was elevated with BRJ intake ($p < 0.05$). Exercise increased MAP and HR and decreased venous volume in the forearm and calf, but there were no differences between CON and BRJ. Thus, these findings suggest that acute BRJ intake does not alter the sympathetic venoconstriction in the non-exercising limbs and MAP response to exercise in healthy young adults, despite the enhanced activity of nitric oxide.

**Keywords:** beetroot juice; exercise; nitric oxide; sympathoexcitation; venoconstriction

**Citation:** Oue, A.; Iimura, Y.; Shinagawa, A.; Miyakoshi, Y.; Ota, M. Effect of Acute Dietary Nitrate Supplementation on the Venous Vascular Response to Static Exercise in Healthy Young Adults. *Nutrients* **2022**, *14*, 4464. https://doi.org/10.3390/nu14214464

Academic Editor: Daniela Rigano

Received: 10 October 2022
Accepted: 21 October 2022
Published: 24 October 2022

**Publisher's Note:** MDPI stays neutral with regard to jurisdictional claims in published maps and institutional affiliations.



## 1. Introduction

Dietary nitrate ($NO_3^-$) supplementation with beetroot juice (BRJ) reduces resting blood pressure (BP) in normotensive and hypertensive populations [1–6]. The hypotensive effect of BRJ is likely to be due to both peripheral and central factors. The peripheral factor underlying the hypotensive effect might be related to the vasodilatory impact of the increase in nitric oxide (NO) bioavailability induced by the stepwise reduction of dietary nitrate ($NO_3^-$) to nitrite ($NO_2^-$) and subsequently to NO (i.e., the $NO_3^- \rightarrow NO_2^- \rightarrow$ NO pathway) [4,7,8]. In addition to the role of NO as a vasodilator, the increase in NO availability related to the $NO_3^- \rightarrow NO_2^- \rightarrow$ NO pathway may also alter efferent sympathetic outflow. Indeed, Notay et al. [9] recently reported that acute dietary $NO_3^-$ supplementation with BRJ decreased muscle sympathetic nerve activity (MSNA) at rest and blunted the MSNA response to sympathoexcitation via static handgrip exercise in young adults.

Veins have high distensibility and contain approximately 60–70% of the total blood volume at rest [10]. Venous tone is controlled by the sympathetic nervous system and changes with physiological stress (e.g., exercise), thereby leading to alterations in venous volume and/or compliance and contributing to the control of circulatory responses (e.g., BP and cardiac output) [10]. For example, venoconstriction and/or decreased venous compliance occurs sympathetically in the non-exercising limb during exercise [11–16]. In addition to sympathetic control, NO is also an important signaling molecule that contributes to the modulation of venous tone [17,18]. Moreover, in our recent study [19], it is suggested that the increased bioavailability of vasodilator NO associated with dietary $NO_3^-$ supplementation with BRJ could contribute to the control of the peripheral vascular tone in not only arteries but also veins under resting conditions, which may perhaps, in part, be attributable to the hypotensive effect of BRJ. However, the effect of BRJ intake on the venous vascular response to exercise-induced sympathoexcitation has not been investigated. Aging and physiological inactivity might cause stiffness of the veins [20,21], which could be a factor in the pathogenesis of hypertension [22]. Considering these findings, it is also important to understand the hypotensive effect of BRJ from the perspective of the venous vascular system.

Therefore, the purpose of this study was to investigate the effect of acute dietary $NO_3^-$ supplementation with BRJ on venous vascular and circulatory responses to static handgrip exercise in humans. Because the scientific evidence indicates that the $NO_3^- \rightarrow NO_2^- \rightarrow NO$ pathway-related increase in NO bioavailability in response to acute dietary $NO_3^-$ supplementation with BRJ causes vasodilation [4,7,8] and attenuates the MSNA response to static handgrip exercise [9], we hypothesized that acute intake of BRJ would attenuate the venoconstriction and BP response to exercise.

## 2. Materials and Methods

### 2.1. Participants

Sixteen healthy individuals (10 men, 6 women, 22.3 ± 1.4 years, 167.4 ± 9.3 cm, 64.7 ± 13.9 kg) volunteered for this study. None of the women were using oral contraceptives and all were in the self-reported follicular phase (3–10 days after the onset of menstruation) during the experiments. This study was approved by the Human Ethics Committee of Toyo University (TU2019-018-TU2020-H-019) and was conducted in accordance with the Declaration of Helsinki. The purpose, procedure, and risks of the study were explained to the participants, and their written and verbal informed consent was obtained. Throughout the study, the participants were instructed to avoid vigorous exercise, caffeine, and alcohol for 24 h before each visit. For 3 days prior to the main experimental protocol and blood sampling protocol, participants were asked to refrain from high $NO_3^-$ foods (e.g., green leafy vegetables and traditional Japanese foods) [23,24]. In addition, because oral bacteria are involved in reducing $NO_3^-$ to $NO_2^-$ in vivo [25], the participants were asked to abstain from using mouthwash.

### 2.2. Experimental Design

Participants visited the laboratory on five occasions. During the first visit, they performed maximal voluntary contraction (MVC) of the left hand using a handgrip dynamometer to determine their 30% MVC. In addition, they were familiarized with the experimental procedure and equipment. Main experiments were carried out at the second and third visits, and blood samplings were performed at the fourth and fifth visits. For the main experiments and blood samplings, participants were assigned in a randomized crossover design to consume BRJ (Beet It®; James White Drinks, Ipswich, UK; 140 mL/day, containing ~8 mmol $NO_3^-$) or placebo control beverage (CON) consisting of prune juice (Sunsweet®; POKKA SAPPORO Food & Beverage Ltd., Nagoya, Japan; 166 mL/day; <0.01 mmol $NO_3^-$). We selected prune juice as the placebo beverage because the $NO_3^-$ in prune juice is at a negligible level and the carbohydrate and fiber contents of prune juice are similar to those of BRJ [26,27]. The amount (166 mL/day) of prune juice was calculated to match the energy contained in 140 mL/day of BRJ. A washout period of at least 7 days separated each supplementation period for both the main experiments and the blood sampling.

### 2.3. Protocol of the Main Experiment

After arrival at the laboratory, participants were instructed to consume BRJ or CON. Two hours later, all participants rested in the supine position for 20 min before data acquisition in an air-conditioned room (26.4 °C ± 0.5 °C). The main experiment comprised two protocols: (1) measurement of the changes in venous volume in the non-exercising limbs during the static handgrip exercise, and (2) measurement of the circulatory parameters during static handgrip exercise. In protocol 1, following the pre-exercise baseline period for 4 min, the static handgrip exercise of the left hand was performed at 30% MVC for 2 min. The cuffs on the right wrist and the right ankle were inflated to 200–220 mmHg at the same time as the start of the baseline period, and the cuffs on the right upper arm and the right thigh were inflated to 30–40 mmHg from the first minute of the baseline period. All cuff inflations were maintained until the end of the exercise. Throughout protocol 1, the changes in venous volume in the non-exercising right forearm and right calf were measured using venous occlusion plethysmography (Hokanson, EC6; D. E. Hokanson, Bellevue, WA, USA). In protocol 2 as well as in protocol 1, static handgrip exercise of the left hand at 30% MVC for 2 min was performed following the 4-min baseline period. Throughout protocol 2, the circulatory parameters were measured in the right middle finger using Finapres NOVA (Finapres Medical Systems BV, Enschede, The Netherlands). Protocol 1 and protocol 2 were performed in random order, and rest periods of at least 20 min were allowed between protocols 1 and 2. In addition, for all protocols, participants controlled their respiratory rate at 10 or 15 breaths per minute, guided by a metronome.

### 2.4. Blood Sampling Protocol

To measure the plasma $NO_3^-$ concentration, venous blood samples were drawn from an antecubital vein in a seated position before and 2 h after BRJ or CON consumption on a different day from the main experimental protocols.

### 2.5. Measurements

Systolic blood pressure (SBP), diastolic blood pressure (DBP), and mean arterial pressure (MAP) were measured noninvasively and continuously from the right middle finger using Finapres NOVA, which was calibrated with the right upper arm cuff and height adjustment. In addition, heart rate (HR) was determined from the BP waveform using the Modelflow software program (Finapres Medical Systems BV, Enschede, The Netherlands).

To assess venoconstriction in the non-exercising forearm and calf during static handgrip exercise of the left hand, the changes in volume in the right forearm and right calf were measured. Inflatable cuffs were wrapped around the right wrist, right upper arm, right ankle, and right thigh, and strain gauges were placed on the sites of maximal thickness in the forearm and calf. Throughout the protocol, the wrist and ankle cuffs were inflated to 200–220 mmHg to arrest the blood circulation of the hand and foot, which have arteriovenous anastomoses, because we wanted to investigate the venoconstriction in the forearm and calf. One minute after the cuff inflation of the wrist and ankle, the cuffs of the upper arm and thigh were inflated to 30–40 mmHg for 3 min, and the volume in the forearm and calf increased until it approached an asymptote, and then static handgrip exercise of the left hand was performed for 2 min. Throughout the protocol (total time, 6 min), the change in volume in the right forearm and the right calf was measured using venous occlusion plethysmography. This method was adopted because when the intravascular pressure of the conduit vein in the limb is maintained at a constant level, the decrease in the limb volume reflects the venoconstriction or the elevated venous vascular tone [15,16]. In addition, the cuff inflation pressure can be considered equivalent to the intravascular pressure of the conduit vein in the limb [28].

Venous blood samples were immediately mixed with EDTA and centrifuged at 3000× $g$ rpm for 10 min. Plasma was placed in microcentrifuge tubes and frozen for the subsequent analysis of plasma $NO_3^-$ and $NO_2^-$ concentrations [29]. The $NO_3^-$ in a sample is reduced by a cadmium column to $NO_2^-$, which reacts with a Griess reagent to form a purple azo dye. The $NO_3^-$ in a sample needs no reaction and thus reacts with the Griess reagent when the cadmium column is bypassed, as well as when it is used. The dye was developed in a 60 °C water bath, the sample was cooled by a 0 °C water bath, and its absorbance at 546 nm was detected using a flow-through UV-Vis spectrophotometer (V-750, JASCO Corporation, Tokyo, Japan). Because the plasma $NO_2^-$ levels were very low and below the quantifiable limit (1 μM) in all participants, only $NO_3^-$ concentration data are presented in this study.

*2.6. Data Analysis and Statistics*

Data are expressed as the mean ± standard deviation. A priori sample size calculation estimated a required sample of 12 participants, assuming a change in venous compliance of 0.031 ± 0.027 mL/dL of tissue/mmHg [30] in a crossover trial with an assigned $\alpha$ of 0.05 and $\beta$ of 0.2. Pre-exercise baseline values of SBP, DBP, MAP, and HR were defined as the mean value obtained from 1 to 3 min of the baseline period. In addition, pre-exercise baseline values of the forearm and calf venous volume were defined as the average of the last 10 s of the 3-min cuff inflation to 30–40 mmHg. The SBP, DBP, MAP, HR, and limb venous volume values during exercise at 2 min were obtained as the average of the last 1 min of exercise. Relative increases in SBP, DBP, MAP, and HR and relative decreases in venous volume in the forearm and calf with exercise at 2 min from the pre-exercise baseline were calculated. The relative decrease in venous volume was used as the index of venoconstriction.

To compare the changes in SBP, DBP, MAP, HR, and forearm venous volume and calf venous volume with exercise between CON and BRJ, a two-way analysis of variance (ANOVA) with repeated measurements (condition × time) was applied. If the main effect of condition (CON and BRJ), that of time (pre-exercise and during exercise), and/or an interaction effect were detected, post hoc analysis using a paired *t*-test was performed. To compare the plasma $NO_3^-$ concentration before and after BRJ or CON supplementation and the plasma $NO_3^-$ concentration between BRJ and CON, a paired *t*-test was used. In addition, we calculated the differences in the absolute MAP at 2 min of exercise between CON and BRJ, in the MAP elevation with exercise from pre-exercise (pressor response to exercise) between CON and BRJ, and in the relative decreases in venous volume in the limbs with exercise between BRJ and CON. Using these values, Spearman's rank correlation coefficients were calculated to examine the relationships between venoconstriction and MAP responses to exercise with acute ingestion of BRJ. Statistical significance was set at $p < 0.05$. All statistical analyses were performed using SPSS version 27 (IBM Corp., Armonk, NY, USA).

## 3. Results

Figure 1 shows the plasma $NO_3^-$ concentration before and after intake of CON and BRJ. The plasma $NO_3^-$ concentration increased in all participants after BRJ ingestion (before, 15 ± 6 μM; after, 574 ± 120 μM; $p < 0.05$). However, there was no significant change in the plasma $NO_3^-$ concentration after CON ingestion (before, 14 ± 3 μM; after, 14 ± 3 μM). Furthermore, there was a significant difference in the plasma $NO_3^-$ concentration between CON and BRJ after their intake ($p < 0.05$).

Static handgrip exercise under both CON and BRJ conditions caused similar increases in SBP, DBP, MAP, and HR (Table 1). ANOVA indicated a significant time effect (all $p < 0.01$), and post hoc testing revealed significant differences in SBP, DBP, MAP, and HR between pre-exercise and during exercise (all $p < 0.05$). However, SBP, DBP, MAP, and HR at pre-exercise or during exercise were similar for CON and BRJ. In addition, the degrees of increases in these parameters during exercise from pre-exercise did not differ between CON and BRJ (Figure 2A–D). In contrast, static handgrip exercise under both CON and BRJ conditions induced similar decreases in venous volume in the non-exercising forearm and calf (Table 1). ANOVA indicated a significant time effect (all $p < 0.01$), and post hoc testing showed significant differences in these parameters between pre-exercise and during static handgrip exercise (all $p < 0.05$). This result meant that sympathetic venoconstriction was obtained during exercise. However, the venous volumes in the forearm and calf pre-exercise or during exercise were similar for CON and BRJ. In addition, the degrees of the decreases in the venous volume (venoconstriction) in the forearm and calf during exercise from pre-exercise did not differ between CON and BRJ (Figure 2E,F)

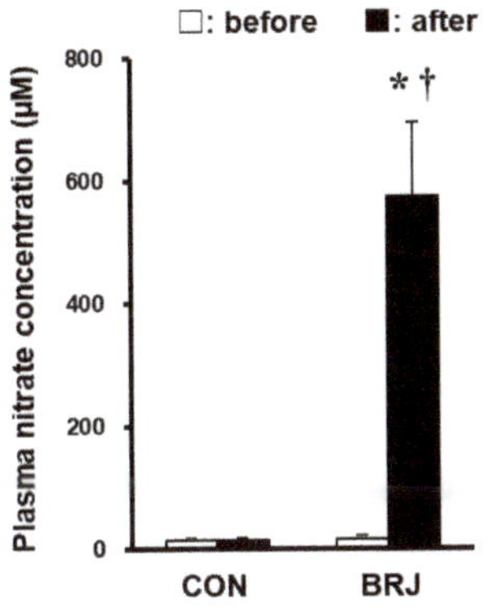


**Figure 1.** Plasma nitrate concentration before and 2h after CON and BRJ consumption. Data are shown as the mean ± standard deviation. CON: $NO_3^-$-depleted prune juice; BRJ: beetroot juice. * $p < 0.05$, significant difference between before and after. † $p < 0.05$, significant difference between CON and BRJ. Data are expressed as values for n = 16 (10 men and 6 women).

**Table 1.** Changes in circulatory parameters and venous volume in non-exercising limbs with static handgrip exercise in CON and BRJ.

| | Pre-Exercise | During Exercise at 2 min | ANOVA *p* |
|---|---|---|---|
| | SBP, mmHg | | |
| CON | 126 ± 13 | 139 ± 15 * | Condition: 0.346<br>Time: 0.001 |
| BRJ | 123 ± 13 | 139 ± 15 * | Condition × Time: 0.346 |
| | DBP, mmHg | | |
| CON | 75 ± 6 | 87 ± 7 * | Condition: 0.445<br>Time: 0.001 |
| BRJ | 73 ± 7 | 86 ± 11 * | Condition × Time: 0.203 |
| | MAP, mmHg | | |
| CON | 95 ± 8 | 110 ± 11 * | Condition: 0.334<br>Time: 0.001 |
| BRJ | 93 ± 9 | 109 ± 14 * | Condition × Time: 0.283 |
| | HR, bpm | | |
| CON | 65 ± 13 | 71 ± 12 * | Condition: 0.339<br>Time: 0.001 |
| BRJ | 64 ± 11 | 70 ± 12 * | Condition × Time: 0.767 |
| | Forearm venous volume, mL/dL of tissue | | |
| CON | 1.63 ± 0.53 | 1.52 ± 0.54 * | Condition: 0.682<br>Time: 0.001 |
| BRJ | 1.69 ± 0.40 | 1.54 ± 0.40 * | Condition × Time: 0.369 |
| | Calf venous volume, mL/dL of tissue | | |
| CON | 2.02 ± 0.76 | 1.90 ± 0.79 * | Condition: 0.502<br>Time: 0.001 |
| BRJ | 2.11 ± 0.70 | 2.01 ± 0.72 * | Condition × Time: 0.607 |

Values are mean ± standard deviation. SBP: systolic blood pressure, DBP: diastolic blood pressure, MAP: mean arterial pressure, HR: heart rate. * $p < 0.05$, significant difference between pre-exercise and during exercise at 2 min. Data are expressed as values for n = 16 (10 men, 6 women).

In the individual data, the difference in the decrease in forearm venous volume (venoconstriction) during exercise between BRJ and CON had no significant relationship with the MAP response to exercise (Figure 3A,B). In contrast, a smaller venoconstriction in the calf during exercise with BRJ rather than CON was significantly associated with a lower MAP elevation with exercise with BRJ but not with absolute MAP at 2 min of exercise (Figure 3C,D).

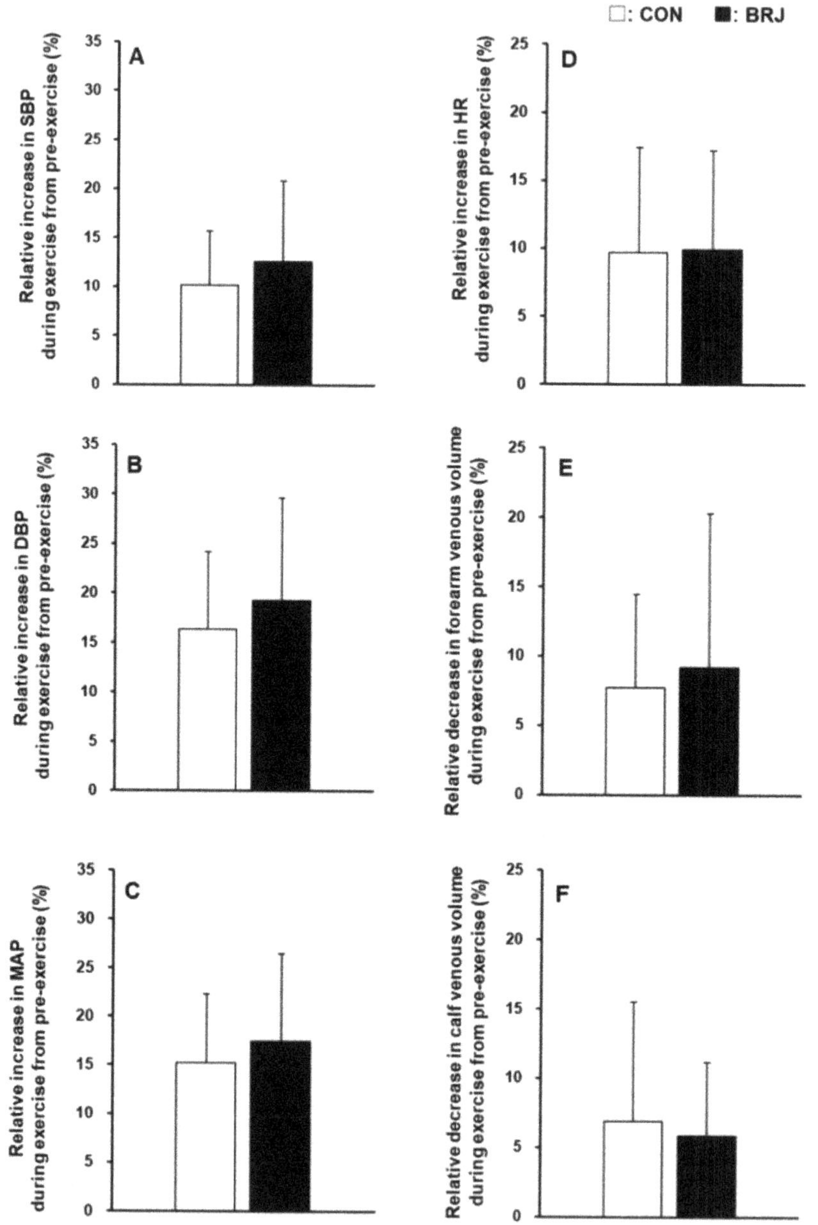


**Figure 2.** Relative changes in circulatory parameters and venous volume in the non-exercising limbs with static handgrip exercise in CON and BRJ. Percentage increases in the systolic blood pressure (SBP; (**A**)), diastolic blood pressure (DBP; (**B**)), mean arterial pressure (MAP; (**C**)), and heart rate (HR, (**D**)) and percentage decreases in venous volume in the forearm (**E**) and calf (**F**) during exercise from pre-exercise. Data are shown as the mean ± standard deviation. CON: $NO_3^-$-depleted prune juice; BRJ: beetroot juice. Data are expressed as values for n = 16 (10 men and 6 women).

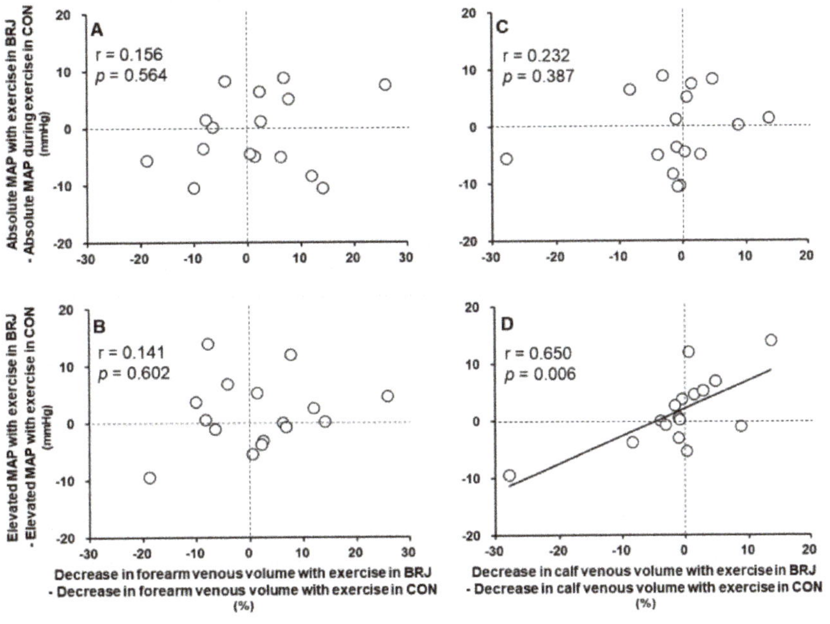


**Figure 3.** Relationship between the mean arterial pressure response and venous vascular response to exercise. Changes in the absolute mean arterial pressure (MAP) at 2 min of exercise (**A**,**C**) and elevated MAP with exercise (**B**,**D**) from BRJ to CON were plotted against changes in decreased venous volume in the non-exercising forearm and calf with exercise from BRJ to CON. Associations were determined with the use of a Spearman's rank correlation coefficient assessment. Data are expressed as values for n = 16 (10 men and 6 women).

## 4. Discussion

The new findings in our study are (1) that the increases in SBP, DBP, MAP, and HR and decreases in the venous volume in the non-exercising forearm and calf during static handgrip exercise did not differ between CON and BRJ, despite the elevated plasma $NO_3^-$ concentration after BRJ ingestion, and (2) that, in the individual data, a greater attenuation in the decrease in calf venous volume with exercise after BRJ ingestion was significantly associated with a lower MAP elevation with exercise after BRJ ingestion but not with the absolute MAP at 2 min of exercise. These results suggest that the enhanced NO bioavailability induced by BRJ supplementation does not alter sympathetic venoconstriction in non-exercising limbs or the pressor response to exercise in healthy young adults. In addition, although BRJ does not cause a decrease in venoconstriction in all subjects, when the attenuation of venoconstriction in the calf during static handgrip exercise after BRJ ingestion is obtained, the degree of the MAP pressor response to exercise could be reduced to a degree dependent on the attenuation of venoconstriction.

In the present study, the venous volume in the non-exercising forearm and calf decreased with static handgrip exercise under both CON and BRJ conditions, and these decreases in the venous volume did not differ between the two groups (Table 1 and Figure 2E,F), despite the increased plasma $NO_3^-$ concentration after BRJ intake. These results suggest that the NO bioavailability increase induced by acute BRJ ingestion does not attenuate the venoconstriction in the non-exercising limbs during exercise, which does not support our hypothesis. This is the first study to investigate the effect of BRJ on the venous vascular response to exercise. Some studies have investigated the effect of NO on the arterial vascular response. For example, sympathetic vasoconstriction in human and animal preparations is inhibited by NO derived from both endothelial nitric oxide synthase [31] and neuronal nitric oxide synthase [31]. In contrast, several studies have reported that the NO bioavailability associated with dietary $NO_3^-$ supplementation (e.g., BRJ) does not alter sympathetic vasoconstrictor responsiveness to exercise [32,33]. Despite differences between arterial and venous vessels and between active and inactive limbs in the previous and present studies, our findings and those of the other studies suggest that exogenous NO augmentation via the $NO_3^- \rightarrow NO_2^- \rightarrow$ NO pathway after acute BRJ consumption might not affect sympathetic vasoconstriction in either the artery or vein during exercise.

As mentioned above, we found no effect of BRJ on the venous vascular response in the non-exercising limb to exercise. Although we have no conclusive explanation, there are several possibilities for this result. First, in our study, the exogenous production of NO by BRJ was perhaps observed significantly because of increased $NO_3^-$ by 574 ± 120 μM 2 h after intake of BRJ, although it may not contribute to the attenuation of venous vascular tone (relaxation of smooth muscle) during exercise for the following reason. Our study participants were healthy young adults who were expected to have normal endothelial function. This meant that the endogenous NO production may have been sufficient to control the venous vascular tone. Indeed, in previous studies investigating the effect of BRJ on arterial endothelial function, flow-mediated dilation (FMD) was improved in participants with impaired endothelial function, including older adults [27,34], overweight or obese men [35], and patients with hypertension [2], but the FMD was very slightly increased [7] or unchanged [4,36] in healthy humans. Second, the inhibition of efferent sympathetic nerve activity induced by the NO increase with BRJ intake might not perhaps be observed in our study because plasma catecholamines during exercise have been reported to not be altered by acute intake of BRJ when compared with placebo [37]. Because the available scientific evidence is limited, we need to further investigate the effects of vasodilator action and inhibition of sympathetic nerve activity induced by the BRJ supplementation-related increase in NO on the venous vascular responses to exercise.

SBP, DBP, and MAP at pre-exercise and during exercise did not differ between CON and BRJ (Table 1). In addition, the degrees of the increases in these parameters with exercise were also similar for CON and BRJ (Figure 2A–D). Consistent with the present findings, previous studies have also reported no effects of acute dietary $NO_3^-$ supplementation on resting BP [38,39] and BP during exercise [9,37] in young normotensive individuals. In contrast, some studies reported a reduction in BP following acute dietary $NO_3^-$ supplementation with BRJ [1,3,4,40,41]. Similarly, there is no consensus on the hypotensive effect of BRJ. Some possibilities may be considered as to why the effect of dietary $NO_3^-$ supplementation on BP differed among studies. First, the individual response to dietary $NO_3^-$ supplementation varies widely [41,42]. Second, eating habits might differ among participants. For example, "traditional" foods found in a Japanese diet appear to be high in $NO_3^-$ [24]. Because chronic intake of a $NO_3^-$-rich diet might mitigate the ability of dietary $NO_3^-$ supplementation to increase the levels of biomarkers of NO synthesis [43], chronic exposure to a $NO_3^-$-rich Japanese diet may have contributed to the lack of a significant effect on BP, even though the participants were requested to refrain from consuming these foods throughout the study period. Finally, aerobic fitness levels might have varied among participants because NO synthase has been reported to be increased by physical activity [44].

Interestingly, in the individual data, a smaller decrease in calf venous volume during exercise with BRJ intake rather than CON was significantly related to a lower MAP elevation with exercise under BRJ conditions (Figure 3D). In other words, if the venoconstriction in the calf during exercise is attenuated by BRJ, the MAP pressor response to exercise may also be lower and in line with the degree of venoconstriction attenuation. In our previous study, in the individual data, a greater increased venous compliance with BRJ tended to be associated with a lower resting BP under BRJ conditions [19]. Based on the present and previous findings, the mechanism for the hypotensive effect of BRJ might be partly attributed to the control of venous vascular tone. Elevated venous stiffness appears to at least somewhat be a factor in the pathogenesis of hypertension [22]. Thus, it is very important to understand the influence of nutritional component(s) with hypotensive effects on venous vascular control, and our present results may boost the development of interventions to improve and maintain vascular health.

This study has several limitations. First, the venous vessel is modulated by an active factor that indicates activation of the sympathetic nerve [12,14–16] and by a passive factor that shows the change in volume-flow dependence [45]. The decrease in limb volume has been reported to reflect the venoconstriction or the elevation of venous tone when the intravascular pressure of the large vein in the limb is maintained at a constant level [15,16]. In addition, the cuff inflation pressure can be considered equivalent to the intravascular pressure of the large vein in the limb [28]. Considering these findings, we believe that the decreases in venous volume in the forearm and calf under cuff pressure of a constant subdiastolic BP of 30–40 mmHg in the present study were caused by sympathetic activation so that this decreased venous volume in limbs reflects the venoconstriction. Second, we did not quantify $NO_2^-$ concentrations. This is important because the conversion of $NO_3^-$ to $NO_2^-$ is necessary for biological effects to occur [4]. However, because there is evidence that a significant increase in the plasma $NO_3^-$ level is accompanied by an increase in $NO_2^-$ concentration in healthy young adults [40,46,47], we believe that the plasma $NO_2^-$ concentration was also elevated after consumption of BRJ in our study. Finally, because the small number of limited participants might have introduced a degree of sampling bias, several variables can interfere with the results and conclusions inferred. Thus, our results cannot be generalized, and the conclusion of our study is applicable to a limited extent to the healthy young adults who participated in this study.

## 5. Conclusions

In this study, we investigated the effect of acute dietary $NO_3^-$ supplementation with BRJ on the venous vascular response and circulatory responses to exercise. Our findings suggest that acute BRJ supplementation leads to an increase in the plasma $NO_3^-$ concentration but does not change venoconstriction in the non-exercising limbs or the MAP response to static exercise in young healthy adults. In addition, although BRJ does not cause a decrease in venoconstriction in all subjects, when venoconstriction during exercise is attenuated after BRJ ingestion, the MAP pressor response to exercise may be reduced to a degree dependent on the attenuation of venoconstriction.

**Author Contributions:** Conceptualization, A.O.; Methodology, A.O., Y.I., A.S., Y.M. and M.O.; Data analysis, A.O., Y.I. and A.S.; Writing—Original Draft Preparation, A.O.; Writing—Review and Editing, A.O., Y.I., A.S., Y.M. and M.O.; Funding Acquisition, A.O. All authors have read and agreed to the published version of the manuscript.

**Funding:** This research was funded by a grant from the Japan Society for the Promotion of Science, Grant-in-Aid for Scientific Research (C) (21K11561).

**Institutional Review Board Statement:** The study was conducted according to the guidelines of the Declaration of Helsinki and approved by the Human Ethics Committee of Toyo University (TU2019-018-TU2020-H-019).

**Informed Consent Statement:** Informed consent was obtained from all participants involved in the study.

**Acknowledgments:** We would like to express our gratitude to everyone who volunteered to participate in this study.

**Conflicts of Interest:** The authors have no financial conflict of interest to declare.

## References


1. Hobbs, D.A.; Kaffa, N.; George, T.W.; Methven, L.; Lovegrove, J.A. Blood pressure-lowering effects of beetroot juice and novel beetroot-enriched bread products in normotensive male subjects. *Br. J. Nutr.* **2012**, *108*, 2066–2074. [CrossRef] [PubMed]
2. Kapil, V.; Khambata, R.S.; Robertson, A.; Caulfield, M.J.; Ahluwalia, A. Dietary nitrate provides sustained blood pressure lowering in hypertensive patients: A randomized, phase 2, double-blind, placebo-controlled study. *Hypertension* **2015**, *65*, 320–327. [CrossRef]
3. Vanhatalo, A.; Bailey, S.J.; Blackwell, J.R.; DiMenna, F.J.; Pavey, T.G.; Wilkerson, D.P.; Benjamin, N.; Winyard, P.G.; Jones, A.M. Acute and chronic effects of dietary nitrate supplementation on blood pressure and the physiological responses to moderate-intensity and incremental exercise. *Am. J. Physiol. Regul. Integr. Comp. Physiol.* **2010**, *299*, R1121–R1131. [CrossRef] [PubMed]
4. Webb, A.J.; Patel, N.; Loukogeorgakis, S.; Okorie, M.; Aboud, Z.; Misra, S.; Rashid, R.; Miall, P.; Deanfield, J.; Benjamin, N.; et al. Acute blood pressure lowering, vasoprotective, and antiplatelet properties of dietary nitrate via bioconversion to nitrite. *Hypertension* **2008**, *51*, 784–790. [CrossRef] [PubMed]
5. Ashor, A.W.; Lara, J.; Siervo, M. Medium-term effects of dietary nitrate supplementation on systolic and diastolic blood pressure in adults: A systematic review and meta-analysis. *J. Hypertens.* **2017**, *35*, 1353–1359. [CrossRef]
6. Gee, L.C.; Ahluwalia, A. Dietary nitrate lowers blood pressure: Epidemiological, pre-clinical experimental and clinical trial evidence. *Curr. Hypertens. Rep.* **2016**, *18*, 17. [CrossRef]
7. Bondonno, C.P.; Yang, X.; Croft, K.D.; Considine, M.J.; Ward, N.C.; Rich, L.; Puddey, I.B.; Swinny, E.; Mubarak, A.; Hodgson, J.M. Flavonoid-rich apples and nitrate-rich spinach augment nitric oxide status and improve endothelial function in healthy men and women: A randomized controlled trial. *Free Radic. Biol. Med.* **2012**, *52*, 95–102. [CrossRef]
8. Heiss, C.; Meyer, C.; Totzeck, M.; Hendgen-Cotta, U.B.; Heinen, Y.; Luedike, P.; Keymel, S.; Ayoub, N.; Lundberg, J.O.; Weitzberg, E.; et al. Dietary inorganic nitrate mobilizes circulating angiogenic cells. *Free Radic. Biol. Med.* **2012**, *52*, 1767–1772. [CrossRef]
9. Notay, K.; Incognito, A.V.; Millar, P.J. Acute beetroot juice supplementation on sympathetic nerve activity: A randomized, double-blind, placebo-controlled proof-of-concept study. *Am. J. Physiol. Heart Circ. Physiol.* **2017**, *313*, H59–H65. [CrossRef]
10. Rothe, C. Venous system: Physiology of the capacitance vessels. In *Handbook of Physiology: The Cardiovascular System III*; Shephard, J.T., Abbound, F.M., Bethesda, M.D., Eds.; American Physiological Society: Rockville, MD, USA, 1983.
11. Ooue, A.; Sato, K.; Hirasawa, A.; Sadamoto, T. Tendon vibration attenuates superficial venous vessel response of the resting limb during static arm exercise. *J. Physiol. Anthropol.* **2012**, *31*, 29. [CrossRef]
12. Ooue, A.; Sato, K.; Hirasawa, A.; Sadamoto, T. Superficial venous vascular response of the resting limb during static exercise and postexercise muscle ischemia. *Appl. Physiol. Nutr. Metab.* **2013**, *38*, 941–946. [CrossRef] [PubMed]
13. Oue, A.; Sato, K.; Yoneya, M.; Sadamoto, T. Decreased compliance in the deep and superficial conduit veins of the upper arm during prolonged cycling exercise. *Physiol. Rep.* **2017**, *5*, e13253. [CrossRef] [PubMed]
14. Oue, A.; Sadamoto, T. Compliance in the deep and superficial conduit veins of the nonexercising arm is unaffected by short-term exercise. *Physiol. Rep.* **2018**, *6*, e13724. [CrossRef] [PubMed]
15. Rowell, L.B.; Brengelmann, G.L.; Detry, J.M.; Wyss, C. Venomotor responses to rapid changes in skin temperature in exercising man. *J. Appl. Physiol.* **1971**, *30*, 64–71. [CrossRef] [PubMed]
16. Seaman, R.G.; Wiley, R.L.; Zechman, F.W.; Goldey, J.A. Venous reactivity during static exercise (handgrip) in man. *J. Appl. Physiol.* **1973**, *35*, 858–860. [CrossRef]
17. Blackman, D.J.; Morris-Thurgood, J.A.; Atherton, J.J.; Ellis, G.R.; Anderson, R.A.; Cockcroft, J.R.; Frenneaux, M.P. Endothelium-derived nitric oxide contributes to the regulation of venous tone in humans. *Circulation* **2000**, *101*, 165–170. [CrossRef]
18. Hamilton, C.A.; Berg, G.; Mcintyre, M.; Mcphaden, A.R.; Reid, J.L.; Dominiczak, A.F. Effects of nitric oxide and superoxide on relaxation in human artery and vein. *Atherosclerosis* **1997**, *133*, 77–86. [CrossRef]
19. Oue, A.; Iimura, Y.; Shinagawa, A.; Miyakoshi, Y.; Ota, M. Acute dietary nitrate supplementation does not change venous volume and compliance in healthy young adults. *Am. J. Physiol. Regul. Integr. Comp. Physiol.* **2022**, *323*, R331–R339. [CrossRef]
20. Hernandez, J.P.; Franke, W.D. Age- and fitness-related differences in limb venous compliance do not affect tolerance to maximal lower body negative pressure in men and women. *J. Appl. Physiol.* **2004**, *97*, 925–929. [CrossRef]
21. Young, C.N.; Stillabower, M.E.; DiSabatino, A.; Farquhar, W.B. Venous smooth muscle tone and responsiveness in older adults. *J. Appl. Physiol.* **2006**, *101*, 1362–1367. [CrossRef]
22. Safar, M.E.; London, G.M. Arterial and venous compliance in sustained essential hypertension. *Hypertension* **1987**, *10*, 133–139. [CrossRef] [PubMed]
23. Blekkenhorst, L.C.; Prince, R.L.; Ward, N.C.; Croft, K.D.; Lewis, J.R.; Devine, A.; Shinde, S.; Woodman, R.J.; Hodgson, J.M.; Bondonno, C.P. Development of a reference database for assessing dietary nitrate in vegetables. *Mol. Nutr. Food Res.* **2017**, *61*, 1600982. [CrossRef]
24. Sobko, T.; Marcus, C.; Govoni, M.; Kamiya, S. Dietary nitrate in Japanese traditional foods lowers diastolic blood pressure in healthy volunteers. *Nitric Oxide* **2010**, *22*, 136–140. [CrossRef] [PubMed]
25. Govoni, M.; Jansson, E.Å.; Weitzberg, E.; Lundberg, J.O. The increase in plasma nitrite after a dietary nitrate load is markedly attenuated by an antibacterial mouthwash. *Nitric Oxide* **2008**, *19*, 333–337. [CrossRef] [PubMed]
26. Berry, M.J.; Justus, N.W.; Hauser, J.I.; Case, A.H.; Helms, C.C.; Basu, S.; Rogers, Z.; Lewis, M.T.; Miller, G.D. Dietary nitrate supplementation improves exercise performance and decreases blood pressure in COPD patients. *Nitric Oxide* **2015**, *48*, 22–30. [CrossRef] [PubMed]

27. Jones, T.; Dunn, E.L.; Macdonald, J.H.; Kubis, H.-P.; McMahon, N.; Sandoo, A. The effects of beetroot juice on blood pressure, microvascular function and large-vessel endothelial function: A randomized, double-blind, placebo-controlled pilot study in healthy older adults. *Nutrients* **2019**, *11*, 1792. [CrossRef]
28. Halliwill, J.R.; Minson, C.T.; Joyner, M.J. Measurement of limb venous compliance in humans: Technical considerations and physiological findings. *J. Appl. Physiol.* **1999**, *87*, 1555–1563. [CrossRef]
29. Green, L.C.; Wagner, D.A.; Glogowski, J.; Skipper, P.L.; Wishnok, J.S.; Tannenbaum, S.R. Analysis of nitrate, nitrite, and [15N]nitrate in biological fluids. *Anal. Biochem.* **1982**, *126*, 131–138. [CrossRef]
30. Oue, A.; Saito, M.; Iimura, Y. Effect of short-term endurance training on venous compliance in the calf and forearm differs between continuous and interval exercise in humans. *Physiol. Rep.* **2019**, *7*, e14211. [CrossRef]
31. Jendzjowsky, N.G.; DeLorey, D.S. Role of neuronal nitric oxide in the inhibition of sympathetic vasoconstriction in resting and contracting skeletal muscle of healthy rats. *J. Appl. Physiol.* **2013**, *115*, 97–106. [CrossRef]
32. Nelson, M.D.; Rosenberry, R.; Barresi, R.; Tsimerinov, E.I.; Rader, F.; Tang, X.; Mason, O.; Schwartz, A.; Stabler, T.; Shidban, S.; et al. Sodium nitrate alleviates functional muscle ischaemia in patients with Becker muscular dystrophy. *J. Physiol.* **2015**, *593*, 5183–5200. [CrossRef]
33. Rosenmeier, J.B.; Fritzlar, S.J.; Dinenno, F.A.; Joyner, M.J. Exogenous NO administration and alpha-adrenergic vasoconstriction in human limbs. *J. Appl. Physiol.* **2003**, *95*, 2370–2374. [CrossRef] [PubMed]
34. Rammos, C.; Hendgen-Cotta, U.B.; Sobierajski, J.; Bernard, A.; Kelm, M.; Rassaf, T. Dietary nitrate reverses vascular dysfunction in older adults with moderately increased cardiovascular risk. *J. Am. Coll. Cardiol.* **2014**, *63*, 1584–1585. [CrossRef] [PubMed]
35. Joris, P.J.; Mensink, R.P. Beetroot juice improves in overweight and slightly obese men postprandial endothelial function after consumption of a mixed meal. *Atherosclerosis* **2013**, *231*, 78–83. [CrossRef] [PubMed]
36. Bahra, M.; Kapil, V.; Pearl, V.; Ghosh, S.; Ahluwalia, A. Inorganic nitrate ingestion improves vascular compliance but does not alter flow-mediated dilatation in healthy volunteers. *Nitric Oxide* **2012**, *26*, 197–202. [CrossRef] [PubMed]
37. de Vries, C.J.; DeLorey, D.S. Effect of acute dietary nitrate supplementation on sympathetic vasoconstriction at rest and during exercise. *J. Appl. Physiol.* **2019**, *127*, 81–88. [CrossRef]
38. Casey, D.P.; Treichler, D.P.; Ganger, C.T.; Schneider, A.C.; Ueda, K. Acute dietary nitrate supplementation enhances compensatory vasodilation during hypoxic exercise in older adults. *J. Appl. Physiol.* **2015**, *118*, 178–186. [CrossRef]
39. Kim, J.-K.; Moore, D.J.; Maurer, D.G.; Kim-Shapiro, D.B.; Basu, S.; Flanagan, M.P.; Skulas-Ray, A.C.; Kris-Etherton, P.; Proctor, D.N. Acute dietary nitrate supplementation does not augment submaximal forearm exercise hyperemia in healthy young men. *Appl. Physiol. Nutr. Metab.* **2015**, *40*, 122–128. [CrossRef]
40. Lee, J.-S.; Stebbins, C.L.; Jung, E.; Nho, H.; Kim, J.-K.; Chang, M.-J.; Choi, H.-M. Effects of chronic dietary nitrate supplementation on the hemodynamic response to dynamic exercise. *Am. J. Physiol. Regul. Integr. Comp. Physiol.* **2015**, *309*, R459–R466. [CrossRef]
41. Wylie, L.J.; Kelly, J.; Bailey, S.J.; Blackwell, J.R.; Skiba, P.F.; Winyard, P.G.; Jeukendrup, A.E.; Vanhatalo, A.; Jones, A.M. Beetroot juice and exercise: Pharmacodynamic and dose-response relationships. *J. Appl. Physiol.* **2013**, *115*, 325–336. [CrossRef]
42. Wilkerson, D.P.; Hayward, G.M.; Bailey, S.J.; Vanhatalo, A.; Blackwell, J.R.; Jones, A.M. Influence of acute dietary nitrate supplementation on 50 mile time trial performance in well-trained cyclists. *Eur. J. Appl. Physiol.* **2012**, *112*, 4127–4134. [CrossRef]
43. Breese, B.C.; Poole, D.C.; Okushima, D.; Bailey, S.J.; Jones, A.M.; Kondo, N.; Amano, T.; Koga, S. The effect of dietary nitrate supplementation on the spatial heterogeneity of quadriceps deoxygenation during heavy-intensity cycling. *Physiol. Rep.* **2017**, *5*, e13340. [CrossRef]
44. Nyberg, M.; Blackwell, J.R.; Damsgaard, R.; Jones, A.M.; Hellsten, Y.; Mortensen, S.P. Lifelong physical activity prevents an age-related reduction in arterial and skeletal muscle nitric oxide bioavailability in humans. *J. Physiol.* **2012**, *590*, 5361–5370. [CrossRef] [PubMed]
45. Noble, B.J.; Drinkhill, M.J.; Myers, D.S.; Hainsworth, R. Reflex control of splanchnic blood volume in anaesthetized dogs. *J. Physiol.* **1998**, *513*, 263–272. [CrossRef] [PubMed]
46. Amano, T.; Okushima, D.; Breese, B.C.; Bailey, S.J.; Koga, S.; Kondo, N. Influence of dietary nitrate supplementation on local sweating and cutaneous vascular responses during exercise in a hot environment. *Eur. J. Appl. Physiol.* **2018**, *118*, 1579–1588. [CrossRef] [PubMed]
47. Stanaway, L.; Rutherfurd-Markwick, K.; Page, R.; Wong, M.; Jirangrat, W.; Teh, K.H.; Ali, A. Acute supplementation with nitrate-rich beetroot juice causes a greater increase in plasma nitrite and reduction in blood pressure of older compared to younger adults. *Nutrients* **2019**, *11*, 1683. [CrossRef] [PubMed]

nutrients

Article

# Effects of Fermentation Time and Type of Tea on the Content of Micronutrients in Kombucha Fermented Tea

Karolina Jakubczyk [1,*], Patrycja Kupnicka [2], Klaudia Melkis [1], Oliwia Mielczarek [1], Joanna Walczyńska [1], Dariusz Chlubek [2] and Katarzyna Janda-Milczarek [1]

1 Department of Human Nutrition and Metabolomics, Pomeranian Medical University in Szczecin, 24 Broniewskiego Street, 71-460 Szczecin, Poland
2 Department of Biochemistry and Medical Chemistry, Pomeranian Medical University, 72 Powstańców Wielkopolskich Street, 70-111 Szczecin, Poland
* Correspondence: karolina.jakubczyk@pum.edu.pl; Tel.: +48-91-441-48-11

**Abstract:** The fermented tea beverage Kombucha is obtained through a series of biochemical and enzymatic reactions carried out by symbiotic cultures of bacteria and yeasts (SCOBY). It contains organic acids, vitamins, amino acids, and biologically active compounds, notably polyphenols, derived mainly from tea. Kombucha exhibits a range of health-promoting properties, including antioxidant or detoxifying effects. This fermented beverage is traditionally brewed with black tea, but other types of tea are used increasingly, which may have significant implications in terms of chemical composition and health-promoting effects. In this preliminary study, we investigated the content of micronutrients (manganese (Mn), copper (Cu), iron (Fe), chromium (Cr) and zinc (Zn)) by the ICP-OES method in Kombucha prepared with black, red, green and white tea at different time points of fermentation (1, 7, 14 days). It should be noted that the composition of separate ingredients such as tea, leaven or sugar has not been studied. Kombucha had the highest content of zinc—0.36 mg/L to 2.08 mg/L, which accounts for between 3% and 26% of the RDA (Recommended Dietary Allowance) for adults, and the smallest amounts of chromium (0.03 mg/L to 0.09 mg/L), which however represents as much as between 75% and 232% of the RDA. It has been demonstrated that the type of tea as well as the day of fermentation have a significant effect on the concentrations of selected minerals. Kombucha can therefore supplement micronutrients in the human diet.

**Keywords:** kombucha; fermentation; microelements

**Citation:** Jakubczyk, K.; Kupnicka, P.; Melkis, K.; Mielczarek, O.; Walczyńska, J.; Chlubek, D.; Janda-Milczarek, K. Effects of Fermentation Time and Type of Tea on the Content of Micronutrients in Kombucha Fermented Tea. *Nutrients* **2022**, *14*, 4828. https://doi.org/10.3390/nu14224828

Academic Editors: Daniela Rigano and Paola Bontempo

Received: 23 September 2022
Accepted: 11 November 2022
Published: 15 November 2022



## 1. Introduction

Kombucha is a low-alcohol beverage made by fermenting a sugared tea infusion with symbiotic cultures of bacteria and yeasts (SCOBY), commonly called "tea fungus". This complex community comprises acetic acid bacteria (AAB) (*Gluconobacter: G. entanii, G. oxydans, Acetobacter: A. xylionoides, A. aceti, A. pasteurianus*), *Komagataeibacter* (*K. intermedius, K. rhaeticus*), lactic acid bacteria (LAB) (*Lactobacillus* and *Leuconostoc*), and yeasts (*Schizosaccharomyces pombe, Zygosaccharomyces bailii, Saccharomyces*). Its dynamics are still not fully understood [1–7].

Kombucha fermentation is a combination of three fermentation processes: alcoholic, lactic and acetic acid. The bacteria present in the tea fungus are responsible for the production of acetic acid, while yeasts, representatives of the osmophilic type, induce the breakdown of sucrose. The resulting product, glucose, is a substrate for both lactic fermentation and alcoholic fermentation [4,8]. Under the influence of lactic acid bacteria, glucose is converted into lactic acid. During alcoholic fermentation, on the other hand, glucose is converted into ethyl alcohol, releasing carbon dioxide. Ethanol produced during the breakdown of glucose is oxidised by acetic acid bacteria to acetic acid and acetaldehyde (Figure 1).

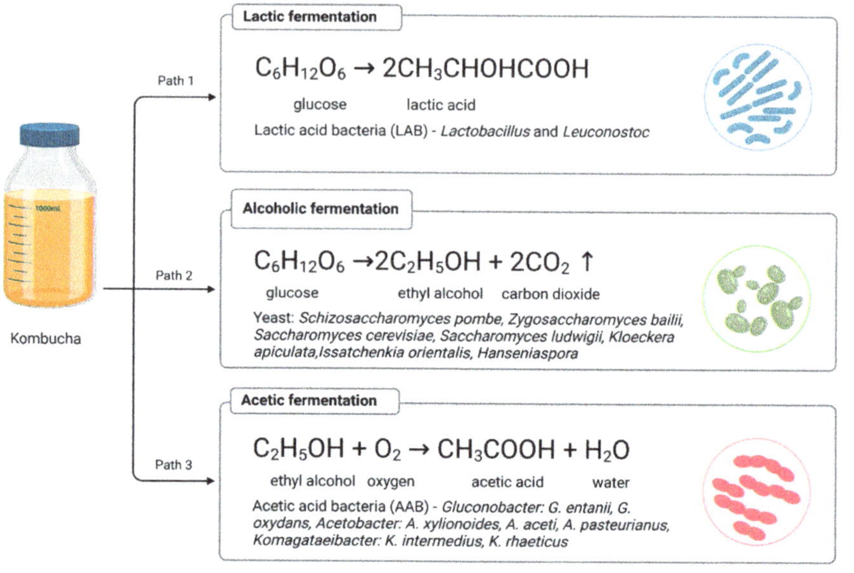


**Figure 1.** Types of fermentation in Kombucha. Created with BioRender.com.

*Acetobacter* are also responsible for the oxidation of glucose to glucuronic and gluconic acid, the key detoxifying agents in Kombucha. The process is also accompanied by the production of cellulose, which is part of the tea fungus [7]. Thus, sweetened tea is transformed into Kombucha by a process involving three types of fermentation, whose activity and dominance changes over time. Initially, the beverage is rich in glucose, followed by alcohol, while the final stage is dominated by organic acids, including acetic acid.

The elementary ingredients of the traditional recipe are black tea and white sugar [9]. Sucrose is the main source of carbon in Kombucha fermentation due to its uncomplicated structure and ability to provide simple carbohydrates for microbial metabolic pathways, as well as its low cost and easy availability [1,10]. However, Kombucha is more and more often made with green, red and white teas or herbal infusions instead of black tea [7], and with coconut sugar, cane sugar, maple syrup or honey instead of white sugar [1,10,11]. The entire Kombucha production process takes place at room temperature over 7–14 days, during which it acquires its distinct chemical and organoleptic characteristics. The flavor of the finished tea beverage is described as mildly sour, fruity, fizzy, resembling that of cider [12–14].

Research to date indicates that the fermented tea beverage contains numerous bioactive substances, originating from the material used, mainly tea, but also resulting from the enzymatic transformations of organic compounds carried out by microorganisms. The bioactive compounds include vitamins (E, K, B, C), amino acids (especially theanine, a derivative of glutamine), polyphenolic compounds, i.e., catechins and flavonoids, and a variety of minerals [4,7,15–17].

It is worth noting that Kombucha has been hailed as a functional fermented beverage with antioxidant, antimicrobial, antioxidant, anti-diabetic properties, reducing cholesterol levels, supporting immune and digestive function, and also stimulating liver detoxification [7,17–19]. However, the presence and the amounts of nutrients in Kombucha, and accordingly its beneficial effects, are determined by a number of factors, such as the parameters of the fermentation process, e.g., time, or the ingredients used [12].

In addition, Kombucha is gaining popularity as one of the novelties offered by fermented food manufacturers. The valuable composition of such foods translates into manifold biological effects in consumers' bodies [20]. As a result of the biochemical transformations of organic compounds by microorganisms, it is possible to obtain foods with not only extended shelf life and microbiological stability, but also a higher nutritional value. The fermentation process largely breaks down the antinutrients in food, e.g., phytates, enhancing the potential for utilizing nutrients of key importance in the human diet. Also, fermented foods may be better tolerated by people with certain food sensitivities and intolerances [21]. Recent scientific reports confirm the interactions between microorganisms and plant products [22,23]. The microbiota present in the product is responsible for "digesting" plant material into absorbable active small molecules, which then induce physiological changes in the body [24–27]. Research findings show that fermentation of tea residues can significantly increase the antioxidant activity (up to 3.25 times), as well as the polyphenol concentrations (5.68 times) of Kombucha. Interestingly, green tea residues showed a stronger effect than those of black tea [26].

Even though Kombucha is gaining increasing recognition and the range of flavors available on the food market continues to expand, the detailed composition and the effects of individual fermentation parameters or type of tea on the properties of this product are not fully understood [12]. A few scientific articles report that Kombucha is a source of micronutrients, but the data are incomplete. Hence, the aim of this study was to evaluate for the first time the content of selected micronutrients in Kombucha beverages made with black, green, white and red tea at different time points of fermentation.

## 2. Materials and Methods

### 2.1. Plant Material

The material consisted of four types of leaf tea (*Camellia sinensis*): black, green, white and red (Pu-ERH) originating in China.

### 2.2. Preparation of Kombucha

The Kombucha cultures in the present article were purchased from a commercial shop. Kombucha bacterium component belongs to the strains of *Acetobacter*, while the yeasts are *Saccharomyces cerevisiae* and *Zygosaccharomyces*. One hundred grams of sugar (100.0 g/L, 10.0%), eight grams of tea (8.0 g/L, 0.8%) and 1 L of hot distilled water (90 °C) were added to the flask. The solution was infused for 10 min in a sterile conical flask. After cooling to 30 °C, the tea decoction was filtered into clean glass bottles and Kombucha pellicle (100.0 g/L, 10.0%) and one hundred milliliters of leaven from a previous culture (100.0 mL/L, 1.0%) were added (Figure 2).

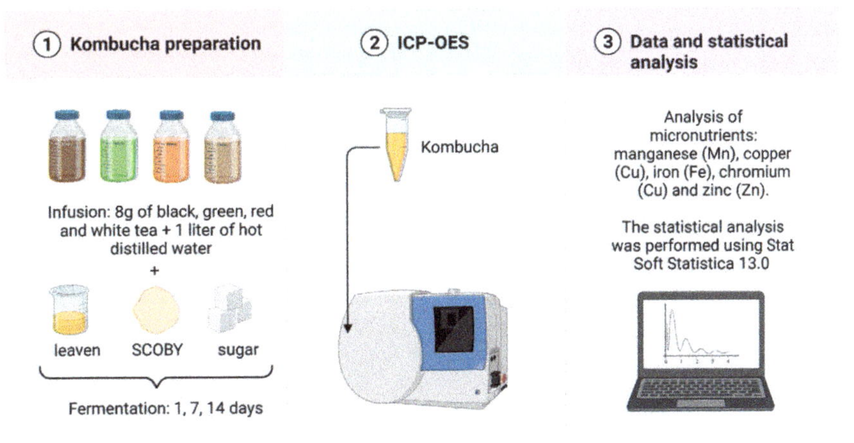


**Figure 2.** Material and methods—Preparation of Kombucha and laboratory analyses. Created with BioRender.com.

### 2.3. Fermentation of Kombucha

Kombucha culture was kept under aseptic conditions. Fermentation was carried out by incubating the Kombucha culture at 28 ± 1 °C for 1, 7 and 14 days. Replicates were prepared so that each replicate was completely collected after its stipulated period of fermentation. The Kombucha obtained was filtered and analyzed.

### 2.4. Determining Elements Content in Infusions

Sample preparation:

The samples were mineralized using the MARS 5 CEM microwave digestion system. The volume of the sample given to research was 0.8 mL. The samples were transferred to clean polypropylene tubes, 2 mL of 65% $HNO_3$ (Suprapur, Merck, Darmstadt, Germany) was added to each vial and each sample was allowed 30 min pre reaction time in the clean hood. After completion of the pre-reaction time, 0.5 mL of non-stabilized 30% $H_2O_2$ solution (Suprapur, Merck, Darmstadt, Germany) was added to each vial. Once the addition of all reagents was complete, the samples were placed in special Teflon vessels and heated in the microwave digestion system for 35 min at 180 °C (15 min ramp to 180 °C and maintained at 180 °C for 20 min). At the end of digestion all samples were removed from the microwave and allowed to cool to room temperature. In the clean hood, samples were transferred to acid-washed 15 mL polypropylene sample tubes. A further 5-fold dilution was performed prior to ICP-OES measurement. A volume of 2 mL was taken from each digest. The samples were spiked with an internal standard to provide a final concentration of 0.5 mg/L Ytrium, 1 mL of 1% Triton (Triton X-100, Sigma, Kawasaki, Japan) and diluted to the final volume of 10 mL with 0.075% nitric acid (Suprapur, Merck, Darmstadt, Germany). Blank samples were prepared by adding concentrated nitric acid (500 μL) to tubes without a sample and subsequently diluted in the same manner described above. Multielement calibration standards (ICP multi-element standard solution IV, Merck, Darmstadt, Germany) were prepared with different concentrations of inorganic elements in the same manner as in blanks and samples. Deionized water (Direct Q UV, Millipore, Burlington, MA, USA, approximately 18.0 MΩ) was used for preparation of all solutions.

Sample determination:

Samples were analyzed using inductively coupled plasma optical emission spectrometry (ICP-OES, ICAP 7400 Duo, Thermo Scientific), which is often utilized to measure the concentrations of mineral nutrients as well as heavy metals and allows simultaneous measurements of many different elements, also in plant samples [28,29]. ICP-OES with a concentric nebulizer and cyclonic spray chamber was used to determine the content of micro and macroelements. The analysis was performed in both radial and axial modes. The wavelengths used in the analysis were: Zn 206.200, Cr 205.560, Mn 257.610, Cu 224.700, Fe 259.940.

Validation was performed by evaluating the following: NIST SRM 8414 reference material (National Institute of Standards and Technology, USA), limit of detection (LOD), and the recovery of internal standard (yttrium). To eliminate possible interference, the emission lines were selected empirically in pilot measurements. This model of validation is often used in ICP-OES studies, also those regarding plant samples [29]. The recovery of Y was within 90–106%. The R2 values for all standard curves were in the range between 0.998 and 1.000.

### 2.5. Statistical Analysis

All determinations were carried out in at least three replicates. Statistical analysis was performed using Stat Soft Statistica 13.0 and Microsoft Excel 2017. Distributions of values for individual parameters were analyzed using the Shapiro-Wilk test. Since the distribution of continuous variables deviated from normal, the Kruskal-Wallis test was used to evaluate the differences between the studied parameters. Spearman's correlation test was used to determine the correlations between the parameters studied. Results were expressed as mean values and standard deviation; however, median values and quartile ranges were used for statistical analyses. Differences were considered significant at $p \leq 0.05$.

## 3. Results

Analysis of Kombucha beverages prepared with black, green, white and red tea at different time points of fermentation revealed the presence of five trace elements (Tables 1–3). The content of identified micronutrients manganese (Mn), copper (Cu), iron (Fe), chromium (Cu) and zinc (Zn) was quantified. It should be noted that the composition of separate ingredients such as tea, leaven or sugar was not studied. The general pattern of mineral concentrations in the Kombucha samples was as follows: Zn > Mn > Fe > Cu > Cr.

**Table 1.** Analysis of reference material Bovine Muscle NIST-SRM 8414.

| Element | Certified [mg/L] | Measured [mg/L] ($n$ = 3) | LOD [mg/L] | %RSD Range |
|---|---|---|---|---|
| Mn | 0.37 ± 0.09 | 0.43 | 0.00026 | 1.0–6.2 |
| Zn | 142 ± 14 | 138 | 0.00065 | 1.5–5.7 |
| Cu | 2.84 ± 0.45 | 3.06 | 0.00186 | 2.4–8.2 |
| Fe | 71.2 ± 9.2 | 76.1 | 0.00022 | 1.8–6.7 |
| Cr | 0.071 ± 0.038 | 0.080 | 0.00044 | 3.9–9.2 |

LOD—limits of detection, %RSD range—relative sample deviation.

**Table 2.** The microminerals content (Mn, Cu) and Recommended Dietary Allowances (RDA) in Kombucha.

| | | Mn (mg/L) | | | | Cu (mg/L) | | | |
|---|---|---|---|---|---|---|---|---|---|
| Kombucha | Time Points-Day | Mean | SD | % RDA Men | % RDA Women | Mean | SD | % RDA Men | % RDA Women |
| BK | 1 [a] | 0.43 [*,b,c,d,g] | ± 0.04 | 18.87 | 24.11 | 0.12 [*,c] | ± 0.03 | 13.00 | 13.00 |
| | 7 [b] | 0.67 [*,a,e] | ± 0.13 | 29.00 | 37.05 | 0.13 [*,c] | ± 0.07 | 14.89 | 14.89 |
| | 14 [c] | 0.68 [*,a,f] | ± 0.03 | 29.48 | 37.66 | 0.25 [*,a,b,f,i] | ± 0.01 | 28.21 | 28.21 |
| GK | 1 [d] | 1.18 [*,a,e,f,j] | ± 0.08 | 51.25 | 65.48 | 0.10 [*,f] | ± 0.06 | 11.26 | 11.26 |
| | 7 [e] | 1.40 [*,b,d,k] | ± 0.08 | 61.03 | 77.98 | 0.09 [*,f] | ± 0.04 | 10.17 | 10.17 |
| | 14 [f] | 1.40 [*,c,d,l] | ± 0.02 | 60.88 | 77.79 | 0.20 [*,c,d,e,l] | ± 0.00 | 22.20 | 22.20 |
| RK | 1 [g] | 0.93 [*,a,i] | ± 0.06 | 40.35 | 51.56 | 0.12 | ± 0.07 | 12.89 | 12.89 |
| | 7 [h] | 0.92 [*,i] | ± 0.06 | 39.86 | 50.93 | 0.07 [*,i] | ± 0.03 | 7.29 | 7.29 |
| | 14 [i] | 1.21 [*,g,h] | ± 0.02 | 52.66 | 67.29 | 0.17 [*,c,h] | ± 0.00 | 18.88 | 18.88 |
| WK | 1 [j] | 0.57 [*,d] | ± 0.07 | 24.97 | 31.91 | 0.19 [*,l] | ± 0.06 | 20.90 | 20.90 |
| | 7 [k] | 0.63 [*,e] | ± 0.08 | 27.42 | 35.04 | 0.12 | ± 0.05 | 13.70 | 13.70 |
| | 14 [l] | 0.71 [*,f] | ± 0.00 | 30.78 | 30.78 | 0.064 [*,c,f,j] | ± 0.00 | 7.13 | 7.13 |

Different numbers (a–l) in the columns represent statistically significant differences * $p < 0.05$ between particular types of Kombucha (1, 7, 14 days of fermentation) and tea: [a]—BK 0, [b]—BK 7, [c]—BK 14, [d]—GK 0, [e]—GK 7, [f]—GK 14, [g]—WK 0, [h]—WK 7, [i]—WK 14, [j]—RK 0, [k]—RK 7, [l]—RK, BK—Kombucha prepared from black tea, GK—Kombucha prepared from green tea, RK—Kombucha prepared from red tea, WK—Kombucha prepared from white tea.

The content of manganese in Kombucha ranged from 0.43 mg/L to 1.40 mg/L and was dependent on both fermentation time and type of tea used (Table 2). Statistically significant differences are presented in Table 1. The lowest values were observed in Kombucha prepared with black and white tea, while the highest were found in green tea Kombucha. Irrespective of the type of tea used, the highest results were observed on day 14 of fermentation. One liter of the beverage can cover from 19% to 61% of manganese requirement for men and 24% to 78% for women.

**Table 3.** The microminerals content (Fe, Cr) and Recommended Dietary Allowances (RDA) in Kombucha.

| | | Fe (mg/L) | | | | Cr (mg/L) | | | |
|---|---|---|---|---|---|---|---|---|---|
| Kombucha | Time Points-Day | Mean | SD | % RDA Men | % RDA Women | Mean | SD | % RDA Men | % RDA Women |
| BK | 1 [a] | 0.23 [*,c] | ± 0.04 | 2.27 | 1.26 | 0.07 | ± 0.02 | 167.19 | 167.19 |
| | 7 [b] | 0.24 [*,c,e] | ± 0.03 | 2.44 | 1.36 | 0.04 [*,c] | ± 0.01 | 99.34 | 99.34 |
| | 14 [c] | 0.36 [*,a,b,f] | ± 0.01 | 3.60 | 2.00 | 0.07 [*,b,f] | ± 0.00 | 174.09 | 174.09 |
| GK | 1 [d] | 0.25 | ± 0.09 | 2.52 | 1.40 | 0.04 [*,f] | ± 0.02 | 95.72 | 95.72 |
| | 7 [e] | 0.27 [*,h] | ± 0.07 | 2.68 | 1.49 | 0.03 [*,f] | ± 0.01 | 76.28 | 76.28 |
| | 14 [f] | 0.25 [*,c,l] | ± 0.01 | 2.49 | 1.39 | 0.09 [*,c,d,e,l] | ± 0.00 | 231.63 | 231.63 |
| RK | 1 [g] | 0.44 [*,h,j] | ± 0.24 | 4.35 | 2.42 | 0.05 | ± 0.01 | 132.21 | 132.21 |
| | 7 [h] | 0.18 [*,b,e,g,i] | ± 0.01 | 1.81 | 1.01 | 0.03 [*,i] | ± 0.02 | 74.53 | 74.53 |
| | 14 [i] | 0.32 [*,h,l] | ± 0.00 | 3.22 | 1.79 | 0.09 [*,h,l] | ± 0.00 | 217.51 | 217.51 |
| WK | 1 [j] | 0.19 [*,g,l] | ± 0.04 | 1.90 | 1.06 | 0.04 | ± 0.03 | 102.34 | 102.34 |
| | 7 [k] | 0.21 [*,l] | ± 0.02 | 2.11 | 1.17 | 0.07 | ± 0.04 | 180.09 | 180.09 |
| | 14 [l] | 0.46 [*,f,i,j,k] | ± 0.00 | 4.6 | 2.56 | 0.06 [*,f,i] | ± 0.00 | 160.32 | 160.32 |

Different numbers (a–l) in the columns represent statistically significant differences * $p < 0.05$ between particular types of Kombucha (1, 7, 14 days of fermentation) and tea: [a]—BK 0, [b]—BK 7, [c]—BK 14, [d]—GK 0, [e]—GK 7, [f]—GK 14, [g]—WK 0, [h]—WK 7, [i]—WK 14, [j]—RK 0, [k]—RK 7, [l]—RK, BK—Kombucha prepared from black tea, GK—Kombucha prepared from green tea, RK—Kombucha prepared from red tea, WK—Kombucha prepared from white tea.

The content of copper in Kombucha depended on both fermentation time and the tea used (Table 2) and ranged from 0.01 mg/L to 0.25 mg/L. Kombucha brewed with red tea had the lowest levels of copper, while that prepared with black tea had the highest. Irrespective of the type of tea used, the highest results were observed on day 14 of fermentation, except for white tea (Table 2). Between 7% and 28% of the requirement for this micronutrient was met for both men and women. With respect to this element, there were no statistically significant differences between Kombuchas made with different types of tea on days 1 and 7 of fermentation. The only differences observed were on day 14, leading to the conclusion that the type of tea does not affect copper content in the early days of fermentation.

The content of iron in Kombucha ranged from 0.18 mg/L to 0.46 mg/L and was dependent on both fermentation time and type of tea used (Table 3). The lowest iron levels were observed in Kombucha brewed with green tea, while the highest were found in white tea Kombucha. One liter of the beverage covers as little as 1.8% to 4.6% of the iron requirement for men, and 1% to 2.5% for women.

The chromium content in Kombucha ranged from 0.03 mg/L to 0.09 mg/L and was dependent on both fermentation time and type of tea used (Table 3). The lowest levels were observed in Kombucha brewed with green tea and red (7 days of fermentation). Irrespective of the type of tea used, the highest results were observed on day 14 of fermentation. The requirement for this element was covered at 75% to 232% for both men and women.

The content of zinc in Kombucha ranged from 0.36 mg/L to 2.08 mg/L and was likewise dependent on both fermentation time and type of tea used (Table 4). The lowest levels were noted in Kombucha brewed with white tea, while the highest were found in black tea Kombucha. Irrespective of the type of tea used, the highest results were observed on day 14 of fermentation. The requirement for zinc was covered at 3% to 19% for men and 5% to 26% for women. The mineral content was almost invariably highest on day 14 of fermentation regardless of the type of tea. There were no statistically significant differences between Kombuchas made with different teas on days 1 and 7 of fermentation, so initially the type of tea does not have a significant effect the content of this element.

**Table 4.** The microminerals content (Zn) and Recommended Dietary Allowances (RDA) in Kombucha.

| | | Zn (mg/L) | | | | |
|---|---|---|---|---|---|---|
| **Kombucha** | **Time Points-Day** | **Mean** | | **SD** | **% RDA Men** | **% RDA Women** |
| BK | 1 [a] | 0.44 [*,a,b] | ± | 0.30 | 3.98 | 5.47 |
| | 7 [b] | 0.74 [*,a] | ± | 0.63 | 6.73 | 9.25 |
| | 14 [c] | 2.08 [*,a,b,f,i] | ± | 0.09 | 18.87 | 25.95 |
| GK | 1 [d] | 0.75 | ± | 0.52 | 6.80 | 9.35 |
| | 7 [e] | 1.11 | ± | 0.76 | 10.13 | 13.92 |
| | 14 [f] | 0.54 [*,c,l] | ± | 0.01 | 4.91 | 6.76 |
| RK | 1 [g] | 0.88 | ± | 0.65 | 8.04 | 11.06 |
| | 7 [h] | 0.38 | ± | 0.29 | 3.44 | 4.73 |
| | 14 [i] | 0.62 [*,c] | ± | 0.00 | 5.63 | 7.74 |
| WK | 1 [j] | 0.36 [*,l] | ± | 0.35 | 3.26 | 4.48 |
| | 7 [k] | 0.36 [*,l] | ± | 0.22 | 3.24 | 4.46 |
| | 14 [l] | 0.99 [*,f,j,k] | ± | 0.00 | 8.99 | 12.36 |

Different numbers (a–l) in the columns represent statistically significant differences * $p < 0.05$ between particular type of Kombucha (1, 7, 14 days of fermentation) and tea: [a]—BK 0, [b]—BK 7, [c]—BK 14, [d]—GK 0, [e]—GK 7, [f]—GK 14, [g]—WK 0, [h]—WK 7, [i]—WK 14, [j]—RK 0, [k]—RK 7, [l]—RK, BK—Kombucha prepared from black tea, GK—Kombucha prepared from green tea, RK—Kombucha prepared from red tea, WK—Kombucha prepared from white tea.

In addition, the content of micronutrients was analyzed independently of the day of fermentation, taking into account the type of tea used to prepare Kombucha. The highest concentrations of Zn, Cu and Cr were found in the beverage made with black tea. In the case of manganese, the highest concentration of this element was observed in Kombucha brewed with green tea, while that of iron in the Kombucha prepared with red tea (Table 5). The only statistically significant differences were observed for manganese. For the other elements, the type of tea did not affect their content.

**Table 5.** The microminerals content in different type of Kombucha.

| Kombucha | (mg/L) | Mn | Zn | Cu | Fe | Cr |
|---|---|---|---|---|---|---|
| BK [a] | mean | 0.582 [*,b,c] | 1.139 | 0.185 | 0.276 | 0.057 |
| | SD | 0.135 | 0.855 | 0.088 | 0.066 | 0.022 |
| GK [b] | mean | 1.329 [*,a,d] | 0.747 | 0.128 | 0.253 | 0.047 |
| | SD | 0.129 | 0.585 | 0.064 | 0.065 | 0.028 |
| RK [c] | mean | 1.034 [*,a,d] | 0.647 | 0.117 | 0.319 | 0.053 |
| | SD | 0.179 | 0.428 | 0.063 | 0.179 | 0.026 |
| WK [d] | mean | 0.629 [*,b,c] | 0.595 | 0.127 | 0.285 | 0.056 |
| | SD | 0.075 | 0.366 | 0.068 | 0.126 | 0.031 |

Different letters (a, b, c, d) in the columns represent statistically significant differences * $p < 0.05$ between particular type of Kombucha (1, 7, 14 days of fermentation) and tea: [a]—BK, [b]—GK, [c]—WK, [d]—RK, BK—Kombucha prepared from black tea, GK—Kombucha prepared from green tea, RK—Kombucha prepared from red tea, WK—Kombucha prepared from white tea.

Additionally, significant positive correlations were found between some micronutrients (Table 6); however, this relationship was variable for different types of Kombucha. Negative significant correlations were also found for Kombucha prepared on the basis of red tea: Zn vs. Cu and Cu vs. Fe (Table 6).

**Table 6.** Spearman's rank correlation between micronutrients for different types of Kombucha.

| **BK:** | **Mn** | **Zn** | **Cu** | **Fe** | **Cr** | **GK** | **Mn** | **Zn** | **Cu** | **Fe** | **Cr** |
|---|---|---|---|---|---|---|---|---|---|---|---|
| Mn | 1.00 | 0.05 | 0.48 * | 0.37 | −0.14 | Mn | 1.00 | −0.08 | 0.37 | −0.24 | 0.18 |
| Zn | 0.05 | 1.00 | 0.66 * | 0.74 * | 0.29 | Zn | −0.08 | 1.00 | 0.00 | 0.78 * | 0.04 |
| Cu | 0.48 * | 0.66 * | 1.00 | 0.52 * | 0.30 | Cu | 0.37 | 0.00 | 1.00 | 0.16 | 0.88 * |
| Fe | 0.37 | 0.74 * | 0.52 * | 1.00 | 0.28 | Fe | −0.24 | 0.78 * | 0.16 | 1.00 | 0.17 |
| Cr | −0.14 | 0.29 | 0.30 | 0.28 | 1.00 | Cr | 0.18 | 0.04 | 0.88 * | 0.17 | 1.00 |
| RK | Mn | Zn | Cu | Fe | Cr | WK | Mn | Zn | Cu | Fe | Cr |
| Mn | 1.00 | −0.19 | 0.50 * | 0.33 | 0.63 * | Mn | 1.00 | 0.82 * | −0.28 | 0.76 * | 0.37 |
| Zn | −0.19 | 1.00 | 0.04 | 0.51 * | 0.25 | Zn | 0.82 * | 1.00 | −0.57 * | 0.91 * | 0.45 |
| Cu | 0.50 * | 0.04 | 1.00 | 0.31 | 0.45 | Cu | −0.28 | −0.57 * | 1.00 | −0.51 * | −0.04 |
| Fe | 0.33 | 0.51 | 0.31 | 1.00 | 0.53 * | Fe | 0.76 * | 0.91 * | −0.51 * | 1.00 | 0.42 |
| Cr | 0.63 * | 0.25 | 0.45 | 0.53 * | 1.00 | Cr | 0.37 | 0.45 | −0.04 | 0.42 | 1.00 |

* $p < 0.05$, BK—Kombucha prepared from black tea, GK—Kombucha prepared from green tea, RK—Kombucha prepared from red tea, WK—Kombucha prepared from white tea.

Regardless of the type of Kombucha, a weak but statistically significant correlation was found between time and the content of Mn (0.279), Zn (0.348) and a moderate, significant relationship between time and the content of Fe (0.423) and Cr (0.447).

Regardless of the day of fermentation, quite strong relationships between the time and concentration of selected mineral compounds were shown in Table 7. Therefore, it can be concluded that both the fermentation process and the time significantly affect the chemical composition of this drink.

**Table 7.** Spearman's rank correlations between fermentation time and microminerals content for different types of Kombucha.

| | **BK** | **GK** | **RK** | **WK** |
|---|---|---|---|---|
| Mn | 0.712 * | 0.712 * | 0.659 * | 0.554 * |
| Zn | 0.712 * | NS | NS | 0.791 * |
| Cu | 0.765 * | 0.712 * | NS | −0.844 * |
| Fe | 0.791 * | NS | NS | 0.805 * |
| Cr | NS | 0.594 * | 0.580 * | NS |

* $p < 0.05$, BK—Kombucha prepared from black tea, GK—Kombucha prepared from green tea, RK—Kombucha prepared from red tea, WK—Kombucha prepared from white tea.

## 4. Discussion

Kombucha fermented tea is becoming increasingly popular, not only for its sensory properties, but also for its health-promoting benefits. In addition, the drink is classified as a functional food or nutraceutical [7,30,31].

Our study confirms that Kombucha can be a source of micronutrients: chromium (Cr), manganese (Mn), copper (Cu), zinc (Zn), and iron (Fe). The general pattern of mineral concentrations in the Kombucha samples was as follows: Zn > Mn > Fe > Cu > Cr. Our analysis was carried out at different time points of fermentation (days 1, 7 and 14). In addition, for the first time, different types of leaf tea were used as the base of the beverage: black, green, red and white tea. We have shown that the content of selected micronutrients is dependent not only on the day of fermentation but also on the type of tea used.

Micronutrients are minerals which, although present in trace amounts, are essential for the normal development and functioning of the human body. Given that cells do not have the ability to synthesize trace elements, they must be supplied with food. A properly balanced diet should contain essential minerals in such quantities that the total supply is adequate to meet demand, as both deficiency and excess can cause a range of dysfunctions and disorders [32,33]. The search for and analysis of different types of new foods can add valuable information about a source of micronutrients.

Manganese plays an important role in development, digestion, reproduction, antioxidant defense, energy production, immune response and regulation of neuronal activity [34]. The adult requirement for this mineral ranges from 1.8 mg (women) to 2.3 mg (men) [35]. Our analysis showed that manganese (Mn) concentrations ranged from 0.43 mg/L to 1.40 mg/L, accounting for 19% to 61% of the requirement for this element for men and 24% to 78% for women. The highest manganese content was found in the green tea Kombucha from day 14 of fermentation.

Copper is a component of superoxide dismutase and thus influences free radical decomposition reactions. Moreover, it is responsible for enzymatic reactions and the synthesis of collagen and neurotransmitters [36]. The RDA for adults is 0.9 mg [35]. The amount of copper (Cu) in our study ranged from 0.06 mg/L to 0.25 mg/L on day 14 of fermentation, representing between 7% and 28% of the RDA for both sexes. The highest content of this element was detected in the beverage made with black tea on day 14 of fermentation.

Iron, an essential mineral for health and life, is responsible for the synthesis of hemoglobin—a protein found in erythrocytes, which carry oxygen molecules from the lungs to peripheral tissues and support immune and nerve functions [37]. To prevent iron deficiencies, a daily intake of 10 to 18 mg of this mineral is recommended, depending on life stage and sex [35]. Our Kombuchas contained between 0.18 mg/L and 0.46 mg/L of iron (Fe), representing between 1.8% and 4.6% of the RDA for men and between 1% and 2.5% for women. The highest iron content was found in the beverage prepared with white tea on day 14 of fermentation. The results of the analysis for each beverage revealed differences in the content of minerals, correlated with fermentation time as well as the tea used, and the differences were statistically significant ($p < 0.05$).

Chromium is an essential nutrient for normal metabolism of glucose, protein and fat. It enhances insulin sensitivity in tissues and also participates in intracellular redox reactions. The recommended daily intake of this mineral for men and women aged 19–50 years is 0.04 mg [35,38]. Our results showed that the chromium content in Kombuchas made with infusions from different types of tea ranged from 0.03 mg/L to 0.09 mg/L, meaning that Kombucha can cover as much as 75% to 232% of the RDA for both men and women. The highest content of this mineral was detected in the product made with green and red tea on day 14 of fermentation. Consuming even small amounts of Kombucha will correct deficiencies of this element.

Zinc (Zn) is a key element in many processes, from cell growth and differentiation, to regulating immune system function and modulating mechanisms related to learning and memory [39]. The recommended daily intake is 11 mg for men and 8 mg for women [35]. Zinc was detected in amounts ranging from 0.36 mg/L to 2.08 mg/L on day 14 of black tea fermentation, which accounts for 3% to 19% of the requirement for men and 5% to 26% for women. In addition, it has been demonstrated that *Acetobacter aceti* bacteria biotransform chromium and zinc and increase their amounts. These properties are used in the treatment of diabetes due to their hypoglycemic effect [40].

Minerals such as Mn, Zn and Cu can find be found in plant protection products, fertilizers, pesticides and fungicides. Their presence in the product may be related to agricultural practices in use and the content of these elements in phytosanitary products [41].

Kombucha, despite its long tradition, is not adequately researched. There are few studies analyzing the mineral content of this beverage, particularly in terms of micronutrients. It should be emphasized, however, none of them take into account the different time points of fermentation and type of tea.

Ivanišová et al. assessed the chemical composition and antioxidant, antimicrobial and sensory properties of Kombucha made with black tea on day 7 of fermentation, and their findings appear to be consistent with the present study. Their Kombucha, however, was prepared in a different way. The infusion was made by boiling 1 liter of water, 5 g of black tea leaves (Darjeeling, India) and 30 g of white sugar, left to steep for 15 min and fermented at 22 degrees Celsius for 7 days [30]. In our study, we used 1 liter of water, 100 g of white sugar and 8 g of tea. In the study by Ivanišová et al., the content of manganese was 1.57 mg/L, copper 0.14 mg/L, iron 0.31 mg/L, zinc 0.53 mg/L and chromium was not detected during the analyses. Our results were: 0.67 mg/L for manganese, 0.13 mg/L for copper, 0.24 mg/L for iron, 0.74 mg/L for zinc and 0.04 mg/L for chromium. The most pronounced differences were noted in the content of manganese and the presence of chromium, but this may be related to the use of material of a different origin [30]. The researchers conclude that the content of minerals valuable for the human body increases with fermentation time. In a comparison of the chemical composition and properties of tea and Kombucha, fermented tea proved to be more valuable, and its antioxidant activity was several times higher. During fermentation, the content of the essential elements Fe, Mn, Zn and Ni increased significantly, too. In addition, the authors of the study emphasize that due to the absence of harmful elements, Kombucha is safe to consume [30].

Jayabalan et al. analyzed the chemical composition of the tea fungus (SCOBY) used to brew black tea Kombucha. Their study also made reference to time points, with tests carried out on days 7, 14 and 21 of fermentation. Biochemical properties, including mineral content, increased throughout fermentation time, reaching maximum values on day 21 [2]. Among micronutrients, the highest concentrations in dried tea fungus were found for zinc and manganese. However, the finished Kombucha drink was not analyzed. The results are consistent with our report.

It is worth noting that similar findings were obtained in our earlier study focusing on fluoride ions. In that case, too, longer fermentation resulted in a higher fluoride concentration in the beverage [14].

Similarly, a comparative analysis of sweet black tea vs. Kombucha showed that fermentation significantly increases the content of selected minerals [42]. Concentrations of Zn, Cu, Fe, Mn, Ni and Co were determined. Tests for certain toxic elements showed that Pb and Cr were present in very small amounts, while Cd was not found. Among micronutrients, the highest concentrations were noted for manganese (0.462 ug/mL), iron (0.353 ug/mL), copper (0.237 ug/mL), and the lowest for zinc (0.154 ug/mL) and chromium (0.001 ug/mL). These results are very similar to our observations, with similar concentrations of Mn, Cu and Fe. The Cr content in our study was slightly higher, but the biggest differences were found for Zn. According to Bauer-Petrovska and Petrushevska-Tozi, the levels of copper, iron, manganese, nickel and zinc increase due to the metabolic activity of Kombucha [42].

Tea as such is an important source of elements in the human diet [43]. What is more, the type or species of plant can significantly affect the mineral content in the beverage. This is related, among other things, to the individual differences between species, the diversity of production processes of the plants concerned and the type of soil in which they grow, the capacity of the plants to store nutrients, the types of pollution, climate or geographical location [43–45].

Brzezicha-Cirocka et al. analyzed 118 black teas, determining the concentrations of 14 elements. In terms of micronutrients, the highest concentration was found for Mn, along with the highest percentage of the RDA (15%) per daily intake of this beverage [46]. Similar results were also obtained by Koch et al. where, for all black teas tested, mineral contents were ranked in the following order: K > Ca > Mg > Mn > Fe > Na > Zn > Cu. It has been shown that mineral composition can be significantly affected by the origin of black tea, defined not only as a country, but also a region or province [45]. Differences in mineral composition have been noted between black and green teas, and the authors pointed out that apart from production-related factors, mineral content can be also influenced by soil conditions, location, rainfall, altitude, genetic characteristics of the plant, and age of the tea leaves [47].

Even though all teas used in this study are derived from the same plant species, *Camellia sinensis*, and come from a single producer in China, they differ considerably in their production processes, as also demonstrated in this paper. Depending on the process followed to obtain the final product, different types of tea can be distinguished. The main types are black tea, white tea, green tea and red tea. Processing treatments affect the color, aroma, taste, intensity, and also the chemical composition of the tea infusion [48]. Black tea is obtained by complete oxidation of tea leaves. After harvesting, the leaves are left to wither completely. During this time, they are also crushed and rolled to speed up the oxidation process. Once they have acquired a sufficiently dark color, they are dried at a high temperature. Red tea is made from green buds and young leaves. Immediately after harvesting, they are heated to inactivate the enzymes. While still moist, the leaves are rolled and then dried in the sun. This type of tea also undergoes fermentation during prolonged storage in high humidity conditions [49]. White tea is obtained from the buds of the tea plant harvested in the spring. Oxidation takes place only where the leaves are damaged, and it is minimal. However, the leaves for white tea are not immediately subjected to a drying process, but are allowed to rest freely, which allows for the activation of enzymatic processes in the leaves, i.e., enzymatic oxidation. Green tea is not oxidized. To make it, withering tea leaves are steamed or pan-fried (Figure 3) [50]. The general principles for obtaining dried tea are very similar, but small differences in production result in distinct flavors and aromas, as well as the chemical composition of the infusion [50].

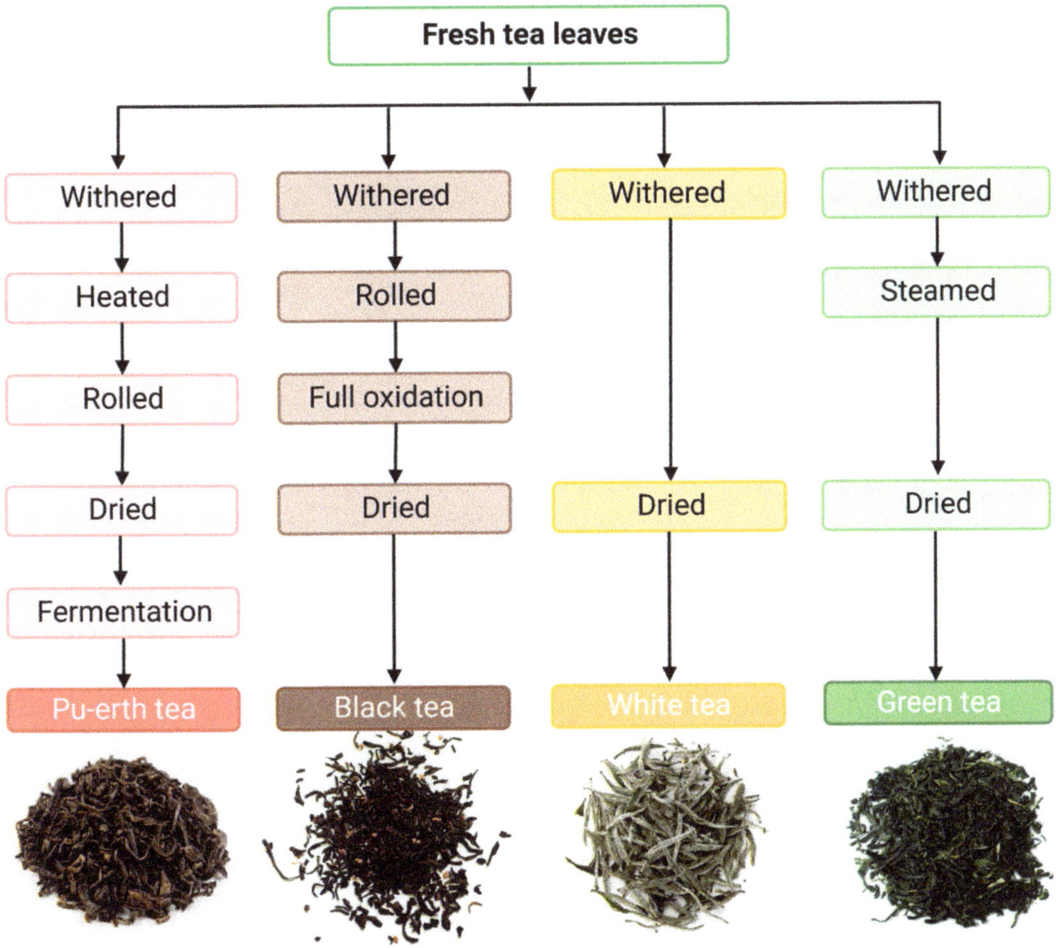


**Figure 3.** Types of teas and their production process. Created with BioRender.com.

In our study, irrespective of the day of fermentation, the content of individual micronutrients in Kombucha varied, with the type of tea being the determining factor. The highest concentrations of Zn, Cu and Cr were observed in the beverage prepared according to the traditional recipe, i.e., using the fully oxidized black tea. In the case of manganese, the highest concentration of this element was observed in Kombucha brewed with green tea, while that of iron in the Kombucha prepared with red tea. White tea, which undergoes the least amount of processing, only drying, gentle rolling and light oxidation, has the lowest concentrations of micronutrients. Hence, one of the factors determining the final mineral composition is the processing of the leaves themselves. Tea oxidation appears to be particularly important, as Kombucha made with black tea contained the highest levels of minerals.

It is also worth noting that the micronutrient content increased with the day of fermentation of the beverage, reaching the maximum level on day 14. Today, a significant number of manufacturers embrace the use of fermentation in food production, due to its positive effects on enhancing biosafety, extending shelf life and functional properties [46]. It appears that in most cases, the content of both macro- and micro-nutrients in foods significantly increases during fermentation [51].

The fermentation process increases the bioavailability of micronutrients and trace elements through the degradation of insoluble metal cation complexes and anti-nutritive substances such as oxalates, tannins and phytates [52]. These compounds are hydrolyzed by enzymes (e.g., phytases) produced by microorganisms such as lactic acid bacteria and yeast that make up SCOBY. In addition, the synthesis of lactic acid during fermentation, causing pH changes, provides the conditions necessary for the activation of microbial enzymes, thus contributing to the intensification of their action [53,54]. This mechanism is confirmed by the results of studies by Castro-Alba et al. which showed a correlation between increased availability of iron, zinc and calcium in fermented quinoa flour (3.6, 4.0 and 3.5 times, respectively) and a reduction in phytate levels [55].

Sometimes, microorganisms use individual elements for their own metabolism as a substrate to initiate the fermentation process or the synthesis of secondary metabolites, including vitamins and polyols [56]. We also observed these changes in our study, but they were strongly related to the type of tea and the day of fermentation, which could be related to the activity of microorganisms [51]. Ivanišová et al. reported a decrease in Ca and Pb concentration in the Kombucha drink. Cultures of Kombucha microorganisms show the ability to detoxify the drink, as these bacteria are considered biosorbents. These SCOBY microorganisms have properties that allow them to accumulate and bind heavy metal contaminants on their cellular structure [30]. The results of Mamisahebei et al. showed that the Kombucha cultures used in the beer brewing process are very effective in removing heavy metals such as arsenic, chromium and copper. The content of Co did not increase in Kombucha, probably due to its inclusion in vitamin B 12, as the B vitamins (mainly B 1, B 6 and B 12) are mainly produced during the fermentation process [57].

Scientific studies have confirmed the potential of LAB strains to improve the bioavailability of minerals. The results confirm that the LAB strains LAB *L. fermentum* B4655, *L. plantarum* B4495, *L. casei* B1922, *L. bulgaricus* CFR2028 i *L. acidophilus* B4496 reduced the content of phytic acid when used to ferment soy milk at 37° C for 24 h. In addition, the results show an increase in Mg and Ca levels in fermented soy milk compared to control [58]. Bahaciu et al. [59] showed that germination and fermentation (Lactobacillus) of soybeans for four days at 25 °C led to an increase of 40.87%, 43.41%, 59.56% and 53.4% for Zn, Mg, Fe and Ca, respectively, which are higher than the values obtained for germination alone. Additionally, fermentation (4 or 10 h at 30 °C) of ground quinoa seeds *with L. plantarum* 299v significantly reduced the phytic acid content and improved the bioavailability of minerals such as Ca, Fe and Zn [60]. This direction should be extended in future scientific research.

The limitations of this preliminary study include the lack of composition analysis of the discrete ingredients such as tea, leaven and sugar. These results will provide a complete understanding of the biochemical processes in Kombucha. Nevertheless, it should be noted that Kombucha, especially when subjected to a longer fermentation process, contains more organic acids, and thus its pH is strongly acidic. Moreover, the produced CO2 can start to accumulate between the drinks (Kombucha) and the biofilm (SCOBY). This can prevent the transfer of nutrients and thus block the continuity of chemical changes in the reaction environment. It should therefore be consumed in limited quantities, diluted with water or fermented for a shorter period of time [7].

## 5. Conclusions

Our findings clearly show that the type of tea used to make Kombucha has a significant impact on the micronutrient content of the final product. In addition, fermentation time also determined the levels of selected minerals. Irrespective of the type of tea, the highest results were observed mainly on day 14 of fermentation. Kombucha had the highest content of zinc (0.36 mg/L to 2.08 mg/L), which accounts for between 3% and 26% of the RDA for adults, and the smallest content of chromium (0.03 mg/L to 0.09 mg/L), which, however, represents as much as between 75% and 232% of the RDA. Black tea proved to be the best source of Zn, Cu and Cr; green tea was rich in Mn; while red tea had the highest iron content. In conclusion, Kombucha, particularly based on black tea, can supplement micronutrients in the human diet.

**Author Contributions:** Conceptualization, J.K.; methodology, J.K.; formal analysis, O.M., J.-M.K. and J.K.; investigation, J.K., P.K. and D.C.; writing—original draft preparation, K.M., J.W. and J.K.; writing—review and editing, J.K. and J.-M.K.; supervision, J.K.; project administration, J.K.; funding acquisition J.K. All authors have read and agreed to the published version of the manuscript.

**Funding:** This research was funded by the Pomeranian Medical University in Szczecin.

**Institutional Review Board Statement:** Not applicable.

**Informed Consent Statement:** Not applicable.

**Conflicts of Interest:** The authors declare no conflict of interest.

## References

1. Nyhan, L.M.; Lynch, K.M.; Sahin, A.W.; Arendt, E.K. Advances in Kombucha Tea Fermentation: A Review. *Appl. Microbiol.* **2022**, *2*, 73–103. [CrossRef]
2. Jayabalan, R.; Malini, K.; Sathishkumar, M.; Swaminathan, K.; Yun, S.-E. Biochemical Characteristics of Tea Fungus Produced during Kombucha Fermentation. *Food Sci. Biotechnol.* **2010**, *19*, 843–847. [CrossRef]
3. Jayabalan, R.; Malbaša, R.V.; Lončar, E.S.; Vitas, J.S.; Sathishkumar, M. A Review on Kombucha Tea—Microbiology, Composition, Fermentation, Beneficial Effects, Toxicity, and Tea Fungus. *Compr. Rev. Food Sci. Food Saf.* **2014**, *13*, 538–550. [CrossRef] [PubMed]
4. Villarreal-Soto, S.A.; Beaufort, S.; Bouajila, J.; Souchard, J.-P.; Taillandier, P. Understanding Kombucha Tea Fermentation: A Review. *J. Food Sci.* **2018**, *83*, 580–588. [CrossRef]
5. Chakravorty, S.; Bhattacharya, S.; Chatzinotas, A.; Chakraborty, W.; Bhattacharya, D.; Gachhui, R. Kombucha Tea Fermentation: Microbial and Biochemical Dynamics. *Int. J. Food Microbiol.* **2016**, *220*, 63–72. [CrossRef]
6. Gaggìa, F.; Baffoni, L.; Galiano, M.; Nielsen, D.S.; Jakobsen, R.R.; Castro-Mejía, J.L.; Bosi, S.; Truzzi, F.; Musumeci, F.; Dinelli, G.; et al. Kombucha Beverage from Green, Black and Rooibos Teas: A Comparative Study Looking at Microbiology, Chemistry and Antioxidant Activity. *Nutrients* **2019**, *11*, 1. [CrossRef]
7. Jakubczyk, K.; Kałduńska, J.; Kochman, J.; Janda, K. Chemical Profile and Antioxidant Activity of the Kombucha Beverage Derived from White, Green, Black and Red Tea. *Antioxidants* **2020**, *9*, 447. [CrossRef]
8. Antolak, H.; Piechota, D.; Kucharska, A. Kombucha Tea—A Double Power of Bioactive Compounds from Tea and Symbiotic Culture of Bacteria and Yeasts (SCOBY). *Antioxidants* **2021**, *10*, 1541. [CrossRef]
9. Greenwalt, C.J.; Steinkraus, K.H.; Ledford, R.A. Kombucha, the Fermented Tea: Microbiology, Composition, and Claimed Health Effects. *J. Food Prot.* **2000**, *63*, 976–981. [CrossRef]
10. Malbaša, R.; Lončar, E.; Djurić, M. Comparison of the Products of Kombucha Fermentation on Sucrose and Molasses. *Food Chem.* **2008**, *106*, 1039–1045. [CrossRef]
11. Hasan, B.J.M.M.; Osman, F.; Muhamad, R.; Sapawi, C.W.N.S.C.W.; Anzian, A.; Voon, W.W.Y.; Hussin, A.M. Effects of Sugar Sources and Fermentation Time on the Properties of Tea Fungus (Kombucha) Beverage. *Int. Food Res. J.* **2019**, *26*, 481–487.
12. Llamas-Arriba, M.; Hernández-Alcántara, A.; Yépez, A.; Aznar, R.; Dueñas, M.; López, P. Functional and nutritious beverages produced by lactic acid bacteria. In *Nutrients in Beverages*; Academic Press: Cambridge, MA, USA, 2019; pp. 419–465. ISBN 978-0-12-816842-4.

13. Kim, J.; Adhikari, K. Current Trends in Kombucha: Marketing Perspectives and the Need for Improved Sensory Research. *Beverages* **2020**, *6*, 15. [CrossRef]
14. Jakubczyk, K.; Gutowska, I.; Antoniewicz, J.; Janda, K. Evaluation of Fluoride and Selected Chemical Parameters in Kombucha Derived from White, Green, Black and Red Tea. *Biol. Trace Elem. Res.* **2021**, *199*, 3547–3552. [CrossRef] [PubMed]
15. Zhenjun, Z.; Yucheng, S.; Huawei, W.; Caibi, Z.; Xianchun, H.; Jian, Z. Flavour Chemical Dynamics during Fermentation of Kombucha Tea. *Emir. J. Food Agric.* **2018**, *30*, 732–741.
16. Jafari, R.; Naghavi, N.S.; Khosravi-Darani, K.; Doudi, M.; Shahanipour, K. Kombucha Microbial Starter with Enhanced Production of Antioxidant Compounds and Invertase. *Biocatal. Agric. Biotechnol.* **2020**, *29*, 101789. [CrossRef]
17. Cardoso, R.R.; Neto, R.O.; Dos Santos D'Almeida, C.T.; do Nascimento, T.P.; Pressete, C.G.; Azevedo, L.; Martino, H.S.D.; Cameron, L.C.; Ferreira, M.S.L.; de Barros, F.A.R. Kombuchas from Green and Black Teas Have Different Phenolic Profile, Which Impacts Their Antioxidant Capacities, Antibacterial and Antiproliferative Activities. *Food Res. Int.* **2020**, *128*, 108782. [CrossRef]
18. Kapp, J.M.; Sumner, W. Kombucha: A Systematic Review of the Empirical Evidence of Human Health Benefit. *Ann. Epidemiol.* **2019**, *30*, 66–70. [CrossRef]
19. Kaewkod, T.; Bovonsombut, S.; Tragoolpua, Y. Efficacy of Kombucha Obtained from Green, Oolong, and Black Teas on Inhibition of Pathogenic Bacteria, Antioxidation, and Toxicity on Colorectal Cancer Cell Line. *Microorganisms* **2019**, *7*, E700. [CrossRef]
20. Castellone, V.; Bancalari, E.; Rubert, J.; Gatti, M.; Neviani, E.; Bottari, B. Eating Fermented: Health Benefits of LAB-Fermented Foods. *Foods* **2021**, *10*, 2639. [CrossRef]
21. Marco, M.L.; Heeney, D.; Binda, S.; Cifelli, C.J.; Cotter, P.D.; Foligné, B.; Gänzle, M.; Kort, R.; Pasin, G.; Pihlanto, A.; et al. Health Benefits of Fermented Foods: Microbiota and Beyond. *Curr. Opin. Biotechnol.* **2017**, *44*, 94–102. [CrossRef]
22. Martínez, Y.; Más, D. Role of herbs and medicinal spices as modulators of gut microbiota. In *Herbs Spices*; IntechOpen: London, UK, 2020. [CrossRef]
23. Dingeo, G.; Brito, A.; Samouda, H.; Iddir, M.; La Frano, M.R.; Bohn, T. Phytochemicals as Modifiers of Gut Microbial Communities. *Food Funct.* **2020**, *11*, 8444–8471. [CrossRef] [PubMed]
24. Adebo, O.A.; Gabriela Medina-Meza, I. Impact of Fermentation on the Phenolic Compounds and Antioxidant Activity of Whole Cereal Grains: A Mini Review. *Molecules* **2020**, *25*, 927. [CrossRef] [PubMed]
25. An, X.; Bao, Q.; Di, S.; Zhao, Y.; Zhao, S.; Zhang, H.; Lian, F.; Tong, X. The Interaction between the Gut Microbiota and Herbal Medicines. *Biomed Pharm.* **2019**, *118*, 109252. [CrossRef]
26. Zhou, D.-D.; Saimaiti, A.; Luo, M.; Huang, S.-Y.; Xiong, R.-G.; Shang, A.; Gan, R.-Y.; Li, H.-B. Fermentation with Tea Residues Enhances Antioxidant Activities and Polyphenol Contents in Kombucha Beverages. *Antioxidants* **2022**, *11*, 155. [CrossRef] [PubMed]
27. Ziemlewska, A.; Nizioł-Łukaszewska, Z.; Bujak, T.; Zagórska-Dziok, M.; Wójciak, M.; Sowa, I. Effect of Fermentation Time on the Content of Bioactive Compounds with Cosmetic and Dermatological Properties in Kombucha Yerba Mate Extracts. *Sci. Rep.* **2021**, *11*, 18792. [CrossRef]
28. Karahan, F.; Ozyigit, I.I.; Saracoglu, I.A.; Yalcin, I.E.; Ozyigit, A.H.; Ilcim, A. Heavy Metal Levels and Mineral Nutrient Status in Different Parts of Various Medicinal Plants Collected from Eastern Mediterranean Region of Turkey. *Biol. Trace Elem. Res.* **2019**, *197*, 316–329. [CrossRef]
29. Senila, M.; Drolc, A.; Pintar, A.; Senila, L.; Levei, E. Validation and Measurement Uncertainty Evaluation of the ICP-OES Method for the Multi-Elemental Determination of Essential and Nonessential Elements from Medicinal Plants and Their Aqueous Extracts. *J. Anal. Sci. Technol.* **2014**, *5*, 37. [CrossRef]
30. Ivanišová, E.; Meňhartová, K.; Terentjeva, M.; Harangozo, Ľ.; Kántor, A.; Kačániová, M. The Evaluation of Chemical, Antioxidant, Antimicrobial and Sensory Properties of Kombucha Tea Beverage. *J. Food Sci. Technol.* **2020**, *57*, 1840–1846. [CrossRef]
31. Watawana, M.I.; Jayawardena, N.; Gunawardhana, C.B.; Waisundara, V.Y. Health, Wellness, and Safety Aspects of the Consumption of Kombucha. *J. Chem.* **2015**, *2015*, e591869. [CrossRef]
32. Gombart, A.F.; Pierre, A.; Maggini, S. A Review of Micronutrients and the Immune System–Working in Harmony to Reduce the Risk of Infection. *Nutrients* **2020**, *12*, 236. [CrossRef]
33. Veronese, N.; Barbagallo, M. Magnesium and Micro-Elements in Older Persons. *Nutrients* **2021**, *13*, 847. [CrossRef] [PubMed]
34. Manganese Metabolism in Humans. Available online: https://pubmed.ncbi.nlm.nih.gov/29293455/ (accessed on 15 April 2022).
35. Jarosz, M.; Szczygła, A. *Normy Żywienia Dla Populacji Polskiej—Nowelizacja*; Instytut Żywności i Żywienia: Warszawa, Poland, 2012; ISBN 978-83-86060-83-2.
36. Copper Deficiency Anemia: Review Article. Available online: https://pubmed.ncbi.nlm.nih.gov/29959467/ (accessed on 15 April 2022).
37. Current Understanding of Iron Homeostasis | The American Journal of Clinical Nutrition | Oxford Academic. Available online: https://academic.oup.com/ajcn/article/106/suppl_6/1559S/4823167 (accessed on 15 April 2022).
38. Żwierełło, W.; Styburski, D.; Maruszewska, A.; Piorun, K.; Skórka-Majewicz, M.; Czerwińska, M.; Maciejewska, D.; Baranowska-Bosiacka, I.; Krajewski, A.; Gutowska, I. Bioelements in the Treatment of Burn Injuries—The Complex Review of Metabolism and Supplementation (Copper, Selenium, Zinc, Iron, Manganese, Chromium and Magnesium). *J. Trace Elem. Med. Biol.* **2020**, *62*, 126616. [CrossRef] [PubMed]
39. Jarząb, S. Characterisation and application of zinc in cosmetology and dietetics. *Aesthetic Cosmetol.* **2021**, *10*, 189–193.

40. Huang, Y.-Y.; Qin, X.-K.; Dai, Y.-Y.; Huang, L.; Huang, G.-R.; Qin, Y.-C.; Wei, X.; Huang, Y.-Q. Preparation and Hypoglycemic Effects of Chromium- and Zinc-Rich Acetobacter Aceti. *World J. Diabetes* **2022**, *13*, 442–453. [CrossRef]
41. Antoniewicz, J.; Jakubczyk, K.; Kupnicka, P.; Bosiacki, M.; Chlubek, D.; Janda, K. Analysis of Selected Minerals in Homemade Grape Vinegars Obtained by Spontaneous Fermentation. *Biol. Trace Elem. Res.* **2022**, *200*, 910–919. [CrossRef]
42. Bauer-Petrovska, B.; Petrushevska-Tozi, L. Mineral and Water Soluble Vitamin Content in the Kombucha Drink. *Int. J. Food Sci. Technol.* **2000**, *35*, 201–205. [CrossRef]
43. Karak, T.; Kutu, F.R.; Nath, J.R.; Sonar, I.; Paul, R.K.; Boruah, R.K.; Sanyal, S.; Sabhapondit, S.; Dutta, A.K. Micronutrients (B, Co, Cu, Fe, Mn, Mo, and Zn) Content in Made Tea (Camellia Sinensis L.) and Tea Infusion with Health Prospect: A Critical Review. *Crit. Rev. Food Sci. Nutr.* **2017**, *57*, 2996–3034. [CrossRef] [PubMed]
44. Milani, R.F.; Silvestre, L.K.; Morgano, M.A.; Cadore, S. Investigation of Twelve Trace Elements in Herbal Tea Commercialized in Brazil. *J. Trace Elem. Med. Biol.* **2019**, *52*, 111–117. [CrossRef]
45. Koch, W.; Kukula-Koch, W.; Czop, M.; Baj, T.; Kocki, J.; Bawiec, P.; Casasnovas, R.O.; Głowniak-Lipa, A.; Głowniak, K. Analytical Assessment of Bioelements in Various Types of Black Teas from Different Geographical Origins in View of Chemometric Approach. *Molecules* **2021**, *26*, 6017. [CrossRef]
46. Brzezicha-Cirocka, J.; Grembecka, M.; Ciesielski, T.; Flaten, T.P.; Szefer, P. Evaluation of Macro- and Microelement Levels in Black Tea in View of Its Geographical Origin. *Biol. Trace Elem. Res.* **2017**, *176*, 429–441. [CrossRef]
47. Street, R.; Szakova, J.; Drabek, O.; Mladkova, L. The Status of Micronutrients (Cu, Fe, Mn, Zn) in Tea and Tea Infusions in Selected Samples Imported to the Czech Republic. *Czech J. Food Sci. UZPI* **2006**, *24*, 62–71. [CrossRef]
48. Matysek-Nawrocka, M.; Cyrankiewicz, P. Substancje biologicznie aktywne pozyskiwane z herbaty, kawy i kakao oraz ich zastosowanie w kosmetykach. *Post. Fitoter.* **2016**, *17*, 139–144.
49. Sady, S.; Sielicka, M.; Pawłowski, T. Ocena potencjału przeciwutleniającego zielonej herbaty z dodatkiem miodu i cukru. *Farm. Współczesna* **2016**, *9*, 169–175.
50. Miazga-Sławińska, M.; Grzegorczyk, A. Herbaty—Rodzaje, właściwości, jakość i zafałszowania. *Kosmos* **2014**, *63*, 473–479.
51. Samtiya, M.; Aluko, R.E.; Puniya, A.K.; Dhewa, T. Enhancing Micronutrients Bioavailability through Fermentation of Plant-Based Foods: A Concise Review. *Fermentation* **2021**, *7*, 63. [CrossRef]
52. Gupta, R.; Gangoliya, S.; Singh, N. Reduction of Phytic Acid and Enhancement of Bioavailable Micronutrients in Food Grains. *J. Food Sci. Technol.* **2013**, *52*, 676–684. [CrossRef]
53. Sharma, R.; Garg, P.; Kumar, P.; Bhatia, S.K.; Kulshrestha, S. Microbial Fermentation and Its Role in Quality Improvement of Fermented Foods. *Fermentation* **2020**, *6*, 106. [CrossRef]
54. Plant Food Anti-Nutritional Factors and Their Reduction Strategies: An Overview | Food Production, Processing and Nutrition | Full Text. Available online: https://fppn.biomedcentral.com/articles/10.1186/s43014-020-0020-5 (accessed on 30 October 2022).
55. Fermentation of Pseudocereals Quinoa, Canihua, and Amaranth to Improve Mineral Accessibility through Degradation of Phytate. Available online: https://www.ncbi.nlm.nih.gov/pmc/articles/PMC6771823/ (accessed on 30 October 2022).
56. Cuvas-Limon, R.B.; Nobre, C.; Cruz, M.; Rodriguez-Jasso, R.M.; Ruíz, H.A.; Loredo-Treviño, A.; Texeira, J.A.; Belmares, R. Spontaneously Fermented Traditional Beverages as a Source of Bioactive Compounds: An Overview. *Crit. Rev. Food Sci.Nutr.* **2021**, *61*, 2984–3006. [CrossRef]
57. Mamisahebei, S.; Khaniki, G.R.J.; Torabian, A.; Nasseri, S.; Naddafi, K. Removal of arsenic from an aqueous solution by pretreated waste tea fungal biomass. *J. Environ. Health Sci. Eng.* **2007**, *4*, 85–92.
58. Rekha, C.R.; Vijayalakshmi, G. Bioconversion of Isoflavone Glycosides to Aglycones, Mineral Bioavailability and Vitamin B Complex in Fermented Soymilk by Probiotic Bacteria and Yeast. *J. Appl. Microbiol.* **2010**, *109*, 1198–1208. [CrossRef]
59. Bahaciu, G.V.; Nicolae, C.G.; Șuler, A.D.; Segal, R. Germinated and Lactic Fermented Soybean Seeds, a Natural Alternative for Healthy Bones. A Scientific Approach. *Bull. Univ. Agric. Sci. Vet. Med. Cluj Napoca Food Sci. Technol.* **2018**, *75*, 8–14. [CrossRef]
60. Castro-Alba, V.; Lazarte, C.E.; Perez-Rea, D.; Sandberg, A.-S.; Carlsson, N.-G.; Almgren, A.; Bergenståhl, B.; Granfeldt, Y. Effect of Fermentation and Dry Roasting on the Nutritional Quality and Sensory Attributes of Quinoa. *Food Sci. Nutr.* **2019**, *7*, 3902–3911. [CrossRef] [PubMed]

nutrients

Article

# The Hypolipidemic Characteristics of a Methanol Extract of Fermented Green Tea and Spore of *Eurotium cristatum* SXHBTBU1934 in Golden Hamsters

**Fuhang Song [1,*], Kai Zhang [1], Jinpeng Yang [2], Annette S. Wilson [3], Caixia Chen [3] and Xiuli Xu [2,*]**

1 School of Light Industry, Beijing Technology and Business University, Beijing 100048, China; zhangkai2030302071@st.btbu.edu.cn
2 School of Ocean Sciences, China University of Geosciences, Beijing 100083, China; yangjinpeng@cugb.edu.cn
3 School of Medicine, University of Pittsburgh, Pittsburgh, PA 15213, USA; aswilson@pitt.edu (A.S.W.); cac321@pitt.edu (C.C.)
* Correspondence: songfuhang@btbu.edu.cn (F.S.); xuxl@cugb.edu.cn (X.X.)

**Abstract:** Fuzhuan brick tea (FBT), a distinctive Chinese dark tea with the predominant fungus of *Eurotium cristatum*, offered significant health benefits to Chinese people. In the current study, the in vivo bioactivities of *E. cristatum* (SXHBTBU1934) fermented green tea and spores of *E. cristatum* fermented on wheat were investigated, respectively. The methanol extract of fermented green tea and spore of *E. cristatum* both showed potent lipid-lowering activity in the blood of a high-fat diet induced hyperlipidemia model in golden hamsters and significantly reduced the accumulation of fat granules in the liver. These results indicated that the key active components were produced by *E. cristatum*. Chemical investigations suggested similar components in the two extracts and led to the identification of a new alkaloid, namely variecolorin P (**1**), along with four known structurally related compounds, (-)-neoechinulin A (**2**), neoechinulin D (**3**), variecolorin G (**4**), and echinulin (**5**). The structure of the new alkaloid was elucidated by HRESIMS, $^{1}$H, $^{13}$C, and 2D NMR analysis. The lipid-lowering activity of these compounds was evaluated using an oleic acid-induced HepG2 cell line model. Compound **1** significantly reduced the lipid accumulation in the HepG2 cell line with an $IC_{50}$ value of 0.127 μM.

**Keywords:** hypolipidemic characteristics; *Eurotium cristatum*; Fuzhuan brick tea; methanol extract; spore; alkaloid

**Citation:** Song, F.; Zhang, K.; Yang, J.; Wilson, A.S.; Chen, C.; Xu, X. The Hypolipidemic Characteristics of a Methanol Extract of Fermented Green Tea and Spore of *Eurotium cristatum* SXHBTBU1934 in Golden Hamsters. *Nutrients* **2023**, *15*, 1329. https://doi.org/10.3390/nu15061329

Academic Editors: Paola Bontempo and Daniela Rigano

Received: 29 January 2023
Revised: 3 March 2023
Accepted: 7 March 2023
Published: 8 March 2023



## 1. Introduction

The continuous ingestion of a high-fat diet and decreased energy expenditure cause the excessive accumulation of fat, which lead to chronic diseases such as hepatic steatosis, cardiovascular diseases, and type 2 diabetes [1–3]. Obesity has become a widespread public health issue all over the world, most of which comes from the excessive accumulation of fat [4]. To treat obesity and avoid related diseases, increasing amounts of studies focused on exploring a new dietary effective functional food, to slow down fatty liver disease and regulate or improve intestinal microbiota with limited side effects [5].

Fuzhuan brick tea (FBT) is a popular beverage in China due to its unique flavor and variety of health-promoting functions. FBT or metabolites isolated from FBT exhibited a variety of bioactivities against metabolic diseases, including regulating the expression of multiple genes which mediated regulation of blood lipids [6–11], preventing obesity [12–14], hyperlipidemia [15] and hyperglycemia [10,16,17], ameliorating colitis [18], anti-bacterial [19], anti-oxidation [20], and modulating gut microbiota [21]. *Eurotium cristatum*, the predominant microorganism in FBT, was considered a probiotic that can alleviate the obesity of rodents induced by a high-fat diet [12]. The theabrownins from dark tea fermented by *E. cristatum* PW-1 were proved to show hypolipidemic activity in high-fat zebrafish [22]. The active metabolites secreted by *E. cristatum* may enhance the human immune system [23]. Further investigation of biomolecules from the spore of *E. cristatum* will promote the development of new functional foods.

Traditionally, FBT is cooked with milk or butter, which suggests that both water-soluble and lipid-soluble constituents could be extracted. As methanol is a good organic solvent to extract most of the polar and nonpolar compounds, in this study, it was used for extraction. The lipid-lowering activities of the methanol extract of *E. cristatum* fermented green tea and spore from *E. cristatum* fermented wheat were investigated, respectively in a high-fat-diet-induced hyperlipidemia model in golden hamsters. Furthermore, the chemical composition and bioactivity of the extracts were investigated.

## 2. Materials and Methods

### 2.1. Fungal Material, FBT, Spore of E. cristatum, and Extraction

Strain SXHBTBU1934 was isolated from FBT, bought from the market of Xixian New District, Shaanxi Province, China, and grown on a potato dextrose agar plate at 28 °C for 7 days. The genomic DNA of SXHBTBU1934 was extracted using DNA quick Plant System (Tiangen). The Internal Transcribed Spacer (ITS) sequence was amplified by using a conventional primer pair of ITS4 (5′-TCCTCCGCTTATTGATATGC-3′) and ITS5 (5′-GGAAGTAAAAGTCGTAACAAGG-3′). PCR products were sequenced by Beijing Qingke Biotechnology Co., Ltd. (Beijing, China). Strain SXHBTBU1934 was identified as *E. cristatum* based on gene sequence analysis of ITS. The strain was deposited at the Beijing Technology and Business University, Beijing, China.

Strain SXHBTBU1934 was inoculated on ten potato dextrose agar plates and cultured for 5 days at 28 °C. In total, 10 mL of distilled water was added to the plate to wash out the spore (3 times), and the spore suspension was combined as seed. In total, 5 mL of the seed was inoculated into twenty 1000 mL conical flasks, each containing 100 g green tea and 30 mL distilled water which was sterilized for 15 min at 121 °C. The inoculated flasks were incubated stationary at 28 °C for 20 days. The fermented tea by SXHBTBU1934 was dried by air. In total, 2 kg of dried fermented tea was soaked in 5 L methanol for 24 h and extracted three times. The organic solvent was evaporated in vacuo at 45 °C to yield a brown crude extract, which was used for HPLC-MS analysis and in vivo experiments.

For comparison, 10 g fermented tea by SXHBTBU1934 was added to 100 mL fresh milk and boiled for 2 min. Then, 1 mL methanol was added to a 2 mL aliquot of the milk extract and centrifuged (10,000 rpm) for 3 min. In total, 1 mL of supernatant was filtered by a 0.45 μm filter for HPLC analysis. The green tea without fermentation was also extracted with methanol.

Spores of *E. cristatum* were collected by Bio-tea Co., Ltd. of Shaanxi Biotech Group (Xianyang, China). *E. cristatum* was fermented on wheat on a large scale and the spores were collected as mentioned above. Briefly, the fermented wheat was washed three times with distilled water and the water was combined to collect the spore suspension, which was then centrifuged at 4000 rpm for 20 min. The supernatant was discarded and the spore pellet was dried by vacuum freeze-drying.

### 2.2. HPLC-MS Analysis of Extracts

The HPLC-MS analysis was performed on an Agilent 1200-MSD HPLC-MS system with reversed-phase Agilent Eclipse XDB-C8 column (4.6 mm i.d. × 150 mm, 5 μm, Agilent, Santa Clara, CA, USA) at 28 °C. Mobile phase A consisted of ultrapure water containing 0.1% methanolic acid, and mobile phase B was 100% acetonitrile. Gradient elution was performed by 10–100% B with a linear change within 15 min, then 100% acetonitrile for 5 min. Absorbance was monitored at 254 nm.

### 2.3. Compound Isolation

The crude extract from the spore of *E. cristatum* was dissolved in MeOH at a concentration of 100 mg/mL. The sample was subjected to preparative HPLC using a reversed-phase

Agilent Eclipse XDB-C8 column (9.4 mm i.d. × 250 mm, 5 μm, Agilent, Santa Clara, CA, USA) at 28 °C, with a gradient elution of MeCN in $H_2O$ from 40–80% within 20 min to yield compounds **1–5**.

### 2.4. Structure Determination

NMR spectra were obtained on a Bruker Avance 500 spectrometer at 25 °C (operating at 500 MHz for $^1$H-NMR, 125 MHz for $^{13}$C-NMR) with residual solvent peaks as references ($CDCl_3$-$d_6$: $\delta_H$ 7.26, $\delta_C$ 77.16). High-resolution electrospray ionization mass spectrometry (HR-ESIMS) measurements were obtained on an Accurate-Mass-Q-TOF LC/MS 6520 instrument (Santa Clara, CA, USA) in the positive ion mode. Optical rotations were measured on a Perkin-Elmer Model 343 polarimeter. Compound structures were elucidated by analyzing $^1$H NMR, $^{13}$C NMR, Heteronuclear Multiple Bond Correlation (HMBC), Heteronuclear Single Quantum Correlation (HSQC), $^1$H-$^1$H Correlation Spectroscopy (COSY), and comparison with the previously reported data. The structure formula was finally confirmed by HR-ESIMS.

### 2.5. Animal Experimental Design

In total, 50 healthy specific pathogen-free male golden hamsters (6 weeks old with a similar body weight of 100 g) were purchased from Beijing Vital River Laboratory Animal Technology Co., Ltd. (Beijing, China). Hamsters were maintained in a temperature-controlled room (23 °C) with a 12 h:12 h light–dark cycle (lights on from 8:00 to 20:00) and allowed one week acclimatization period [24]. Deionized water was provided continuously for hamsters to drink freely. All procedures using hamsters in this study were conducted by the recommendations in the Guide for the Care and Use of Laboratory Animals of the People's Republic of China. Golden hamsters were allowed one week of the adaptation period. Then, the hamsters were divided into two groups, which included 10 and 40 hamsters, respectively. In total, 10 hamsters were fed a normal diet as the control group (ND), and 40 hamsters in the experiment group were fed a high-fat diet (HFD) for rapid body weight growth. After 2 weeks, 30 hamsters from the HFD group with higher body weight (weight gain > 8 g) were selected and divided into 3 groups: high-fat diet control group (HFD), high-fat diet with methanol extract of fermented green tea intervention group (MET), and high-fat diet with the spore intervention group (ST). The dried methanol extract of fermented green tea and the dried spore of *E. cristatum* were both dissolved or suspended in 0.5% boxymethylcellulose sodium (80 mg/mL) for the in vivo experiment. The hamsters in MET and ST groups were intragastrically administered 400 mg of dissolved fermented green tea extract or spore suspension per kg body weight (5 mL/kg body weight), once daily for 2 weeks. The hamsters in the ND and HFD groups were administered 0.5% boxymethylcellulose sodium in the same way. After the last administration, all of the hamsters fasted for 12 h [25,26].

### 2.6. Serum and Liver Collection/Biochemical Analyses

Blood was collected from the retro-orbital plexus after the hamsters fasted for 12 h. The blood samples were centrifuged at 3000 rpm for 5 min to separate the serum for further biochemical analyses. Sections of fresh liver were put into a glass homogenizer and homogenized with 9 times (*w*/*w*) of physiological saline. The homogenized liver was centrifuged at 3000 rpm for 5 min and the supernatant was collected for further analysis. Total cholesterol (TC), triglyceride (TG), high-density lipoprotein cholesterol (HDL-C), and low-density lipoprotein cholesterol (LDL-C) levels in serum and liver, glucose, insulin, leptin, and free fatty acids (FFA) levels in serum were analyzed using commercial kits following the manufacturer's instructions. Free fatty acid (FFA) assay kit, mouse insulin ELISA kit, and mouse leptin ELISA kit were purchased from Beijing Solarbio Science & Technology Co, Ltd., Beijing, China. Mouse glucose assay kit, total cholesterol (TC) content assay kit, triglyceride (TG) content assay kit, HDL-cholesterol assay kit, and LDL-cholesterol assay kit were purchased from BioSino Bio-Technology & Science Inc., Beijing, China.

### 2.7. Cell Culture and Treatment

Human hepatoma HepG2 cells were obtained from the Beijing Union Medical Cell Resource Center (Basic Medical Cell Center, Institute of Basic Medical Sciences, Chinese Academy of Medical Sciences). The cells were cultured in DMEM containing 10% FBS and 1% penicillin–streptomycin and incubated with 5% $CO_2$ at 37 °C. Cells were allowed to grow to approximately 80% confluency and were sub-cultured at a ratio of 1:3 [27]. In this study, to simulate the process of nonalcoholic fatty liver disease in vitro, 0.2 mM oleic acid (OA) was added to the culture medium for 48 h to induce human HepG2 to establish a nonalcoholic fatty liver cell model [28].

### 2.8. Cell Viability Assay

HepG2 cells were seeded into 96-well plates after achieving logarithmic growth and cultured overnight at 37 °C with 5% $CO_2$. After treatment with different concentrations of compounds (dissolved in DMSO) for 24 h, 10 μL CCK-8 solution (Cell Counting Kit-8, Sigma-Aldrich (St. Louis, MO, USA)) was added to each well and the plate was incubated for 2 h in the incubator. Then, the absorbance at 450 nm was measured using a microplate reader. The normal control group was treated with the same amount of DMSO as the experiment group [29]. Cell viability was calculated by: cell viability (%) = $(OD_{drug}/OD_{control}) \times 100\%$.

### 2.9. Bodipy Staining

After treatment with compounds for 24 h in a 96-well plate, the supernatant of the HepG2 cells was discarded and cells were washed twice with PBS, then the cells were fixed with 4% paraformaldehyde at room temperature for 20 min. After removing the fixative reagent, the cells were washed twice with PBS. HepG2 cells were stained with 100 μL Bodipy (2 μg/mL, Thermo Fisher Scientific (Waltham, MA, USA)) in the dark at 25 °C for 30 min [30]. The fluorescence was measured using a fluorescence plate reader (Ex = 500 nm, Em = 550 nm, POLARstar, BMG Labtech (Offenburg, Germany)) [29]. The inhibition rate was calculated by comparison with the control group which was just treated with DMSO.

### 2.10. Oil Red O Staining

HepG2 cells were treated with different concentrations of compounds for 24 h, then were washed twice with PBS and fixed with 4% paraformaldehyde for 15–20 min. The cells were washed with PBS and stained with Oil Red O solution (50 μL per well, Sigma-Aldrich) at 25 °C for 1 h. After washed with PBS for 3–5 times, the lipid droplets and cell morphology were observed by a light microscope and photographed [29]. The area of lipid droplets was statistically compared for $IC_{50}$ calculation with the control group which was treated with DMSO using Image J software.

### 2.11. Statistical Analysis

Statistical analyses were performed using GraphPad Prism 8.0.2 (GraphPad Software, San Diego, CA, USA). The statistical significance of differences between groups was calculated by ordinary one-way ANOVA. The experimental data are presented as mean ± standard deviations (SD). Statistical significance is denoted by * for $p < 0.05$, ** for $p < 0.01$, *** for $p < 0.001$, and **** for $p < 0.0001$.

## 3. Results

### 3.1. Spores of *E. cristatum* Share Similiar Components as *E. cristatum* Fermented Green Tea

As FBT is traditionally cooked with butter or milk, we investigated the difference between the milk and methanol extracts of *E. cristatum*-fermented green tea, as well as the methanol extract of the spores. The extracts were analyzed by HPLC with a reversed-phase column. As shown in Figure 1, after fermentation by *E. cristatum*, milk, and methanol extracts showed several low-polarity HPLC peaks between the retention times of 13–18 min compared with the extract of green tea without fermentation. For the methanol extract of spore, more peaks with retention times between 7 and 13 min were found. As this extract shows an abundance of compounds, it was selected for further investigation for compound isolation.

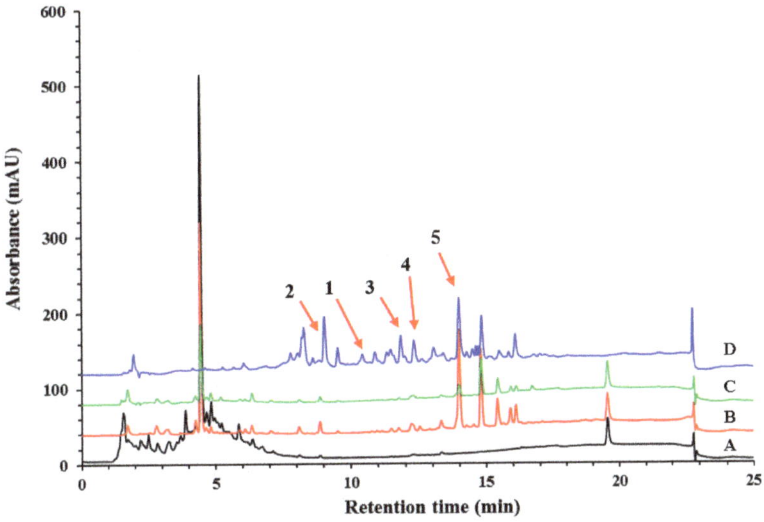


**Figure 1.** Component comparison of methanol extract of spores, methanol extracts of green tea with and without fermentation, and milk extract of fermented green tea. **A**: methanol extract of green tea (black); **B**: methanol extract of fermented green tea (red); **C**: milk extract of fermented green tea (green); **D**: methanol extract of spore (blue). **1** (variecolorin P); **2** ((-)-neoechinulin A); **3** (neoechinulin D); **4** (variecolorin G); **5** (echinulin).

### *3.2. Methanol Extract of Fermented Green Tea, as Well as Spore Suspension Alleviated HFD-Induced Body Weight and Ratio of Liver Weight to Body Weight*

The methanol and milk extracts of fermented green tea shared similar HPLC peaks, while the methanol extract of spores of *E. cristatum* contained more different peaks than the other two. Therefore, we selected the methanol extract of fermented green tea and spore for further investigation. The in vivo effects of the methanol extract of fermented green tea and the spore suspension on body weight and liver weight were studied in a high-fat induced hyperlipidemic model in golden hamsters. After 2 weeks of treatment with a high-fat diet, the body weight of golden hamsters in the HFD group increased to 139.3 ± 6.9 g, which showed a significant difference compared with the NC group (128.8 ± 5.6 g). Interestingly, after following 2 weeks of 400 mg/kg of the methanol extract of fermented green tea or spore suspension intervention, body weight was significantly controlled in groups of MET (155.7 ± 13.8 g, $p < 0.01$) and ST (158.5 ± 10.4 g, $p < 0.05$) (Figure 2A) while the body weight of the high-fat diet control group increased to 176.9 ± 9.2 g. The ratios of liver weight to body weight of hamsters in the MET and ST groups were significantly decreased to 0.0408 ± 0.0035 and 0.0411 ± 0.0031, respectively, $p < 0.001$ compared with the HFD group (0.0524 ± 0.0078). These results showed that the methanol extract of fermented green tea and spore suspension of *E. cristatum* were both able to inhibit the increase of body weight and the ratio of liver weight to body weight in golden hamsters.

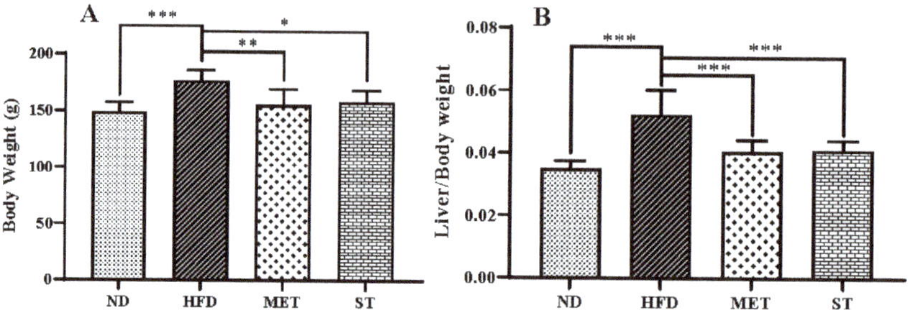


**Figure 2.** Effects of methanol extract of fermented green tea and spore suspension on body weight of hamsters for 2 consecutive weeks. (**A**): body weight; (**B**): ratio of liver weight to body weight. ND: normal diet treated group; HFD: high-fat diet; MET: high-fat diet with methanol extract of fermented green tea intervention group; ST: high-fat diet with the spore intervention group. Data are expressed as mean ± SD ($n = 10$). * $p \leq 0.05$; ** $p \leq 0.01$; and *** $p \leq 0.001$.

### 3.3. *Effect of Methanol Extract of Fermented Green Tea and Spore Suspension of* E. cristatum *on Lipid Levels in Serum and Liver*

The TC, TG, HDL-C, and LDL-C levels in the serum of golden hamsters for each group are shown in Figure 3. After four weeks of HFD induction, serum lipid levels, including TC, TG, HDL-C, and LDL-C, significantly increased from 4.30 ± 0.50, 1.94 ± 0.60, 2.68 ± 0.21, and 0.74 ± 0.22 mmol/L to 11.45 ± 2.08, 6.18 ± 1.92, 3.87 ± 0.25 and 2.89 ± 0.74 mmol/L, respectively (all groups with $p < 0.0001$). Methanol extract from fermented green tea treatment caused a significant decrease in TC (8.89 ± 2.01 mmol/L, $p < 0.01$), TG (3.60 ± 1.53 mmol/L, $p < 0.01$), and LDL-C (2.02 ± 0.50 mmol/L, $p < 0.01$) levels in comparison with HFD control (11.45 ± 2.08 mmol/L, 6.18 ± 1.92 mmol/L, and 2.89 ± 0.74 mmol/L, respectively). Interestingly, treatment with spores of *E. cristatum* also reduced the TC, TG, and LDL-C levels to 7.86 ± 1.40 mmol/L ($p < 0.0001$), 3.60 ± 1.60 mmol/L ($p < 0.01$), and 1.59 ± 0.36 mmol/L ($p < 0.0001$), respectively. However, neither of the treatments showed a significant effect on serum HDL-C levels. The TC, TG, HDL-C, and LDL-C levels in the livers of golden hamsters for each group were also investigated. As shown in Figure 4, methanol extract of fermented green tea and spore suspension of *E. cristatum* decreased TC (from 0.64 ± 0.17 to 0.45 ± 0.14 and 0.46 ± 0.14 mmol/100 g liver, respectively, both $p < 0.05$) and LDL-C (from 0.57 ± 0.13 to 0.39 ± 0.14 and 0.42 ± 0.10 mmol/100 g liver, with $p < 0.01$ and $p < 0.05$, respectively) levels in liver compared with the HFD control. Both interventions did not show significant effects on TG and HDL-C levels in the liver of golden hamsters.

To detect the effect of these samples on lipids accumulation in the liver, microscopic observations were performed by the hematoxylin and eosin (H&E) staining method. The liver fat granules in groups treated with methanol extract from fermented green tea and spore suspension of *E. cristatum* were significantly reduced compared with the HFD group (Figure 5). These results indicated that the methanol extract of fermented green tea and spore suspension can both reduce the accumulation of lipid droplets in the hamster's liver after 2 weeks intervention.

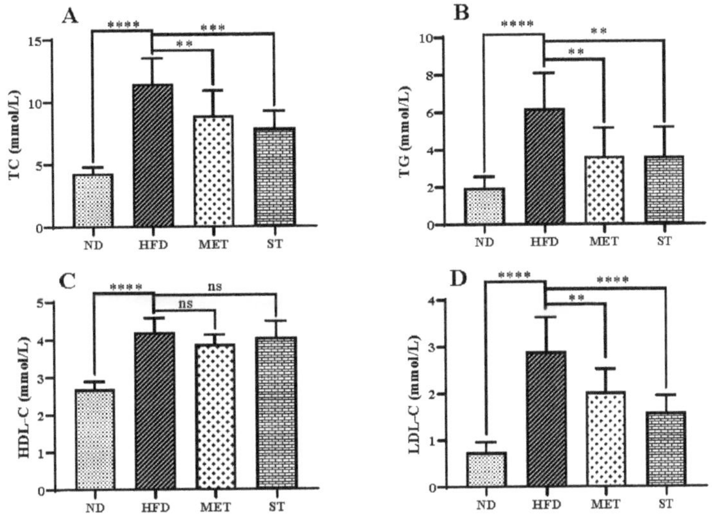


**Figure 3.** Effects of methanol extract of fermented green tea and spore suspension on TC (**A**), TG (**B**), HDL-C (**C**), LDL-C (**D**) levels in the serum of hamsters for 2 consecutive weeks of intervention. Statistical significance of differences between groups was determined using Ordinary one-way ANOVA. Data are expressed as mean ± SD ($n$ = 10). ** $p \leq 0.01$; *** $p \leq 0.001$; and **** $p \leq 0.0001$. "ns" indicates not significant ($p > 0.05$).

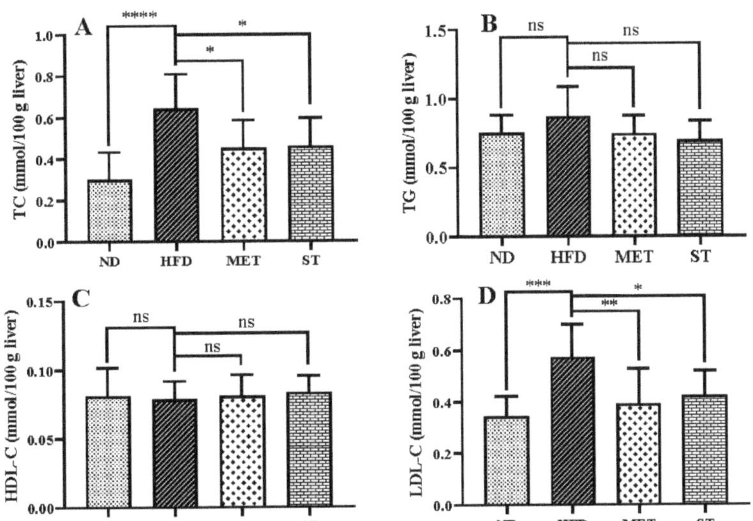


**Figure 4.** Effects of methanol extract of fermented green tea and spore suspension on TC (**A**), TG (**B**), HDL-C (**C**), LDL-C (**D**) levels in the liver of hamsters for 2 consecutive weeks of intervention. Statistical significance of differences between groups was determined using Ordinary one-way ANOVA. Data are expressed as mean ± SD ($n$ = 10). * $p \leq 0.05$; ** $p \leq 0.01$; *** $p \leq 0.001$; and **** $p \leq 0.0001$. "ns" indicates not significant ($p > 0.05$).

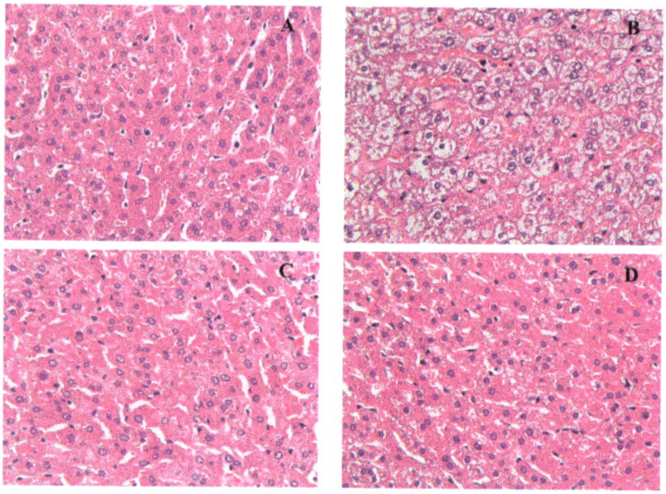


**Figure 5.** Effects of methanol extract of fermented green tea and spore suspension on fat granules in livers of hamsters for 2 consecutive weeks of intervention. (**A**): normal diet treated group (ND); (**B**): high-fat diet group (HFD): (**C**): high-fat diet with methanol extract of fermented green tea intervention group (MET); (**D**): high-fat diet with the spore intervention group (ST).

### 3.4. Methanol Extract of Fermented Green Tea and Spore Suspension Improve Diabetes-Related Biomarkers in Serum

The effects of these samples on biomarker levels in serum were evaluated. When hamsters were treated with HFD, the levels of serum glucose, insulin, leptin, and FFA were all significantly increased compared with the ND group, indicating that intake of HFD caused damage to the normal liver function of hamsters (Figure 6). After two weeks of intervention with methanol extract from fermented green tea and spore suspension of *E. cristatum*, the levels of blood glucose, insulin, and FFA were reduced significantly, while the average levels of leptin did not show a significant decrease.

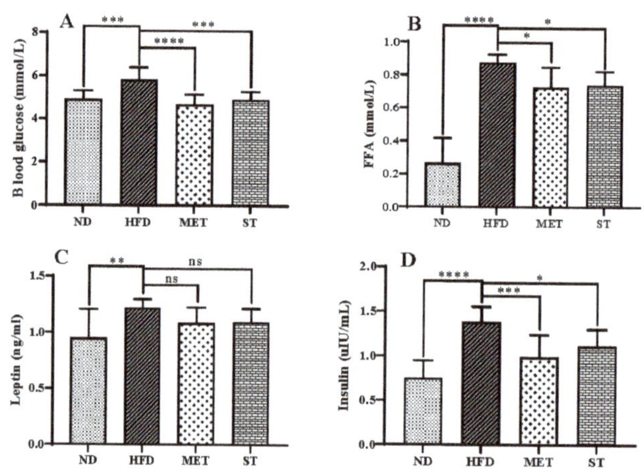


**Figure 6.** Effects of methanol extract of fermented green tea and spore suspension on diabetes related biomarkers in serum. (**A**): blood glucose; (**B**): FFA; (**C**): Leptin; (**D**): Insulin. Statistical significance of differences between groups was determined using Ordinary one-way ANOVA. Data are expressed as mean ± SD ($n$ = 10). * $p \leq 0.05$; ** $p \leq 0.01$; *** $p \leq 0.001$ and **** $p \leq 0.0001$. "ns" indicates not significant ($p > 0.05$).

### 3.5. Isolation and Structure Elucidation of Compounds from Spore

Both the methanol extract of fermented green tea and spore suspension from fermented wheat reduced the body weight, and serum lipid levels in HFD induced hyperlipidemia model in hamsters. To identify the key components displaying these bioactivities, the chemical components of spore extract were investigated. The spore was extracted by methanol and then purified by preparative HPLC to yield compounds **1–5** (Figure 7).

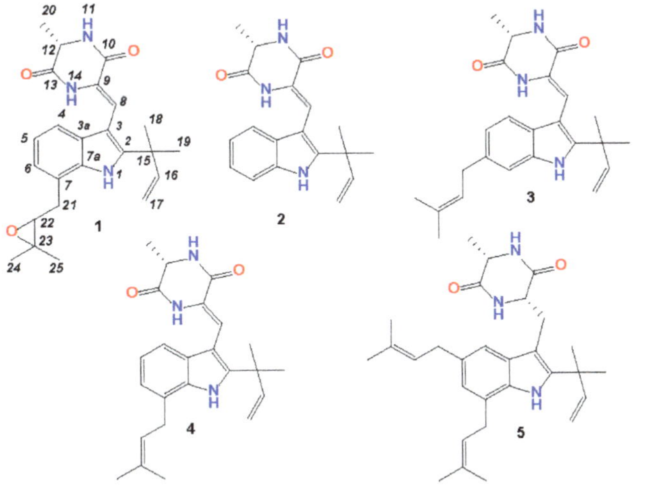

**Figure 7.** Structures of compounds **1–5** isolated from spore.

Compound **1** was isolated as a light yellow amorphous powder. The molecular formula of **1** was determined to be $C_{24}H_{29}N_3O_3$ based on its high-resolution electrospray ionization mass spectrum (HR-ESIMS) ($m/z$ $[M + H]^+$ 408.2289, calculated for $C_{24}H_{30}N_3O_3$, 408.2289), accounting for 12 degrees of unsaturation (Figure S1). The $^1H$, $^{13}C$ and HSQC NMR spectra of **1** (Figures S3–S5, Table 1) showed the presence of one doublet methyl group [$\delta_H$ 1.60/$\delta_C$ 21.0 (20-Me)], four singlet methyl groups [$\delta_H$ 1.55/$\delta_C$ 27.4 (18-Me), $\delta_H$ 1.55/$\delta_C$ 27.5 (19-Me), $\delta_H$ 1.51/$\delta_C$ 19.2 (C-24), $\delta_H$ 25.0/$\delta_C$ 140 (C-25)], one $sp^3$ methylene group [$\delta_H$ 3.24 and 2.92/$\delta_C$ 33.7 (C-21)], one $sp^3$ oxygenated methine [$\delta_H$ 3.01/$\delta_C$ 64.4 (C-22)], one 1,2,3-trisubstituted benzene moiety [$\delta_C$ 126.3 (C-3a), $\delta_H$ 7.19 (d, 8.0)/$\delta_C$ 117.8 (C-4), $\delta_H$ 7.08 (dd, 8.0, 8.0)/$\delta_C$ 121.1 (C-5), $\delta_H$ 6.99 (8.0)/$\delta_C$ 122.7 (C-6), $\delta_C$ 122.2 (C-7), $\delta_C$ 134.2 (C-7a)], one terminal double bond [$\delta_H$ 6.07/$\delta_C$ 144.3 (C-16), $\delta_H$ 5.15 and 5.17/$\delta_C$ 112.9 (C-17)], one olefinic double bond [$\delta_H$ 7.22/$\delta_C$ 112.7 (C-8), $\delta_C$ 112.4 (C-9)], one $sp^3$ oxygenated quaternary carbon [$\delta_C$ 60.4 (C-23)], two $sp^2$ quaternary carbons [$\delta_C$ 144.6 (C-2), $\delta_C$ 103.2 (C-3)], as well as two carbonyl carbons [$\delta_C$ 160.1 (C-10), $\delta_C$ 165.8 (C-13)]. All these data indicated that compound **1** owns a skeleton of indole-containing diketopiperazine alkaloids. $^1H$-$^1H$ COSY spectrum (Figure 8 and Figure S6) indicated the moieties of C-4/C-5/C-6, C-12/C-20, and C-21/C-22. In the HMBC spectra, the correlations from H-4 to C-3, C-6, and C-7a, from H-6 to C-7a, from H-NH-1 to C-2, C-3, and C-3a confirmed the indole moiety. The long range HMBC crossing peaks from $H_3$-18 and $H_3$-19 to C-2, C-15, and C-16 revealed the isopentene moiety (C15/C16/C17/C18/C19) and the connection between C-2 and C-15. The methyl diketopiperazine and the linkage between C-3 and C-8 were suggested by the HMBC correlations from $H_3$-20 to C-12 and C-13, from H-NH-14 to C-8, C-9, C-10, C-12, and C-13, and from H-8 to C-2, C-3a, and C-10. The oxygenated side chain was indicated by the HMBC correlations from $H_3$-24 and $H_3$-25 to C-22 and C-23. C-21 attachment to C-7 was revealed by the HMBC correlations from H-6 to C-21 and from H-21 to C-6, C-7, and C-7a. Combined with molecular formula and chemical shifts for C-22 ($\delta_C$ 64.4) and C-23 ($\delta_C$ 60.4), there should be an epoxy bond between C-2 and C-3.

Thus, the structure of **1** was determined as shown in Figure 8 and named variecolorin P. The alanine moiety in **1** was determined as L by comparing its optical rotation ($[a]_D^{25}$ -30) with the reported analog variecolorin M ($[a]_D^{25}$ -25) [31].

**Table 1.** $^1$H (500 MHz) and $^{13}$C NMR (125 MHz) data of Compound **1** (in $CDCl_3$).

| **Position** | **1** | | |
|---|---|---|---|
| | $\delta_C$ | $\delta_H$ (*J* in Hz) | **HMBC** |
| 1 | | 9.64 s | 2, 3, 3a, 7a, |
| 2 | 144.6 | | |
| 3 | 103.2 | | |
| 3a | 126.3 | | |
| 4 | 117.8 | 7.19 d (8.0) | 3, 6, 7a |
| 5 | 121.1 | 7.08 dd (8.0, 8.0) | 3a, 7 |
| 6 | 122.7 | 6.99 d (8.0) | 4, 7a, 21 |
| 7 | 122.2 | | |
| 7a | 134.2 | | |
| 8 | 112.2 | 7.22, s | 2, 3a, 10 |
| 9 | 124.0 | | |
| 10 | 160.1 | | |
| 12 | 51.8 | 4.29, qd (7.0, 1.5) | 10, 13, 21 |
| 13 | 165.8 | | |
| 14 | | 7.47, s | 8, 9, 10, 12, 13 |
| 15 | 39.3 | | |
| 16 | 144.3 | 6.07 dd (17.5, 10.5) | 2, 15, 18,19 |
| 17a | 112.9 | 5.15 d (17.5) | |
| 17b | | 5.17 d (10.5) | |
| 18 | 27.4 | 1.55 s | 2, 15, 16 |
| 19 | 27.5 | 1.55 s | 2, 15, 16 |
| 20 | 21.0 | 1.60 d (7.0) | 12, 13 |
| 21a | 33.7 | 2.92 d (15.0, 9.5) | 6, 7, 7a, 22, 23 |
| 21b | | 3.24 dd (15.0) | 6, 7, 7a, 22, 23 |
| 22 | 64.4 | 3.01 d (9.5) | 21 |
| 23 | 60.4 | | |
| 24 | 19.2 | 1.51 s | |
| 25 | 25.0 | 1.40 s | |

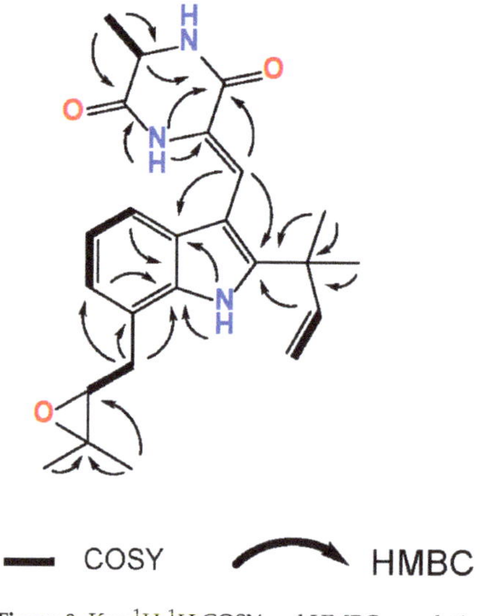


**Figure 8.** Key $^1$H-$^1$H COSY and HMBC correlations for compound **1**.

Four known indole-containing diketopiperazine analogs, (-)-neoechinulin A (**2**) [32], neoechinulin D (**3**) [33], variecolorin G (**4**) [34] and echinulin (**5**) [35] were also isolated and characterized by comparing their molecular weight and NMR data with those reported in the literature. The HPLC peaks of these four compounds as well as compound **1** are labeled in Figure 1.

### *3.6. Compound 1 Attenuated OA-Induced Lipid Accumulation in HepG2 Cells*

All of the compounds were assessed using an in vitro cell-based model in which NAFLD was simulated by inducing excessive oleic acid influx into the HepG2 cell line. In the primary screening, with Bodipy staining, compounds **1**, **2**, and **5** attenuated the accumulation of lipids in HepG2 cells with high cell viability at 100, 100, and 25 μg/mL (Table 2). As a new compound, compound **1** was further tested in the Oil Red O assay and showed a significant inhibitory effect against the lipid accumulation with $IC_{50}$ of 0.127 μM.

**Table 2.** Inhibition of lipid accumulation in HepG2 and cytotoxicity of compounds **1–5**.

| Compound | Concentration (μg/mL) | Inhibitory Rate (%) | Cell Viability (%) |
|---|---|---|---|
| **1** | 100 | 37.30 | 110.29 |
| **2** | 100 | 6.19 | 93.97 |
| **3** | 100 | −67.35 | 11.15 |
| **4** | 100 | −15.27 | 78.06 |
| **5** | 25 | 5.46 | 94.54 |

## 4. Discussion

People with metabolic syndrome normally have conditions including increased blood pressure, abnormal serum levels of blood sugar, cholesterol or triglyceride, and excess body fat. These conditions increase the risk of heart disease, stroke, and type 2 diabetes [36,37]. FBT, a kind of Chinese traditional beverage, is considered to originate in the 16th century (the Ming Dynasty of China) [38]. Chemical investigation of FBT revealed that there are many specific compounds in FBT produced by *E. cristatum* compared with other teas [7–9,39–41]. The in vivo health benefits of water crude extract of FBT have been investigated [14,15,21,42]. Limited studies focused on the in vivo evaluation of organic solvent extracts of FBT and spore of *E. cristatum* [10,43]. Therefore, we studied the in vivo health benefits of methanol extracts from fermented green tea and spore of *E. cristatum* on a high-fat-diet-induced hyperlipidemia model in hamsters. Furthermore, the chemical components of the spores were analyzed, isolated, and their bioactivity against hyperlipidemia was evaluated.

Our results showed that both the methanol extract from fermented green tea and the spore suspension of *E. cristatum* successfully alleviated the HFD-induced body weight in hamsters. In particular, the ratios of liver weight to body weight were also reduced after two weeks of treatment with interventions compared to HFD control. A study of water extract of FBT on a high-fat diet (HFD)-induced obese mouse assay reported that FBT dramatically ameliorated obesity [14]. *E. cristatum* is non-pathogenic and could survive in the mouse intestine. It can prevent HFD-induced obesity in C57BL/6J mice [12], which is similar to our results. Hyperlipidemia, also known as high cholesterol, means an increase of triacylglycerol (TG), total cholesterol (TC), low-density lipoprotein cholesterol (LDL-C), and a decrease of high-density lipoprotein cholesterol (HDL-C) in blood. HDL absorbs cholesterol from blood and transports it to the liver and improves the metabolism of cholesterol. High levels of HDL lower the risk for cardiovascular diseases [44]. LDL, made up of an outer rim of lipoprotein with a cholesterol center, can accumulate in the walls of blood vessels. High levels of LDL raise the incidence of metabolic diseases [45]. Our study revealed that the methanol extract and spore of *E. cristatum* both can significantly decrease the levels of TC, TG, and LDL-C in serum, and TC and LDL-C in the liver with no significant effects on HDL-C level in serum and liver. A previous study on a hexane extract of FBT [43] reported consistent results with our study. As for the spore of *E. cristatum*, our study reported that a methanol extract of spores of *E. cristatum* showed effects on TG, TC, HDL-C, and LDL-C in serum and liver in hamsters. Non-alcoholic fatty liver disease (NAFLD) is a leading cause of cirrhosis and hepatocellular carcinoma induced by unhealthy diets [46], which is mainly characterized by fat deposition in the liver [47,48]. Liu and co-workers evaluated the hypoglycemic effect of the water extract of spore of *E. cristatum* using a Hep-G2 cell hypoglycemic model and observed the evident increase in glucose consumption [10]. Our results indicated that both the methanol extracts of fermented green tea and spores of *E. cristatum* reduced the accumulation of fatty granules in the livers of hamsters. FBT was reported to have no significant effect on blood glucose homeostasis in HFD mice, while high doses of FBT significantly reduced blood glucose levels in fasting animals [14]. The current study showed that the methanol extract of fermented green tea and spores reduced the blood glucose in hamsters. High insulin levels led to decreased lipid utilization, which will increase the lipid accumulation and aggravate obesity and hyperlipidemia [49]. Zhou and co-workers reported the levels of insulin were decreased after the compound Fuzhuan brick intervention. Compared with glucose, free fatty acids (FFA, or non-esterified fatty acids, NEFA) account for a greater energy flux through the circulation [50,51]. However, a high concentration of FFA induces hepatic toxicity and ectopic lipid deposition, which plays a key role in the pathogenesis of NAFLD [52–54]. This study revealed that FBT and spore inhibited the increase of FFA concentration induced by a high-fat diet for the first time. Leptin, a hormone secreted by adipose tissue, can regulate food intake, increase energy release, inhibit the synthesis of adipocytes, and then reduce body weight by participating in the regulation of glucose, fat, and energy metabolism [55]. A high-fat diet feeding increased leptin and affected the body's glucose and lipid metabolism [56]. A previous study reported that a water extract of dark tea significantly reduced leptin mRNA in the liver, but no significant difference in fasting blood leptin was observed. Similarly, our results showed that two weeks of administration of FBT and spore did not reduce serum leptin. These results demonstrated that real health beneficial components in FBT are mostly contributed by the predominant fungus of *E. cristatum*, while raw tea provided nutrition during the fermentation process. The chemical compositions of FBT were greatly changed during the fermentation process by *E. cristatum* [39,40]. Different classes of metabolites, including polyphenols, phenolic acids, flavones, and their glycosides, terpenoids, alkaloids, steroids, tannins, and fatty acids, have been characterized in FBT [38]. However, limited studies focused on the identification of small molecules with bioactivities related to metabolic syndrome [10,43]. Our study investigated the methanol extract and compounds isolated from the spore of *E. cristatum* with hypolipidemic activity, which indicated the potential developmental value of the spore for functional food and pharmaceutical prospects of compound variecolorin P.

## 5. Conclusions

Our results indicated that the methanol extract of fermented green tea and spores of *E. cristatum* share similar secondary metabolites. Furthermore, both of them improved the hypolipidemic characteristics and obesity in a golden hamster model. Chemical investigation of spores of *E. cristatum* led to the identification of a new indole-containing diketopiperazine alkaloid, variecolorin P (**1**), together with four known analogs, including, (-)-neoechinulin A (**2**), neoechinulin D (**3**), variecolorin G (**4**), and echinulin (**5**). Variecolorin P inhibited the accumulation of lipids in HepG2 cells with $IC_{50}$ of 0.127 μM. Further studies on variecolorin P will reveal the potential value of FBT as a functional food for metabolic diseases.

**Supplementary Materials:** The following supporting information can be downloaded at: https://www.mdpi.com/article/10.3390/nu15061329/s1, Figures S1–S6: HRESIMS, 1D and 2D NMR of compound **1**.

**Author Contributions:** Funding acquisition, F.S.; Investigation, F.S., K.Z. and J.Y.; Methodology, F.S.; Project administration, X.X.; Resources, F.S.; Supervision, F.S.; Writing—original draft, F.S.; Writing—review and editing, F.S., K.Z., J.Y., A.S.W., C.C. and X.X. All authors have read and agreed to the published version of the manuscript.

**Funding:** This work was funded by grants from the Research Foundation for Advanced Talents of Beijing Technology and Business University (19008021176).

**Institutional Review Board Statement:** The animal study protocol was approved by the Beijing Viewsolid Biotechnology Co. LTD (protocol code VS212601479 and date of 21 February 2021).

**Informed Consent Statement:** Not applicable.

**Data Availability Statement:** The data are available from the corresponding author.

**Conflicts of Interest:** The authors declare no conflict of interest.

## References


1. Makri, E.; Goulas, A.; Polyzos, S.A. Epidemiology, pathogenesis, diagnosis and emerging treatment of nonalcoholic fatty liver disease. *Arch. Med. Res.* **2021**, *52*, 25–37. [CrossRef] [PubMed]
2. Rohm, T.V.; Meier, D.T.; Olefsky, J.M.; Donath, M.Y. Inflammation in obesity, diabetes, and related disorders. *Immunity* **2022**, *55*, 31–55. [CrossRef] [PubMed]
3. Piche, M.-E.; Tchernof, A.; Despres, J.-P. Obesity phenotypes, diabetes, and cardiovascular diseases. *Circ. Res.* **2020**, *126*, 1477–1500. [CrossRef]
4. Klein, S.; Gastaldelli, A.; Yki-Jarvinen, H.; Scherer, P.E. Why does obesity cause diabetes? *Cell Metab.* **2022**, *34*, 11–20. [CrossRef] [PubMed]
5. Martel, J.; Ojcius, D.M.; Chang, C.-J.; Lin, C.-S.; Lu, C.-C.; Ko, Y.-F.; Tseng, S.-F.; Lai, H.-C.; Young, J.D. Anti-obesogenic and antidiabetic effects of plants and mushrooms. *Nat. Rev. Endocrinol.* **2017**, *13*, 149–160. [CrossRef]
6. Lim, H.J.; Lim, T.J.; Lee, J.H.; Lee, J.H.; Kim, M.O.; Park, J.Y.; Kim, J.T.; Kim, M.J.; Jang, S.H.; Choi, S.H. Anti-obesity effects of dark tea extracts by down-regulation of C/EBPalpha and PPARgamma. *In Vivo* **2022**, *36*, 1753–1760. [CrossRef]
7. Luo, Z.M.; Du, H.X.; Li, L.X.; An, M.Q.; Zhang, Z.Z.; Wan, X.C.; Bao, G.H.; Zhang, L.; Ling, T.J. Fuzhuanins A and B: The B-ring fission lactones of flavan-3-ols from Fuzhuan brick-tea. *J. Agric. Food. Chem.* **2013**, *61*, 6982–6990. [CrossRef]
8. Luo, Z.M.; Ling, T.J.; Li, L.X.; Zhang, Z.Z.; Zhu, H.T.; Zhang, Y.J.; Wan, X.C. A new norisoprenoid and other compounds from Fuzhuan brick tea. *Molecules* **2012**, *17*, 3539–3546. [CrossRef]
9. Ling, T.J.; Wan, X.C.; Ling, W.W.; Zhang, Z.Z.; Xia, T.; Li, D.X.; Hou, R.Y. New triterpenoids and other constituents from a special microbial-fermented tea-Fuzhuan brick tea. *J. Agric. Food. Chem.* **2010**, *58*, 4945–4950. [CrossRef]
10. Liu, G.; Duan, Z.; Wang, P.; Fan, D.; Zhu, C. Purification, characterization, and hypoglycemic properties of eurocristatine from *Eurotium cristatum* spores in Fuzhuan brick tea. *RSC Adv.* **2020**, *10*, 22234–22241. [CrossRef]
11. Fu, D.; Ryan, E.P.; Huang, J.; Liu, Z.; Weir, T.L.; Snook, R.L.; Ryan, T.P. Fermented *Camellia sinensis*, Fu Zhuan Tea, regulates hyperlipidemia and transcription factors involved in lipid catabolism. *Food Res. Int.* **2011**, *44*, 2999–3005. [CrossRef]
12. Kang, D.; Su, M.; Duan, Y.; Huang, Y. *Eurotium cristatum*, a potential probiotic fungus from Fuzhuan brick tea, alleviated obesity in mice by modulating gut microbiota. *Food Funct.* **2019**, *10*, 5032–5045. [CrossRef] [PubMed]
13. Chen, G.J.; Xie, M.H.; Dai, Z.Q.; Wan, P.; Ye, H.; Zeng, X.X.; Sun, Y. Kudingcha and Fuzhuan brick tea prevent obesity and modulate gut microbiota in high-fat diet fed mice. *Mol. Nutr. Food Res.* **2018**, *62*, e1700485. [CrossRef]
14. Liu, D.M.; Huang, J.A.; Luo, Y.; Wen, B.B.; Wu, W.L.; Zeng, H.L.; Liu, Z.H. Fuzhuan brick tea attenuates high-fat diet-induced obesity and associated metabolic disorders by shaping gut microbiota. *J. Agric. Food Chem.* **2019**, *67*, 13589–13604. [CrossRef] [PubMed]
15. Li, Q.; Liu, Z.; Huang, J.; Luo, G.; Liang, Q.; Wang, D.; Ye, X.; Wu, C.; Wang, L.; Hu, J. Anti-obesity and hypolipidemic effects of Fuzhuan brick tea water extract in high-fat diet-induced obese rats. *J. Sci. Food Agric.* **2013**, *93*, 1310–1316. [CrossRef]
16. Xiang, X.; Xiang, Y.; Jin, S.; Wang, Z.; Xu, Y.; Su, C.; Shi, Q.; Chen, C.; Yu, Q.; Song, C. The hypoglycemic effect of extract/fractions from Fuzhuan Brick-Tea in streptozotocin-induced diabetic mice and their active components characterized by LC-QTOF-MS/MS. *J. Food Sci.* **2020**, *85*, 2933–2942. [CrossRef] [PubMed]
17. Xu, W.; Zhou, Y.; Lin, L.; Yuan, D.; Peng, Y.; Li, L.; Xiao, W.; Gong, Z. Hypoglycemic effects of black brick tea with fungal growth in hyperglycemic mice model. *Food Sci. Hum. Wellness* **2022**, *11*, 711–718. [CrossRef]
18. Liu, B.; Yang, T.; Zeng, L.N.; Shi, L.M.; Li, Y.; Xia, Z.G.; Xia, X.P.; Lin, Q.L.; Luo, F.J. Crude extract of Fuzhuan brick tea ameliorates DSS-induced colitis in mice. *Int. J. Food Sci. Technol.* **2016**, *51*, 2574–2582. [CrossRef]
19. Keller, A.C.; Weir, T.L.; Broeckling, C.D.; Ryan, E.P. Antibacterial activity and phytochemical profile of fermented *Camellia sinensis* (fuzhuan tea). *Food Res. Int.* **2013**, *53*, 945–949. [CrossRef]
20. Zhang, Q.-A.; Zhang, X.-L.; Yan, Y.-Y.; Fan, X.-H. Antioxidant evaluation and composition analysis of extracts from uzhuan brick tea and its comparison with two instant tea products. *J. AOAC Int.* **2017**, *100*, 653–660. [CrossRef]

21. Zhang, X.; Wu, Q.; Zhao, Y.; Aimy, A.; Yang, X. Consumption of post-fermented Jing-Wei Fuzhuan brick tea alleviates liver dysfunction and intestinal microbiota dysbiosis in high fructose diet-fed mice. *RSC Adv.* **2019**, *9*, 17501–17513. [CrossRef]
22. Xiao, Y.; Li, M.; Wu, Y.; Zhong, K.; Gao, H. Structural characteristics and hypolipidemic activity of theabrownins from dark tea fermented by single species *Eurotium cristatum* PW-1. *Biomolecules* **2020**, *10*, 204. [CrossRef] [PubMed]
23. Yan, Z.F.; Lin, P.; Kook, M.; Yi, T.H.; Li, C.T. Immune activation effects of *Eurotium cristatum* on T cells through NF-B signaling pathways in humans. *Food Agric. Immunol.* **2017**, *28*, 388–402. [CrossRef]
24. Mangiapane, E.H.; McAteer, M.A.; Benson, G.M.; White, D.A.; Salter, A.M. Modulation of the regression of atherosclerosis in the hamster by dietary lipids: Comparison of coconut oil and olive oil. *Br. J. Nutr.* **1999**, *82*, 401–409. [CrossRef]
25. Zhang, B.; Zhang, C.; Zhang, X.; Li, N.; Dong, Z.; Sun, G.; Sun, X. Atorvastatin promotes AMPK signaling to protect against high fat diet-induced non-alcoholic fatty liver in golden hamsters. *Exp. Ther. Med.* **2020**, *19*, 2133–2142. [CrossRef] [PubMed]
26. Liu, S.; Xiao, C.; Peng, C.; Liang, L.; He, X.; Zhao, S.; Zhang, G. Probiotic strains improve high-fat diet-induced hypercholesterolemia through modulating gut microbiota in ways different from atorvastatin. *Food Funct.* **2019**, *10*, 6098–6109. [CrossRef]
27. Li, N.; Sun, Y.R.; He, L.B.; Huang, L.; Li, T.T.; Wang, T.Y.; Tang, L. Amelioration by *Idesia polycarpa* Maxim. var. vestita Diels. of oleic acid-induced nonalcoholic fatty liver in HepG2 cells through antioxidant and modulation of lipid metabolism. *Oxid. Med. Cell Longev.* **2020**, *2020*, 1208726. [CrossRef]
28. Xia, H.; Zhu, X.; Zhang, X.; Jiang, H.; Li, B.; Wang, Z.; Li, D.; Jin, Y. Alpha-naphthoflavone attenuates non-alcoholic fatty liver disease in oleic acid-treated HepG2 hepatocytes and in high fat diet-fed mice. *Biomed. Pharmacother.* **2019**, *118*, 109287. [CrossRef]
29. Im, A.R.; Kim, Y.H.; Lee, H.W.; Song, K.H. Water extract of *Dolichos lablab* attenuates hepatic lipid accumulation in a cellular nonalcoholic fatty liver disease model. *J. Med. Food* **2016**, *19*, 495–503. [CrossRef]
30. Li, J.; Xie, S.; Teng, W. Sulforaphane attenuates nonalcoholic fatty liver disease by inhibiting hepatic steatosis and apoptosis. *Nutrients* **2021**, *14*, 76. [CrossRef]
31. Wei, X.; Feng, C.; Wang, S.Y.; Zhang, D.M.; Li, X.H.; Zhang, C.X. New indole diketopiperazine alkaloids from soft coral-associated epiphytic fungus *Aspergillus* sp. EGF 15-0-3. *Chem. Biodivers.* **2020**, *17*, e2000106. [CrossRef] [PubMed]
32. Aoki, T.; Kamisuki, S.; Kimoto, M.; Ohnishi, K.; Takakusagi, Y.; Kuramochi, K.; Takeda, Y.; Nakazaki, A.; Kuroiwa, K.; Ohuchi, T.; et al. Total synthesis of (-)-neoechinulin A. *Synlett* **2006**, *2006*, 677–680. [CrossRef]
33. Bovio, E.; Garzoli, L.; Poli, A.; Luganini, A.; Villa, P.; Musumeci, R.; McCormack, G.P.; Cocuzza, C.E.; Gribaudo, G.; Mehiri, M.; et al. Marine fungi from the sponge *Grantia compressa*: Biodiversity, chemodiversity, and biotechnological potential. *Mar. Drugs* **2019**, *17*, 220. [CrossRef]
34. Wang, W.-L.; Lu, Z.-Y.; Tao, H.-W.; Zhu, T.-J.; Fang, Y.-C.; Gu, Q.-Q.; Zhu, W.-M. Isoechinulin-type alkaloids, variecolorins A-L, from halotolerant Aspergillus variecolor. *J. Nat. Prod.* **2007**, *70*, 1558–1564. [CrossRef]
35. Miyake, Y.; Ito, C.; Itoigawa, M.; Osawa, T. Antioxidants produced by *Eurotium herbariorum* of filamentous fungi used for the manufacture of karebushi, dried bonito (Katsuobushi). *Biosci. Biotechnol. Biochem.* **2009**, *73*, 1323–1327. [CrossRef]
36. Huang, P.L. A comprehensive definition for metabolic syndrome. *Dis. Model. Mech.* **2009**, *2*, 231–237. [CrossRef] [PubMed]
37. Naqvi, S.S.Z.H.; Imani, S.; Hosseinifard, H.; Wen, Q.-L.; Shahzad, M.N.; Ijaz, I.; Deng, Y.; Guo, M.; Xu, Y. Associations of serum low-density lipoprotein and systolic blood pressure levels with type 2 diabetic patients with and without peripheral neuropathy: Systemic review, meta-analysis and meta-regression analysis of observational studies. *BMC Endocr. Disord.* **2019**, *19*, 125. [CrossRef]
38. Chen, G.; Peng, Y.; Xie, M.; Xu, W.; Chen, C.; Zeng, X.; Liu, Z. A critical review of Fuzhuan brick tea: Processing, chemical constituents, health benefits and potential risk. *Crit. Rev. Food Sci. Nutr.* **2021**. [CrossRef]
39. Zhu, Y.F.; Chen, J.J.; Ji, X.M.; Hu, X.; Ling, T.J.; Zhang, Z.Z.; Bao, G.H.; Wan, X.C. Changes of major tea polyphenols and production of four new B-ring fission metabolites of catechins from post-fermented Jing-Wei Fu brick tea. *Food Chem.* **2015**, *170*, 110–117. [CrossRef]
40. Jia, W.; Shi, Q.; Shi, L.; Qin, J.; Chang, J.; Chu, X. A strategy of untargeted foodomics profiling for dynamic changes during Fu brick tea fermentation using ultrahigh-performance liquid chromatography-high resolution mass spectrometry. *J. Chromatogr. A* **2020**, *1618*, 460900. [CrossRef]
41. Lu, Y.; He, Y.J.; Zhu, S.H.; Zhong, X.H.; Chen, D.; Liu, Z.H. New acylglycosides flavones from Fuzhuan brick tea and simulation analysis of their bioactive effects. *Int. J. Mol. Sci.* **2019**, *20*, 494. [CrossRef] [PubMed]
42. Zhou, X.; Ge, B.; Zhang, X.; Wang, K.; Zhou, C.; Fu, D. Metabolomics analysis reveals the effects of compound Fuzhuan brick tea (CFBT) on regulating dyslipidemia and metabolic disorders in mice induced by high-fat diet. *Nutrients* **2022**, *14*, 1128. [CrossRef] [PubMed]
43. Song, F.; Dai, W.; Li, H.; Zhang, X.; Xu, X.; Ma, L.; Wang, L. Characterization of hypolipidemic phenol analogues from fermented tea by *Eurotium cristatum*. *Foods* **2022**, *12*, 49. [CrossRef] [PubMed]
44. Ali, K.M.; Wonnerth, A.; Huber, K.; Wojta, J. Cardiovascular disease risk reduction by raising HDL cholesterol—Current therapies and future opportunities. *Br. J. Pharmacol.* **2012**, *167*, 1177–1194. [CrossRef]
45. Wadhera, R.K.; Steen, D.L.; Khan, I.; Giugliano, R.P.; Foody, J.M. A review of low-density lipoprotein cholesterol, treatment strategies, and its impact on cardiovascular disease morbidity and mortality. *J. Clin. Lipidol.* **2016**, *10*, 472–489. [CrossRef]
46. Powell, E.E.; Wong, V.W.-S.; Rinella, M. Non-alcoholic fatty liver disease. *Lancet* **2021**, *397*, 2212–2224. [CrossRef]
47. Cobbina, E.; Akhlaghi, F. Non-alcoholic fatty liver disease (NAFLD)—Pathogenesis, classification, and effect on drug metabolizing enzymes and transporters. *Drug Metab. Rev.* **2017**, *49*, 197–211. [CrossRef]

48. Stefan, N.; Haering, H.-U.; Cusi, K. Non-alcoholic fatty liver disease: Causes, diagnosis, cardiometabolic consequences, and treatment strategies. *Lancet Diabetes Endocrinol.* **2019**, *7*, 313–324. [CrossRef]
49. Li, J.; Wu, H.; Liu, Y.; Yang, L. High fat diet induced obesity model using four strains of mice: Kunming, C57BL/6, BALB/c and ICR. *Exp. Anim.* **2020**, *69*, 326–335. [CrossRef]
50. Frayn, K.N. Non-esterified fatty acid metabolism and postprandial lipaemia. *Atherosclerosis* **1998**, *141* (Suppl. S1), S41–S46. [CrossRef]
51. Hara, T.; Kashihara, D.; Ichimura, A.; Kimura, I.; Tsujimoto, G.; Hirasawa, A. Role of free fatty acid receptors in the regulation of energy metabolism. *Biochim. Biophys. Acta Mol. Cell Biol. Lipids* **2014**, *1841*, 1292–1300. [CrossRef] [PubMed]
52. Wu, X.; Zhang, L.; Gurley, E.; Studer, E.; Shang, J.; Wang, T.; Wang, C.; Yan, M.; Jiang, Z.; Hylemon, P.B.; et al. Prevention of free fatty acid-induced hepatic lipotoxicity by 18 beta-glycyrrhetinic acid through lysosomal and mitochondrial pathways. *Hepatology* **2008**, *47*, 1905–1915. [CrossRef] [PubMed]
53. Feldstein, A.E.; Werneburg, N.W.; Canbay, A.; Guicciardi, M.E.; Bronk, S.F.; Rydzewski, R.; Burgart, L.J.; Gores, G.J. Free fatty acids promote hepatic lipotoxicity by stimulating TNF-alpha expression via a lysosomal pathway. *Hepatology* **2004**, *40*, 185–194. [CrossRef] [PubMed]
54. Fiorenza, C.G.; Chou, S.H.; Mantzoros, C.S. Lipodystrophy: Pathophysiology and advances in treatment. *Nat. Rev. Endocrinol.* **2011**, *7*, 137–150. [CrossRef]
55. Seoane-Collazo, P.; Martinez-Sanchez, N.; Milbank, E.; Contreras, C. Incendiary leptin. *Nutrients* **2020**, *12*, 472. [CrossRef]
56. Lee, S.; Kweon, O.-K.; Kim, W.H. Associations between serum leptin levels, hyperlipidemia, and cholelithiasis in dogs. *PLoS ONE* **2017**, *12*, e0187315. [CrossRef]

nutrients

MDPI

Article

# Thiolated Chitosan as an Intestinal Absorption Carrier with Hesperidin Encapsulation for Obesity Treatment

Tzu-Chien Chen [1], Yu-Yu Ho [1], Rui-Chian Tang [2], Yong-Chen Ke [1], Jhih-Ni Lin [1], I-Hsuan Yang [1] and Feng-Huei Lin [1,3,*]

1 Department of Biomedical Engineering, College of Medicine and College of Engineering, National Taiwan University, No. 49, Fanglan Rd., Taipei 10672, Taiwan; d07528022@ntu.edu.tw (T.-C.C.); paulho56@gmail.com (Y.-Y.H.); lala851206@gmail.com (Y.-C.K.); febe200630@gmail.com (J.-N.L.); tony910028@gmail.com (I.-H.Y.)
2 Department of Biochemical Science and Technology, College of Life Science, National Taiwan University, No. 1, Sec. 4, Roosevelt Rd., Taipei 10617, Taiwan; b06b02006@ntu.edu.tw
3 Institute of Biomedical Engineering and Nanomedicine, National Health Research Institutes, No. 35, Keyan Rd., Zhunan, Miaoli County 35053, Taiwan
* Correspondence: double@ntu.edu.tw; Tel.: +886-2-27327474

**Abstract:** Obesity is characterized as abnormal or excessive fat accumulation harmful to one's health, linked to hormonal imbalances, cardiovascular illness, and coronary artery disease. Since the disease stems mainly from overconsumption, studies have aimed to control intestinal absorption as a route for treatment. In this study, chitosan-thioglycolic acid (CT) was developed as a physical barrier in the gastrointestinal tracts to inhibit nutrient uptake. CT exhibits a superior mucoadhesive property compared to chitosan both in vitro and in vivo for the ability to form disulfide bonds with the intestinal mucosa. For CT as a potential drug delivery platform, hesperidin, a herb for bodyweight control in traditional Chinese medication, is encapsulated in CT and can be released consistently from this absorption barrier. In animal studies, CT encapsulated with hesperidin (CTH) not only results in a weight-controlling effect but limits adipose accumulation by hindering absorption, suggesting a potential role in obesity treatment. Neither CT nor CTH exhibit cytotoxicity or produce adverse immunological reactions in vivo.

**Keywords:** obesity; chitosan; thiol; hesperidin

**Citation:** Chen, T.-C.; Ho, Y.-Y.; Tang, R.-C.; Ke, Y.-C.; Lin, J.-N.; Yang, I.-H.; Lin, F.-H. Thiolated Chitosan as an Intestinal Absorption Carrier with Hesperidin Encapsulation for Obesity Treatment. *Nutrients* **2021**, *13*, 4405. https://doi.org/10.3390/nu13124405

Academic Editor: Daniela Rigano

Received: 20 October 2021
Accepted: 3 December 2021
Published: 9 December 2021



## 1. Introduction

According to the WHO definitions, obesity and overweight refer to the excessive or abnormal fat accumulation that affects health [1]. The American Medical Association (AMA) officially defined obesity as a disease characterized by psychological symptoms, hormonal imbalance, and lack of neurotransmitters [2]. In 2014, there were more than 1.9 billion overweight adults, of whom 600 million were obese. It is estimated that 60% of adult males worldwide will be overweight or obese in 2030 [3], and the underlying childhood obesity problem is becoming gradually serious [4]. The problem of obesity is one of the most serious public health issues of the 21st century.

The etiology of obesity is caused by an energy imbalance. Energy imbalance is mainly caused by genes, diet, lifestyle, diseases, and drugs [5]. As well as congenital genetic defects, obesity may also be caused by Prader Willi syndrome or Alstrom syndrome [6,7]. Besides diseases such as Cushing's syndrome and hypothyroidism, drugs such as steroids and insulin can also cause obesity [8,9].

Accordingly, studies have aimed to control intestinal absorption as a route of treatment, in which bariatric surgeries play a major role. There are two types of bariatric surgery, namely invasive and semi-invasive. The invasive methods include sleeve gastrectomy (SG), which changes the structure of the stomach to limit a patient's appetite [10]; Furthermore,

Roux-en-Y Gastric Bypass (RYGB) creates bypasses of the duodenum to prevent excess nutrients from being absorbed [11]. Nevertheless, these invasive surgeries irreversibly alter the gastrointestinal tracts, leaving only a small number of patients willing to undergo the surgeries. A semi-invasive method, EndoBarrier, utilizes an endoscopically delivered polymer film to block the absorption of nutrients to achieve weight control for patients with severe type 2 diabetes or morbidly obese patients. However, Endobarrier may cause abnormal proliferation where the device is installed. Additionally, it is required to be removed annually by endoscopy in a second operation, which may be uncomfortable for the patient [12–14]. As a result, this experiment aims to develop a polymer that can inhibit nutrient absorption in the small intestine to achieve bodyweight control.

Chitosan, a natural polysaccharide, has been studied for almost two decades in biomedical research because of its biodegradation, biodistribution, and low toxicity [15]. Additionally, the amino group on chitosan gives the material the ability to be easily modified [16,17]. Despite having been evaluated in a number of trials, the effectiveness of chitosan as a dietary supplement to reduce bodyweight remains in dispute [18]. Therefore, we used chitosan as the potential material to create a new type of absorption barrier material that is both safe and effective. The thiol group immobilizing strategy was selected for chitosan modification, for we hypothesized that the spontaneous disulfide bonds formed between thiomers and mucin on the intestine would enable improved adhesion. Moreover, it has been shown that the disulfide bonds formed by thiomers and mucin can be reversed [19,20]. As a consequence, the material may not be fixed permanently to the small intestinal mucosa.

The 1,4-butanediol diglycidyl ether (BDDE) was used in this study to crosslink chitosan with thioglycolic acid to synthesize chitosan-TGA (CT). To test the potential of CT to become a mucoadhesive drug-delivery platform, hesperidin was encapsulated in the CT as CTH. Hesperidin is a traditional Chinese medicinal material extracted from tangerine peel. It can reduce cholesterol, low-density lipoprotein, and triglycerides as shown in previous studies [21,22]. Therefore, we expected that hesperidin mixed with CT could produce a better weight-control effect.

CT was characterized by Fourier transform infrared spectroscopy (FTIR) and nuclear magnetic resonance (NMR) spectroscopy to identify functional groups and the synthetic material structure. WST-1 assay and live/dead staining were used to determine cell viability and biocompatibility. The mucoadhesive property was verified using u-slides seeded with IEC-6 cells and FITC-labeled CT (CTF). The obesity model was developed by C57BL/6 mice fed a high-fat diet. Finally, tissue H&E staining, complete blood count, and serum biochemical analysis were used to evaluate the impact of daily feeding materials on the animal body or tissue.

## 2. Materials and Methods

### 2.1. Synthesis of Chitosan-TGA (CT)

To synthesize the thiolated chitosan, 1 g of chitosan powder was mixed into 50 mL of 0.1 M acetic acid solution, and 0.37 mL of thioglycolic acid (TGA) was added and the mix stirred evenly. Then, 0.098 mL of BDDE was dissolved in 50 mL of isopropanol and mixed into the solution mentioned above. After stirring for three hours in the dark at room temperature, the solution was dialyzed in ddH2O using a dialysis membrane (MWCO: 12,000 Da) for 72 h. The solution was further lyophilized and stored at room temperature for the following experiments.

### 2.2. FTIR Spectroscopy Analysis

The FTIR was used for the determination of functional groups within the candidate materials. A total reflection FTIR spotlight 200i (Perkin Elmer, Waltham, MA, USA) was used to conduct the measurements. The sample was scanned 16 times with a resolution of 4 $cm^{-1}$. The sample was analyzed by transmittance mode in the range of 500–4000 $cm^{-1}$. Absorption peaks of each functional group were analyzed.

### 2.3. NMR Spectroscopy Analysis

An amount of 8 mg chitosan or CT was dissolved in 1 mL of 0.1 M CD3COOD/D2O solution and filtered with a 0.22 μm filter. The inspection was performed by the National Taiwan University Precious Instrument Center. The 1H-NMR was measured with a 500 MHz superconducting magnet on a Bruker Avance III 500 spectrometer.

### 2.4. Cytotoxicity of CT

The cytotoxicity of the CT was evaluated by WST-1 assay and live/dead staining. L929 cells were seeded into the 96 wells, at a cell density of 7000 per well. CT was dissolved in 0.1 M acetic acid aqueous solution at a concentration of 3 mg/mL and mixed with a medium in a ratio of 1:5. The positive control was 0.2 g/mL ZDEC extract, while the negative control was 0.2 g/mL $Al_2O_3$ extract. After co-culturing the cells with the extracts for 24 h, the WST-1 assay was added for 1.5 h of reaction. Finally, an ELISA reader was used to detect the absorbance wavelength at 450 nm to measure cell activity. Live/dead staining was used to evaluate cytotoxicity. The L929 cell line was first seeded in a 24-well plate at a density of 20,000 cells/well. After one day of incubation, 0.5 mL of 1.5% chitosan or CT was added in a serum-free MEM medium to each well. After another 24 h of incubation, the cell culture medium was removed, and the cells were thoroughly washed by phosphate-buffered saline (PBS) 3 times. Triton X-100, as a positive control, was added 5 min before staining. Then, a staining solution containing calcein-AM (2 μM) and ethidium homodimer-1 (EthD-1) (4 μM) was added and allowed to react in the wells for 30 min. After washing with PBS, the cells were observed using a fluorescence microscope to evaluate the cytotoxicity of the chitosan and CT.

### 2.5. In Vitro Mucus Adhesion Test

The mucus adhesion test was adopted to gauge the enhanced mucoadhesive properties of CT. IEC-6 cells. Intestine epithelial cells, which could secrete mucosa layers to mimic the microenvironment of the gastrointestinal tract were utilized. IEC-6 was seeded in the μ-Slides (from ibidi μ-Slide I 0.8 Luer) at a cell density of 400,000 cells/mL overnight. The cells were first stained with Hoechst 33,342 (1 μg/mL) within the medium for 15 min. Then, 6 mg/mL of chitosan-FITC (CF) or CT-FITC (CTF) in 0.1 M acetic acid solution was diluted with the medium at a ratio of 1:1 and incubated with IEC-6 on the μ-slides for 1 h. The microfluidic device generated a constant pumping of culture medium into the slides at a flow rate of 3 mL/h. Finally, at the times of 0, 1, and 2 h, a fluorescent microscope was used to observe the fluorescence intensity of the materials bonded with the cells to evaluate the adhesion of the materials.

### 2.6. Oral Glucose Tolerance Test (OGTT) and Intraperitoneal Glucose Tolerance Test (IPGTT)

Eight-week-old C57BL/6 male mice were starved for 18 h and 6 mice were placed in each group including the control group, the positive control group, and the CT group. Except for the control group, positive control, and CT groups mice were gavaged with PBS or CT (250 mg/kg) respectively 1 h before oral administration of glucose solution (3 g/kg mice) in OGTT. The IPGTT, positive control, and CT groups were gavaged with PBS or CT (250 mg/kg) respectively 1 h before intraperitoneal injections of glucose solution (3 g/kg mice). Mice without glucose gavage served as the control group. Glucose levels were measured using a glucose meter (Accu-Chek Instant, Roche, Basel, Switzerland) every 30 min for 120 min. Moreover, the incremental area under the curve (iAUC) was used to diagnose the response of OGTT.

### 2.7. In Vivo Adhesion Test of CT

The adhesion properties of CF or CTF in vivo were evaluated by the IVIS Luminar II in vivo imaging system (Perkin Elmer, Waltham, MA, USA). C57BL/6 male mice were gavaged with CF or CTF with a dose of 250 mg/kg mice. The gastrointestinal tracts were harvested at each time point of 1, 2, 4, and 24 h.

### 2.8. In Vivo Study

C57BL/6 (4-week-old) male mice were used to test the long-term treatment of CT or CTH. The operations following lasted eight weeks. The mice in the control group were fed with a normal diet (ND) and were gavaged with phosphate-buffered saline (PBS) daily. The mice in the high-fat diet (HFD) group were fed a 60 percent high-fat diet from Dyets (DYET# 112252) and gavaged with PBS. The mice in the HFD + CT group or HFD + CTH group were fed with the same 60 percent high-fat diet but gavaged with different material experiment solutions (250 mg/kg) daily. Every week, the body weight and food intake were measured. A full-body scan employing micro computerized tomography (SKYSCAN, Bruker, Billerica, MA, United States) with a resolution of 35 μm was done at the end of the in vivo investigation to examine the effects of material gavaging on adipose tissue distribution.

### 2.9. Statistics

The results with at least three replicates were presented as "mean ± standard deviation." To assess the statistical significance of the experiment, a one-way ANOVA with multiple comparison tests was used. At a $p$-value of less than 0.05, differences were judged significant $p < 0.05$, *; $p < 0.01$, **; $p < 0.001$, ***.

## 3. Results

### 3.1. Material Molecular Structure Analysis

FTIR, which could measure each functional group's own discrete vibrational energy, was used to identify the functional groups of chitosan and CT (Figure 1a). Both chitosan and CT exhibited similar characteristics of OH absorption at 3600–3200 $cm^{-1}$ and CH stretching at 3000–2850 $cm^{-1}$. As our expectation, there was an absorption band for the thiol group at 2550 $cm^{-1}$ for CT. Apparently, these results indicated that the amino groups in chitosan were successfully modified with thiol groups. By analyzing the 1H-NMR spectrum, the structure of CT could be verified. As seen in Figure 1b for chitosan, H1 appeared at 2.0 ppm, H2, H3, H4, H5, H6 were distributed between 3.4 ppm and 3.8 ppm, which was comparable to the NMR spectroscopy of chitosan in previous studies [23]. In the 1H-NMR spectrum of CT, BDDE, the cross-linking agent, was detected at 3.4 ppm. CT also exhibited an obvious characteristic peak at 3.2 ppm, confirming that a thiol group was grafted onto the molecule (Figure 1b). In the EDS analysis, sulfur signals were detected on the CT surface (Figure 1c). The collaborating results confirmed that CT was successfully synthesized by conjugating TGA with chitosan by BDDE crosslinker.

### 3.2. Cytotoxicity and Mucus Adhesion Test of CT In Vitro

Following the specifications of ISO10993, the cytotoxicity of CT was evaluated by the WST-1 cell viability assay and live/dead staining. To determine the cytotoxicity of chemicals in vitro, the WST-1 assay, a tetrazolium salt, which is converted into a colored dye by mitochondrial dehydrogenase enzymes, is the most commonly used test. The control group consisted of conventional culture media; the positive control consisted of ZDEC extract, and the negative control group consisted of $Al_2O_3$ extract. The experimental group comprised chitosan or CT dissolved in the medium. Results showed that the materials were biocompatible and had negligible toxicity in L929 cells (Figure 2a). Live/Dead assay was performed to evaluate whether the material would showcase obvious cytotoxicity toward the L929 cell line as shown in (Figure 2b). The cell-permeant dye calcein AM can be used to determine the viability of most eukaryotic cells. Ethidium homodimer 1 dye (EthD-1) is highly positively charged. Therefore, it can be used to detect nucleic acids in solutions or to stain dead cells selectively. In both the chitosan and CT group, cell morphology was not significantly altered, and most of the cells were alive. Consequently, the above results indicated that chitosan and CT did not exhibit cytotoxicity and possessed desirable biocompatibility.

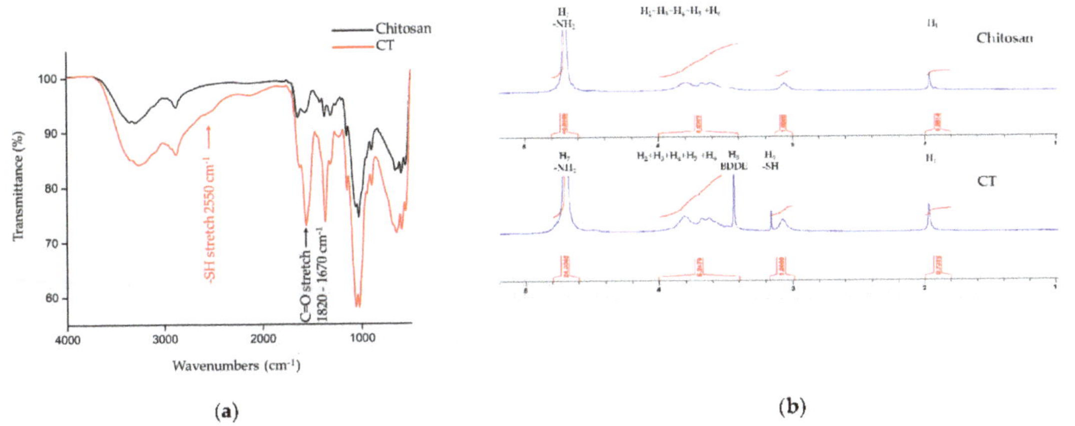


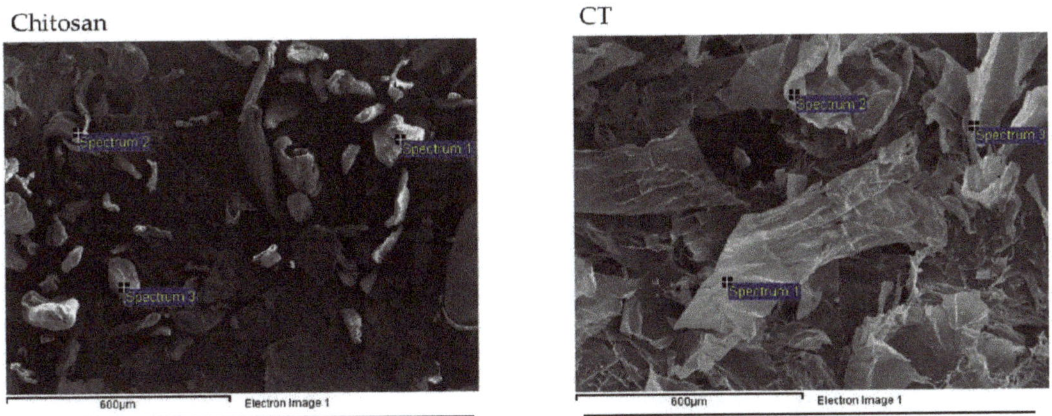


| Chitosan–EDS | C | O | S |
|---|---|---|---|
| Spectrum 1 | 62.91 | 37.09 | 0 |
| Spectrum 2 | 61.07 | 38.93 | 0 |
| Spectrum 3 | 60.89 | 39.11 | 0 |

(c)

| CT–EDS | C | O | S |
|---|---|---|---|
| Spectrum 1 | 51.58 | 44.79 | 3.63 |
| Spectrum 2 | 56.01 | 40.61 | 3.38 |
| Spectrum 3 | 52.55 | 41.99 | 5.46 |

(d)

**Figure 1.** Material molecular structure analysis of CT. (**a**) The FTIR result of chitosan and CT. (**b**) $^1$Hydrogen NMR spectra confirmed the structure of chitosan and CT. EDS results showed that there were sulfur signals on the surface of CT (**d**) but not on chitosan (**c**).

To test the mucoadhesive properties of the thiolated material, a μ-slide containing IEC-6 cells (small intestine epithelial cell) was used, followed by a constant flow of medium. For clear observation, nuclei were stained with Hoechst 33342, and candidate materials were labeled with FITC, demonstrated in blue fluorescence for nucleus and green fluorescence for the material respectively. At the beginning of the test, most of the materials were presented in the μ-slide, causing a strong background and obvious overlapping of signals from nuclei and materials. After 2 h of medium washing, a portion of the materials was flushed away, resulting in a decrease in green fluorescence. Compared with CF, CTF had a stronger signal remaining, indicating the ability to extend its retention time on the mucus with the thiolated modification (Figure 2c). Utilizing ImageJ software, the FITC fluorescence intensity was calculated to quantitate the remaining materials on the mucus, where the initial fluorescence intensity (0 h) was normalized to 100%. As shown in

Figure 2d, the signal of CF was reduced to 40%, whereas the signal of CTF only decreased to 77%, verifying the efficacy of thiolation.

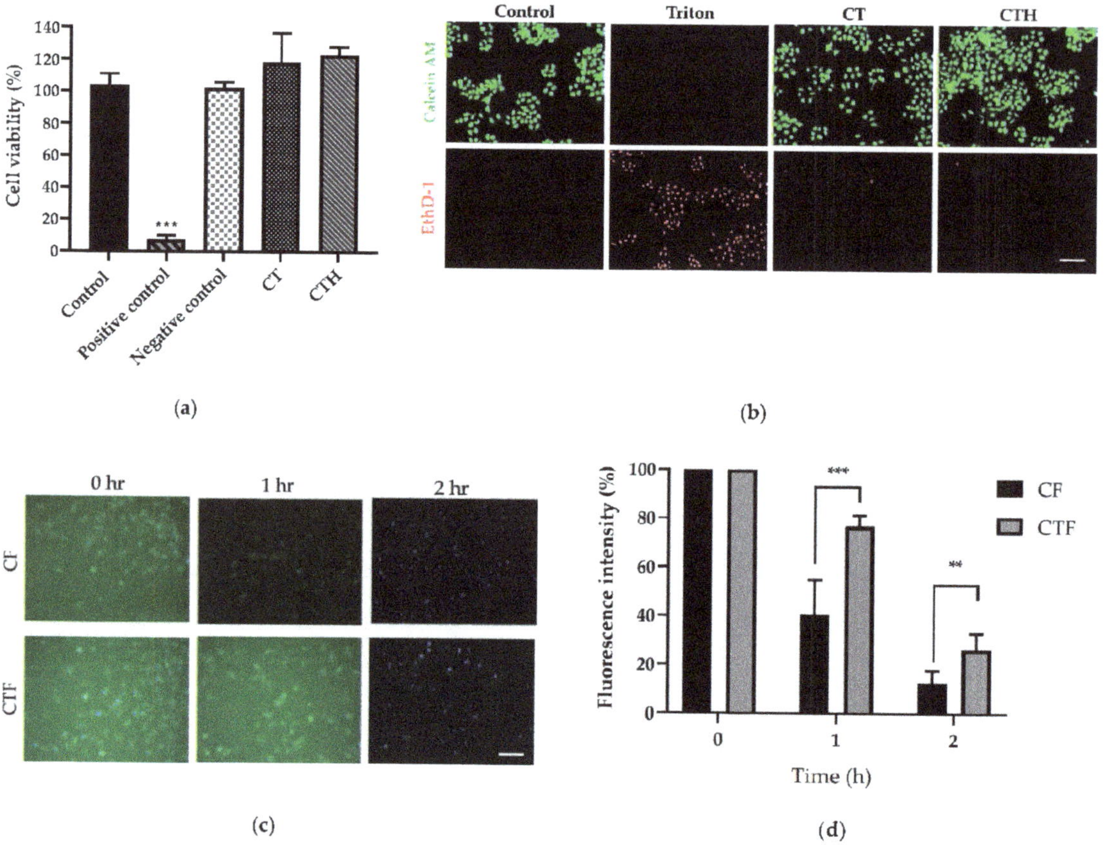


**Figure 2.** Evaluation of biocompatibility and mucoadhesion of CT. (**a**) WST-1 test to evaluate cell viability of CT and CTH with L929 cell. ($n = 6$, *** $p < 0.001$) (**b**) Cytotoxicity of CT and CTH tested using live/dead staining. Living cells were stained with calcein AM and dead cells with homodimer-1 (EthD-1) (scale bar = 100 µm). (**c**) In vitro adhesion evaluation of chitosan -FITC (CF) and CT-FITC (CTF). The fluorescence microscope was used to observe material adhesion on mucus. (**d**) Fluorescence quantification of the in vitro adhesion test. One-way ANOVA ($n = 3$, ** $p < 0.01$ *** $p < 0.001$).

### *3.3. Functional Barrier Test and Mucosal Adhesion In Vivo*

An in vivo investigation of the therapeutic effects of CT on intestinal absorption was necessary. In addition, as obesity could impair glucose tolerance and insulin function [24], an oral glucose tolerance test (OGTT) was performed to determine whether the materials would block glucose absorption in the intestine after forming a physical barrier. C57BL/6 male mice (8 weeks old) were first gavaged with CT to form an intestinal coating. Afterwards, glucose solutions were administered to each group orally except for the control group. Judging from the glucose response curves, the CT group had a much lower level of blood glucose at 326 mg/dL than the positive control group (387 mg/dL) after 30 min. In both positive control and the CT group, blood sugar levels returned to their baseline levels after 120 min due to the action of insulin (Figure 3a). As for quantification, iAUC was significantly reduced in the CT group by 40.27% (Figure 3b), indicating that CT indeed formed a barrier coating on the small intestine to inhibit glucose absorption.

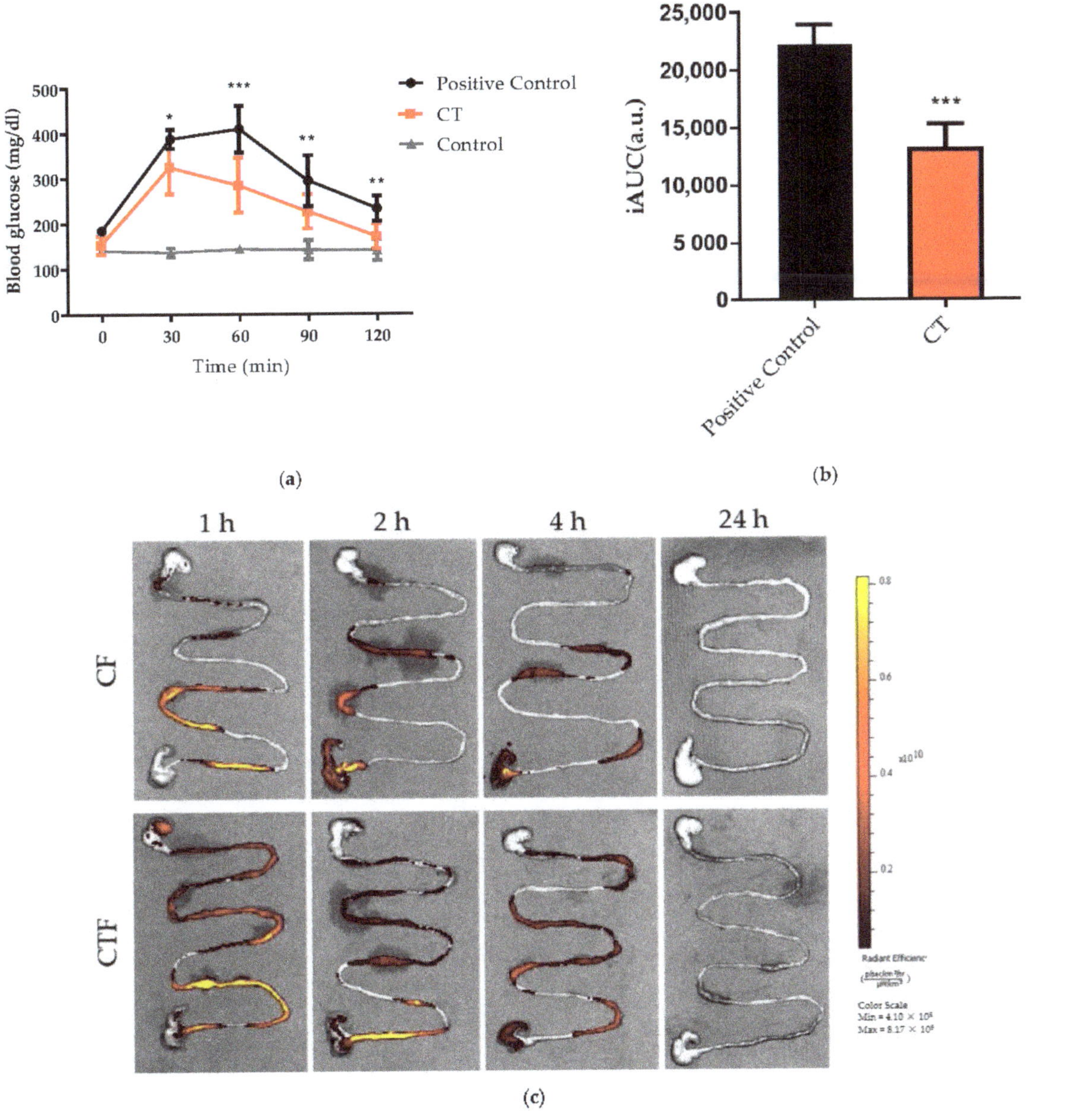


**Figure 3.** Oral glucose tolerance and in vivo adhesion evaluation. (**a**) Blood glucose response post-CT gavage (250 mg/kg). Two-way ANOVA with multiple comparisons ($n = 6$, * $p < 0.1$ ** $p < 0.01$ *** $p < 0.001$). (**b**) iAUC of the OGTT curves. One-way ANOVA with multiple comparisons ($n = 6$, * $p < 0.1$ ** $p < 0.01$ *** $p < 0.001$ compared to positive control) (**c**) Gastrointestinal tracts of mice were harvested from the stomach to cecum after gavage with CF or CTF (250 mg/kg) at each time point followed by IVIS in vivo imaging systems.

The adhesion properties of CF and CTF on the small intestinal mucosa were verified in C57BL/6 male mice. The gastrointestinal tracts were harvested after fluorescence material gavaging at 1, 2, 4, and 24 h respectively. Using an IVIS imaging system, the fluorescence signal was detected in the gastrointestinal tract from the stomach to the cecum. The signals of fluorescence intensity quickly began to differ between CF and CTF after 1 h. The signal of CF deteriorated notably over the course of four hours, CTF, while on the other hand, it still had strong fluorescent signals (Figure 3c). After 24 h, neither CF nor CTF showed any fluorescence. Accordingly, we assumed that the material would not adhere permanently to the gastrointestinal tract. Compared with the CF, CTF formed better adhesion to the intestinal tract of mice in fluorescence imaging, as indicated by an in vitro adhesion test.

### 3.4. CT as a Drug Release Hydrogel with a Long-Term Effect in an HFD Model

There were four groups in the long-term evaluation of materials: the normal diet group (ND), the high-fat diet group (HFD), and the high-fat diet plus CT group (HFD + CT). In addition, hesperidin is a flavanone glycoside that has been known to be effective in relieving symptoms such as hyperlipidemia and hyperglycemia [21]; therefore, a high-fat diet plus a hesperidin-loaded CT group (HFD + CTH) was added to evaluate whether CT could be used as an anti-obesity drug delivery system. PBS was given to the control group as well as the group with HFD for eight weeks. A daily dose of 250 mg/kg mice of material was given to both the CT and CTH groups for eight weeks. During the experiment, the bodyweight of mice was measured once a week, and the rate at which their weight changed was recorded.

In comparison to the HFD group, the weight gains of the CT group and CTH group were significantly lower after 35 days of treatment (Figure 4a), with the CTH group showing slightly better results than the CT group. The difference in weight changes between each group could also be observed clearly in Figure S1. Compared with the HFD group, the HFD + CT group had a significant reduction in weight gain of 26.28% and the HFD + CTH group had a greater reduction of 40.91%. Computed tomography (CT) scan, a non-invasive medical imaging technique, was used to visualize the accumulation of adipose tissue. In both the CT and CTH groups, there was a significant decrease in adipose distribution compared to the HFD group (Figure 4b).

In the following eight-week period, animals were sacrificed, and the epididymal white adipose tissue (eWAT) and subcutaneous white adipose tissue (sWAT) were weighed (Figure 4c). The HFD group, HFD + CT, and the HFD + CTH group showcased a significant reduction in adipose tissue as compared to the HFD group after dissection. eWAT was 50.15 percent lower in the HFD + CT group than in the HFD group, whereas eWAT of the HFD + CTH group was 56.8 percent lower than in the HFD group (Figure 4d). As shown in Figure 4d, the sWATs of groups HFD + CT and HFD + CTH were significantly lower than that of the HFD group by 32.14 percent and 57.85 percent. The results above indicate that both the CT and CTH groups could successfully reduce eWAT and sWAT; However, CTH was proved to be more successful in reducing eWAT.

### 3.5. The Result of H&E Staining

A large amount of lipid was accumulated in the liver of the HFD group, whereas the control group had much less lipid accumulation in liver H&E staining despite HFD, CT, and CTH having significantly reduced fatty liver symptoms, according to the results in Figure 5a. Furthermore, no differences were found in the H&E staining of the stomach or the small intestines (Figure S2). The H&E staining results of eWAT (Figure 5b) showed that the adipocyte diameters and volumes in the HFD group were larger than the control, as well as the HFD + CT, and HFD + CTH groups. It is speculated that CT or CTH suppresses nutrient absorption, thereby slowing lipid synthesis.

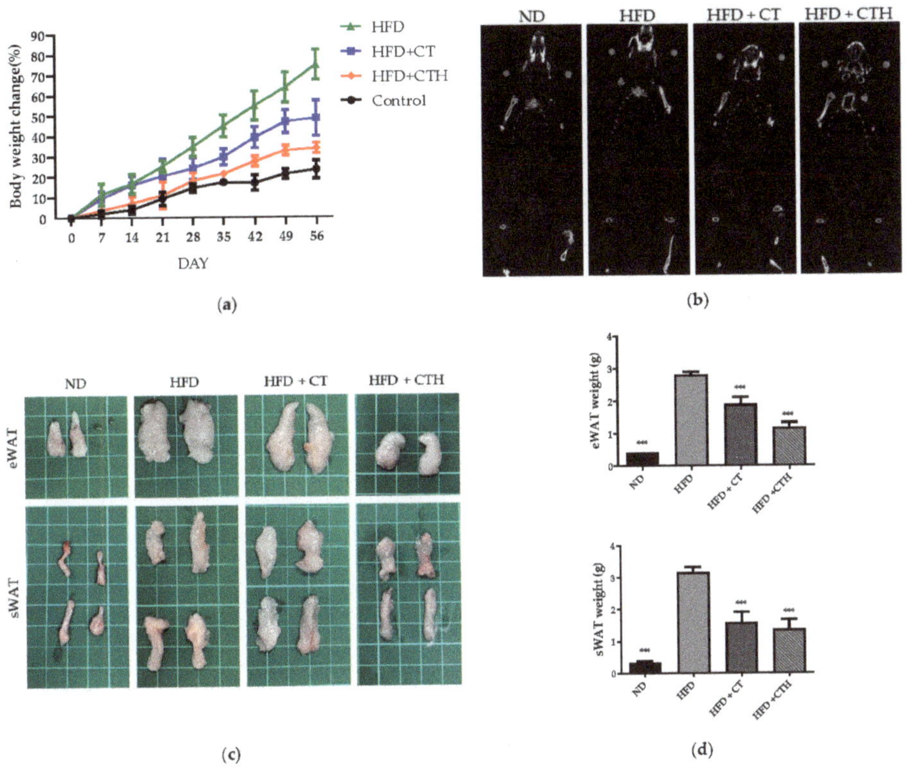


**Figure 4.** Evaluation of the bodyweight and fat accumulation (**a**) Eight-week weight record for normal diet group (ND), high-fat diet group (HFD), high-fat diet with CT group (HFD + CT), and high-fat diet with CT combined hesperidin (HFD + CTH). Two-way ANOVA with multiple comparisons ($n$ = 6, *** $p < 0.001$ for HFD + CT vs. HFD). (**b**) CT images of mice with a resolution voxel spacing of 35 μm. (**c**) Representative images of epididymal white adipose tissue (eWAT) and subcutaneous white adipose tissue (sWAT). (**d**) Weight of eWAT and sWAT. One-way ANOVA with multiple comparisons compared with HFD ($n$ = 6 for each group, *** $p < 0.001$).

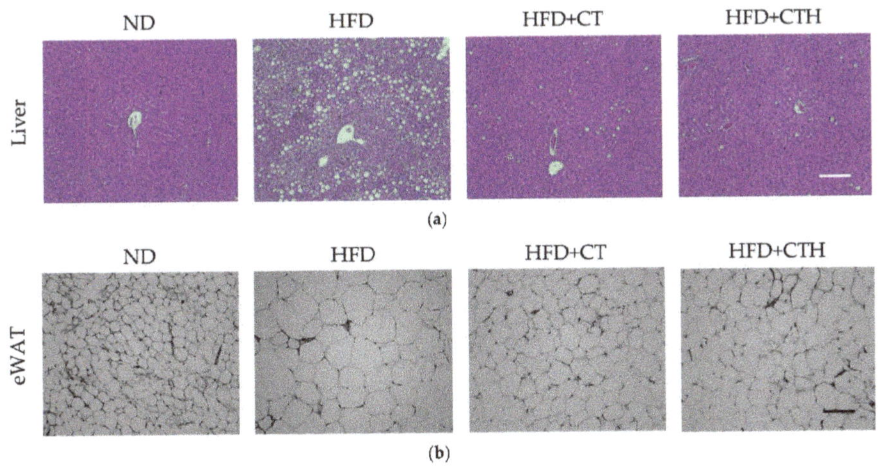


**Figure 5.** Results of hematoxylin and eosin stain. HE staining for (**a**) liver and (**b**) epididymal white adipose tissue (eWAT) (scale bar = 100 μm).

### 3.6. Serum Biochemical Analysis and Complete Blood Count

C57b/6 mice were sacrificed by carbon dioxide inhalation, and blood samples were taken by cardiac puncture using a 26G needle for the biochemical analysis of serum and complete blood count. In the serum biochemical analysis, aspartate aminotransferase (AST) and alanine aminotransferase (ALT) were widely used as enzyme biomarkers of liver injury. Furthermore, biochemical markers associated with lipid metabolisms, such as the total cholesterol (TC), triglycerides (TG), high-density lipoproteins (HDL), and low-density lipoproteins (LDL), were also investigated. Finally, kidney dysfunction can be evaluated by testing blood urea nitrogen (BUN) and creatinine (Cre) in serum biochemical analysis (Table 1).

**Table 1.** Serum biochemical assessment.

| | ND | HFD | HFD + CT | HFD + CTH |
|---|---|---|---|---|
| TC (mg/dL) | 89.40 ± 12.41 | 158.60 ± 8.46 *** | 145.80 ± 19.63 *** | 144.10 ± 9.31 *** |
| TG (mg/dL) | 93.32 ± 26.73 | 128.20 ± 21.83 * | 116.60 ± 13.57 * | 105.3 ± 7.55 |
| HDL (mg/dL) | 75.6 ± 14.52 | 140.90 ± 7.89 *** | 123.80 ± 14.61 *** | 122.70 ± 8.59 *** |
| LDL (mg/dL) | 8.63 ± 3.93 | 15.84 ± 2.20 *** | 13.96 ± 2.91 ** | 13.93 ± 2.29 ** |
| AST (U/L) | 108.1 ± 26.14 | 286.3 ± 91.52 *** | 153.4 ± 51.6 ### | 88.24 ± 28.7 ### |
| ALT (U/L) | 35.62 ± 6.37 | 40.63 ± 8.70 | 37.64 ± 8.26 | 33.73 ± 11.73 |
| BUN (mg/dL) | 20.86 ± 3.91 | 19.48 ± 2.61 | 18.35 ± 2.57 | 17.97 ± 2.33 |
| Crea (mg/dL) | 0.25 ± 0.05 | 0.23 ± 0.05 | 0.21 ± 0.03 | 0.23 ± 0.05 |

One-way ANOVA with multiple comparisons ($n = 6$ for each group, * $p < 0.05$, ** $p < 0.01$, *** $p < 0.005$ compared to ND, ### $p < 0.005$ compared to HFD).

The results of the complete blood count (CBC) including white blood cells (WBC), neutrophil (NEUT), eosinophil (EO), basophil (BASO), monocyte (MO), lymphocyte (LYMPH), red blood cell (RBC), hemoglobin (HGB), hematocrit (HCT), mean corpuscular volume (MCV), mean corpuscular hemoglobin (MCH), reticulocyte count (RET), mean corpuscular hemoglobin concentration (MCHC), and platelets (PLT) are within the normal range compared with the control group (Table 2). According to the CBC results, the gavaged CT or CT with hesperidin will not cause adverse immune reactions in the body, making the materials safe for experimental use in animals.

**Table 2.** Complete blood count.

| | ND | HFD | HFD + CT | HFD + CTH |
|---|---|---|---|---|
| RBC (M/μL) | 10.25 ± 0.62 | 9.975 ± 0.261 | 10.71 ± 0.213 | 10.48 ± 0.376 |
| HGB (g/dL) | 14.57 ± 0.85 | 14.25 ± 0.212 | 15.41 ± 0.3586 | 15.10 ± 0.916 |
| HCT (%) | 55.71 ± 3.75 | 53.05 ± 1.061 | 53.62 ± 1.45 | 54.50 ± 3.830 |
| MCV (fL) | 52.98 ± 1.033 | 53.20 ± 0.282 | 50.06 ± 1.276 | 51.93 ± 1.790 |
| MCH (pg) | 14.62 ± 0.25 | 14.30 ± 0.141 | 14.39 ± 0.209 | 14.40 ± 0.360 |
| MCHC (g/dL) | 27.61 ± 0.516 | 26.90 ± 0.141 | 28.77 ± 0.9447 | 27.70 ± 0.264 |
| RET (K/μL) | 429.4 ± 106.8 | 514.8 ± 18.24 | 490.3 ± 79.54 | 502.8 ± 113.9 |
| PLT (K/μL) | 1124 ± 259.0 | 802.0 ± 114.6 | 1258 ± 140.30 | 928.7 ± 328.7 |
| WBC (K/μL) | 4.631 ± 2.186 | 3.650 ± 0.480 | 6.193 ± 1.229 | 2.547 ± 0.140 |
| NEUT (K/μL) | 1.181 ± 1.015 | 0.755 ± 0.304 | 1.324 ± 0.57 | 0.450 ± 0.155 |
| LYMPH (K/μL) | 3.263 ± 1.568 | 2.570 ± 0.08485 | 7.523 ± 2.474 | 1.887 ± 0.2228 |
| MONO (K/μL) | 0.1589 ± 0.1776 | 0.190 ± 0.084 | 0.257 ± 0.150 | 0.1300 ± 0.079 |
| EO (K/μL) | 0.026 ± 0.018 | 0.1267 ± 0.006 | 0.233 ± 0.105 | 0.076 ± 0.035 |

## 4. Discussion

Obesity is defined as the presence of excessive body fat that poses a health risk, including an increased risk of chronic diseases like cancer [25,26]. According to estimates, these preventable diseases will result in a combined medical costs of $48–66 billion/year in the US and £1.9–2 billion/year in the UK by 2030 [27]. Natural materials such as chitosan are gradually being developed, modified, and tested due to the increase in overweight or obese people in the world. In this study, we used chitosan as main material and BDDE as a crosslinking agent and combined it with thioglycolic acid as a CT. BDDE is a bifunctional compound containing epoxy groups on both ends that can be used to crosslink nucleophilic molecules such as hydroxyl group, amino group, and sulfhydryl group through a ring-opening mechanism. Ellman's assay result also revealed sulfhydryl groups on CT (Figure S3). Our research examined the possibility that chitosan hydroxyls and amines could form ether bonds with free thiols on TGA. In combination with the above results, CT was confirmed by functional group analysis (Figure 1a,b) and the sulfur signal on the surface was determined in the EDS result (Figure 1c,d).

Intestines contained a lot of mucosa of epithelium and connective tissue. The epithelial and mucosal tissue, which provided the first defense system, contained redox pairs, such as glutathione, cysteine, and thioredoxin reductase [28]. Due to these redox pairs, the mucosa contained enough thiol groups to ensure optimal adhesion between mucoadhesive materials and the mucosa. Moreover, the mucoadhesive material in this study will be excreted, thus eliminating potential toxicity concerns or the risk of accumulating bioactive molecules. From in vivo verification (Figure 3), the results demonstrate that there is a good glucose blocking effect after CT gavaging. In vitro glucose barrier functional tests have also established similar results (Figure S4). However, in the IPGTT, the glucose response did not differ, suggesting that the decreased glucose response is due to a localized barrier coating instead of a systemic effect (Figure S5). It is also evident from the results of IVIS that CT can increase the time of stay in the gastrointestinal tract but will not accumulate permanently.

Hesperidin has anti-inflammatory, anti-oxidative, and anti-cancer properties that contribute to lowering cholesterol levels, blood pressure, and obesity [29]. To be more specific, hesperidin modulates AMPK and PPAR signaling pathways in obesity treatment, which affects the antioxidant index, anti-apoptosis, and NF-κB signaling pathways indirectly [30]. CT added with hesperidin (CTH) not only had the effect of CT blocking gastrointestinal absorption, with the effect of hesperidin, it could also make further improvements in terms of weight loss. However, there have been many strategies proposed to help people lose weight and maintain their weight, including caffeine, ephedra, capsaicin, and green tea, since they have the potential to increase energy expenditure and counter the decline in metabolic rate associated with weight increase [31]. In the future, we will try adding auxiliary additives to CT materials to assist their effectiveness

## 5. Conclusions

In this study, we crosslinked chitosan with BDDE and grafted TGA onto the chitosan, creating chitosan-TGA (CT). The mucoadhesive CT had better mucoadhesive properties than unmodified chitosan. The prepared hesperidin-containing CT (CTH) and CT had no cytotoxicity in vitro and no systemic toxicity in vivo.

At the same time, CT could be excreted from the body after 24 h of feeding materials and will not accumulate in the body. Hesperidin mixed with CT (CTH) not only effectively reduced the body weight by 40.91% compared with the HFD group, but also greatly reduced the accumulation of body fat. From the results, it could be shown that the developed CTH could be used as a biosafety mucosal gel for obesity treatment. CT as a barrier to inhibit absorption can also be used as a drug carrier for different natural Chinese herbal medicines to treat obesity or nonalcoholic fatty liver disease (NAFLD) in the future.

**Supplementary Materials:** The following are available online at https://www.mdpi.com/article/10.3390/nu13124405/s1, Figure S1: Representative images of the control group, the HFD group, HFD + CT group and the HFD + CTH group, Figure S2: H&E staining of stomach and small intestine, Figure S3: Ellman's assay of CT, Figure S4: The glucose barrier functional test, and Figure S5: IPGTT.

**Author Contributions:** Conceptualization, T.-C.C. and R.-C.T.; Data curation, T.-C.C., Y.-Y.H., Y.-C.K., J.-N.L. and I.-H.Y.; Formal analysis, T.-C.C.; Funding acquisition, F.-H.L.; Investigation, R.-C.T. and Y.-C.K.; Methodology, Y.-Y.H.; Project administration, T.-C.C. and F.-H.L.; Resources, F.-H.L.; Validation, T.-C.C. and Y.-Y.H.; Visualization, F.-H.L.; Writing–original draft, T.-C.C. and R.-C.T.; Writing—review & editing, T.-C.C., R.-C.T. and F.-H.L. All authors have read and agreed to the published version of the manuscript.

**Funding:** This research was funded by the National Health Research Institutes in grant numbers BN-110-GP-07, BN-110-PP-01, and BN-110-GP-09.

**Institutional Review Board Statement:** The study protocol was approved by the Institutional Animal Care and Use Committee of the National Taiwan University (IACUC, No. 20201042).

**Conflicts of Interest:** The authors declare no conflict of interest.

## References

1. Jiang, S.Z.; Lu, W.; Zong, X.F.; Ruan, H.Y.; Liu, Y. Obesity and hypertension. *Exp. Ther. Med.* **2016**, *12*, 2395–2399. [CrossRef]
2. Stoner, L.; Cornwall, J. Did the American Medical Association make the correct decision classifying obesity as a disease? *Australas. Med. J.* **2014**, *7*, 462. [CrossRef] [PubMed]
3. Kelly, T.; Yang, W.; Chen, C.-S.; Reynolds, K.; He, J. Global burden of obesity in 2005 and projections to 2030. *Int. J. Obes.* **2008**, *32*, 1431–1437. [CrossRef] [PubMed]
4. Ng, M.; Fleming, T.; Robinson, M.; Thomson, B.; Graetz, N.; Margono, C.; Mullany, E.C.; Biryukov, S.; Abbafati, C.; Abera, S.F. Global, regional, and national prevalence of overweight and obesity in children and adults during 1980–2013: A systematic analysis for the Global Burden of Disease Study 2013. *Lancet* **2014**, *384*, 766–781. [CrossRef]
5. Jiménez, E.G. Obesity: Etiologic and pathophysiological analysis. *Endocrinol. Y Nutr. (Engl. Ed. )* **2013**, *60*, 17–24.
6. Angulo, M.; Butler, M.; Cataletto, M. Prader-Willi syndrome: A review of clinical, genetic, and endocrine findings. *J. Endocrinol. Investig.* **2015**, *38*, 1249–1263. [CrossRef]
7. Bray, G.A. Epidemiology, risks and pathogenesis of obesity. *Meat Sci.* **2005**, *71*, 2–7. [CrossRef] [PubMed]
8. Kanneganti, T.-D.; Dixit, V.D. Immunological complications of obesity. *Nat. Immunol.* **2012**, *13*, 707–712. [CrossRef]
9. Lumeng, C.N.; Saltiel, A.R. Inflammatory links between obesity and metabolic disease. *J. Clin. Investig.* **2011**, *121*, 2111–2117. [CrossRef] [PubMed]
10. Himpens, J.; Dobbeleir, J.; Peeters, G. Long-term results of laparoscopic sleeve gastrectomy for obesity. *Ann. Surg.* **2010**, *252*, 319–324. [CrossRef] [PubMed]
11. Schauer, P.R.; Ikramuddin, S.; Gourash, W.; Ramanathan, R.; Luketich, J. Outcomes after laparoscopic Roux-en-Y gastric bypass for morbid obesity. *Ann. Surg.* **2000**, *232*, 515. [CrossRef]
12. Schouten, R.; Rijs, C.S.; Bouvy, N.D.; Hameeteman, W.; Koek, G.H.; Janssen, I.M.; Greve, J.-W.M. A multicenter, randomized efficacy study of the EndoBarrier Gastrointestinal Liner for presurgical weight loss prior to bariatric surgery. *Ann. Surg.* **2010**, *251*, 236–243. [CrossRef]
13. Patel, S.R.; Hakim, D.; Mason, J.; Hakim, N. The duodenal-jejunal bypass sleeve (EndoBarrier Gastrointestinal Liner) for weight loss and treatment of type 2 diabetes. *Surg. Obes. Relat. Dis.* **2013**, *9*, 482–484. [CrossRef] [PubMed]
14. Forner, P.M.; Ramacciotti, T.; Farey, J.E.; Lord, R.V. Safety and effectiveness of an endoscopically placed duodenal-jejunal bypass device (EndoBarrier®): Outcomes in 114 patients. *Obes. Surg.* **2017**, *27*, 3306–3313. [CrossRef] [PubMed]
15. Kean, T.; Thanou, M. Biodegradation, biodistribution and toxicity of chitosan. *Adv. Drug Deliv. Rev.* **2010**, *62*, 3–11. [CrossRef] [PubMed]
16. Bagheri-Khoulenjani, S.; Taghizadeh, S.; Mirzadeh, H. An investigation on the short-term biodegradability of chitosan with various molecular weights and degrees of deacetylation. *Carbohydr. Polym.* **2009**, *78*, 773–778. [CrossRef]
17. Hsu, S.-h.; Whu, S.W.; Tsai, C.-L.; Wu, Y.-H.; Chen, H.-W.; Hsieh, K.-H. Chitosan as scaffold materials: Effects of molecular weight and degree of deacetylation. *J. Polym. Res.* **2004**, *11*, 141–147. [CrossRef]
18. Jull, A.B.; Mhurchu, C.N.; Bennett, D.A.; Dunshea-Mooij, C.A.; Rodgers, A. Chitosan for overweight or obesity. *Cochrane Database Syst. Rev.* **2008**, *3*, CD003892. [CrossRef] [PubMed]
19. Leitner, V.M.; Walker, G.F.; Bernkop-Schnürch, A. Thiolated polymers: Evidence for the formation of disulphide bonds with mucus glycoproteins. *Eur. J. Pharm. Biopharm.* **2003**, *56*, 207–214. [CrossRef]
20. Bernkop-Schnürch, A. Thiomers: A new generation of mucoadhesive polymers. *Adv. Drug Deliv. Rev.* **2005**, *57*, 1569–1582. [CrossRef]
21. Garg, A.; Garg, S.; Zaneveld, L.; Singla, A. Chemistry and pharmacology of the citrus bioflavonoid hesperidin. *Phytother. Res.* **2001**, *15*, 655–669. [CrossRef]
22. Jiao, Q.; Xu, L.; Jiang, L.; Jiang, Y.; Zhang, J.; Liu, B. Metabolism study of hesperetin and hesperidin in rats by UHPLC-LTQ-Orbitrap MS n. *Xenobiotica* **2020**, *50*, 1311–1322. [CrossRef]
23. Fernandes Queiroz, M.; Melo, K.R.T.; Sabry, D.A.; Sassaki, G.L.; Rocha, H.A.O. Does the use of chitosan contribute to oxalate kidney stone formation? *Mar. Drugs* **2015**, *13*, 141–158. [CrossRef] [PubMed]

24. Matsumoto, K.; Miyake, S.; Yano, M.; Ueki, Y.; Yamaguchi, Y.; Akazawa, S.; Tominaga, Y. Glucose tolerance, insulin secretion, and insulin sensitivity in nonobese and obese Japanese subjects. *Diabetes Care* **1997**, *20*, 1562–1568. [CrossRef]
25. Anderson, A.S.; Key, T.J.; Norat, T.; Scoccianti, C.; Cecchini, M.; Berrino, F.; Boutron-Ruault, M.-C.; Espina, C.; Leitzmann, M.; Powers, H. European code against cancer 4th edition: Obesity, body fatness and cancer. *Cancer Epidemiol.* **2015**, *39*, S34–S45. [CrossRef]
26. Marmot, M.; Atinmo, T.; Byers, T.; Chen, J.; Hirohata, T.; Jackson, A.; James, W.; Kolonel, L.; Kumanyika, S.; Leitzmann, C. *Food, Nutrition, Physical Activity, and the Prevention of Cancer: A Global Perspective*; World Cancer Research Fund/American Institute for Cancer Research: Washington, DC, USA, 2007.
27. Wang, Y.C.; McPherson, K.; Marsh, T.; Gortmaker, S.L.; Brown, M. Health and economic burden of the projected obesity trends in the USA and the UK. *Lancet* **2011**, *378*, 815–825. [CrossRef]
28. Circu, M.L.; Aw, T.Y. Intestinal redox biology and oxidative stress. *Proc. Semin. Cell Dev. Biol.* **2012**, *23*, 729–737. [CrossRef] [PubMed]
29. Xiong, H.; Wang, J.; Ran, Q.; Lou, G.; Peng, C.; Gan, Q.; Hu, J.; Sun, J.; Yao, R.; Huang, Q. Hesperidin: A therapeutic agent for obesity. *Drug Des. Dev. Ther.* **2019**, *13*, 3855. [CrossRef]
30. Zhou, Z.; Kandhare, A.D.; Kandhare, A.A.; Bodhankar, S.L. Hesperidin ameliorates bleomycin-induced experimental pulmonary fibrosis via inhibition of TGF-beta1/Smad3/AMPK and IkappaBalpha/NF-kappaB pathways. *EXCLI J.* **2019**, *18*, 723. [PubMed]
31. Diepvens, K.; Westerterp, K.R.; Westerterp-Plantenga, M.S. Obesity and thermogenesis related to the consumption of caffeine, ephedrine, capsaicin, and green tea. *Am. J. Physiol.-Regul. Integr. Comp. Phys.* **2007**, *292*, R77–R85. [CrossRef]

*nutrients*

MDPI

*Article*

# Co-Microencapsulation of Anthocyanins from Cornelian Cherry (*Cornus mas* L.) Fruits and Lactic Acid Bacteria into Antioxidant and Anti-Proliferative Derivative Powders

**Iuliana Maria Enache [1], Mihaela Aida Vasile [1], Oana Crăciunescu [2], Ana Maria Prelipcean [2], Anca Oancea [2], Elena Enachi [1], Viorica Vasilica Barbu [1], Nicoleta Stănciuc [1] and Camelia Vizireanu [1,*]**

1 Faculty of Food Science and Engineering, Dunarea de Jos University of Galati, 800008 Galati, Romania
2 National Institute for Research and Development of Biological Sciences in Romania, 060031 Bucharest, Romania
* Correspondence: camelia.vizireanu@ugal.ro

**Abstract:** Driven by the modern lifestyle, the consumers are interested in healthier and balanced diets, including both probiotics and natural antioxidants. The beneficial health effects of probiotics are mainly due to their capacity of modulating the human intestinal microbiota, although achieving at least a 6 log of viable cells at the targeted site is still challenging. Therefore, in this study, an attempt to improve the bioaccessibility of *Lacticaseibacillus casei* (*L. casei*) 431® through a co-microencapsulation by complex coacervation and freeze-drying, using an extract from Cornelian cherry and two biopolymeric matrices, whey protein isolates and casein (WPI-CN) and inulin (WPI-I), was studied. The powders showed a comparable anthocyanin content of around 32.00 mg cyanidin-3-rutinoside (C3R)/g dry matter (DM) and a microbial load of about 10 Log CFU/g DM. A high stability of the lactic acid bacteria was assessed throughout 90 days of storage, whereas the anthocyanins' degradation during storage followed a first order kinetic model, allowing the estimation of a half-time of 66.88 ± 1.67 days for WPI-CN and 83.60 ± 2.46 days for WPI-I. The in vitro digestion showed a high stability of anthocyanins in the simulated gastric juice, whereas the release in the simulated intestinal juice was favored in the variant with inulin (up to 38%). The use of casein permitted to obtain finer spherosomes, with smaller diameters, whereas a double encapsulation was obvious in both variants, thus explaining the high resistance in the gastric environment. The anti-proliferative effect against the human colon cancer cell line (HT-29) was also demonstrated. No cytotoxicity has been found for the concentrations between 1 and 25 μg/mL for the WPI-I variant, whereas a cell proliferation effect was observed at low concentrations of 1–5 μg/mL.

**Keywords:** *Cornus mas*; anthocyanins; *Lacticaseibacillus casei* (*L. casei*); co-microencapsulation; anti-proliferative effect

**Citation:** Enache, I.M.; Vasile, M.A.; Crăciunescu, O.; Prelipcean, A.M.; Oancea, A.; Enachi, E.; Barbu, V.V.; Stănciuc, N.; Vizireanu, C. Co-Microencapsulation of Anthocyanins from Cornelian Cherry (*Cornus mas* L.) Fruits and Lactic Acid Bacteria into Antioxidant and Anti-Proliferative Derivative Powders. *Nutrients* **2022**, *14*, 3458. https://doi.org/10.3390/nu14173458

Academic Editor: LaVerne L. Brown

Received: 31 July 2022
Accepted: 21 August 2022
Published: 23 August 2022



## 1. Introduction

Food coloring agents reached a worldwide turnover of 8000 tons per year, as they are used in different food applications, such as carbonated drinks, salads, juices, ice creams, and sweets [1]. Food color is a decisive factor in the consumer's choice; therefore, an estimation in this regard suggested that 62–90% of consumer's assessment is based on color evaluation [2]. Driven by the consumer's preferences, which turns the activity into an economically profitable one, the producers are focused on identifying natural alternatives to prevent and eliminate the shortcomings of synthetic dyes, especially those related to possible negative side effects on health, such as obesity, cancer, diabetes, hypertension, dementia, or eye diseases [3]. Several criteria are used for the classification of different natural dyes, such as their origin (vegetal, animal, bacterial, fungal, etc.) or their hue (red, yellow, purple, blue, green, etc.), while the flavonoids (anthocyanins), isoprenoids (carotenoids), and nitrogen-heterocyclic derivatives (betalains) are the most used natural dyes, in a chemical structure's dependent manner [2]. When considering the long-term use of natural colorants, some significant benefits in the human's health should be considered, such as increasing the resistance of low-density lipoproteins to lipid peroxidation, the protection of proteins against oxidation, the chelation of transition metals that promote oxidative reactions, and the inhibition of the enzymes involved in the oxidative stress [4].

Among all the natural sources, plants, mainly fruits and vegetables, are naturally rich sources of pigments, with particular characteristics, such as solubility in water or lipids and solvents, significantly different both in structure and metabolic pathways [5]. However, when considering the use of natural pigments, some crucial steps should be considered, such as extraction methods to obtain a maximum extraction yield, purification, and stabilization to provide an appropriate environment for the bioactive to limit the prone to degrade at different parameters. It is well known that bioactives are prone to degradation to light, heat, oxygen, and temperature and can interact with each other or with different components of the food matrix, via polymerization or co-pigmentation reactions. Another critical step in the use of natural pigments is the low bioavailability, which has a crucial importance in biological activity maintenance [6]. However, the remarkable progress in the use of encapsulation technologies, both at scientific and application levels, provides an immense opportunity for industry to provide stable, health-promoting compounds for different fields [7]. The need for natural colorant encapsulation is given by the significant technological and biological advantages, related to their easier handling, increased chemical and thermal stability, preservation, or masking of their aroma, taste and/or scent, controlled and targeted release, and increased bioavailability [7]. In order to respond to the complex demands of the industry, there is a need to develop technological alternatives that address the multifunctional properties of an ingredient by incorporating probiotics [8]. The use of probiotics in food and nutraceuticals is limited due to their low resistance to temperature, light, or oxygen, while their ability to remain metabolically active under the gastrointestinal conditions so as to exert their beneficial effects are still challenging [9]. Numerous studies revealed the prebiotic function of anthocyanins, due to their ability to improve the diversity of the intestinal microbiota [10,11], by different mechanisms, hence allowing the growth of beneficial bacteria and inhibiting potentially harmful ones, and increasing the concentration of short-chain fatty acids [10,12].

Cornelian cherries (*Cornus mas* L.) are well known for their rich content in bioactive compounds, including organic molecules, carbohydrates, fatty acids, vitamins, and minerals [13]. The organic molecules can be divided into five structural groups: anthocyanins, iridoids, phenolic acids, flavonoids, and tannins [14]. The phytochemical profile of Cornelian cherry makes the fruits a purposive food in the food industry [15], being used to produce different drinks, syrups, gels, jams, and other traditional products [16].

The *Lacticaseibacillus casei* (*L. casei*) group is represented by *L. casei*, *L. paracasei*, and *L. rhamnosus* species, being some of the most well studied and applied probiotic species of lactobacilli, especially due to their commercial, industrial, and applied health-promoting potential [17]. Usually, *L. casei* is used in the milk fermentation processes, as starter or adjunct culture, to deliver yogurt or cheeses, and it has been associated with important health-promoting benefits, such as regulation of intestinal microbiota [18], tumor inhibition, pro-apoptotic and anti-proliferative effects [19], and with the production of bioactive peptides in fermented milk [20]. *L. casei* 431® is a probiotic strain associated with several health benefits, including boosting the immune system's health, maintaining a healthy respiratory tract and bowel function. *L. casei* 431® shows a high stability being able to survive under the gastrointestinal tract and it has been successfully used in food and dietary supplements since 1995 [21].

The aim of our study was to design and test two synbiotic co-microencapsulated delivery systems, containing bioactive compounds from Cornelian cherry (CM) and *L. casei* 431® into biopolymeric whey protein isolated-based materials, including casein and inulin, through complex coacervation and freeze-drying in order to develop potential candidates for food and pharmaceuticals applications. The extract was obtained from freeze-dried

CM fruits using a liquid-solid ultrasound-assisted extraction and co-microencapsulated in whey protein isolate-casein (WPI-CN) and whey protein isolate-inulin (WPI-I) with *L. casei* 431® by complex coacervation and freeze-drying. The two biopolymeric combinations were selected based on their properties, in order to increase the co-microencapsulation efficiencies, both for polyphenolic compounds and probiotic cells and to create a protective environment able to allow a good resistance of co-microencapsulated bacteria in the gastrointestinal stress and a controlled release of anthocyanins. The selected microencapsulation matrices have unique properties, both from technological and nutritional points of view. For example, milk proteins are rich sources of essential amino acids and have excellent water binding, gelling, emulsifying, and foaming properties. Moreover, both whey proteins and casein have slow digestion patterns and immune benefits [22,23]. Inulin, a natural dietary soluble fiber consisting of a mixture of oligo- and/or polysaccharides of β (2→1) linked D-fructose units with a terminal glucose residue linked by α (1→2) bond [19], is frequently used as ingredient for functional food, given its prebiotic potential, technological properties, and beneficial effects on health [24].

Therefore, the two powders were tested for anthocyanins and lactic acid bacteria encapsulation efficiency, phytochemical profile, in terms of total polyphenolic content (TPC), total flavonoids content (TFC), total monomeric anthocyanin (TAC), antioxidant activity, cells viability, color, structure and morphology, in vitro bioavailability of anthocyanins, and storage stability. The powders were also tested for anti-proliferative activity on HT-29 human colon cancer cell line and cyto-compatibility on L929 fibroblasts cells.

The obtained results highlighted the effectiveness of the customized co-microencapsulation designs, in terms of polyphenolic compounds, whey proteins, casein, and inulin in preserving the bioactive chemical stability and cells viability during shelf life, whereas a controlled release of anthocyanins in gastrointestinal environment was suggested.

## 2. Materials and Methods

### *2.1. Chemicals*

Acetic acid, hydrochloric acid, aluminum chloride, 2,2-Diphenyl-1-picrylhydrazyl (DPPH), ethanol, Folin–Ciocâlteu reagent, gallic acid, inulin from chicory, methanol, pancreatin (Kreon), pepsin from gastric porcine, potassium acetate, Trizma hydrochloride, 6-Hydroxy-2,5,7,8-tetramethylchromane-2-carboxylic acid (Trolox), and inulin were purchased from Sigma Aldrich (Taufkirchen, Germany), while sodium bicarbonate was purchased from Honeywell, Fluka (Selze, Germany) and whey protein isolate 894 (WPI) from Fonterra (Clandeboye, New Zealand). *L. casei* 431® strain was purchased from Chr. Hansen (Hoersholm, Denmark). For the cell viability studies of the *L. casei* 431®, de Man, Rogosa, and Sharpe agar (MRS agar) (Merck (Darmstadt, Germany) was used. The human colon cancer cell line (HT-29, American Type Culture Collection—ATCC), fibroblasts from the NCTC cell line clone L929 (European Collection of Authenticated Cell Cultures, Sigma-Aldrich, Darmstadt, Germany), Dulbecco's Modified Eagle Medium (DMEM), fetal calf serum (FCS), Medium Minim Essential (MEM), L-glutamine, and a mixture of antibiotics (penicillin-streptomycin-neomycin) were used for antiproliferative activity and cytotoxicity.

### *2.2. Fresh Fruits Processing*

CM fruits were purchased from a local market (Galati, Romania) in October 2018. The fresh fruits were manually removed from their seeds, washed, and dried with a paper towel and freeze-dried at −42 °C under a pressure of 10 Pa for 48 h. The freeze-dried CM fruits were collected and packed in dark glass containers, hermetically sealed, and stored at 4 °C until analysis.

### *2.3. Phytochemical Extractions*

The anthocyanins' extraction was performed by liquid-solid ultrasonic-assisted method, using ethanol solution as a solvent (70%). An amount of 50 g of freeze-dried CM ground powder was mixed with 400 mL of ethanol solution, followed by ultrasound-assisted extraction at 40 °C for 30 min (100 W and 40 KHz). The resulting liquid containing anthocyanins was centrifuged at 6000× *g* for 20 min and concentrated under reduced pressure at 40 °C (AVC 2-18, Christ, UK) to remove the solvent.

### 2.4. Co-Microencapsulation of the Anthocyanins and Lactic Acid Bacteria

A volume of 100 mL of biopolymer solutions, obtained by dissolving WPI and CN (WPI-CN) and WPI and I (WPI-I), in a ratio of 1:1 (*w*:*w*) was mixed with the extract in a ratio of 2.5%. The solutions were allowed to mix on a magnetic stirrer until complete hydration at 650 rpm, and the pH was adjusted at 4.6. The inoculation of the mixture with *L. casei* 431® lyophilized culture (with an initial CFU of $10^{11}$/g) was performed as described by Enache et al. [25]. The resulting freeze-dried powders were collected, packed in metallized bags, and kept at 4 °C until further analysis.

### 2.5. Phytochemical Profile of the Extract and Freeze-Dried Powders

The extract and freeze-dried powders were analyzed for TPC by Folin–Ciocalteu method, TAC using pH differential method, TFC by aluminum chloride method, and antioxidant activity using DPPH method, as previously described [25]. Briefly, for TPC evaluation, a volume of 0.2 mL of corresponding extract was diluted with 15.8 mL of distillated water, followed by addition of 1 mL of Folin–Ciocalteu reagent and 3 mL of 20% of $Na_2CO_3$. The mixtures were vigorously agitated and allowed to react for 60 min in the dark, followed by the absorbance reading at λ = 765 nm (Jenway Scientific Instruments, Essex, UK) The TPC was expressed as mg gallic acid equivalents (mg GAE) per g dry matter (DM extract or powder), using a gallic acid calibration curve.

For TFC evaluation, a volume of 0.250 mL of the extract or powder extract was mixed with 0.075 mL of sodium nitrite (5%), 0.15 mL aluminum chloride (10%), and 0.5 mL sodium hydroxide (1 M). The absorbance of the mixture was measured at 510 nm against the suitable blank. The TF was expressed in mg catechin equivalents (CE) per g of DM (extract or powder), using a catechin calibration curve.

The DPPH antiradical scavenging potential, a volume of 3.9 mL of DPPH solution (0.1 M in methanol) and 0.1 mL of the extract and powders extract were thoroughly mixed and allowed to react for 30 min, in the dark. The absorbance was read at the wavelength of 515 nm, using a blank with methanol instead of sample. The scavenging percentage of DPPH was expressed as mMol Trolox/g DM using a calibration curve.

### 2.6. Viability of L. casei 431®

Sterile physiological serum (0.9 g NaCl%, *w*/*v*) and the pour plate technique were used to estimate the number of viable cells of *L. casei* 431®. The standardized method [26] was used, involving the estimation of the viable cell number, by the number of colony-forming units (CFU) when cultivated on the MRS-agar plates (medium at pH 5.7) for 48 h of aerobic incubation at 37 °C. The counts were expressed as CFU/g DM.

### 2.7. Co-Microencapsulation Efficiency

The co-microencapsulation efficiency of the anthocyanins was estimated as described earlier by Enache et al. [25]. The method involves two steps for the evaluation of the surface anthocyanins content (*SAC*) and total anthocyanins content (*TAC*). Two different medium were used to dissolve the powders, namely methanol:acetic acid:distillated water mixture (25:4:21, *v*/*v*/*v*) for *TAC* and methanol:ethanol in a ratio of 1:1 (*v*/*v*) for *SAC*. An amount of 0.2 g of co-microencapsulated powder was dissolved in 5 mL of appropriate mixture, with and without homogenization, and sonicated to destroy the microparticles for *TAC* and *SAC*, respectively. The total monomeric anthocyanin content in the resulting medium was determined by the pH differential method [27], whereas the results were expressed as mg of cyanidin-3-rutinoside equivalents (C3R) per g of DM of powder (mg/g DM). The co-microencapsulation efficiency (*cME*) was calculated using Equation (1):

$$cME(\%) = \frac{TAC - SAC}{TAC} \times 100 \quad (1)$$

For lactic acid bacteria co-microencapsulation efficiency, the protocol described by Colín-Cruz et al. [28] was used. In brief, the quantification of the viable bacteria was performed by pour plate technique. In order to determine the co-microencapsulation efficiency (%), Equation (2) was used:

$$EE(\%) = \frac{log\ (N)}{log\ (N_0)} \quad (2)$$

where $N$ is the number of viable cells (CFU/g) in the powder, and $N_0$ is the number of viable cells in the mixture before freeze-drying [25].

*2.8. Structure and Morphology of the Powders*

The structure and morphology of the powders were measured using a LSM 710 Carl Zeiss Confocal Laser scanning microscope (Carl Zeiss, Oberkohen, Germany), by using four types of lasers, such as a diode laser (405 nm), Ar-laser (458, 488, 514 nm), DPSS laser (diode pumped solid state—561 nm), and HeNe laser (633 nm). The images were captured and rendered with the black edition of the ZEN 2012 SP1 software (Carl Zeiss, Oberkohen, Germany). The powders' fluorescence was assessed both in their unlabeled (native) and labeled with the Red Congo (40 µM) fluorophore.

*2.9. In Vitro Simulated Digestion of Anthocyanins*

In vitro digestion of the anthocyanins from the powders was performed using a static digestion model, which involves the addition of the powders in simulated gastric juice (SGJ) at pH 2.0 and simulated intestinal juice (SIJ) at pH 7.7, as described earlier by Enache et al. [25].

*2.10. Colorimetric Analysis*

A CR 410 Chroma Meter (Konica Minolta, Tokyo, Japan) colorimeter was used to appreciate the color of the co-microencapsulated powders. The three coordinates, respectively: *L** (brightness, 0 for blackness, 100 for whiteness), *a** (positive value for redness, negative value for greenness), and *b** (positive value for yellowness, negative value for blueness) were used in this study.

*2.11. Anti-Proliferative Activity and Cytocompatibility of the Powders*

2.11.1. Cell Culture and Treatment

The anti-proliferative activity of the powders was tested on human colon cancer cell line HT-29 (ATCC), whereas the cyto-compatibility was tested on mouse fibroblasts from NCTC clone L929 cell line (ECACC). For the estimation of anti-proliferative activity, the cells were cultured in Dulbecco's Modified Eagle Medium (DMEM) supplemented with 20% (*v/v*) fetal calf serum (FCS), 2 mM L-glutamine and 1% (*v/v*) antibiotic mixture (penicillin-streptomycin-neomycin), in a humidified atmosphere with 5% $CO_2$, at 37 °C, until sub-confluence. For the cyto-compatibility tests, the cell was cultivated on Minimum Essential Medium (MEM) supplemented with 10% (*v/v*) fetal calf serum (FCS), 2 mM L-glutamine, and 1% (*v/v*) antibiotic mixture (penicillin-streptomycin-neomycin) in a humidified atmosphere with 5% $CO_2$, at 37 °C, until sub-confluence. For testing the effect of the co-microencapsulated powders, serial dilutions of stock a solution (1 mg/mL) were used, followed by sterile filtration using 0.22 µm membrane filters (Millipore, Burlington, MA, USA). For both experiments, the cells were seeded in 24-well microplates, at a density of $4 \times 10^4$ cells/mL, followed by an incubation at 37 °C, in a humidified atmosphere with

5% $CO_2$, for 24 h. Further, the culture medium was replaced with fresh medium, containing different concentrations of samples ranging from 10 to 1000 μg/mL and incubated at 37 °C under standard conditions, for 24 h and 48 h, respectively. The cells incubated in the culture medium without the sample were used as the control culture.

2.11.2. Cells' Viability

For the cells' viability estimation, the Neutral Red (NR) assay was used, as previously described [29]. Briefly, the method involved the removing the specific culture medium at the end of each incubation period, followed by addition of 50 μg/mL NR solution and incubation at 37 °C, for 3 h. After cell washing, the incorporated dye was released in 1% (*v/v*) acetic acid solution in 50% (*v/v*) ethanol by gentle shaking, for 15 min. The number of viable cells was directly proportional with the amount of dye taken up. The optical density was measured at 540 nm in a Sunrise microplate reader (Tecan, Grödig, Austria). The results were reported as percentage relative to the control culture, considered 100% viable and the effective concentration was calculated as the sample concentration that decreased the cell viability down to 50% (EC50).

2.11.3. Cell Morphology

Prior to imaging, the cultured cells were washed, fixed with methanol, and Giemsa stained. The cell morphology of the cells incubated in the presence of the samples was observed by light microscopy, after 48 h of cultivation. The micrographs were acquired with an optical microscope Axio Observer D1 equipped with digital camera (Carl Zeiss, Germany).

*2.12. Storage Stability*

The stability of the phytochemicals and lactic acid bacteria cells in the co-microencapsulated powders was tested after 21 and 90 days of storage in the absence of light at 4 °C.

*2.13. Kinetics of Anthocyanins and Antioxidant Activity Degradation during Storage*

The degradation kinetics of anthocyanin and antioxidant activity during storage was obtained by first order kinetic model, as described by Azarpazhooh et al. [30], by estimating the rate constant against the time. The half-life ($t_{1/2}$) values were calculated using Equations (3) and (4).

$$C_t = C_0 exp(kt) \tag{3}$$

$$t_{1/2} = -\frac{ln\ 0.5}{k} \tag{4}$$

where, $C_0$ is the initial anthocyanin contents and $C_t$ is the anthocyanin contents after time $t$ (days), while $k$ is the first-order kinetic constant [30].

*2.14. Statistical Analysis*

Unless otherwise stated, the data reported in this study represent the average of triplicate analyses and were reported as mean ± standard deviation (SD). The statistical analysis was performed on control-sample pairs using two-paired, two-sample equal variance Student's *t*-test. Significant differences were considered at $p < 0.05$.

## 3. Results

*3.1. Phytochemical Characterization of the Extract*

The combined liquid-solid solvent-ultrasound assisted extraction method allowed to obtain an extract with TAC of 34.01 ± 2.66 mg C3R/g DM, TPC of 27.00 ± 0.46 mg GAE/g DM, and TFC of 14.31 ± 1.70 mg CE/g DM, leading to an antioxidant activity of 178.43 ± 4.34 mMol/g DW. Gąstoł et al. [31] reported a total polyphenolic CM fruit content of 45.6 mg GAE/g fresh weight (FW), suggesting a higher polyphenols level compared to apple, pear, and plum fruits. Our results for TPC are higher than those reported by Okan et al. [32], who suggested a concentration of 19.87 ± 0.27 mg GAE/g DM in CM fruits. A high level of TAC was found in our study, related to the color of the fruits, when compared to the data from literature. For example, Dzydzan et al. [33] reported that the red CM fruits contain anthocyanins in a concentration of 20.67 mg/g DM. TFC values varied between 0.5 and 207 mg quercetin equivalents (QE)/g DM for different varieties cultivated in Turkey [33]. However, the extraction efficiency of the bioactive from different matrices is highly affected by the extraction method. Therefore, Kutlu et al. [34] used microwave and ohmic heating-assisted microwave methods to extract the phenolic compounds from Cornelian cherry, suggesting that a preliminary ohmic treatment prior to the extraction enhanced the TPC yield 1.1- and 5.4-fold compared to that of the microwave and maceration, respectively. These authors reported a lower TAC value of 0.65 mg C3G/g DM, whereas a comparable value for the TPC was suggested (29.4 ± 0.67 mg GAE/g DM) by using a combination of ohmic heating assisted with microwave extraction.

### 3.2. Co-Microencapsulation Efficiency and Phytochemical Profile of the Powders

In terms of the co-microencapsulation efficiency of anthocyanins, values of 77.97 ± 0.57% for WPI-CN and 79.03 ± 0.72% for WPI-I were determined using Equation (1). The co-microencapsulation efficiency of the lactic acid bacteria was of 90% for both of the powders. The phytochemical profiles of the powders are given in Table 1.

**Table 1.** Co-microencapsulation efficiency, phytochemical content, and *L. casei* 431® content of the co-microencapsulated powders.

| Analyzed Parameter | WPI-CN | WPI-I |
|---|---|---|
| Anthocyanins' co-microencapsulation efficiency (%) | 77.97 ± 0.57 | 79.03 ± 0.72 |
| *Lactobacillus casei* 431® co-microencapsulation efficiency (%) | 90.03 ± 0.56 [a] | 90.07 ± 0.34 [a] |
| Total monomeric anthocyanins content (mg C3R/g DM) | 32.14 ± 0.97 [a] | 32.29 ± 0.26 [a] |
| Total polyphenolic content (mg GAE/g DM) | 9.67 ± 0.12 [a] | 9.79 ± 0.15 [a] |
| Total flavonoids content (mg CE/g DM) | 5.59 ± 0.51 [a] | 5.94 ± 0.24 [a] |
| Antioxidant activity (mMol Trolox/g DM) | 50.05 ± 0.94 [a] | 47.62 ± 1.43 [b] |
| *Lactobacillus casei* 431® (CFU/g DM) | $8.02 \times 10^9$ [a] | $8.10 \times 10^9$ [a] |

WPI-CN—Co-microencapsulated extract from Cornelian cherry fruits with *L. casei* 431® in whey protein isolates and casein. WPI-I—Co-microencapsulated extract from Cornelian cherry fruits with *L. casei* 431® in whey protein isolates and inulin. For each tested parameter and powder, values that are on the same row that do not have the same lowercase letters (([a]) and ([b])) are statistically different at $p < 0.05$ based on the Tukey method and the 95% confidence interval.

It can be observed that no significant differences were found in the bioactive content between the two co-microencapsulation matrices, whereas WPI-CN revealed a higher antioxidant activity. Oancea et al. [35] microencapsulated anthocyanins from cherry skins in whey protein isolate and reported an encapsulation efficiency of 70.30 ± 2.20%, while Tao et al. [36] suggested that the encapsulation efficiency of blueberry anthocyanins using different ratios of WPI, acacia gum, maltodextrin, and β-cyclodextrin reached 82%. However, the biopolymeric combination used in our study allowed to enhance the microencapsulation efficiency of lactic acid bacteria. Neuenfeldt et al. [8] suggested values ranging between 82 and 84% for *L. rhamnosus* co-microencapsulation with blueberry extract and Arabic gum, inulin, and maltodextrin.

### 3.3. Structural and Morphological Analysis of the Powders

The native samples of powder (Figure 1a,b) revealed irregular, polygonal, large, and thin formations with autofluorescence between 500 and 540 nm, in which anthocyanins and lactic acid bacteria are trapped in the matrix network. If in the WPI-I powder, the spherosomes are scarce and the scale-like formations are predominant (Figure 1b); when CN was used as an encapsulating adjuvant (WPI-CN), the spherosomes were numerous, smaller (with diameters between 17.42–34.03 μm) with an agglutination tendency (Figure 1a). The *L. casei* 431® formed a compact biofilm in the structure of the scales in both variants.

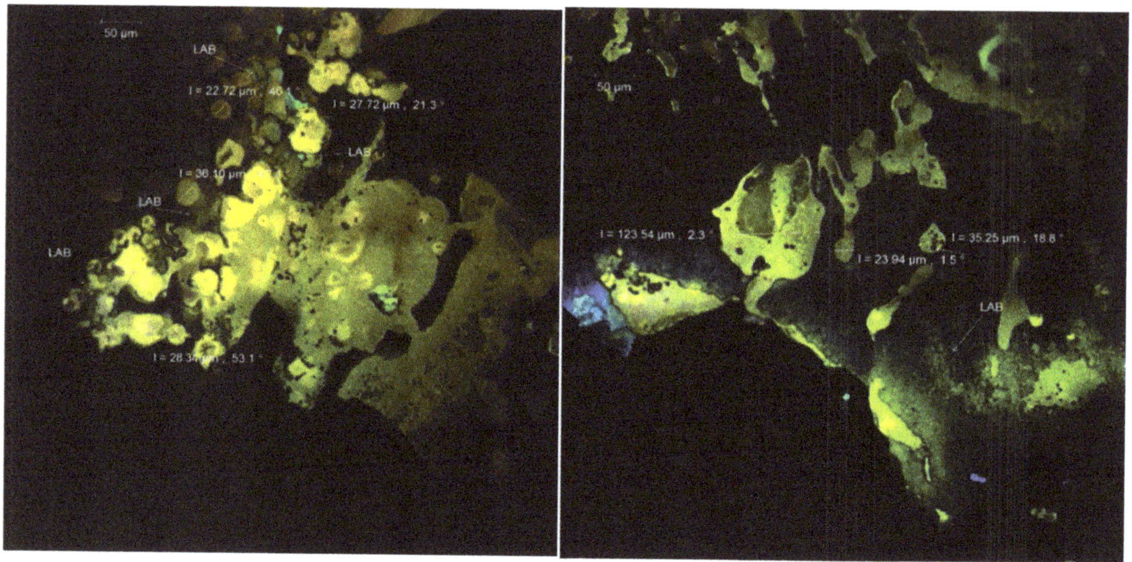


(**a**) (**b**)

**Figure 1.** Confocal laser scanning microscopy images of the co-microencapsulated extract from Cornelian cherry fruits with *L. casei* 431® in whey protein isolates and casein (**a**) and whey protein isolates and inulin (**b**) in native state.

The fluorescent labeling with Congo Red of the samples (Figure 2a,b) suggested a double encapsulation of bioactives in the spherosomes in both experimental variants.

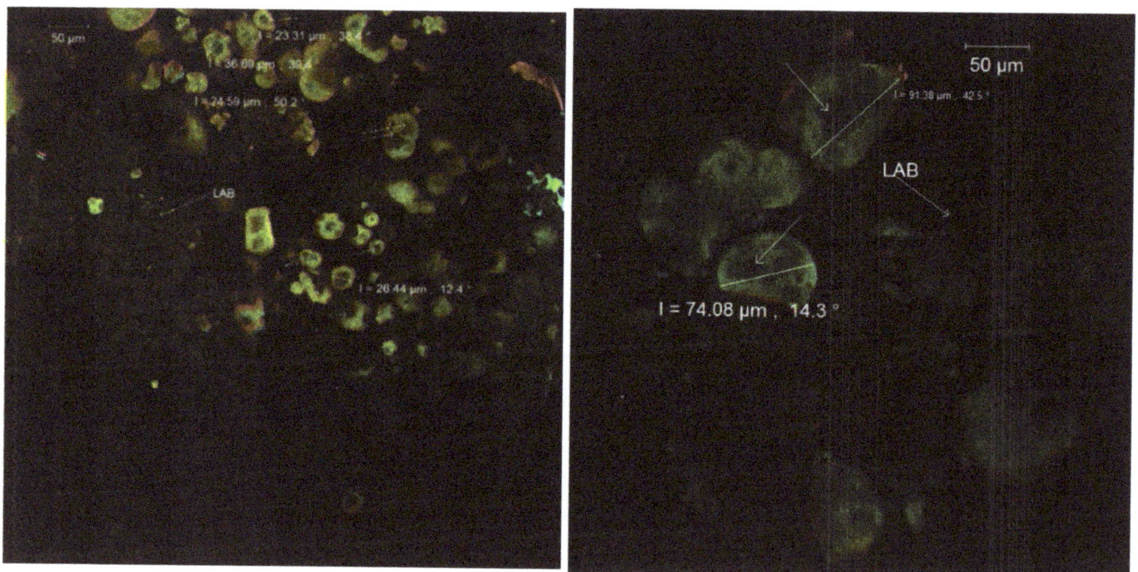


(**a**) (**b**)

**Figure 2.** Confocal laser scanning microscopy images of co-microencapsulated extract from Cornelian cherry fruits with *L. casei* 431® in whey protein isolates and casein (**a**) and whey protein isolates and inulin (**b**) stained with fluorophores.

The microstructural difference between the two powders consisted of the fact that the WPI-CN was softer, with uniform spherosomes with smaller diameters, whereas in the WPI-I, the spherosomes were bigger and reached dimensions of 74.08–91.38 μm.

### 3.4. In Vitro Digestion of Anthocyanins

The TAC content in SGJ increased slightly with 8.75% in the WPI-CN and with 7.10% in the WPI-I, with no significant differences ($p > 0.05$) in the influence of the gastric digestion time tested (120 min) on TAC. These results highlight the protective effects of both biopolymeric matrices upon the anthocyanin compounds in SGJ. In SIJ, the controlled release of the anthocyanins was found for both of the co-microencapsulated powders (Figure 3).

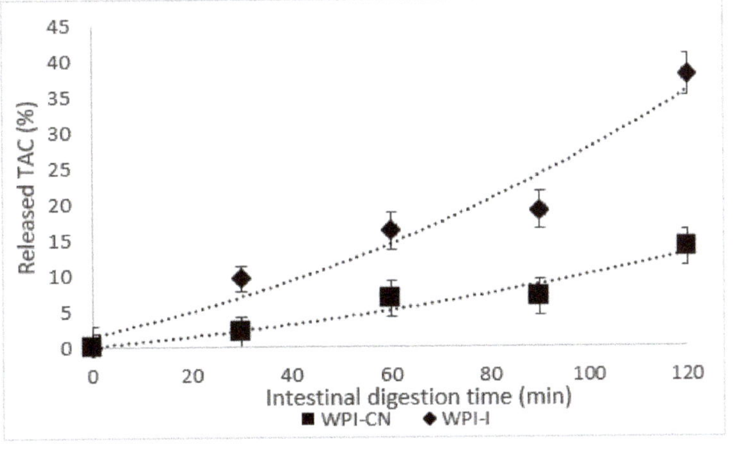


**Figure 3.** The controlled release of total anthocyanins in the simulated intestinal environment (WPI-CN—co-microencapsulated extract from Cornelian cherry fruits with *L. casei* 431® in whey protein isolates and casein, WPI-I—co-microencapsulated extract from Cornelian cherry fruits with *L. casei* 431® in whey protein isolates and inulin). The values represent mean ± SD ($n = 3$).

The mixture of WPI-CN was less effective in releasing the anthocyanins, thus at the end of the intestinal digestion, about 14% of the anthocyanins present in the microcapsules were released. By comparison, the WPI-I matrix allowed the release of about 38% of TAC, which may be correlated with a higher absorption capacity (Figure 3). It is important to mention that both of the biopolymeric combinations had a similar protective effect for TAC in the gastric environment, whereas the WPI-I allowed a better release of the anthocyanins in the intestinal environment.

The bioaccessibility of TAC took into account the number of compounds released from the powder after the gastrointestinal digestion that could become available for the absorption into the systemic circulation [37]. In this study, the TAC bioaccessibility was calculated by dividing the TAC (mg C3R/g DM) in the digested samples after gastrointestinal digestion and the TAC (mg C3R/DM) in powder before gastrointestinal digestion. A higher bioaccessibility of TAC of 55% was found in the WPI-CN powder, when compared with 49% in the WPI-I.

### 3.5. Colorimetric Analysis

Color represents one of the most important properties underlying the quantification of the freshness of food. The colorimetric properties of the co-microencapsulated powders are shown in Table 2.

**Table 2.** Colorimetric analysis of the powders.

| Variants | $L^*$ | $a^*$ | $b^*$ | $c^*$ | h* | ΔE |
|---|---|---|---|---|---|---|
| WPI-CN | 37.17 ± 0.24 [a] | 21.45 ± 0.18 [a] | 6.67 ± 0.01 [a] | 22.47 ± 0.17 [a] | 0.30 ± 0.01 [a] | 43.43 ± 0.29 [a] |
| WPI-I | 35.35 ± 0.06 [b] | 21.30 ± 0.12 [a] | 6.00 ± 0.02 [b] | 22.13 ± 0.12 [a] | 0.27 ± 0.01 [b] | 41.71 ± 0.11 [b] |

WPI-CN—Co-microencapsulated extract from Cornelian cherry fruits with *L. casei* 431® in whey protein isolates and casein. WPI-I—Co-microencapsulated extract from Cornelian cherry fruits with *L. casei* 431® in whey protein isolates and inulin. For each parameter tested, the values that are on the same column that do not have the same lowercase letters (([a]) and ([b])) are statistically different at $p < 0.05$ based on the Tukey method and the 95% confidence interval.

The parameter $a^*$ describes the tendency of the powder to red, which can be associated with the anthocyanins released from proanthocyanidins [38] and the formation of pigments derived from anthocyanins that stabilize the red flavilium [39]. From Table 2, it can be observed that the powders showed comparable redness and yellowness coefficients, which is in good agreement with the results for the TAC content. The $L^*$ parameter showed that the two variants of co-microencapsulated powders had a tendency to light, due to the use of the microencapsulated white materials.

### 3.6. Viability of Lactic Acid Bacteria

A critical parameter when evaluating the co-microencapsulation process of lactic acid bacteria is the guarantee of the probiotic survivability during processing stages and to keep the cells viable during storage [40]. A minimum amount of $10^6$ CFU/g of viable probiotics in the ready-to-eat product is recommended to obtain beneficial health effects [41]. The cells viability of the *L. casei* 431® was tested immediately after the freeze-drying, as well as after 21 and 90 days of storage at 4 °C, in the dark. The powders showed an initial viable cells value of 9.9 Log CFU/g DM. No significant reduction in the viable counts were found after 21 days of storage (9.8 log CFU/g DM), whereas after 90 days, *L. casei* 431® retained up to 97% viable cells in WPI-I and up to 93% in WPI-CN (9.18 log CFU/g DM and 9.58 log CFU/g DM, respectively). Therefore, the co-microencapsulated lactic acid bacteria cells remained viable in both powders during 90 days of storage, at 4 °C. Based on the obtained results, it is fair to conclude that the unique combination of biopolymers and bioactives from Cornelian cherry extract provided a protective environment for *L. casei* 431®. Neuenfeldt et al. [8] suggested that at room temperature (25 °C), the free bacteria remained viable for approximately 30 days of storage, with a cell reduction to 4.06 log CFU/g, whereas the microencapsulated *L. rhamnosus* using different blueberry extract concentrations showed viability for over 90 days, with counts above 6.0 log CFU/g.

### 3.7. Storage Stability of the Bioactive Compounds in the Co-Microencapsulated Powders

The TAC, TPC, TFC, and antioxidant activity were determined after 21 and 90 days, in order to evaluate the stability of co-microencapsulated compounds (Table 3).

**Table 3.** Storage stability of bioactive in co-microencapsulated variants during storage.

| Variants | Storage (Days) | Antioxidant Activity (mMol/g DM) | TPC (mg GAE/g DM) | TFC (mg CE/g DM) | TAC (mg C3R/g DM) |
|---|---|---|---|---|---|
| WPI-CN | 0 | 50.05 ± 0.94 [a] | 9.67 ± 0.12 [a] | 5.59 ± 0.51 [a] | 32.14 ± 0.97 [a] |
| | 21 | 49.73 ± 2.35 [b] | 9.65 ± 0.07 [a] | 5.65 ± 0.31 [a] | 14.62 ± 1.12 [b] |
| | 90 | 46.33 ± 1.01 [a] | 9.57 ± 0.08 [a] | 5.85 ± 0.65 [a] | 10.64 ± 0.66 [b] |
| WPI-I | 0 | 47.62 ± 1.43 [b] | 9.79 ± 0.15 [a] | 5.94 ± 0.24 [a] | 32.29 ± 0.26 [a] |
| | 21 | 46.03 ± 0.24 [a] | 9.75 ± 0.02 [a] | 5.81 ± 0.41 [a] | 17.26 ± 0.81 [a] |
| | 90 | 45.87 ± 1.11 [a] | 9.54 ± 0.02 [a] | 5.67 ± 0.33 [a] | 13.35 ± 0.37 [a] |

WPI-CN—Co-microencapsulated extract from Cornelian cherry fruits with *L. casei* 431® in whey protein isolates and casein. WPI-I—Co-microencapsulated extract from Cornelian cherry fruits with *L. casei* 431® in whey protein isolates and inulin. Values that are on the same row that do not have the same lowercase letters (([a]) and ([b])) are statistically different at $p < 0.05$ based on the Tukey method and the 95% confidence interval.

From Table 3, it can be observed that TAC were the most sensitive compounds, with a significant 67% reduction in WPI-CN and 56% in WPI-I. No significant differences were found in TPC and TFC, whereas a positive correlation between the decrease in TAC and antioxidant activity was found for both samples.

The obtained results allowed the use of the first order kinetic model for the estimation of the half-life for anthocyanins content and antioxidant activity in the powders. The $t_{1/2}$ values for the powders were significantly different, with 66.88 ± 1.67 days for WPI-CN and 83.60 ± 2.46 days for WPI-I, suggesting a higher stability of anthocyanins in the WPI-I matrices. Concomitantly, the $t_{1/2}$ for the antioxidant activity were of 752.43 ± 2.56 days and 3009.75 ± 4.57 days, respectively, suggesting an excellent stability of biologically active compounds in the two experimental variants. The results obtained are consistent with Moser et al. [39], who concluded that grape juice encapsulated with maltodextrin and soy or whey proteins had high stability during the 150 days of storage.

In addition, Azarpazhooh et al. [30] studied the degradation kinetic of TAC in a microencapsulated pomegranate peel extract under different storage conditions (temperature of 4 °C and 25 °C and relative humidity of 52 and 75%), suggesting a higher stability of TAC at lower temperature. For example, the microencapsulated TAC in a 10% maltodextrin with calcium alginate (0.1%) revealed a half time of 115.52 days at 4 °C and a relative humidity of 75%.

### 3.8. *In Vitro Anti-Proliferative Activity of the Co-Microencapsulated Powders*

The anti-proliferative effect of the co-microencapsulated powders was assessed on the human intestinal cells (HT-29) by NR assay. The obtained data indicated an anti-proliferative effect of both powders, in a concentration-dependent manner (Figure 4).

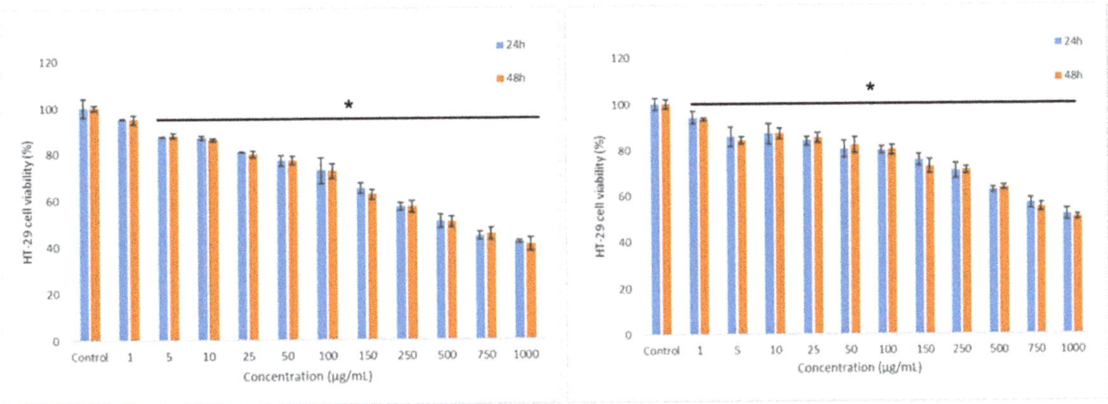


(**a**) (**b**)

**Figure 4.** The anti-proliferative effect of co-microencapsulated powders containing *L. casei* 431® in whey protein isolates and inulin (**a**) and in whey protein isolates and casein (**b**) on human intestinal cells HT-29 assessed by NR method at 24 and 48 h. The results were expressed as percent relative to the control culture (untreated), considered 100% viable. The values represent mean ± SD ($n$ = 3), * $p < 0.05$, compared to control.

The effect of powders on cell viability decreased under 80% at concentrations higher than 100 µg/mL. The EC50 values were calculated in the range of 750–1000 µg/mL, with WPI-I powder having the most prominent anti-proliferative activity. Microscopy investigation of the tumoral cells morphology confirmed the NR quantitative data (Figure 5). The micrographs showed a concentration-dependent decrease of the cell density for both variants of co-microencapsulated powders. Even though treated cells maintained their normal aggregative phenotype, the cell viability and proliferation were severely affected, at concentrations higher than 250 and 750 μg/mL of WPI-I and WPI-CN variants, respectively.

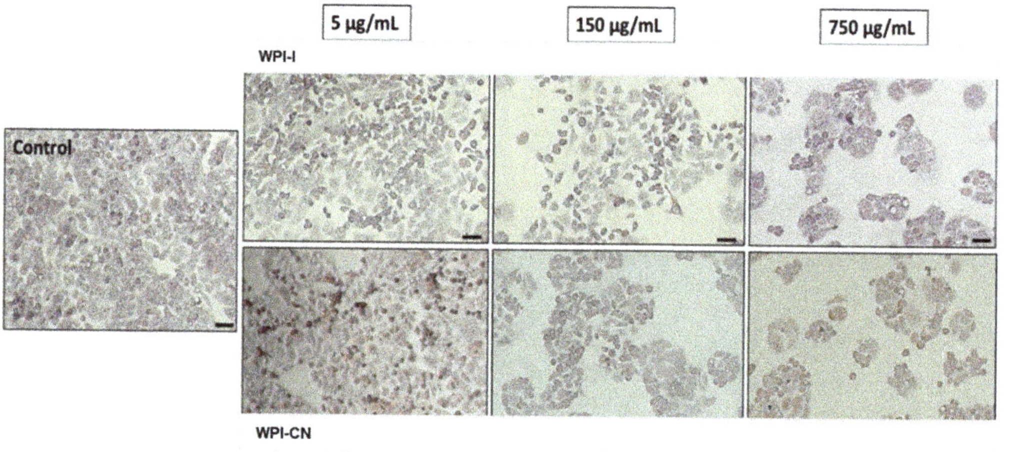


**Figure 5.** Selection of light micrographs showing the cell morphology of HT-29 tumoral cells cultivated in the presence of different concentrations of powders, for 48 h. Scale bar = 50 μm.

Previous studies have shown the dose-dependent anti-tumoral effect of CM juice through the inhibitory effect on the viability of various cancer cell lines (Hep-G2, Caco-2, HT-29, CT-26, MCF-7), including HT-29 human colon cancer cells [42]. Recently, the anti-proliferative effect was found to be correlated with the polyphenols content and iridoid presence, as a major monoterpenoid constituent of CM [43]. For example, Blagojević et al. [43] suggested that different CM fruits genotypes exhibited high antioxidant capacity and anti-proliferative effect on HT-29 cells. However, these authors reported significant lower inhibitory concentrations in the range of 9.14 mg/mL to 13.97 mg/mL, as a function of genotype. The bioactives from CM extract considered as major contributors to anti-proliferative activity were delphinidin and cyanidin galactosides, loganic acid, and sweroside [43]. The customized designs used in our study allowed enhancing the anti-proliferative activity of the CM extracts, with a potential to inhibit HT-29 cell viability with 50% in the range of 500–1000 μg/mL.

### 3.9. *In Vitro Cytocompatibility of the Co-Microencapsulated Powders*

Cell viability was evaluated in L929 fibroblast cell culture by NR assay. The results showed that both co-microencapsulates variants were cytocompatible in a concentration-dependent manner (Figure 6). From Figure 6, it can be seen that cell viability values after 48 h of cultivation were higher than 80% at powders' concentrations ranging from 1 μg/mL to 25 μg/mL for WPI-I variant (Figure 6a) and from 1 μg/mL to 10 μg/mL for WPI-CN variant (Figure 6b). Additionally, WPI-I variant stimulated the cell proliferation at low concentrations, ranging from 1 to 5 μg/mL, after 24 h of treatment. At concentrations between 50 and 250 μg/mL, the WPI-I variant induced a decrease of cell viability (60–80% values), with a significant decrease down to 35% at higher concentration (750—1000 μg/mL).

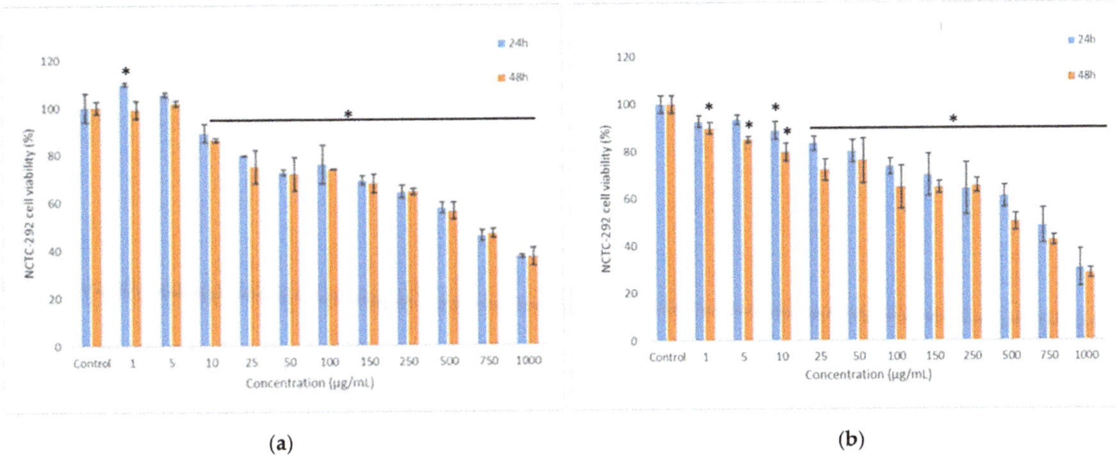


**Figure 6.** Cell viability of L929 fibroblasts cultivated in the presence of co-microencapsulated extract from cornelian cherry fruits with *L. casei* 431® in whey protein isolates and inulin (**a**) and in whey protein isolates and casein (**b**) for 24 h and 48 h, respectively, determined by NR assay. The results were expressed as percent relative to the untreated culture (control), considered 100% viable. The values represent mean ± SD ($n$ = 3), * $p < 0.05$, compared to control.

A wider concentration range between 1–50 μg/mL was found for the WPI-CN variant, suggesting a good cyto-compatibility (cell viability >80%). For the WPI-CN variant, a moderate cytotoxicity (60–80% cell viability) was found at concentration range of 100–250 μg/mL, after 48 h of cultivation and even 500 μg/mL at 24 h of cultivation. At higher concentrations of WPI-CN powder, the cell viability decreased down to 30%.

Light microscopy images showed that L929 cells cultivated in the presence of co-microencapsulated variants maintained their normal fusiform phenotype, specific for fibroblast cells, similar to the control culture, at low concentrations ranging between 1 and 50 μg/mL (Figure 7).

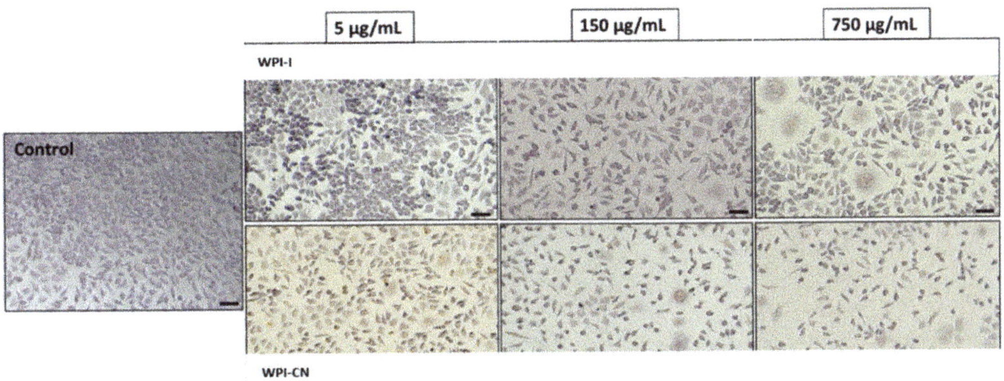


**Figure 7.** Selection of light micrographs showing the cell morphology of L929 cells cultivated in the presence of different concentrations of powders, for 48 h. Scale bar = 50 μm.

The cells were homogeneously distributed on the surface and their density was comparable to the control culture. At concentrations of 50–250 μg/mL, the cell density slightly decreased, while at higher concentrations, some structural modifications of cells, such as vesicle formation were observed in cells treated with both variants. Thus, these qualitative observations of light microscopy confirmed the NR quantitative data of cytocompatibility.

Popović et al. [44] used freeze-dried CM fruit to obtain free and microencapsulated extracts by using inclusion complexes with β-cyclodextrin. These authors tested the cytotoxic profile of β-cyclodextrin, Cornelian cherry water/ethanol extract, and β-cyclodextrin encapsulate of 50% ethanolic-water extract on HT-29 cell lines. A significant higher cytotoxic profile ranging from 312.5 to 2500 μg/mL was suggested, with a uniform cytotoxic response and low dose-dependent increase of cytotoxic activity in evaluated concentration range for the microencapsulated variant.

However, when considering the anti-proliferative and cytotoxic effects of the powders in different food, pharmaceutical, and nutraceutical applications, the safety dose should be further investigated in appropriate toxicity studies.

## 4. Conclusions

In this study, the prebiotic effect of Cornelian cherry fruits extract on the *Lacticaseibacillus casei* (*L. casei*) 431® was tested through the formulation of two variants of co-microencapsulation based on whey proteins isolate with casein and inulin, by a complex coacervation and freeze-drying. Both powders revealed a good retention of anthocyanins, with a microencapsulation efficiency of 78–79% for anthocyanins and 90% for lactic acid bacteria. Two pink-reddish powders were obtained, with a significant content of bioactive compounds and viable cells, of $10^9$ CFU/g DW, with different structure and morphologies particularities. The design of the synbiotic delivery systems based on whey protein isolate and casein led to soft microparticles, with uniform spherosomes and small diameters, whereas the unique combinations of biopolymers allowed a double microencapsulation, therefore explaining the in vitro resistance of anthocyanins, especially in the acidic environment. The in vitro digestion revealed the protective effect of the encapsulating matrices on the anthocyanins in the gastric environment, whereas the combination of whey protein isolates and inulin was more effective for anthocyanins' release in the intestinal simulated juice. During storage, anthocyanins were more sensitive to the environmental conditions, while a clear cell viability of at least 9 Log CFU/g DW in both powders was observed. The inulin-containing variant showed the most potent anti-proliferative effect, whereas both powders were cytocompatible, but to a different extent. In conclusion, our results demonstrated the potential to develop multifunctional ingredients, provided by both the bioactive compounds from plant sources, Cornelian cherries, and the probiotic lactic acid bacteria, for different applications, such as adding or using these types of powders in nutraceuticals or in food products.

**Author Contributions:** N.S., A.O., and C.V: conceptualization; I.M.E., A.M.P., E.E., M.A.V., V.V.B., and O.C.: investigation, methodology, data curation, original draft preparation; C.V., A.O. and N.S.: supervision, editing; C.V.: supervision, editing; C.V., A.O., and N.S.: supervision, conceptualization, editing. All authors have read and agreed to the published version of the manuscript.

**Funding:** This work was financially supported by the Internal Grant financed by Dunarea de Jos University of Galati, Romania, Contract no. 3637/30.09.2021. The Romanian Ministry of Research, Innovation, and Digitization within the National Program Nucleu, project no. 25N/2019-19270102, also supported this work.

**Institutional Review Board Statement:** Not applicable.

**Informed Consent Statement:** Not applicable.

**Data Availability Statement:** Data will be made available on reasonable request.

**Acknowledgments:** The authors affiliated to Dunarea de Jos University of Galati are grateful for the technical support offered by the Integrated Center for Research, Expertise, and Technological Transfer in Food Industry, and Romanian Center for Modeling Aquaculture Recirculating Systems (MoRas).

**Conflicts of Interest:** The authors declare no conflict of interest.

## References


1. Khan, I.S.; Ali, M.N.; Hamid, R.; Ganie, S.A. Genotoxic effect of two commonly used food dyes metanil yellow and carmoisine using *Allium cepa* L. as indicator. *Toxicol. Rep.* **2020**, *7*, 370–375. [CrossRef] [PubMed]
2. Viera, I.; Pérez-Gálvez, A.; Roca, M. Green Natural Colorants. *Molecules* **2019**, *24*, 154. [CrossRef] [PubMed]
3. Boeing, H.; Bechthold, A.; Bub, A.; Ellinger, S.; Haller, D.; Kroke, A.; Leschik-Bonnet, E.; Müller, M.J.; Oberritter, H.; Schulze, M.; et al. Critical review: Vegetables and fruit in the prevention of chronic diseases. *Eur. J. Nutr.* **2012**, *51*, 637–663. [CrossRef] [PubMed]
4. Vinha, A.F.; Rodrigues, F.; Nunes, M.A.; Oliveira, M.B.P.P. Natural pigments and colorants in foods and beverages. In *Polyphenols: Properties, Recovery, and Applications*; Galanakis, C.M., Ed.; Elsevier: Amsterdam, The Netherlands, 2018; pp. 363–391.
5. Buchweitz, M. Natural solutions for blue colors in food. In *Handbook on Natural Pigments in Food and Beverages: Industrial Applications for Improving Food Color*; Reinhold, C., Schweiggert, R.M., Eds.; Woodhead Publishing: Sawston, UK, 2016; pp. 355–384.
6. Manzoor, M.F.; Ahmad, N.; Ahmed, Z.; Siddique, R.; Zeng, X.A.; Rahaman, A.; Wahab, A. Novel extraction techniques and pharmaceutical activities of luteolin and its derivatives. *J. Food Biochem.* **2019**, *43*, 12974. [CrossRef]
7. Ghosh, S.; Sarkar, T.; Das, A.; Chakraborty, R. Micro- and nanoencapsulation of natural colors: A holistic view. *Appl. Biochem. Biotechnol.* **2021**, *193*, 3787–3811. [CrossRef]
8. Neuenfeldt, N.H.; Almeida Farias, C.A.; de Oliveira Mello, R.; Sasso Robalo, S.; Smanioto Barin, J.; Picolli da Silva, L.; Müller, E.I.; Moraes Flores, E.M.; Teixeira Barcia, M.; Ragagnin de Menezes, C. Effects of blueberry extract co-microencapsulation on the survival of *Lactobacillus rhamnosus*. *LWT* **2022**, *155*, 112886. [CrossRef]
9. Quintana, G.; Simoes, M.G.; Hugo, A.; Alves, P.; Ferreira, P.; Gerbino, E.; Simoes, P.N.; Gomez-Zavaglia, A. Layer-by-layer encapsulation of *Lactobacillus delbrueckii* subsp. *bulgaricus* using block-copolymers of poly (acrylic acid) and pluronic for safe release in gastro-intestinal conditions. *J. Funct. Foods* **2017**, *35*, 408–417. [CrossRef]
10. Li, J.; Zhang, C.; Liu, H.; Liu, J.; Jiao, Z. Profiles of sugar and organic acid of fruit juices: A comparative study and implication for authentication. *J. Food Qual.* **2020**, *2020*, 7236534. [CrossRef]
11. Pan, P.; Lam, V.; Salzman, N.; Huang, Y.; Yu, J.; Zhang, J.; Wang, L. Black raspberries and their anthocyanin and fiber fractions alter the composition and diversity of gut microbiota in F-344 rats. *Nutr. Cancer* **2017**, *69*, 943–951. [CrossRef]
12. Garcia-Mazcorro, J.F.; Lage, N.N.; Mertens-Talcott, S.; Talcott, S.; Chew, B.; Dowd, S.E.; Kawas, J.R.; Noratto, G.D. Effect of dark sweet cherry powder consumption on the gut microbiota, short-chain fatty acids, and biomarkers of gut health in obese db/db mice. *PeerJ* **2018**, *6*, e4195. [CrossRef]
13. Bayram, H.M.; Ozturkcan, S.A. Bioactive components and biological properties of cornelian cherry (*Cornus mas* L.): A comprehensive review. *J. Funct. Foods* **2020**, *75*, 104252. [CrossRef]
14. Szczepaniak, O.M.; Ligaj, M.; Kobus-Cisowska, J.; Maciejewska, P.; Tichoniuk, M.; Szulc, P. Application for novel electrochemical screening of antioxidant potential and phytochemicals in *Cornus mas* extracts. *CyTA J. Food* **2019**, *17*, 781–789. [CrossRef]
15. Kazimierski, M.; Regula, J.; Molska, M. Cornelian cherry (*Cornus mas* L.)—Characteristics, nutritional and pro-health properties. *Acta Sci. Pol. Technol. Aliment.* **2019**, *18*, 5–12. [PubMed]
16. Dinda, B.; Kyriakopoulos, A.M.; Dinda, S.; Zoumpourlis, V.; Thomaidis, N.S.; Velegraki, A.; Dinda, M. *Cornus mas* L. (cornelian cherry), an important European and Asian traditional food and medicine: Ethnomedicine, phytochemistry and pharmacology for its commercial utilization in drug industry. *J. Ethnopharmacol.* **2019**, *193*, 670–690. [CrossRef]
17. Hill, D.; Sugrue, I.; Tobin, C.; Hill, C.; Stanton, C.; Ross, R.P. The *Lactobacillus casei* group: History and health related applications. *Front. Microbiol.* **2018**, *9*, 2107. [CrossRef]
18. Sidira, M.; Galanis, A.; Ypsilantis, P.; Karapetsas, A.; Progaki, Z.; Simopoulos, C.; Kourkoutas, Y. Effect of probiotic-fermented milk administration on gastrointestinal survival of *Lactobacillus casei* ATCC 393 and modulation of intestinal microbial flora. *J. Mol. Microbiol. Biotechnol.* **2010**, *19*, 224–230.
19. Tiptiri-Kourpeti, A.; Spyridopoulou, K.; Santarmaki, V.; Aindelis, G.; Tompoulidou, E.; Lamprianidou, E.E.; Saxami, G.; Ypsilantis, P.; Lampri, E.S.; Simopoulos, C.; et al. *Lactobacillus casei* exerts anti-proliferative effects accompanied by apoptotic cell death and up-regulation of TRAIL in colon carcinoma cells. *PLoS ONE* **2016**, *11*, e0147960. [CrossRef]
20. Abdel-Hamid, M.; Romeih, E.; Gamba, R.R.; Nagai, E.; Suzuki, T.; Koyanagi, T.; Enomoto, T. The biological activity of fermented milk produced by *Lactobacillus casei* ATCC 393 during cold storage. *Int. Dairy J.* **2019**, *91*, 1–8. [CrossRef]
21. Trachootham, D.; Chupeerach, C.; Tuntipopipat, S.; Pathomyok, L.; Boonnak, K.; Praengam, K.; Promkam, C.; Santivarangkn, C. Drinking fermented milk containing *Lactobacillus paracasei* 431 (IMULUSTM) improves immune response against H1N1 and cross-reactive H3N2 viruses after influenza vaccination: A pilot randomized triple-blinded placebo controlled trial. *J. Funct. Foods* **2017**, *33*, 1–10. [CrossRef]
22. Matalanis, A.; Jones, O.G.; MacClements, D.J. Structured biopolymer based delivery systems for encapsulation, protection, and release of lipophilic compounds. *Food Hydrocol.* **2011**, *25*, 1865–1880. [CrossRef]
23. McIntosh, G.H.; Royle, P.J.; Le Leu, R.K.; Regester, G.O.; Johnson, M.A.; Grinstead, R.L.; Kenward, R.S.; Smithers, G.W. Whey proteins as functional food ingredients? *Int. Dairy J.* **1996**, *8*, 425–434. [CrossRef]
24. Tarifa, M.C.; Piqueras, C.M.; Genovese, D.B.; Brugnoni, L.I. Microencapsulation of *Lactobacillus casei* and *Lactobacillus rhamnosus* in pectin and pectin-inulin microgel particles: Effect on bacterial survival under storage conditions. *Int. J. Biol. Macromol.* **2021**, *179*, 457–465. [CrossRef] [PubMed]

25. Enache, I.M.; Vasile, A.M.; Enachi, E.; Barbu, V.; Stănciuc, N.; Vizireanu, C. Co-microencapsulation of anthocyanins from black currant extract and lactic acid bacteria in biopolymeric matrices. *Molecules* **2020**, *25*, 1700. [CrossRef] [PubMed]
26. *ISO 8261 IDF122:2001*; Milk and Milk Products—General Guidance for the Preparation of Test Samples, Initial Suspensions and Decimal Dilutions for Microbiological Examination. ISO: Geneva, Switzerland, 2001.
27. AOAC Official Method 2005.02. Total Monomeric Anthocyanin Pigment Content of Fruit Juices, Beverages, Natural Colorants, and Wines—pH Differential Method. In *Official Methods of Analysis*, 18th ed.; Association of Officiating Analytical Chemists: Washington, DC, USA, 2005; Method 935.14 and 992.24.
28. Colín-Cruz, M.A.; Pimentel-González, D.J.; Carrillo-Navas, H.; Alvarez-Ramírez, J.; Guadarrama-Lezama, A.Y. Co-encapsulation of bioactive compounds from blackberry juice and probiotic bacteria in biopolymeric matrices. *LWT* **2019**, *110*, 94–101. [CrossRef]
29. Voicescu, M.; Ionescu, S.; Manoiu, V.S.; Anastasescu, M.; Craciunescu, O.; Moldovan, L. Synthesis and biophysical characteristics of riboflavin/HSA protein system on silver nanoparticles. *Mat. Sci. Eng. C* **2019**, *96*, 30–40. [CrossRef]
30. Azarpazhooh, E.; Sharayei, P.; Zomorodi; Ramaswamy, H.S. Physicochemical and phytochemical characterization and storage stability of sreeze-dried encapsulated pomegranate peel anthocyanin and in vitro evaluation of its antioxidant activity. *Food Bioproc. Technol.* **2019**, *12*, 199–210. [CrossRef]
31. Gąstoł, M.; Krośniak, M.; Derwisz, M.; Dobrowolska-Iwanek, J. Cornelian cherry (*Cornus mas* L.) juice as a potential source of biological compounds. *J. Med. Food* **2013**, *16*, 728–732. [CrossRef]
32. Okan, O.T.; Serencam, H.; Baltas, N.; Can, Z. Some edible forest fruits their in vitro antioxidant activities, phenolic compounds and some enzyme inhibition effects. *Fresenius Environ. Bull.* **2019**, *28*, 6090–6098.
33. Dzydzan, O.; Bila, I.; Kucharska, A.Z.; Brodyak, I.; Sybirna, N. Antidiabetic effects of extracts of red and yellow fruits of cornelian cherries (*Cornus mas* L.) on rats with streptozotocin-induced diabetes mellitus. *Food Funct.* **2019**, *16*, 6459–6472. [CrossRef]
34. Kutlu, N.; Isci, A.; Sakiyan, O.; Yilmaz, A.M. Extraction of phenolic compounds from cornelian cherry (*Cornus mas* L.) using microwave and ohmic heating assisted microwave methods. *Food Bioproc. Technol.* **2020**, *14*, 650–664. [CrossRef]
35. Oancea, A.M.; Hasan, M.; Vasile, A.M.; Barbu, V.; Enachi, E.; Bahrim, G.; Râpeanu, G.; Silvi, S.; Stănciuc, N. Functional evaluation of microencapsulated anthocyanins from sour cherries skins extract in whey proteins isolate. *LWT* **2018**, *95*, 129–134. [CrossRef]
36. Tao, Y.; Wang, P.; Wang, J.; Wu, Y.; Han, Y.; Zhou, J. Combining various wall materials for encapsulation of blueberry anthocyanin extracts: Optimization by artificial neural network and genetic algorithm and a comprehensive analysis of anthocyanin powder properties. *Powder Technol.* **2017**, *311*, 77–87. [CrossRef]
37. Paucar-Menacho, L.M.; Peñas, E.; Hernandez-Ledesma, B.; Frias, J.; Martínez-Villaluenga, C. A comparative study on the phenolic bioaccessibility, antioxidant and inhibitory effects on carbohydrate-digesting enzymes of maca and mashua powders. *LWT* **2020**, *131*, 109798. [CrossRef]
38. Kaderides, K.; Mourtzinos, I.; Goula, A.M. Stability of pomegranate peel polyphenols encapsulated in orange juice industry by-product and their incorporation in cookies. *Food Chem.* **2020**, *310*, 125849. [CrossRef]
39. Moser, P.; Telis, V.R.N.; de Andrade Neves, N.; García-Romero, E.; Gómez-Alonso, S.; Hermosín-Gutiérrez, I. Storage stability of phenolic compounds in powdered BRS Violeta grape juice microencapsulated with protein and maltodextrin blends. *Food Chem.* **2017**, *214*, 308–318. [CrossRef]
40. Champagne, C.P.; Raymond, Y.; Guertin, N.; Belanger, G. Effects of storage conditions, microencapsulation and inclusion in chocolate particles on the stability of probiotic bacteria in ice cream. *Int. Dairy J.* **2015**, *47*, 109–117. [CrossRef]
41. Tripathi, M.K.; Giri, S.K. Probiotic functional foods: Survival of probiotics during processing and storage. *J. Funct. Foods* **2014**, *9*, 225–241. [CrossRef]
42. Tiptiri-Kourpeti, A.; Fitsiou, E.; Spyridopoulou, K.; Vasileiadis, S.; Iliopoulos, C.; Galanis, A.; Vekiari, S.; Pappa, A.; Chlichlia, K. Evaluation of antioxidant and antiproliferative properties of *Cornus mas* L. fruit juice. *Antioxidants* **2019**, *8*, 377. [CrossRef]
43. Blagojević, B.; Agić, D.; Serra, A.M.; Matić, S.; Matovina, M.; Bijelić, S.; Popović, B.M. An in vitro and in silico evaluation of bioactive potential of cornelian cherry (*Cornus mas* L.) extracts rich in polyphenols and iridoids. *Food Chem.* **2021**, *335*, 127619. [CrossRef]
44. Popović, B.M.; Blagojević, B.; Latković, D.; Cetojević-Simin, D.; Kucharska, A.Z.; Parisi, F.; Lazzara, G. A one-step enhanced extraction and encapsulation system of cornelian cherry (*Cornus mas* L.) polyphenols and iridoids with β-cyclodextrin. *LWT* **2021**, *141*, 110884. [CrossRef]

MDPI
St. Alban-Anlage 66
4052 Basel
Switzerland
www.mdpi.com

*Nutrients* Editorial Office
E-mail: nutrients@mdpi.com
www.mdpi.com/journal/nutrients

www.ingramcontent.com/pod-product-compliance
Lightning Source LLC
Chambersburg PA
CBHW061808210926
51850CB00006B/649
* 9 7 8 3 0 3 6 5 9 0 9 6 7 *